Game of Life Cellular Automata

T0135353

Andrew Adamatzky

Editor

Game of Life Cellular Automata

 Springer

Editor
Prof. Andrew Adamatzky
University of the West of England
Bristol BS16 1QY
United Kingdom
andrew.adamatzky@uwe.ac.uk

ISBN 978-1-4471-6154-7   ISBN 978-1-84996-217-9 (eBook)
DOI 10.1007/978-1-84996-217-9
Springer London Dordrecht Heidelberg New York

British Library Cataloguing in Publication Data
A catalogue record for this book is available from the British Library

Printed on acid-free paper

Springer is part of Springer Science+Business Media (www.springer.com)

Preface

In 1970 Martin Gardner spelled out the rules of a new solitaire game forged by John Horton Conway.[1] An unparalleled combination of functional simplicity with behavioural complexity made Conway's Game of Life the most popular cellular automaton of all time. We commemorate the Game of Life's 40th birthday with a unique collection of works authored by renowned mathematicians, computer scientists, physicists and engineers. The superstars of science, academy and industry present their visions of the Game of Life cellular automaton, its extensions and modifications, and spatially-extended systems inspired by the Game.

The book covers hot topics in theory of computation, pattern formation, optimization, evolution, non-linear sciences and mathematics. Academics, researchers, hobbyists and students interested in the Game of Life theory and applications will find this monograph a valuable guide to the field of cellular automata and excellent supplementary reading.

Bristol, UK   *Andrew Adamatzky*

---

[1] "...each cell of the checkerboard (assumed to be an infinite plane) has eight neighboring cells, four adjacent orthogonally, four adjacent diagonally. The rules are:

1. Survivals. Every counter with two or three neighboring counters survives for the next generation.
2. Deaths. Each counter with four or more neighbors dies (is removed) from overpopulation. Every counter with one neighbor or none dies from isolation.
3. Births. Each empty cell adjacent to exactly three neighbors — no more, no fewer — is a birth cell. A counter is placed on it at the next move.

...all births and deaths occur simultaneously."
Martin Gardner, The fantastic combinations of John Conway's new solitaire game "life". Scientific American 223 (October 1970): 120–123.

Contents

Contributors

Derek Abbott Department of Electrical and Electronic Engineering, the University of Adelaide, Adelaide 5005, Australia, dabbott@eleceng.adelaide.edu.au

Susumu Adachi National Institute of Information and Communications Technology, Japan, sadachi_nict@yahoo.co.jp

Andrew Adamatzky University of the West of England, Bristol BS16 1QY, UK, andrew.adamatzky@uwe.ac.uk

Ramón Alonso-Sanz ETSI Agronomos (Estadistica), Polytechnic University of Madrid, Madrid 28040, Spain, ramon.alonso@upm.es

Carter Bays Department of Computer Science and Engineering, University of South Carolina, Columbia, SC 29208, USA, bays@sc.edu

Geoffrey Chu Department of Computer Science and Software Engineering, University of Melbourne, Parkville, 3052, Melbourne, Australia, gchu@csse.unimelb.edu.au

Claudio Conti Institute for Complex Systems (ISC-CNR), Department of Physics, University Sapienza, Piazzale Aldo Moro 2, 00185 Rome, Italy, claudio.conti@roma1.infn.it

Héctor Francisco Coronel Brizio Departamento de Inteligencia Artificial, Facultad de Física e Inteligencia Artificial, Universidad Veracruzana, Xalapa Veracruz, Mexico, hcoronel@uv.mx

David Eppstein Computer Science Department, University of California, Irvine, CA 92697-3435, USA, eppstein@uci.edu

Kellie Michele Evans Department of Mathematics, California State University, Northridge, CA 91330-8313, USA, kellie.m.evans@csun.edu

Nazim Fatès INRIA Nancy Grand-Est – LORIA, Nancy, France, nazim.fates@loria.fr

Adrian P. Flitney School of Physics, University of Melbourne, Parkville 3010, Australia, aflitney@unimelb.edu.au

Nick Gotts Integrated Land Use Systems, Macaulay Land Use Research Institute, Aberdeen AB15 8QH, UK, n.gotts@macaulay.ac.uk

Adam P. Goucher Chesterfield, Derbyshire, UK, apgoucher@gmx.com

Alejandro Raúl Hernández-Montoya Departamento de Inteligencia Artificial, Facultad de Física e Inteligencia Artificial, Universidad Veracruzana, Xalapa Veracruz, Mexico, alhernandez@uv.mx

Nathaniel Johnston Department of Mathematics and Statistics, University of Guelph, Guelph, ON, Canada N1G 2W1, njohns01@uoguelph.ca

Alexis Kirke Interdisciplinary Centre for Computer Music Research, University of Plymouth, Drake Circus, Plymouth PL4 8AA, UK, alexis.kirke@plymouth.ac.uk

Jia Lee College of Computer Science, ChongQing University, Chong-Qing, China, lijia315yu@gmail.com

Maurice Margenstern Laboratoire d'Informatique Théorique et Appliquée, Université de Metz, I.U.T. de Metz, 57045 Metz Cedex, France, margens@univ-metz.fr

Genaro J. Martínez Instituto de Ciencias Nucleares, Universidad Nacional Autónoma de México, Mexico, Mexico and Department of Computer Science, University of the West of England, Bristol BS16 1QY, UK, genaro.martinez@uwe.ac.uk

Harold V. McIntosh Departamento de Aplicación de Microcomputadoras, Instituto de Ciencias, Universidad Autónoma de Puebla, Puebla, Mexico, mcintosh@servidor.unam.mx

Eduardo R. Miranda Interdisciplinary Centre for Computer Music Research, University of Plymouth, Drake Circus, Plymouth PL4 8AA, UK, eduardo.miranda@plymouth.ac.uk

Kenichi Morita Department of Information Engineering, Graduate School of Engineering, Hiroshima University, Higashi-Hiroshima 739-8527, Japan, morita@iec.hiroshima-u.ac.jp

Mark D. Niemiec Phoenix, AZ, USA, mniemiec@gmail.com

Nick Owens Department of Electronics, University of York, York, UK, ndlo100@york.ac.uk

Ferdinand Peper National Institute of Information and Communications Technology, Nano ICT Group, 588-2 Iwaoka, Nishi-ku, Kobe 651-2492, Japan, peper@nict.go.jp

Karen Elizabeth Petrie School of Computing, University of Dundee, Dundee, DD1 4HN, Scotland, UK, karenpetrie@computing.dundee.ac.uk

Marcus Pivato Department of Mathematics, Trent University, Peterborough, Ontario K9J 7B8, Canada, marcuspivato@trentu.ca

Paul Rendell Department of Computer Science, University of the West of England, Bristol BS16 1QY, UK, paul@rendell-attic.org

Manuel Enríque Rodríguez Achach Departamento de Física, Facultad de Física e Inteligencia Artificial, Universidad Veracruzana, Xalapa Veracruz, Mexico, manurodriguez@uv.mx

Emmanuel Sapin Department of Computer Science, University of the West of England, Bristol BS16 1QY, UK, emmanuelsapin@hotmail.com

Susan Stepney Department of Computer Science, University of York, UK, susan.stepney@cs.york.ac.uk

Jeffrey Ventrella ventrella.com, Jeffrey@ventrella.com

Robert Wainwright Iona College, Department of Mathematics, 715 North Ave., New Rochelle, NY 10801, USA, RWainwright@iona.edu

Neil Yorke-Smith American University of Beirut, Lebanon, nysmith@aub.edu.lb and Artificial Intelligence Center, SRI International, Menlo Park, CA 94025, USA, nysmith@ai.sri.com

Chapter 1
Introduction to Cellular Automata and Conway's Game of Life

Carter Bays

Although cellular automata has origins dating from the 1950s, widespread popular interest was not created until John Conway's "Game of Life" cellular automaton was initially revealed to the public in a 1970 Scientific American article [2]. The single feature of his "game" that probably caused this intensive interest was undoubtedly the discovery of "oscillators" (periodic forms) and "gliders" (translating oscillators).

1.1 A Brief Background

Cellular Automata (CA) can be constructed in one, two, three or more dimensions and can best be explained by giving an example utilizing Conway's rule. Start with an infinite grid of squares. Each individual square has 8 touching neighbors; typically these neighbors are treated the same (a "Moore neighborhood"), whether they touch a candidate square on a side or at a corner. We now fill in some of the squares; we shall say that these squares are "alive". Discrete time units called generations evolve; at each generation we apply a "rule" to the current configuration in order to arrive at the configuration for the next generation; in our example we shall use the rule below.

- If a live cell is touching 2 or 3 live cells (called "neighbors"), then it remains alive next generation, otherwise it dies.
- If a non-living cell is touching exactly 3 live cells, it comes to life next generation.

Figure 1.1 depicts the evolution of a simple configuration of filled-in ("live") cells for the above rule.

There are many notations for describing CA rules; these can differ depending upon the type of CA. For CA of more than one dimension, and in our present discussion, we shall utilize the following notation, which is standard for describing CA in two dimensions with Moore neighborhoods.

A. Adamatzky (ed.), *Game of Life Cellular Automata*, DOI 10.1007/978-1-84996-217-9_1, © Springer-Verlag London Limited 2010

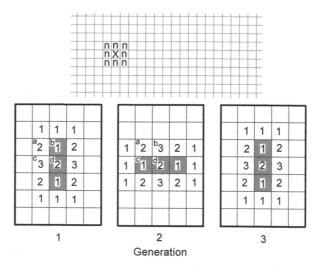

Fig. 1.1 *Top*: Each cell in a grid has 8 neighbors. The cells containing "*n*" are neighbors of the cell containing the "*X*". Any cell in the grid can be either "dead' or "alive". *Bottom*: Here we have outlined a specific area of what is presumably a much larger grid. At the *left* we have installed an initial shape. *Shaded cells* are alive; all others are dead. The number within each cell gives the quantity of live neighbors for that cell. (Cells containing no numbers have zero live neighbors.) Depicted are three generations, starting with the configuration at generation 1. Generations 2 then 3 show the result when we apply the following cellular automata rule: "Live cells with exactly 2 or 3 live neighbors remain alive (otherwise they die); dead cells with exactly 3 live neighbors come to life (otherwise they remain dead)". Let us now evaluate the transition from generation 1 to generation 2. In our diagram, cell "*a*" is dead. Since it does not have exactly 3 live neighbors, it remains dead. Cell "*b*" is alive, but it needs exactly 2 or 3 live neighbors to remain alive; since it only has 1, it dies. Cell "*c*" is dead; since it has exactly 3 live neighbors, it comes to life. And cell "*d*" has 2 live neighbors; hence it will remain alive. And so on. Notice that the form repeats every two generations. Such forms are called oscillators

We write a rule as

$$E_1, E_2, \ldots / F_1, F_2, \ldots$$

where the E_i ("environment") specify the number of live neighbors required to keep a living cell alive, and the F_i ("fertility") give the number required to bring a non-living cell to life. The E_i and F_i will be listed in ascending order; hence if $i > j$ then $E_i > E_j$, etc.

Thus the rule for the CA given above is 2,3/3. This rule, discovered by John Horton Conway, was examined in several articles in Scientific American and elsewhere, beginning with the seminal article in 1970 [2]. It is popularly known as Conway's "game of life". Of course it is not really a "game" in the usual sense, as the outcome is determined as soon as we pick a starting configuration.

Note that the shape in Fig. 1.1 repeats, with a period of two. A repeating form such as this is called an oscillator. Stationary forms can be considered oscillators with a period of one. In Figs. 1.2 and 1.3 we show several oscillators that

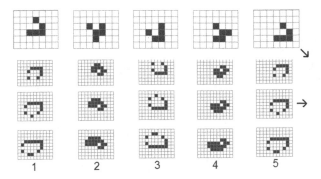

Fig. 1.2 Here we see a few of the small gliders that exist for 2,3/2. The form at the *top* — the original "glider" — was discovered by John Conway in 1968. The remaining forms were found shortly thereafter. Soon after Conway discovered rule 2,3/2 he started to give his various shapes rather whimsical names. That practice continues to this day. Hence, the name "glider" was given only to the simple shape at the *top*; the other gliders illustrated were called (from *top* to *bottom*) "lightweight spaceship", "middleweight spaceship" and "heavyweight spaceship". The numbers give the generation; each of the gliders shown has a period of four. The exact movement of each is depicted by its shifting position in the various small enclosing grids

move across the grid as they change from generation to generation. Such forms are called translating oscillators, or more commonly, gliders. Conway's rule popularized the term; in fact a flurry of activity began during which a great many shapes were discovered and exploited. These shapes were named whimsically — "blinker" (Fig. 1.1), "boat", "beehive" and an unbelievable myriad of others. Most translating oscillators were given names other than the simple moniker "glider" — there were "lightweight spaceships", "puffer trains", etc.

Of course rule 2,3/3 is not the only CA rule (even though it is the most interesting). Configurations under some rules always die out, and other rules lead to explosive growth. (We say that rules with expansive growth are unstable.)

We can easily find gliders for many unstable rules; for example Fig. 1.4 illustrates some simple constructs for rule 2/2. Note that it is practically impossible NOT to create gliders with this rule!

Hence we shall only look at gliders for rules that stabilize (i.e. exhibit bounded growth) and eventually yield only zero or more oscillators. We call such rules GoL (game of life) rules. "Stability" can be a rather murky concept, since there may be some carefully constructed forms within a GoL rule that grow without bounds.

Typically, such forms would never appear in random configurations. Hence, we shall informally define a GoL rule as follows:

1. All neighbors must be touching the candidate cell and all are treated the same (a Moore neighborhood).

Fig. 1.3 Conway's rule 2,3/2 is rich with interesting forms — stationary or translating (i.e. gliders), as well as weird constructs that can expand forever. Here are but two of many hundreds that have been discovered. The glider at the *top* has a period of 5. The large form at the *left* is called a "wickstretcher" and grows forever. It is depicted at the *right* after 83 generations. The portion at the *top* of the wickstretcher moves up, while the blob at the *bottom* remains in place (though it does exhibit some turbulence). The "wick" in the *middle* increases in length and undulates downward like movie marquee lights

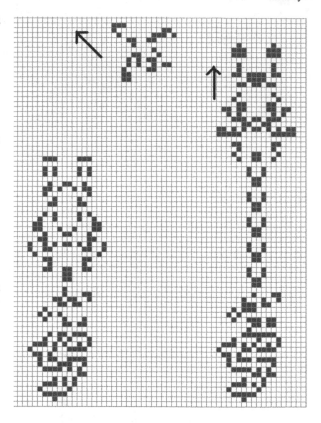

2. There must exist at least one translating oscillator (a glider).
3. Random configurations must eventually stabilize.

The above definition is a bit simplistic; for a more formal definition of a GoL rule refer to [1].

1.2 The Original Glider Gun

Conway's rule 2,3/3 is the original GoL rule and is unquestionably the most famous CA rule known. A challenge put forth by Conway was to create a configuration that would generate an ever increasing quantity of live cells. This challenge was met by William Gosper in 1970 — back when computing time was expensive and computers were slow by today's standards. He devised a form that spit out a continuous stream of gliders — a "glider gun" so to speak.

Interestingly, his gun configuration was displayed not as nice little squares, but as a rather primitive typewritten output (Fig. 1.5); this emphasizes the limited resources available in 1970 for seeking out such complex structures. Soon a "cottage industry" developed — all kinds of intricate initial configurations were discovered and exploited. Such research continues to this day and has pretty much

Fig. 1.4 Gliders exist under a large number of rules, but almost all such rules are unstable. For example the rule 2/2 exhibits rapid unbounded growth, and almost any starting configuration will yield "gliders"; e.g. just two live cells will produce two gliders going off in opposite directions. But almost any small form will quickly grow without bounds. The form at the *bottom left* expands to the shape at the *right* after only 10 generations. The generation is given with each form

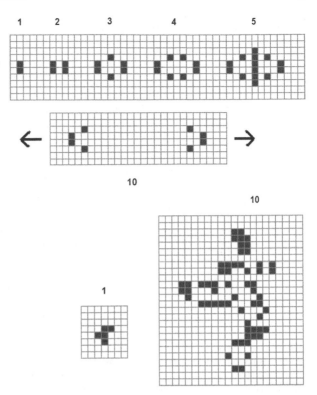

guaranteed that Conway's Game of Life will always be the most famous CA rule.

1.3 Other GoL Rules in the Square Grid

The rule 2,4,5/3 is also a GoL rule and sports the glider shown in Fig. 1.6. It has not been seriously investigated and will probably not reveal the vast array of interesting forms that exist under 2,3/3. Interestingly, 2,3/3,8 appears to be a GoL rule which not surprisingly supports many of the constructs of 2,3/3. This ability to add terms of high neighbor counts onto known GoL rules, obtaining other GoL rules, seems to be easy to implement — particularly in higher dimensions or in grids with large neighbor counts such as the triangular grid, which has a neighbor count of 12.

1.4 Why Treat All Neighbors the Same?

By allowing only Moore neighborhoods in two (and higher) dimensions we greatly restrict the number of rules that can be written. And certainly we could consider

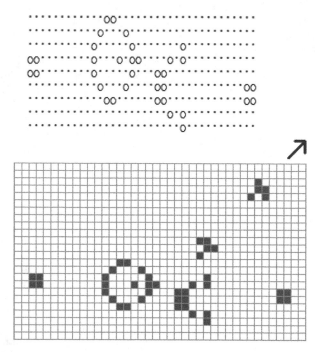

Fig. 1.5 A fascinating challenge was proposed by Conway in 1970 — he offered $50 to the first person who could devise a form for 2,3/2 that would generate an infinite number of living cells. One such form could be a "glider gun" — a construct that would create an endless stream of gliders. The challenge was soon met by William Gosper, then a student at MIT. His glider gun is illustrated here. At the *top*, testifying to the primitive computational power of the time, is an early illustration of Gosper's gun. At the *bottom* we see the gun in action, sending out a new glider every thirty generations (here it has sent out two gliders). Since 1970 there have been numerous such "guns" that generate all kinds of forms — some gliders and some stationary oscillators. Naturally in the latter case the generator must translate across the grid, leaving its intended stationary debris behind

"specialized" neighborhoods — e.g. treat as neighbors only those cells that touch on sides, or touch only the left two corners and nowhere else, or touch anywhere, but state in our rule that two or more live neighbors of a subject cell must not touch each other, etc. Consider the following rule for finding the next generation.

1. A living cell dies.
2. A dead cell comes to life if and only if its left side touches a live cell.

If we start, say, with a single cell we will obtain a "glider" of one cell that moves to the right one cell each generation! Such rules are easy to construct, as are more complex glider-producing positional rules. So we shall not investigate them further.

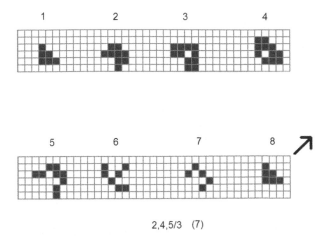

2,4,5/3 (7)

Fig. 1.6 There are a large number of interesting rules that can be written for the square grid but Rule 2,3/2 is undoubtedly the most fascinating — however it is not the only GoL rule. Here we depict a glider that has been found for the rule 2,4,5/3. And since that rule stabilizes, it is a valid GoL rule. Unfortunately it is not as interesting as 2,3/2 because its glider is not as likely to appear in random (and other) configurations — hence limiting the ability of 2,4,5/3 to produce interesting moving configurations. Note that the period is 7, indicated in *parentheses*

References

1. Bays, C.: A note on the Game of Life in hexagonal and pentagonal tessellations. Complex Syst. **15**, 245–252 (2005)
2. Gardner, M.: The fantastic combinations of John Conway's new solitaire game 'life'. Sci. Am. **223**, 120–123 (1970)
3. Preston, K., Jr., Duff, M.J.B.: Modern Cellular Automata. Plenum Press, New York (1984)
4. Wolfram, S.: A New Kind of Science. Wolfram Media, Champaign (2002)

Part I
Historical

Chapter 2
Conway's Game of Life:
Early Personal Recollections

Robert Wainwright

When the October 1970 issue of Scientific American arrived, I had no idea the extent to which Martin Gardner's article in that issue would affect my life. As long as I can remember, my custom would be to seek out the Mathematical Games column in search for Gardner's latest topic with the usual reader challenges. My first reaction to that particular article introducing a new pastime titled "The fantastic combinations of John Conway's new solitaire game 'life'" was only mildly interesting. A couple of days later, still curious about the outcome of random patterns, I located an old checkerboard and a small jarful of pennies to investigate this new game.

The simplicity and unpredictability of Life was intriguing and I realized that using coins was too cumbersome and left no record of the succession of generations. At that time, as a systems analyst for a large firm in Manhattan, I had access to an IBM mainframe computer and the following week wrote a program to "play" Life. Gardner had posed several challenges in his column and I set about to check them. My primary interest, however, concerned tracking the outcome of large areas randomly populated with "bits".

Since these computer runs required significant mainframe capacity, they were submitted for overnight processing. Initially, the jobs were aborted by the operators who thought the output was some sort of program error. After a few weeks, a summary of these "random broth" runs formed the basis of my first correspondence to Gardner about Life.

Late in October, I was delighted to receive a response (my first ever) from Gardner. In his letter he thanked me for solving one of the challenges and was awaiting confirmation from other readers. He mentioned that this was a common strategy for verifying the validity of material from readers responding to his monthly challenges. I also learned that Gardner did not work in an office at Scientific American headquarters on Madison Avenue but rather out of his home in nearby Westchester County. The geographic proximity offered an opportunity to personally meet with him, and on several occasions we did so to discuss developments readers were sending.

During one of our meetings, he mentioned a telegram (Fig. 2.1) he had received from a William Gosper at MIT claiming to have solved the biggest challenge of all,

Fig. 2.1 Seminal telegram from Gosper to Gardner

finding a finite pattern that endlessly replicates. The telegram contained coordinates for a small set of starting "bits" which would evolve into a glider gun. Gardner had no way of knowing whether or not Gosper's claim was valid and asked if I could possibly verify this for him. Around mid-November, I input this starting configuration into the program which produced several dozen generations confirming that Gosper's claim was indeed true. This information pleased Gardner who in turn notified Conway of the discovery.

Gardner also said that this particular column had generated an unprecedented volume of reader response including many discoveries of which some were new to Conway himself. He felt that a second column would soon be necessary and had to convince the magazine editors to agree to this. About this time, I suggested possibly starting a newsletter to serve as a clearing house to handle the large number of inquiries. Both Gardner and Conway agreed to this idea. In February 1971, Gardner wrote a column about cellular automata which presented more of the technical background upon which Life was based.

In March 1971, with Gardner's encouragement including a list of about 150 reader names and addresses, LIFELINE, a quarterly newsletter for enthusiasts of John Conway's Game of Life, was initiated (Fig. 2.2). This was mentioned in the April column along with details for an annual subscription of one dollar. Over the first year, a growing base of readers sent in more discoveries which provided new material for the newsletter (Fig. 2.3). Toward the end of 1971, during a weekend trip to Boston, I met Gosper and a few others including Ed Fredkin who headed the AI Lab there. The lab's computing capability which included a large circular CRT display was truly amazing. This was the first time I observed Life patterns rapidly

```
┌─────────────────────────────────────────────────────────────────────────┐
│ A QUARTERLY NEWSLETTER FOR ENTHUSIASTS OF JOHN CONWAY'S GAME OF LIFE       │
│                                                                           │
│   O      OOOOO OOOOO OOOOO O     OOOOO O   O OOOOO                         │
│   O        O   O     O     O         O  OO O O                            │
│   O        O   OOO   OOO   O         O   O O O OOO                        │
│   O        O   O     O     O         O   O  OO O                          │
│   OOOOO OOOOO O       OOOOO OOOOO OOOOO O   O OOOOO                        │
│                                                                           │
│ NUMBER 1                                                      MARCH 1971   │
└─────────────────────────────────────────────────────────────────────────┘
```

· Editor and Publisher - Robert T. Wainwright ·

What you are now reading is the prototype issue of LIFELINE, a newsletter for enthusiasts of John Horton Conway's game of 'Life'.
Scientific American having already devoted two full Mathematical
Games columns to this subject can not, obviously, continue to
provide the space required to report adequately on all the new
developments still occurring. Many readers (the writer included)
have expressed an interest to have some means by which they may
continue to exchange new developments. My own prior investment of
time and effort motivates me to establish this newsletter and I will
maintain it in proportion to the degree of interest expressed by you,
the 150 correspondents of Martin Gardner's October 1970 and February
1971 columns.

This first newsletter is compiled from information contained in your
letters to Martin Gardner and from experiments conducted by the
writer. Subsequent newsletters will necessarily depend upon the
extent of your response to LIFELINE. A subscription form is provided
for you and anyone you choose who would be interested in keeping
abreast of new Life developments. I will attempt to provide an
interesting mix of information in a free format and solicit your
comments and suggestions on how this could best be done.

John Conway first presented his game of Life to Martin Gardner early
last year. At that time he had followed the life histories of all
but one of the pentominoes, all but one of the hexominoes, and all
but seven of the heptominoes. By now we all know the fate of the
notorious R-pentomino which, in its first generation, becomes a hexomino (the one who's fate was unknown to Conway). This apparently
confused a number of readers who wondered how Conway could have
known about all the hexominoes as stated on page 122 of the October
column.

This leaves us with the seven 'unknown' heptominoes shown here
which Conway arbitrarily labeled B, C, D, E, F, H, and I.

```
┌─────────────────────────────────────────────────────────────────────────┐
│                  Conway's seven 'unknown' heptominoes                      │
│                                                                           │
│      B       C       D       E       F       H       I                     │
│                                                                           │
│    O OO     OOO     O       OOO     OO      OO      OO                      │
│    OOO      OOO     OOO     OO      O       O       O                       │
│    O        O       O O     OO      O       OOO     OO                      │
│                     O               OOO     O       OO                      │
└─────────────────────────────────────────────────────────────────────────┘
```

Heptomino B whose first generation appears in the 29th generation of
the R-pentomino eventually becomes three blocks, one ship, and two
gliders after 148 generations - so its history is known. This was
confirmed by Mr. Hugh W. Thompson of Lefrak City, New York.

Fig. 2.2 First issue of LIFELINE

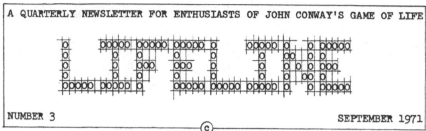

A QUARTERLY NEWSLETTER FOR ENTHUSIASTS OF JOHN CONWAY'S GAME OF LIFE

NUMBER 3   SEPTEMBER 1971

© . Editor and Publisher - Robert T. Wainwright .
1280 Edcris Road,Yorktown Heights,N.Y.10598
914-245-5150

New inquiries to LIFELINE are still coming in but (fortunately) the
rate is decreasing. With the readership base almost established at
more than 500, I can afford to devote more attention to the many new
developments which are occurring at a rate very nearly proportional
to time (since October 1970) squared. In order to maintain some degree of continuity in these series of newsletters, I will continue
to follow the general outline of previous issues. While doing so I
will point out new and the more interesting developments as well as
answer (and pose) some new questions about this incredible game.

When I first outlined
this issue in late
August, it appeared
that well over two-
thirds of the new de-
velopments were in
the area of Life deal-
ing with transfinite
objects. In early
September, I received
some information from
the group at the
M.I.T. Artificial
Intelligence Labora-
tory (Gosper, et.al.)
which convinced me
there are still many
surprises in the
finite kingdom of Life.
After reading this
issue, I am sure you
will agree that LIFE-
LINE Number Three will
very likely be remem-
bered for the extra-
ordinary achievements
reported.

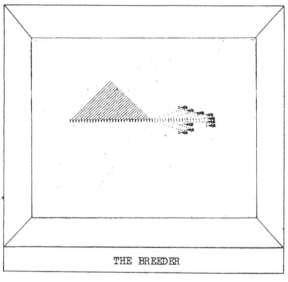

THE BREEDER

In this issue I will cover first, the area of Life dealing with
finite objects and events, then activity associated with one-dimen-
sionally infinite objects ('wicks') which includes fuses, next activity associated with two-dimensionally infinite objects (agars),
and finally developments in other selected areas of cellular automina.

The expanded classification system for finite Life objects makes it
more convenient to identify and place new discoveries. However, Class
V which includes Life events needs some further refinement. I will
go thru each subclass in turn discussing new developments. These are
definitely not in any order of significance.

Fig. 2.3 Number 3 of LIFELINE

Primary Class	Subclassified According To:		Subclass	Example*	Size
AN EXPANDED CLASSIFICATION SYSTEM FOR FINITE LIFE OBJECTS					
I-Still Lifes	Degree of symmetry	2-way orthogonal, 2-way diagonal, 90° rotational	I.A	block	4
		2-way orthogonal, 180° rotational	I.B.1	beehive	6
		2-way diagonal, 180° rotational	I.B.2	ship	6
		90° rotational	I.B.3	spiral	20
		1-way orthogonal	I.C.1	hat	9
		1-way diagonal	I.C.2	boat	5
		180° rotational	I.C.3	snake	6
		none	I.D	fishhook	7
II-Oscillators	Type of periodic activity	flip-flops (all period two)	II.A	blinker	3
		billiard table configurations	II.B	pinwheel	35
		inductors	II.C	tumbler	16
		pulsators	II.D	pentadecthalon	12
		shuttles	II.E	queen bee	20
		miscellaneous	II.F	M.I.T. osc.	18
III-Spaceships	Motion of spaceship	orthogonal	III.A	lt.wt.s.s.	8
		diagonal	III.B	glider	5
IV-Propagators	Nature of activity	spaceship guns	IV.A	the Gun	36
		puffer trains	IV.B	?	?
V-(Unstable (all objects not in above)	Final census	dies (θ)	V.A	bit	1
		Class I	V.B	latent block	3
		Class II	V.C	blinker pred.	4
		Class III	V.D.1	R-pentomino	5
		Class III (only)	V.D.2	glider pred.	5
		Class IV	V.E	Gun pred.	26
		unknown	V.F	acorn**	7

* smallest known object (for Classes II,III,IV-minimum phase)
** 'apparently' unknown - see text page six

Fig. 2.4 An early classification system

evolving rather than manually paging through mainframe output one page (generation) at a time. Fredkin suggested that Life might actually be the basis for a model describing how subatomic particles behaved.

Around the middle of 1972, Conway came to New York to meet with Gardner. During his visit, I had the fortunate opportunity to meet him and hear firsthand about how his idea for Life developed. He said that he was excited to learn of Gosper's discovery and could not believe the amount of interest Gardner's columns had generated. At that time, he posed a second challenge called The Grandfather Problem which asked: "Is there a configuration which has a father but no grandfather?". This challenge along with an offer of another $50 prize was included in the newsletter. Conway's second prize generated even more interest in LIFELINE, which by then had grown to nearly one thousand subscribers.

The initial society of Life enthusiasts were like a group of taxonomists, giving names to the wide variety of forms that were tumbling out of the S32/B3 rule (Fig. 2.4). This is just the opposite of what goes on in science. Ordinarily one starts off with a set of data and then attempts to determine what underlying principles or laws control these results. Life players had the underlying principle already (Conway's rule); they sought to discover the universe it implied.

Late in 1973, near the end of its third year, LIFELINE ceased publication. It had become too great a burden and time consuming to continue due to priorities of family, career, and other personal matters which had been long neglected.

Like many others at that time, I wondered if Life was just a superficial game or was there something of real significance implied in its deceptively simple rules. Gardner, in an earlier Scientific American article,[1] wrote the following concerning simplicity in nature:

> A closely related question is whether the natural laws themselves are simple or complicated. Most biologists, particularly those working with the brain and nervous system, are impressed by the complexity of life. In contrast, although quantum theory has become enormously more complicated with the discovery of weird new particles and interactions, most physicists retain a strong faith in the ultimate simplicity of basic laws. This was especially true of Albert Einstein who wrote: 'Our experience justifies us in believing that nature is the realization of the simplest conceivable mathematical ideas'.

It is remarkable how such a simple system of genetic rules can lead to such complex results. It may even be argued as Fredkin suggested earlier that the configurations so far examined correspond roughly to the subatomic level in the real universe. If a two-state cellular automaton can produce such varied and esoteric phenomena from these simple rules, how much more so in our own universe?

---

[1] Gardner, Martin, Mathematical Games. Scientific American, August 1969.

Chapter 3
Conway's *Life*

Harold V. McIntosh

The rules of Conway's *Life*, a two dimensional cellular automaton, are explained. Some of its characteristics and typical behavior are described. An algorithm to calculate periodic states using de Bruijn diagrams is explained.[1]

3.1 Introduction

The game of *Life* is an invention of John Horton Conway, a British mathematician, which was widely publicized in the early seventies through Martin Gardner's monthly column "Mathematical Recreations" in *Scientific American*. Programmers at numerous computer centers set out to explore the game with the machines at their disposal, those with access to visual display equipment having somewhat of an advantage, particularly if their equipment was interactive. Nevertheless many others apparently found pencil and paper adequate to obtain significant results.

Sufficient interest existed to maintain a quarterly newsletter, which was published by Robert T. Wainwright, for almost three years. The third issue was certainly one of the highlights of the series, reporting a wide variety of constructions, of glider guns, glider collisions, puffer trains and so on. A sufficient number of artifacts were found to enable Conway to demonstrate a universal constructor, which has been one of the principal goals of the theory ever since John von Neumann became interested in automatic factories.

The factories which he envisioned don't just make particular things; they work from a description of whatever they are supposed to build. Presumably one of them could be given its own description, making it universal in accordance with Alan Turing's sense of the theory of computability.

Gradually computer capacity and the theoretical understanding of the subject became saturated, although there were attempts in later years to apply some of the precepts of information theory to the game. A new surge of interest was provoked

---

[1]The paper is written in July 4, 1988.

A. Adamatzky (ed.), *Game of Life Cellular Automata*,
DOI 10.1007/978-1-84996-217-9_3, © Springer-Verlag London Limited 2010

from another direction when detailed computer experiments with nonlinear differential equations began to illuminate some of the more abstruse theories of topological dynamics. Quite independently of the fact that cellular automata define the continuous mappings within one esoteric approach to the subject, Stephen Wolfram thought that he saw an application of Stephen Smale's theory of strange attractors to one dimensional cellular automata. He created a classification into four categories for the occasion.

The early eighties saw intensive activity on Wolfram's part, exploring the evolution of linear cellular automata of one kind or another, with the intention of relating their behavior to the theory of computational complexity as well as the topologist's classification of the asymptotic behavior of dynamical systems. Some insights that were gained from this approach, such as the use of the Bruijn diagrams to discover the periods of cycles of evolution, can be applied to the original game of *Life*, and so resolve some of the questions which were left unanswered during the original excitement.

It is also true that the computers with which *Life* can be studied have improved over the years, both in terms of capacity and of availability. Results which were previously out of reach can now be computed, with the prospect of obtaining still further results as the performance of computers improves. Nevertheless one of the lessons to be learned is the exponential growth of all such computations; what can be calculated will always remain relatively modest.

Attempts to modify Conway's rule have not met with much success, although variants have been considered. The apparent 2^{2^9} different choices of a rule is illusory; rules without rotational and reflective symmetry are not acceptable, nor are rules without a quiescent state. Given restrictions such as totality, thousands or millions of candidates remain—many, but not entirely impossible, to sort through.

Some promising three dimensional analogues of *Life* have been found, but working with higher dimensions is clearly a formidable proposition.

3.2 The Rules of the Game

Automata are characterized by having states, and rules for deciding whether and how to change their states from time to time. The cells making up cellular automata all change their own states simultaneously, according to the states of their neighbors in the lattice in which they are situated. Consequently cellular automata are classified by the basic number of states their cells can display, and the exact form of their neighborhood.

With just two states per cell, binary automata are the simplest. The distinction between "living" and "dead" or "inactive" gave the published version of *Life* a readily understandable ecological interpretation, then a very popular concern.

As for their neighborhoods, the first consideration is dimension and the second radius. One dimensional automata are very restricted in the way that connections can be eventually be established between the cells; perhaps a reason that they were generally disregarded before Wolfram's interest. Two dimensional automata are more satisfactory, although von Neumann had to exercise considerable ingenuity to be able to construct crossing wires which would not short circuit one another. Two dimensional automata seem to suffice; more dimensions would increase the complexity of all computations without necessarily providing any better insight, granted that the conclusions to be drawn from them are highly symbolic anyway.

Even though a two dimensional network has been chosen, there are choices between square, triangular and hexagonal lattices, as well as the exact constitution of the neighborhood. Von Neumann used a square lattice, taking single neighbors on each side in both the horizontal and the vertical direction, but he needed a large number of states to accomplish his purposes (28 plus a quiescent state). Conway settled for fuller neighborhoods, restricted to binary cells. Thus *Life* uses diagonal neighbors as well as lateral neighbors; a neighborhood contains altogether nine cells, eight of them bordering the central cell.

Finally, the rule of evolution has to be chosen; Conway's final choice was

- a live cell survives if it has two or three live neighbors
- a new cell is born whenever there are three live neighbors
- all other cells either die or remain inactive

Fanciful interpretations assuming some strange kind of microbe provide a useful metaphor for discussing the game, although not to be taken too seriously.

Conway's criterion was that the rule should neither lead to populations which quickly died out, nor which expanded without end from limited beginnings. Through Martin Gardner, he challenged the readers of *Scientific American* to find out whether long term growth was nevertheless possible. Two mechanisms were suggested.

The first was by the construction of a "glider gun." Early experimentation with the rule had revealed an aggregate of five cells which moved across the *Life* field in a diagonal direction, advancing one square every four generations. If some mass of live cells could be found which emitted gliders periodically, it would produce the growth he was seeking. Evidence for the existence of some such configuration was the fact that gliders were frequent evolutionary products in random fields that had been observed. William Gosper at MIT soon found one glider gun; then another, structurally different one, was found. Since then no more have been reported.

The second possibility was that there were more complicated objects than gliders which could also move about a *Life* field, leaving stable debris behind. Belief in their existence was encouraged by the fact that certain larger structures, which were christened "space ships," had been found which moved horizontally or vertically, and that numerous small stable structures had been observed in the evolution of random fields.

Interestingly enough, a variety of "puffer trains" were also found, but more interesting still was the discovery that there were puffer trains that left gliders as debris in just such a fashion that they could collide to form glider guns.

This last combination, a mass whose periphery increases with time, violated Conway's concept of a stable ecology in the worst possible way; nevertheless it was a carefully engineered and delicate construct, and not a combination likely to be found by chance. The behavior of the rules which he had discarded was evidently much cruder.

Fig. 3.1 Some small still
lifes

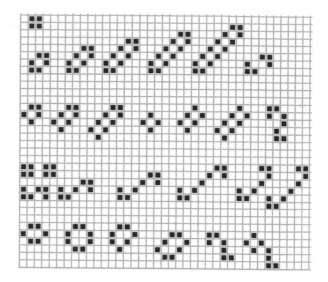

3.3 Still Lifes

Since finite automata eventually evolve into a cycle of states which repeat indefinitely thereafter, cycles are one of the first things to look for; the same principle
holds for infinite automata as well, even though such repetition is not obligatory.
The shortest possible cycle, with period 1, would be a configuration of states which
never changes; they are called still lifes.

A field in which all cells are inactive remains so; such quiescence was deliberately included in the definition of *Life*. Examination of the evolution of random *Life*
fields reveals many still lifes formed from small clusters of live cells; the simplest is
a 2 × 2 square in which each cell has exactly three neighbors, and thus persists from
one generation to the next. Cells outside the cluster have at most two live neighbors,
which is never enough for any additional cells to be formed. Thus the conditions for
a still life are met.

An empirical search of random fields reveals configurations that are easily
formed, especially if a very limited sample is taken. It needs to be supplemented
by a more thorough procedure, so it is not surprising that much of the response to
Gardner's article consisted in carefully recording the evolution of all the small objects that it was possible to form. Gradually a consensus was built up concerning the
catalog of still lifes up to twelve pixels, including the existence of several families
of still lifes and rules for generating them.

Figure 3.1 shows the 21 symmetry classes of connected still lifes containing eight
live cells or less. The symmetry of *Life*'s evolutionary rule ensures that objects differing by rotation or reflection undergo a similarly rotated or reflected evolution, so
only one representative of each class has to be shown.

Fig. 3.2 Alternators due to overcrowding

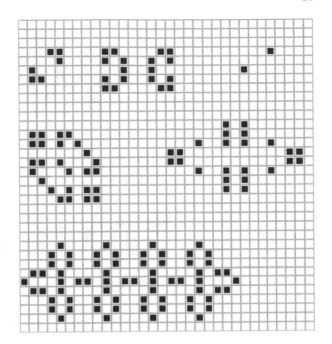

3.4 Period Two

The structures called blinkers, consisting of three live cells in a row, alternate in form between three horizontal cells and three vertical cells. Next to square clusters of four blocks, they are the most common debris remaining after evolution has run its course. They exemplify structures with period two, which, like the smaller still lifes, often arise during the evolution of random fields.

Alternators, or period two oscillators, tend to be rather fragile. Nevertheless, they are easily constructed, once the basic technique has been grasped. There are really two different kinds of alternators, depending on whether cells disappear from isolation or from overcrowding. Of course, it is possible for a large configuration to incorporate both kinds of alternators.

Figure 3.2 shows an assortment of alternators which depend on overcrowding. The neighbors of the deceased remain in place, giving birth to a new cell which they promptly crush, repeating the cycle indefinitely.

Figure 3.3 shows some typical examples of the converse situation, in which the field contains only isolated cells or pairs of cells, which will surely disappear. However, their mutual placement is sufficient to reproduce a similar configuration in the next generation. Then, under the proper conditions of symmetry, this second generation recreates the first, establishing a permanent cycle. Extremely large and intricate networks can be built up from a few basic configurations.

Fig. 3.3 Alternators due to isolation

3.5 Gliders

Gliders were probably the most unexpected characteristic of Conway's *Life*, and in the end the one responsible for its great complexity and utility. Conway quickly realized that gliders were the substitute for von Neumann's wires as the mechanism for transporting information from one place to another in the *Life* field. They form an essential ingredient of his demonstration of a self-replicating configuration for *Life*, which does not detract from their interest for other applications.

Figure 3.4 shows the four phases of a glider, which would constitute a cycle except for the displacement. Consequently gliders can never have a finite period in an infinite *Life* field; an exception to the theorem that finite automata must eventually retrace a cycle of states, except for the finiteness condition. Allowed to evolve on a finite torus, gliders will have periods sufficiently long for them to traverse the length of the torus and return to their starting position.

A variety of glider precursors were reported in Wainwright's newsletter; in view of the asymmetry of the glider itself, it would seem that the principal requirement for the precursor is that it or significant segments of it also be asymmetric. For all of their interest and importance, gliders are not hard to find in random *Life* fields; it is reported that they were first observed wiggling their way across a video display.

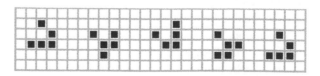

Fig. 3.4 The cycle of evolution of a glider

Fig. 3.5 The three primitive space ships

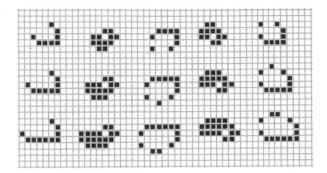

Their prevalence is consistent with the fact that they are formed from only five live cells, since the smaller stable figures are by far the most common residues of long term evolution.

The search for other moving configurations eventually turned up the three space ships shown in Fig. 3.5. They and their symmetric images are capable of horizontal or vertical motion. In principle the central section of the space ships could be stretched out to arbitrary lengths, but in practice the designs shown are not quite correct; they are always accompanied by some additional "sparks," which promptly die out.

The sparks are too big to die out in the longer space ships, but they can be suppressed by stacking a series of smaller space ships alongside the longer ones to produce a flotilla which will travel together smoothly.

Certain other combinations produce the puffer trains which Conway had foreseen. However, gliders and the three small space ships are the only primitive structures which have ever been found capable of self propulsion.

3.6 Oscillators

Compared with the case with which still lifes or alternators can be formed, it is quite difficult to form configurations with longer periods, so that they are not often found as transients during the evolution of random fields nor among the debris which remains when they have stabilized. Nevertheless their construction is not impossible, and oscillators of every small period have been found. As with alternators, they tend to fall into different categories depending upon the mechanism involved.

One scheme is to create vacancies and then replacements for them, a common aspect of the two classes of alternators. If the replacement can be deferred, a longer cycle that of period two can be created. Figure 3.6 shows a cycle of length 3 created in this manner.

Another mechanism is to note that some configurations are naturally expansive, such as those which evolve from a long row of live cells. However, as the center of the row expands laterally, the end of the row contracts longitudinally; the competition between these two tendencies can result in an alternation which leads to a cycle

Fig. 3.6 A period 3 oscillator

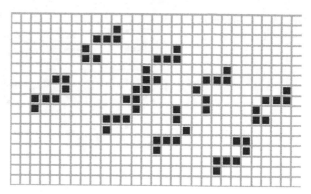

Fig. 3.7 A period 15 oscillator based on expansion and contraction

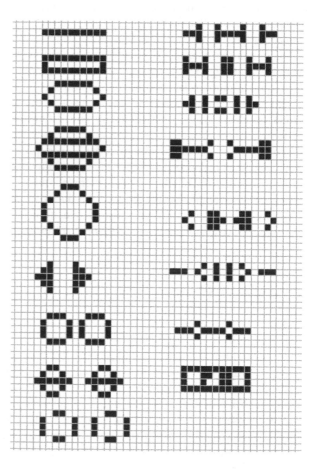

of long period. Experimentation when *Life* was still new led to the discovery that a row of ten live cells would reproduce itself after fifteen generations, as shown in Fig. 3.7.

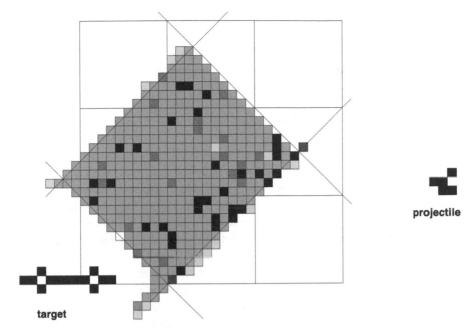

projectile

target

Fig. 3.8 Map of period 15 oscillator collisions with a glider

Another configuration which behaves similarly is formed by two solid 3×3 squares staggered diagonally, touching at their corners.

Yet another arrangement which can be exploited for arbitrarily large assemblages of live cells is to surround figures which are known to be highly expansive with other highly stable figures which are known to dampen expansion. Under favorable circumstances the central figures will reconstitute themselves from the debris, phoenix like, to initiate a new cycle. By repeating itself periodically the combination becomes an oscillator whose period depends roughly on the distance from the center to the periphery.

Many experiments have been performed to observe collisions of moving objects with each other, still lifes, or oscillators. Figure 3.8 shows the possible initial positions for collisions between a glider and the period fifteen oscillator. Most are messy, but dark squares show where the glider will simply be eaten, having struck inessential trash in the oscillator. For just one arrangement, the glider reflects. By bouncing in between two oscillators, a composite oscillator with an arbitrarily long period is possible. Collisions which turn a glider by 90 degrees rather than 180 degrees are extremely rare.

3.7 Glider Guns

The approach of combining "eaters" with naturally expansive configurations led a group of students in Ontario to large numbers of interesting oscillators of diverse periods; over a hundred of them according to reports. However, a slightly different approach led the group at MIT to two different glider guns.

A variant on the line of ten cells is a pair of lines of fifteen cells, both are similar in evolving through a cycle of expansion and contraction, but the fifteen cell configuration leaves behind enough debris to destroy itself rather than continuing through its second cycle. The clever realization that the debris could be broken up by a strategically placed square block to act as an eater led to another shuttle based oscillator.

Thanks to a curiosity as to the consequences of collisions and near collisions, a pair of shuttles were set in motion in a way that led them to just barely graze each other. Altogether many distances of approach and relative phases were tested before a combination was found which produced pure gliders every time the shuttles met. That became the glider gun which Conway had conjectured. It won the prize offered in Martin Gardner's column, and led the way to quite an interesting series of further discoveries.

Besides investigating all kinds of shuttle collisions, there has always been a search for additional shuttles. Although it would seem that there is an impossibly large number of possibilities to sort through, a certain faith that relatively small structures will be significant can be combined with the fact that the evolution of all the different small clusters of live cells has been tracked. Thus there is a certain data base from which one can work.

No other reasonable explanation seems to exist for the fact that a second glider gun was discovered, consisting of another self reversing shuttle in which eaters could be inserted to clean up unwanted scraps of its field.

Pairs of shuttles of the second type, this time colliding at right angles, when correctly combined with each other, again produced a glider gun; it is generally referred to as the "new gun." It is indeed fortunate that the new gun exists and was discovered, because it turns out to be essential for some of the advanced constructions which have been made.

It is difficult to say whether any additional shuttles or glider guns remain to be found. The intense search which revealed the ones that are now known should have discovered others, if they existed; but there are many starting points.

Figure 3.9 shows the glider gun in one of its phases, together with the stream of emitted gliders. Variations on the theme are possible, in which the confining blocks are removed and a cascade of shuttles is allowed to generate parallel glider streams. However their separation is not convenient for certain applications, which is why it is fortunate that the new gun was discovered.

Figure 3.10 shows the new gun in one of its phases, together with the stream of emitted gliders. It has a period of 46, in contrast to the regular glider gun whose period is 30. Consequently the gliders which it produces are spaced further apart; every second glider produced by the period 46 gun is 23 cells. Successive gliders are out of phase by two generations, having been produced by shuttles moving in directions opposite to those of the previous pass.

Fig. 3.9 Glider gun, together with a stream of gliders

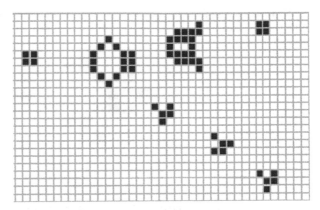

Fig. 3.10 The new gun, with a freshly generated glider

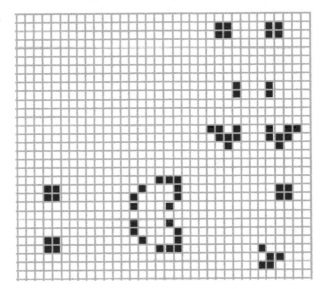

3.8 Puffer Trains

Shuttle precursors tend to turn back on themselves, so the natural tendency is to place eaters at strategic locations to absorb the debris which prevents them from becoming oscillators. There are other small configurations which advance steadily, reproducing portions of themselves, until their leading edge is overwhelmed by advancing debris from behind. There are two different approaches to salvaging the wavefront which have been found effective.

One of them is to place eaters behind them in the hope that they will both control the debris, and somehow reproduce themselves. Two versions of one rather slow diagonal puffer train have been found by using this technique.

The other approach uses space ships as escorts, feasible if the velocity of the advancing front in the same as the velocity of the space ship. The periodic sparks cast off by the space ship will sometimes interact to control the expanding debris

Fig. 3.11 A "smokeless" puffer train escorted by a space ship

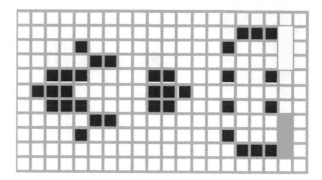

without affecting the integrity of the space ship. Figure 3.11 shows a clean example, in which the pair of space ships and the heptomino advance one space every second generation, but with an overall period of twelve.

Gliders frequently inhabit the cloud lagging behind a puffer train, making it worthwhile to adjust the escort to allow them to escape, yet cleaning up the rest of the cloud. Given that there are known glider collisions producing all the smaller artifacts, the way is open to use multiple puffer trains to create gliders whose collisions produce other desirable objects in their turn.

3.9 Life on a Torus

Although *Life* was intended for an infinite square lattice, it is possible to deform the lattice into a torus by identifying cells which have a fixed separation from one another. This procedure is equivalent to working with configurations which have a spatial periodicity. Since it transforms the infinite lattice into a finite structure, there is some hope of discovering its properties by exhaustive enumeration.

Limitations of both computer space, but principally speed, restrict the dimensions of the cases which can be treated, the present limit being an area encompassing approximately 32 cells. Even so, much data can be gathered, allowing definitive, conclusions about some of the simpler configurations to be reached, and revealing some interesting patterns which were not previously known.

The simplest mapping of the infinite plane onto a torus is to identify all cells differing by a constant horizontal or vertical distance. Conversely, a rectangle with opposite edges identified may be mapped onto the plane by making a new copy of the rectangle whenever crossing one of its boundaries, instead of reentering the opposite side.

More extensive planar areas may be explored by making alternative identifications of the opposite edges of a rectangle, the equivalent of forming a Möbius strip or a Klein bottle. However, it is easy enough to accumulate voluminous data from the simplest mapping, raising the doubt that collecting still further data would contribute further insight. It is true that the feasible dimensions reveal gliders but none of the space ships, much less shuttles, glider guns, puffer trains, or any of the other

Fig. 3.12 A sampling of still
lifes on a 4 × 8 torus

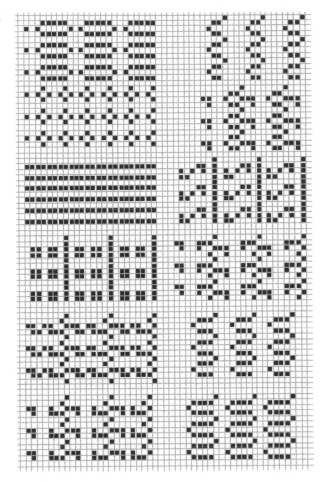

interesting objects. However, their study provides sufficient insight to form fairly
general conclusions.

Figure 3.12 shows some of the still lifes for a 4 × 8 tours, found through an
exhaustive search. Some of them belong to recognized families.

The extreme case of an 1 × *n* torus does not seem very interesting at first, but it
has two advantages. Values of *n* as large as 34 have been completely analyzed, some
of the most interesting detail emerging in the vicinity of *n* = 30. So, it would have
been missed for even a 2 × *n* torus. Better still, the 1 × *n* automaton is really one
dimensional, so subdiagrams of the de Bruijn diagrams used in shift register theory
suffice to calculate explicitly all the configurations of any given period.

Once the relevance of the de Bruijn diagrams has been seen, they can be applied
to 2 × *n* tori, 3 × *n* tori, and so on. The amount of calculation they require rapidly
becomes overwhelming, but their theoretical implications are always valid.

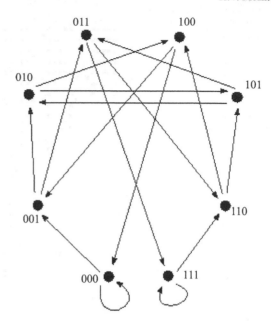

Fig. 3.13 A three stage
binary de Bruijn diagram

3.10 Cycles of Finite Periodicity

One of the advantages of working with *Life* on a finite torus is that there is an algorithm by which the cycles of finite periodicity can be determined. In one dimension the procedure actually finds all the periods on an infinite lattice, but in applying the procedure twice to get to two dimensions the first stage has to remain finite. This is probably a fundamental limitation, given the fact that it is known that many properties of a two dimensional lattice are recursively insoluble, in Turing's sense. Thus the most to be expected are results valid for structures of finite periodicity.

A de Bruijn diagram is a very simple thing, merely a diagrammatic technique to keep track of sequences of digits of a given length. A binary de Bruijn diagram of three stages contains all possible sequences xyz of three bits, eight sequences altogether. Suppose that they are placed around the circumference of the unit circle, forming an octagon. Arrows are drawn to link the sequences according to how they are initial or terminal segments of one another, arrows are to be drawn from all vertices wxy to vertices xyz; for example 001 links to 010, also to 011.

In arithmetic terms, links always run from vertex i to vertices $2i$ and $2i + 1$, modulo 8. Figure 3.13 shows the octagon with its links. With respect to the rules of *Life*, the relevance of the de Bruijn diagram is that it shows how to build up long chains of cells out of overlapping neighborhoods; precisely, it shows which neighborhoods can overlap and which can not.

If the number of stages is just one less than the length of the neighborhoods, links in the diagram match the neighborhoods exactly, classifiable as "good" or "bad" according to the neighborhood. For still lifes, a "good" neighborhood evolves into its central cell. If all the "bad" links are discarded, paths along the surviving links describe chains of cells whose cores remain unchanged in the next generation.

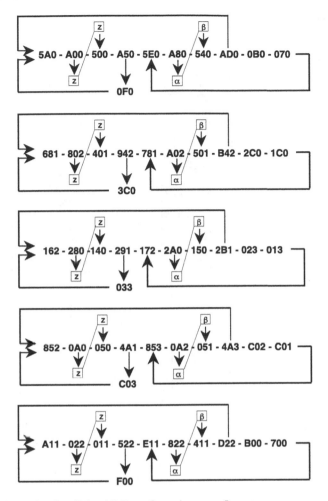

Fig. 3.14   Strip overlap for all the .1(1,0) configurations on a 5 × *n* torus

Shrinkage due to the loose ends can be avoided by discarding all links which are not parts of closed loops. Every possible still life can be read off from the remaining diagram. No ingenuity is involved; the procedure is an algorithm. Long and cumbersome it may be, but less so than testing every configuration; indeed the only configurations actually tested are just the neighborhoods by which the evolutionary rule is defined.

Cycles with longer periods require the longer neighborhoods of the iterated rule of evolution; non-binary automata require the de Bruijn diagram of the corresponding number base. That is the secret to working with *Life* on 2 × *n* tori; a vertical section of two cells is taken to be a single cell of four states. Then the de Bruijn diagram for a four state linear cellular automaton provides the information; similarly a 3 × *n* torus becomes an eight state linear automaton.

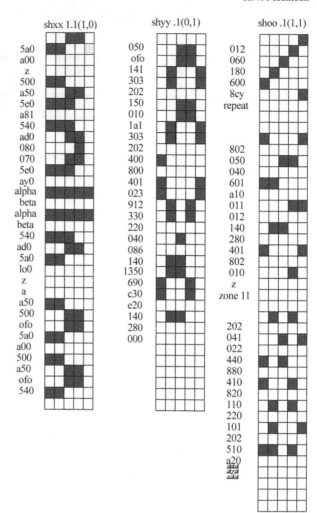

Fig. 3.15 Sample shifting configurations on a $5 \times n$ torus

Fortunately there is an alternative to treating vertical slices as a single cell with a huge number of states. A vertical strip of three cells is sufficiently wide to calculate the central cell of the next generation, and even to compare it with any of the nine members of its neighborhood. Not only still lifes, but shifting patterns can be detected. The first step is to generate the diagram for this single vertical strip. Next, the vertical strips have to be fitted together to fill up the plane.

General fitting seems to be a difficult proposition, but it is easy enough to work out sequences three cells wide which fit into a common horizontal strip, by selecting closed loops in the diagram whose length is equal to the selected width. All such loops are used as nodes in a new diagram whose links indicate that the two left cells of one match the two right cells of the others when they are overlapped. Following a path through the second diagram builds up vertical sections which eventually fill out the plane.

Figure 3.14 shows the strip overlap diagram for the some configurations on a $5 \times n$ torus which correspond to configurations which move horizontally by one cell in each generation.

Figure 3.15 shows some examples of creepers and crawlers on a $5 \times n$ torus, gotten from strip diagrams similar to those of Fig. 3.14.

Chapter 4
Life's Still Lifes

Harold V. McIntosh

The de Bruijn diagram describing those decompositions of the neighborhoods of a one dimensional cellular automaton which conform to predetermined requirements of periodicity and translational symmetry shows how to construct extended configurations satisfying the same requirements. Similar diagrams, formed by stages, describe higher dimensional automata, although they become more laborious to compute with increasing neighborhood size. The procedure is illustrated by computing some still lifes for Conway's game of *Life*, a widely known two dimensional cellular automaton.[1]

4.1 Introduction

Public attention was drawn to cellular automata by Martin Gardner's monthly column *Mathematical Recreations*, a regular feature of *Scientific American* for many years. The October, 1970, issue [2] featured the game of *Life*, which had been invented about that time by the British mathematician John Horton Conway. Sufficient interest was aroused by the game for it to be followed up in several later columns, and to support a newsletter [8] for nearly three years. Gardner's columns have now been collected into one of the compilations that are regularly published by W.H. Freeman and Company [3], while Conway's own version of the game is available in the recent Academic Press book [1] *Winning Ways*.

However, there had been much previous interest in cellular automata, beginning at least with the work [5] of Warren McCulloch and Walter Pitts on neural nets, later including John von Neumann's investigations [7] into self reproduction and automatic factories. Interest still continues, a recent example being Stephen Wolfram's examination [10] of one dimensional automata from the point of view of chaos in complex systems theory.

---

[1] This paper is written in September 10, 1988.

A. Adamatzky (ed.), *Game of Life Cellular Automata*,
DOI 10.1007/978-1-84996-217-9_4, © Springer-Verlag London Limited 2010

One of the fundamental expectations in the theory of automata is that the automaton will eventually settle down into a fixed cycle of states, which will then characterize its long term behavior. Some modification of this principle must be expected for infinite automata, nevertheless the search for states of low period constituted an important part of the activity inspired by the announcement of *Life*. Someone with a crystallographer's frame of mind might well have undertaken a classification of all such states, beginning with those of period 1, which Conway called "still lifes." This article describes how such a classification can be obtained.

4.2 Cellular Automata

Mathematically, an automaton consists of a set of states, together with a set of mappings of the state set into itself. Each mapping is identified with a signal, which is supposed to cause a change of state. Signals can therefore be considered as inputs to the automaton, which in turn could be considered as a neural net, an electronic circuit, or some other structure. In that case, outputs might also considered, and altogether the groundwork has been laid for some kind of fundamental theory of computation, or at least of computing devices. Much of the theory of automata proceeds in that direction.

Cellular automata are those for which a large number of similar automata — the cells — are connected together in some regular pattern, and for which the signals are the information which each cell has concerning some of its neighbors, most likely including self-awareness. From time to time the cells change their state, according to this knowledge. McCulloch and Pitts would have the connectivity of the cells modelling some physiological system, but lacking definite structures to follow, the tendency has been to use crystallographic lattices of low dimension. Von Neumann worked with two dimensions, which was also the arena for Conway's game.

Life presupposed binary cells occupying a two dimensional square lattice, the neighborhood of each cell consisting of itself, four lateral neighbors, and four diagonal neighbors; a total of nine cells altogether. Many other combinations are possible, but Conway chose one of them, as well as a particular rule of transition, for his game after discarding many alternatives. Adopting his picturesque ecological metaphor, binary cells are either dead or alive; in each generation,

- new cells are born to three live "parents"
- old cells survive if they have two or three live neighbors
- all other cells either die or remain dead

There are 2^9, or 512, different combinations of dead and live neighbors. Each combination can evolve in its own way, giving the enormous number of 2^{2^9} different rules, or games, which Conway could have chosen; nevertheless that one choice has lived up to his expectations of finding an interesting game. Part of the choice consisted in selecting a symmetric rule; it is reasonable to suppose that a square lattice would evolve similarly if it were rotated or reflected, as well as if any configuration were shifted to on side by a given distance. Beyond this, the rule depends on *numbers* of live neighbors, not on particular groupings.

Whatever the reasons for choosing one rule in preference to another, the analysis which follows is applicable to all cellular automata; so *Life* just happens to be a particularly interesting special case. Consequently its results are available for comparison and checking against other rules.

4.3 Still Life

For any given rule of evolution, some cells will retain their existing state, while others change; generally there is no correlation between the two alternatives, giving the automaton a different appearance from generation to generation. Still, it could happen that there are particular combinations which remain immobile. One such, deliberately included in Conway's choice of a rule, is that if a cell and all its neighbors are dead, it remains dead. Automata which follow this requirement are said to have a quiescent state; live cells cannot appear spontaneously, but only near other live cells.

A quiescent state is not usually considered to be a still life; the latter term is reserved for collections of live cells for which cells neither die nor are born with succeeding generations.

It is curious that the simplest possible approach, if managed properly, suffices to enumerate the still lifes for an automaton. To begin with, neighborhoods can be classified as "good" or "bad" according to how their cell evolves. For the moment, a good neighborhood is one whose cell remains fixed; a cell that changes state is in a bad neighborhood. Obviously we are only interested in good neighborhoods.

The next step checks the neighborhoods of the neighbors; but not all pairs of neighborhoods overlap consistently. Only good neighborhoods that overlap well need be considered. The direct approach proceeds along some path, fitting good neighborhoods together until an inconsistency results. By backtracking and considering alternative neighborhoods, it might be possible to generate a whole region which is unaffected by evolution. Trying again and again until all the possibilities are exhausted would eventually produce a complete list of static regions.

The whole plane can never be covered by this process; but it would be possible to stop when a quiescent border was reached, or even if the region began to repeat itself after a certain distance. So, at the very least, it should be possible to find all the still lifes covering a fixed area. Of course, the quantity of computation required grows exponentially with the area to be covered.

Giving first priority to the compatibility of overlapping neighborhoods, later rejecting those whose evolution is not satisfactory, places the computation on a firmer foundation. In either case, there is a simple diagram from which the compatibility of neighborhoods can be ascertained.

4.4 De Bruijn Diagram

The representation of overlapping sequences and the establishment of some of their properties is facilitated by using a diagram which is often called the de Bruijn diagram, or its associated connectivity matrix. The concepts involved have had a fairly long history [6]; sometimes the diagram is given other names.

Basically, a diagram is prepared whose nodes represent short segments taken from a sequence; an example would be a string of three binary numbers, eight nodes corresponding to all the possible sequences. Links are drawn in the diagram according to the ways that the first member of the sequence can be discarded, and a new final member appended; it is natural to label the links by the longer segments, including the discarded and adjoined elements.

In this binary example, 0 could be discarded from the sequence 011; then 0 adjoined to produce 110 or else 1 to produce 111. Accordingly 011 would be linked to each of the nodes 110 and 111, but no others. The first link would be labelled 0110, the second 0111.

In the case of *Life*, and for automata in general, we are interested in dissecting the neighborhood of a cell into two overlapping pieces, each of which overlaps an appropriate partner among adjoining neighborhoods. In the following sample,

a	b	c	d	e
f	g	h	i	j
k	l	m	n	o

the cells g, h, and i have the respective neighborhoods and partial neighborhoods

Cell	Neighborhood	Left half	Right half
g	$\begin{array}{ccc}a&b&c\\f&g&h\\k&l&m\end{array}$	$\begin{array}{cc}a&b\\f&g\\k&l\end{array}$	$\begin{array}{cc}b&c\\g&h\\l&m\end{array}$
h	$\begin{array}{ccc}b&c&d\\g&h&i\\l&m&n\end{array}$	$\begin{array}{cc}b&c\\g&h\\l&m\end{array}$	$\begin{array}{cc}c&d\\h&i\\m&n\end{array}$
i	$\begin{array}{ccc}c&d&e\\h&i&j\\m&n&o\end{array}$	$\begin{array}{cc}c&d\\h&i\\m&n\end{array}$	$\begin{array}{cc}d&e\\i&j\\n&o\end{array}$

The de Bruijn diagram for *Life* and other automata based on the same neighborhood has 64 nodes, due to six binary cells forming each overlapping half. Eight links emanate from each node, since three binary cells are discarded and three added to advance from one neighborhood to the next. Links and neighborhoods correspond exactly; there are 512 altogether. Octal notation readily labels the nodes; if

$$\alpha = 4a + 2f + k$$
$$\beta = 4b + 2g + l$$
$$\gamma = 4c + 2h + m$$

then the link $\alpha\beta\gamma$ joins the node $\alpha\beta$ to the node $\beta\gamma$.

The large number of nodes and links would make such a diagram laid out on a small sheet of paper overly crowded; a better representation would be the connectivity matrix of the diagram, or even a simple listing in which each line contained its own node followed by a list of the nodes to which it was linked.

The choice of a de Bruijn diagram whose links are the neighborhoods in Conway's *Life* means that any path through the diagram represents a possible row of cells in the automaton, surrounded by their respective neighborhoods. The nodes are partial neighborhoods; the lack of a link between a particular pair shows that they cannot be overlapped to form a complete neighborhood.

Insofar as the links represent neighborhoods, they can be considered to reflect the properties of their neighborhoods as well. By dropping the links corresponding to bad neighborhoods, any remaining paths through the diagram can only represent a good row, which in the present context would be a row of a still life. In other words, there exists a diagram from which all the still life rows can be read off just by following paths through the diagram.

Building up still life rows is only the first stage of construction; a new second stage de Bruijn diagram governs the overlapping of rows to cover the plane.

4.5 First Stage

The maximal first stage de Bruijn diagram has 64 nodes connected by 512 links. The nodes are representable by a pair of octal numbers or equivalently, by a single number modulo 64. In that case, it is easy to describe the connectivity matrix M_{ij}, whose elements are defined by

$$M_{ij} = \begin{cases} 1 & i \to j \\ 0 & \text{otherwise} \end{cases}$$

This matrix is shown in greater detail in Fig. 4.1.

By inspection, $Trace(M) = 8$ and M^2 is a matrix solidly filled with 1's. Therefore $Trace(M^2) = 64$ and M satisfies the minimal equation

$$M^3 = 8M^2$$

from which its characteristic equation can be obtained. It is evident from calculating traces of its powers that there are many loops of all possible lengths in the diagram. There are also numerous Hamiltonian loops, these latter passing through all 64 different nodes, giving the longest possible consecutive sequences of neighborhoods that can be formed without repeating one of them.

Dropping links from the complete de Bruijn diagram will break some loops, maybe even isolate certain nodes completely. Normally such artifacts would be discarded; if a node had no exit links, it would mean that there was a partial neighborhood for which no right border existed so that the central cell would remain constant

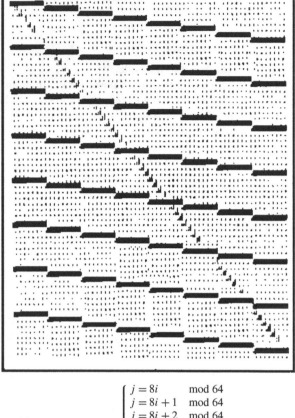

$$M_{ij} = 1 \quad \begin{cases} j = 8i & \mod 64 \\ j = 8i + 1 & \mod 64 \\ j = 8i + 2 & \mod 64 \\ j = 8i + 3 & \mod 64 \\ j = 8i + 4 & \mod 64 \\ j = 8i + 5 & \mod 64 \\ j = 8i + 6 & \mod 64 \\ j = 8i + 7 & \mod 64 \end{cases}$$

$$= 0 \quad \text{otherwise}$$

Fig. 4.1 The full first stage 64×64 de Bruijn matrix

in the next generation. Such partial neighborhoods are barriers to extending a region of still lifes, therefore inappropriate to an infinite plane.

Such barriers would terminate any recursion reaching them in the direct approach to still lifes; the advantage of using de Bruijn diagrams is just that their occurrence is clearly located in the context of the finite number of distinct sequences of neighborhoods which exist.

Although emphasis has been placed on still lifes, it should be noted that the evolved cell can be compared to any of the other cells in the neighborhood from which it arose, not just the original cell. The simplest comparisons are with the shifted cells; by symmetry the interesting ones would be the longitudinal neighbor

Fig. 4.2 The de Bruijn
matrix for *Life*'s still lifes

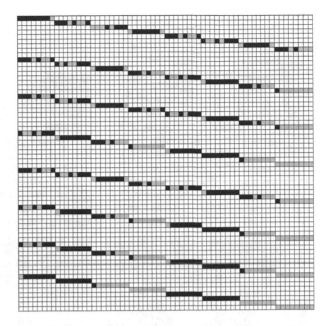

in the direction of progressive overlapping, the transversal neighbor perpendicular
to the direction of extension, and the diagonal neighbor. Using the symbol (x, y, t)
to designate a pattern in which a cell matches the one x cells to the right and y cells
up after t generations, the fundamental neighborhood is large enough to detect the
patterns $(0, 0, 1)$, $(0, 1, 1)$, $(1, 0, 1)$, and $(1, 1, 1)$. The first of these are the still lifes,
the others might be called gliders. Strictly, Conway's glider corresponds to the pattern $(1, 1, 4)$, his space ships to $(0, 2, 4)$ or $(2, 0, 4)$.

In principle any Boolean combination of the evolved cell with the cells of its
neighborhood could form the basis of a pattern. For example, a pattern could consist
of configurations which vanished after a single generation.

Figure 4.2 shows the connection matrix of the first stage de Bruijn diagram for
still lifes, from which its resemblance to the complete diagram can be judged. A better understanding can be formed by inspecting several of the lower powers of the
matrix, as shown in Fig. 4.3.

The block structure of M is evident by the power M^3; the fact that M^2 is still
fairly sparse but M^3 is not indicates the presence of an important component of
period 3. This situation has been strongly evident to those who have examined large
empirical collections of still lifes, among which the length three is a magic number.
In terms of the properties of positive matrices, an explanation would be that the
second largest eigenvalue was triply degenerate (in absolute value).

Although there is probably no real "explanation" of why 3 is magic, those who
have worked with *Life* have noticed that the still lifes for a 3×3 torus are just
those configurations with four live cells, an easy requirement to satisfy. The same
rule works for a general $3 \times N$ torus, although an additional family of still lifes is
also possible. Apparently breaking the periodicity in the cross direction still leaves

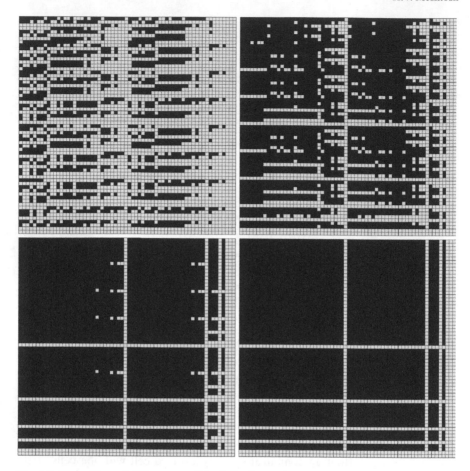

Fig. 4.3 Second through fifth powers of the de Bruijn matrix

ample opportunity to form still lifes. It would be interesting to know whether this phenomenon persists for other rules or for other lattices.

The ergodic set of the diagram has 57 nodes, the remaining seven never appearing in any still life. This is not surprising, since it is foreordained that the live central cell in neighborhood built either by extending the nodes 73, 76, or 77 to the right, or the nodes 37, 67, or 77 to the left will die. The node 57 cannot be extended to the right, nor 75 to the left, but each can be extended in the other direction. They, too, have to be removed from the diagram.

Usually the exclusion of nodes is more subtle, requiring much higher powers of the connectivity matrix to establish the pattern clearly.

Although relatively few nodes are lacking from the full de Bruijn diagram, about half the links are missing. In general such numbers are explained by the fact that there are two classes of neighborhoods, the good and the bad, and that they occur in roughly equal numbers. Thus there is a 50% probability of dropping any given link from the diagram.

Eight links must be dropped to leave a node without exit links; since $(\frac{1}{2})^8 = \frac{1}{256}$, there is about 0.4% chance of finding one such node, which means that not too many bare nodes will be found among 64. The fact that seven are missing has to be considered as a fluctuation, due to some special characteristic of Conway's game. Experience with other rules tends to confirm this evaluation.

Similarly rough estimates apply to the chance of finding loops; we expect 8^N loops of length N (with repeated subloops included) but cutting any link ruins the loop, so that there is one chance in 2^N of finding an intact loop of that length. Thus the expectation is that the number of loops will quadruple with any increment in length. The dominant eigenvalue of the de Bruijn matrix gives the most accurate estimate of this factor, which is generally close to 4.0.

The following table presents the number of loops originating with the quiescent state, the total number of loops, and the maximum number possible. Bear in mind that the trace of the connection matrix counts a loop once for each node which it contains, but not multiple traverses of portions of a loop, so the crude data does not reflect the usual number of cycles as they are commonly counted.

Width	$(0,0)$ element	Trace	Maximum trace
1	1	3	8
2	1	23	64
3	4	156	512
4	24	499	4,096
5	103	2,613	32,768
6	455	12,320	262,144
7	1,114	57,235	2,097,152
8	10,708	279,523	...
9	51,006	...	...

This data shows that the number of loops is multiplied by approximately four for each increment in length, and that there is a strong component of period 3; as confirmed by experimental tallies of the number of still lifes.

4.6 Second Stage

The first stage de Bruijn diagram shows which rows of cells can form a still life (or other pattern). The second stage selects a width, then determines all possible rows of that length. A new de Bruijn diagram, whose links are these rows, describes strips of fixed width but arbitrary height embodying the desired pattern.

It is natural to ask why a fixed width is required. Amongst other representations, the admissible rows can be defined by regular expressions. From the expressions defining the link sequences (three rows high) can be extracted those defining the node sequences (two rows high); compatible rows would be described by intersections of regular expressions. It would seem that one has an instance of Post's correspondence problem, in trying to relate the two classes of regular expressions. Keeping the width fixed and finite eliminates this uncertainty from the calculation.

The simplest criterion for a row is that it be periodic, which means that it would be formed from loops of the first stage diagram of a given length. Other boundary conditions could be imposed, but they would seem to be rather artificial unless the context determining them were included. One possibility, somewhat contrary to the spirit of the generality being described, would be to build up a collection of still lifes with given boundaries. Matching such strips to get even wider strips would reveal some, but not all, of the still lifes having the composite width.

A very important exception to the recommendation to eschew such constructions is when the quiescent state forms the boundary. If the de Bruijn diagram contains a link connecting the quiescent state to itself, isolated patterns can be formed. If in turn the strips themselves are bounded by the quiescent state, there will be free standing figures which can exist in complete isolation from any others.

If "0" is the quiescent state, we need to concentrate on the $(0, 0)$ element of the connection matrix. In any event, suppose that the following illustration represents two consecutive rows of a strip of width 6,

a	b	c	d	e	f
g	h	i	j	k	l
m	n	o	p	q	r
s	t	u	v	w	x

This time the cells are the rows $ghijkl$ and $mnopqr$ with the respective neighborhoods and partial neighborhoods

Cell	Neighborhood	Upper half	Lower half
$ghijkl$	a b c d e f / g h i j k l / m n o p q r	a b c d e f / g h i j k l	g h i j k l / m n o p q r
$mnopqr$	g h i j k l / m n o p q r / s t u v w x	g h i j k l / m n o p q r	m n o p q r / s t u v w x

It is typical of the second stage that there are many broken loops and even isolated nodes, in contrast to the first stage where they are relatively infrequent. The same informal probabilistic arguments given for the first stage explain why. The full diagram for a strip of width N would have 2^{2N} nodes with 2^{3N} links, thus 2^N links per node. But instead of a single neighborhood, half of whose links might be bad, there are N cells, whose "halves" are compounded, reducing the number of good

Fig. 4.4 Second stage
de Bruijn diagram for width 1

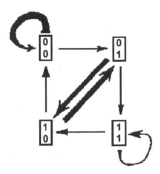

links for the whole row by a factor of $\frac{1}{2}^N$; the two effects compensate, leaving us with an estimated single link per node. Fluctuations can just as readily leave a node without a link as provide it with a pair of links, so that the surviving core of loops — the ergodic set — will still have a certain amount of variety.

One dreads to think of what would happen if this reasoning remained valid for a third stage, as in a three dimensional cellular automaton.

Although the simplest cases are almost trivial, they are also easy to understand, which will help to fix our ideas and understand the general case.

Choosing width 1 means working with constant rows; such a system is a one dimensional linear cellular automaton evolving according to Wolfram's Rule 22. Our table predicts three still life neighborhoods out of eight possible; the pertinent de Bruijn diagram is shown in Fig. 4.4. There is just one still life, discounting the quiescent state; it consists of alternating rows of live and dead cells.

Although this is the simplest case, it already shows some structure, namely that the de Bruijn diagram can consist of two disconnected pieces. It also shows how links and nodes can be dropped from the full de Bruijn diagram. Showing the actual diagram as a subset of the full diagram would be more instructive if a larger diagram were used, but showing complete diagrams for greater widths is too cumbersome for the printed page.

Width 2 shows slightly more variety. It is equivalent to working with a $(4, 1)$ cellular automaton (4 states, first neighbors).

The unrefined de Bruijn diagram has 11 nodes with 23 links, which reduces to 7 nodes with 19 links and finally to the cyclic core with 7 nodes and 9 links shown in Fig. 4.5. Inspection shows that the de Bruijn diagram decomposes into three disjoint pieces. The first contains just the quiescent state linked to itself; of course we can ascribe any periodicity to the quiescent state that we wish, making it a common feature of all diagrams.

The second is really double the configuration of width 1, in which rows of live cells alternate with rows of dead cells. There is just one new configuration, the third piece, a sort of grillwork in which the minimum length of each tier of vertical bars is two cells. Both these latter configurations have to extend indefinitely, since there is no transition bringing either of them to the quiescent state.

In contrast to narrower strips, width 3 is striking both for the complexity of its still lifes and the simplicity of the rule generating some of them. The de Bruijn

Fig. 4.5 Second stage
de Bruijn diagram for width 2

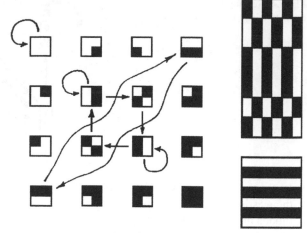

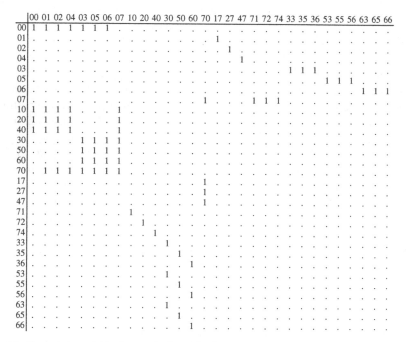

	00	01	02	04	03	05	06	07	10	20	40	30	50	60	70	17	27	47	71	72	74	33	35	36	53	55	56	63	65	66
00	1	1	1	1	1	1	1	1	.	.	.	.	.	.	.	.	.	.	.	.	.	.	.	.	.	.	.	.	.	.
01	.	.	.	.	.	.	.	.	.	.	.	.	.	.	.	1	.	.	.	.	.	.	.	.	.	.	.	.	.	.
02	.	.	.	.	.	.	.	.	.	.	.	.	.	.	.	.	1	.	.	.	.	.	.	.	.	.	.	.	.	.
04	.	.	.	.	.	.	.	.	.	.	.	.	.	.	.	.	.	1	.	.	.	.	.	.	.	.	.	.	.	.
03	.	.	.	.	.	.	.	.	.	.	.	.	.	.	.	.	.	.	.	.	.	1	1	1	.	.	.	.	.	.
05	.	.	.	.	.	.	.	.	.	.	.	.	.	.	.	.	.	.	.	.	.	.	.	.	1	1	1	.	.	.
06	.	.	.	.	.	.	.	.	.	.	.	.	.	.	.	.	.	.	.	.	.	.	.	.	.	.	.	1	1	1
07	.	.	.	.	.	.	.	.	.	.	.	.	.	.	1	.	.	.	1	1	1	.	.	.	.	.	.	.	.	.
10	1	1	1	1	.	.	.	1	.	.	.	.	.	.	.	.	.	.	.	.	.	.	.	.	.	.	.	.	.	.
20	1	1	1	1	.	.	.	1	.	.	.	.	.	.	.	.	.	.	.	.	.	.	.	.	.	.	.	.	.	.
40	1	1	1	1	.	.	.	1	.	.	.	.	.	.	.	.	.	.	.	.	.	.	.	.	.	.	.	.	.	.
30	.	.	.	.	1	1	1	1	.	.	.	.	.	.	.	.	.	.	.	.	.	.	.	.	.	.	.	.	.	.
50	.	.	.	.	1	1	1	1	.	.	.	.	.	.	.	.	.	.	.	.	.	.	.	.	.	.	.	.	.	.
60	.	.	.	.	1	1	1	1	.	.	.	.	.	.	.	.	.	.	.	.	.	.	.	.	.	.	.	.	.	.
70	.	1	1	1	1	1	1	1	.	.	.	.	.	.	.	.	.	.	.	.	.	.	.	.	.	.	.	.	.	.
17	.	.	.	.	.	.	.	.	.	.	.	.	.	.	1	.	.	.	.	.	.	.	.	.	.	.	.	.	.	.
27	.	.	.	.	.	.	.	.	.	.	.	.	.	.	1	.	.	.	.	.	.	.	.	.	.	.	.	.	.	.
47	.	.	.	.	.	.	.	.	.	.	.	.	.	.	1	.	.	.	.	.	.	.	.	.	.	.	.	.	.	.
71	.	.	.	.	.	.	.	.	1	.	.	.	.	.	.	.	.	.	.	.	.	.	.	.	.	.	.	.	.	.
72	.	.	.	.	.	.	.	.	.	1	.	.	.	.	.	.	.	.	.	.	.	.	.	.	.	.	.	.	.	.
74	.	.	.	.	.	.	.	.	.	.	1	.	.	.	.	.	.	.	.	.	.	.	.	.	.	.	.	.	.	.
33	.	.	.	.	.	.	.	.	.	.	.	1	.	.	.	.	.	.	.	.	.	.	.	.	.	.	.	.	.	.
35	.	.	.	.	.	.	.	.	.	.	.	.	1	.	.	.	.	.	.	.	.	.	.	.	.	.	.	.	.	.
36	.	.	.	.	.	.	.	.	.	.	.	.	.	1	.	.	.	.	.	.	.	.	.	.	.	.	.	.	.	.
53	.	.	.	.	.	.	.	.	.	.	.	1	.	.	.	.	.	.	.	.	.	.	.	.	.	.	.	.	.	.
55	.	.	.	.	.	.	.	.	.	.	.	.	1	.	.	.	.	.	.	.	.	.	.	.	.	.	.	.	.	.
56	.	.	.	.	.	.	.	.	.	.	.	.	.	1	.	.	.	.	.	.	.	.	.	.	.	.	.	.	.	.
63	.	.	.	.	.	.	.	.	.	.	.	1	.	.	.	.	.	.	.	.	.	.	.	.	.	.	.	.	.	.
65	.	.	.	.	.	.	.	.	.	.	.	.	1	.	.	.	.	.	.	.	.	.	.	.	.	.	.	.	.	.
66	.	.	.	.	.	.	.	.	.	.	.	.	.	1	.	.	.	.	.	.	.	.	.	.	.	.	.	.	.	.

Fig. 4.6 Second stage de Bruijn matrix for width 3 — connected component of 00

diagram is too complicated to show; however its two disconnected pieces can be exhibited in matrix form.

The first, consisting of 30 nodes, is the least regular, but is the connected component of the quiescent state. If pairs of octal numbers are used to index the nodes, each member of the pair translates directly into a three-cell cross section of the strip. The pertinent connectivity matrix is shown in Fig. 4.6.

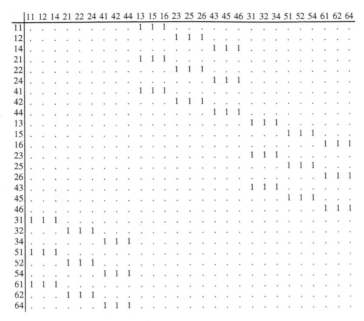

	11	12	14	21	22	24	41	42	44	13	15	16	23	25	26	43	45	46	31	32	34	51	52	54	61	62	64
11	.	.	.	.	.	.	.	.	.	1	1	1	.	.	.	.	.	.	.	.	.	.	.	.	.	.	.
12	.	.	.	.	.	.	.	.	.	.	.	.	1	1	1	.	.	.	.	.	.	.	.	.	.	.	.
14	.	.	.	.	.	.	.	.	.	.	.	.	.	.	.	1	1	1	.	.	.	.	.	.	.	.	.
21	.	.	.	.	.	.	.	.	.	1	1	1	.	.	.	.	.	.	.	.	.	.	.	.	.	.	.
22	.	.	.	.	.	.	.	.	.	.	.	.	1	1	1	.	.	.	.	.	.	.	.	.	.	.	.
24	.	.	.	.	.	.	.	.	.	.	.	.	.	.	.	1	1	1	.	.	.	.	.	.	.	.	.
41	.	.	.	.	.	.	.	.	.	1	1	1	.	.	.	.	.	.	.	.	.	.	.	.	.	.	.
42	.	.	.	.	.	.	.	.	.	.	.	.	1	1	1	.	.	.	.	.	.	.	.	.	.	.	.
44	.	.	.	.	.	.	.	.	.	.	.	.	.	.	.	1	1	1	.	.	.	.	.	.	.	.	.
13	.	.	.	.	.	.	.	.	.	.	.	.	.	.	.	.	.	.	1	1	1	.	.	.	.	.	.
15	.	.	.	.	.	.	.	.	.	.	.	.	.	.	.	.	.	.	.	.	.	1	1	1	.	.	.
16	.	.	.	.	.	.	.	.	.	.	.	.	.	.	.	.	.	.	.	.	.	.	.	.	1	1	1
23	.	.	.	.	.	.	.	.	.	.	.	.	.	.	.	.	.	.	1	1	1	.	.	.	.	.	.
25	.	.	.	.	.	.	.	.	.	.	.	.	.	.	.	.	.	.	.	.	.	1	1	1	.	.	.
26	.	.	.	.	.	.	.	.	.	.	.	.	.	.	.	.	.	.	.	.	.	.	.	.	1	1	1
43	.	.	.	.	.	.	.	.	.	.	.	.	.	.	.	.	.	.	1	1	1	.	.	.	.	.	.
45	.	.	.	.	.	.	.	.	.	.	.	.	.	.	.	.	.	.	.	.	.	1	1	1	.	.	.
46	.	.	.	.	.	.	.	.	.	.	.	.	.	.	.	.	.	.	.	.	.	.	.	.	1	1	1
31	1	1	1	.	.	.	.	.	.	.	.	.	.	.	.	.	.	.	.	.	.	.	.	.	.	.	.
32	.	.	.	1	1	1	.	.	.	.	.	.	.	.	.	.	.	.	.	.	.	.	.	.	.	.	.
34	.	.	.	.	.	.	1	1	1	.	.	.	.	.	.	.	.	.	.	.	.	.	.	.	.	.	.
51	1	1	1	.	.	.	.	.	.	.	.	.	.	.	.	.	.	.	.	.	.	.	.	.	.	.	.
52	.	.	.	1	1	1	.	.	.	.	.	.	.	.	.	.	.	.	.	.	.	.	.	.	.	.	.
54	.	.	.	.	.	.	1	1	1	.	.	.	.	.	.	.	.	.	.	.	.	.	.	.	.	.	.
61	1	1	1	.	.	.	.	.	.	.	.	.	.	.	.	.	.	.	.	.	.	.	.	.	.	.	.
62	.	.	.	1	1	1	.	.	.	.	.	.	.	.	.	.	.	.	.	.	.	.	.	.	.	.	.
64	.	.	.	.	.	.	1	1	1	.	.	.	.	.	.	.	.	.	.	.	.	.	.	.	.	.	.

Fig. 4.7 Second stage de Bruijn matrix for width 3 — connected component of 11

In all diagrams and connectivity matrices a certain amount of symmetry must be present, because the rule of evolution remains the same even though the row from which the diagram is derived is longitudinally shifted or even reflected. The symmetry can manifest in various forms according to whether the still life has the same symmetry as the row, or whether there are several equivalent patterns arising from one another via the symmetry operations. For longer rows whose lengths have various distinct divisors, intermediate cases of partial symmetry can arise.

The second component, consisting of 27 nodes, is disconnected from the quiescent state, but is extremely regular. To begin with, the rule of formation is extremely elegant; given any neighborhood containing exactly four live cells, three cells are to be chosen for the new margin so that the overlapping neighborhood also has four live cells; a construction clearly unique to *Life*. Its connectivity matrix constitutes Fig. 4.7.

Close inspection shows that sections of odd parity — with a single live cell — alternate with sections of even parity — having two live cells — in the ratio of two odd sections to one even section; thus the importance of the height being a multiple of three. Moreover it is pairs of sections which count — the half neighborhoods — so the sequence even–odd, odd–odd, and odd–even must be rigorously followed. Consequently the de Bruijn matrix is imprimitive of degree 3, explaining the magical properties of the number 3.

Further inspection shows that the matrix is the tensor product of a cyclic shift matrix with a full de Bruijn matrix for three states and two stages, which is a consequence of the transversal symmetry. In other words, it doesn't matter which section of a given parity is chosen, but from there on further choices must all be consistent.

Fig. 4.8 Typical still lifes for
a strip of width 3

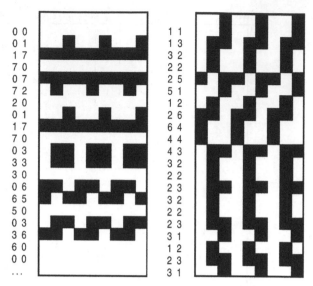

0 0	1 1
0 1	1 3
1 7	3 2
7 0	2 2
0 7	2 5
7 2	5 1
2 0	1 2
0 1	2 6
1 7	6 4
7 0	4 4
0 3	4 3
3 3	3 2
3 0	2 2
0 6	2 3
6 5	3 2
5 0	2 2
0 3	2 3
3 6	3 1
6 0	1 2
0 0	2 3
...	3 1

To better visualize the still lifes of width three, Fig. 4.8 shows typical samples, one for the component of 00, the other for the component of 11. The former contains three fundamental motifs; horizontal bars which must have castellations whenever they are not surrounded by similar bars; rows of blocks, and extended snakes, using the standard terminology of *Life*. The motifs may follow in any sequence, and with any phase relative to one another.

Going on to widths of 4 or wider, more and more complicated diagrams are encountered. However, a new phenomenon becomes visible with width 4, which is that the quiescent state may have nontrivial connected components in both the first stage and the second stage. Since the quiescent state is always self-connected, an arbitrary number of additional quiescent states may always be inserted wherever a single one was encountered. Consequently the live cells from any configuration so found continue to form a still life when moved to an environment which is not periodic.

Furthermore, with fewer diagrams, visual presentations for slightly wider strips are still possible.

Strips of width 4 show the first nontrivial component of the quiescent state; the second order diagram for this component has 24 nodes, as shown in Fig. 4.9. This is the smallest diagram in which completely freestanding figures occur; they are separable from one another or others by arbitrarily long quiescent intervals, either horizontally or vertically.

Of course, the full de Bruijn diagram for width 4 could also be exhibited; but it isn't since the intricacy of the diagrams increases exponentially with width, making their explicit representation increasingly cumbersome and less visual. Eventually the only feasible representation consists of a table, whose entries list the possible successors of each node in turn.

Fig. 4.9 Freestanding strip
of width 4

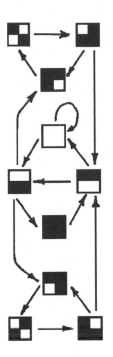

4.7 Comments

It is disconcerting to see the degree to which the detailed labor of hundreds of people, and untold hours of computer time, has been reduced to an algorithm. Even though the necessary computations can be realized quite quickly, it is no less impressive to foresee the stupendous amount of time which would be required to obtain certain additional results which initially seem quite simple. For instance, it is out of the question to use the same approach directly to calculate any structures of period 2, much less gliders and space ships which require period 4.

Much of the activity reported in Wainwright's newsletter [8] involved tracking the evolution of diverse small figures, a byproduct of which was a gradually increasing catalog of small still lifes. Many of them grouped themselves into families, whose general structure could be readily perceived, and many people seem to have become quite skilled at designing still lifes and other predictable patterns. Moreover, after having reviewed large collections of still lifes, one develops an eye for flaws and a feeling for what constitutes a proper still life. Which implies that there must be some pattern present which can be recognized.

Golomb's book [4] on shift register sequences disseminated the use of de Bruijn diagrams to characterize long sequences of overlapping symbols, although an application to automata does not seem to have been published until Wolfram's article [9] in 1984. Nowadays they can be seen as a tool for quickly obtaining the personalities of arbitrary automata, subject to the limitations imposed by exponential growth with respect to any of the parameters involved — dimension of the automaton, length of the period, number of states, and so on. Perhaps an even more elaborate theory of higher dimensional de Bruijn diagrams will eventually result.

References

1. Berlekamp, E.R., Conway, J.H., Guy, R.K.: Winning Ways for Your Mathematical Plays, vol. 2, Chap. 25. Academic Press, San Diego (1982). ISBN 0-12-091152-3
2. Gardner, M.: Mathematical Games — The fantastic combinations of John Conway's new solitaire game "life". Sci. Am. **October** 120–123 (1970)
3. Gardner, M.: Wheels, Life, and Other Mathematical Amusements. Freeman, New York (1983). ISBN 0-7167-1589-9 pbk
4. Golomb, S.W.: Shift Register Sequences. Holden-Day, San Francisco (1967)
5. McCulloch, W.S., Pitts, W.: A logical calculus of the ideas immanent in nervous activity. Bull. Math. Biophys. **5**, 115–133 (1943)
6. Ralston, A.: De Bruijn sequences — A model example of the interaction of discrete mathematics and computer science. Math. Mag. **55**, 131–143 (1982)
7. von Neumann, J.: Theory of Self-reproducing Automata. Edited and completed by Burks, A.W. University of Illinois, Champaign (1966)
8. Wainwright, R.T. (ed.): *Lifeline*, a quarterly newsletter with 11 issues published between March 1971 and September 1973
9. Wolfram, S.: Computation theory of cellular automata. Commun. Math. Phys. **96**, 15–57 (1984)
10. Wolfram, S. (ed.): Theory and Applications of Cellular Automata. World Scientific Press, Singapore (1986). ISBN 9971-50-124-4 pbk

Chapter 5
A Zoo of *Life* Forms

Harold V. McIntosh

A catalog is presented, of those *Life* forms on strips of widths up to nine whose translational behavior during a single generation can be inferred from two levels of de Bruijn diagrams.[1]

5.1 Introduction

Previously, in Chap. 4 on *Life*'s Still Lifes, we presented an algorithm capable of revealing all *Life* forms of the type (m, n, t), which move uniformly m cells in the x-direction and n cells in the y-direction after t generations. Strictly speaking, the algorithm only yields those forms which can be found in strips of finite width or, which are spatially periodic. Although finding all the forms which could fill the whole plane without periodicity is an undecidable proposition, with ingenuity many other interesting forms can nevertheless be found. However, that is another story.

Life possesses the characteristic that many of its forms are surrounded by arbitrarily long quiescent stretches, which makes them freestanding. The same algorithm yields all isolated forms of this nature. In either event the amount of computation required to obtain forms of even modest extent is quite considerable. Forms of long period, likewise all those of period two, are inaccessible to present computer power. Consequently this presentation is limited to forms whose characteristics can be ascertained within a single generation.

Still lifes, creepers and crawlers can be determined; the latter are not gliders because they are not freestanding; rather many are fuses whose quiescent surroundings extend infinitely in only one direction. Nor do they move by reflection and translation. Still, the word "glider" has acquired a generic connotation referring to any moving configuration and is often used where it is not strictly appropriate.

---

[1] The paper is written in October 26, 1988; revised July 20, 1992.

A. Adamatzky (ed.), *Game of Life Cellular Automata*,
DOI 10.1007/978-1-84996-217-9_5, © Springer-Verlag London Limited 2010

The algorithm involves two levels of de Bruijn diagrams. The first level diagrams have a maximum of 64 nodes, typically with four links each, except those which do not belong to the ergodic set of the diagram; those often have none, or participate in chains leading to end nodes which have no continuation. In any event it is awkward to present the de Bruijn diagram in its preferred form as a set of chords of a circle, so a matrix form is used instead.

Even the matrix presentation is unwieldy, so the style actually adopted consists of listing the nodes of the ergodic set on a line together with the nodes to which they are linked. Each line will have a maximum length determined by the number of outgoing links in the full de Bruijn diagram, which in turn will be a fraction of the length of the rows of the full connectivity diagram.

Having formed the first level de Bruijn diagram, the second level can be constructed. All the loops in the first level with a chosen length are candidates to be links in the second level diagram, that length now becoming the width of a periodic strip. Again, links will be discarded for not joining nodes within the ergodic set.

The following sections are laid out according to the behavior that can be discerned after a single generation of evolution, that is, still lifes, followed by longitudinal, transversal, and diagonal gliders. Only one instance of each symmetry class is presented, meaning that further patterns can be gotten by planar rotations or reflections of the ones shown. The list could also be extended by exhibiting the precursors of a completely quiescent field, or of a completely live field; but all such results have been omitted for reasons of space.

Likewise six is the maximum width of the periodic strips shown; diagrams for wider strips would not fit on a single page without some change in the style of presentation. Since the strips are periodic, they are subject to further reflective and rotational symmetries; the tables have been further compressed by showing only symmetry classes. Link superscripts such as L, R, rot, or F imply that the next node is to be rotated to the left, right, arbitrarily, or reflected, before continuing. Only one node of any given symmetry class is shown.

5.2 (0, 0, 1) — Still Lifes

5.2.1 First Level de Bruijn Matrix

At the first level, there are 284 neighborhoods leading to still life for the central cell.

If neighborhoods are excluded which fail to connect to a neighbor either to the left or to the right, or which belong to non-branching chains with such a termination, 8 neighborhoods can be discarded, leaving 276.

If similar reasoning is applied to neighborhoods which cannot be continued upwards or downwards, the number is further reduced to 259 neighborhoods, serving as links connecting 57 nodes. The ratio of links to nodes is therefore 4.54.

The 25 links which can be discarded according to this reasoning have been marked with asterisks in the table above. All of them have been included in the following analysis; the final figures are not affected, but the intermediate results would be slightly different had these links been removed at the outset.

	0	1	2	3	4	5	6	7		0	1	2	3	4	5	6	7
00	00	01	02	03	04	05	06	.	40	00	01	02	.	04	.	.	07
01	10	11	12	.	14	.	.	17	41	10	.	.	13	.	15	16	17
02	.	.	.	23	.	25	26	27	42	.	21	22	23	24	25	26	.
03	.	31	32	33	34	35	36	.	43	30	31	32	.	34	.	.	.
04	40	41	42	.	44	.	.	47	44	40	.	.	43	.	45	46	47
05	50	.	.	53	.	55	56	57	45	.	51	52	53	54	55	56	57
06	.	61	62	63	64	65	66	.	46	60	61	62	.	64	.	.	.
07	70	71	72	.	74	.	.	.	47	70	.	.	.	.	.	.	.
10	00	01	02	.	04	.	.	07	50	00	.	.	03	.	05	06	07
11	10	.	.	13	.	15	16	17	51	.	11	12	13	14	15	16	17
12	.	21	22	23	24	25	26	.	52	20	21	22	.	24	.	.	.
13	30	31	32	.	34	.	.	.	53	30	.	.	.	.	.	.	.
14	40	.	.	43	.	45	46	47	54	.	41	42	43	44	45	46	47
15	.	51	52	53	54	55	56	57	55	50	51	52	53	54	55	56	57
16	60	61	62	.	64	.	.	.	56	60	.	.	.	.	.	.	.
17	70	.	.	.	.	.	.	.	57	.	.	.	.	.	.	.	.
20	00	01	02	.	04	.	.	07	60	00	.	.	03	.	05	06	07
21	10	.	.	13	.	15	16	17	61	.	11	12	13	14	15	16	17
22	.	21	22	23	24	25	26	.	62	20	21	22	.	24	.	.	.
23	30	31	32	.	34	.	.	.	63	30	.	.	.	.	.	.	.
24	40	.	.	43	.	45	46	47	64	.	41	42	43	44	45	46	47
25	.	51	52	53	54	55	56	57	65	50	51	52	53	54	55	56	57
26	60	61	62	.	64	.	.	.	66	60	.	.	.	.	.	.	.
27	70	.	.	.	.	.	.	.	67	.	.	.	.	.	.	.	.
30	00	.	.	03	.	05	06	07	70	.	01	02	03	04	05	06	07
31	.	11	12	13	14	15	16	17	71	10	11	12	13	14	15	16	17
32	20	21	22	.	24	.	.	.	72	20	.	.	.	.	.	.	.
33	30	.	.	.	.	.	.	.	73	.	.	.	.	.	.	.	.
34	.	41	42	43	44	45	46	47	74	40	41	42	43	44	45	46	47
35	50	51	52	53	54	55	56	57	75	50	51	52	53	54	55	56	57
36	60	.	.	.	.	.	.	.	76	.	.	.	.	.	.	.	.
37	.	.	.	.	.	.	.	.	77	.	.	.	.	.	.	.	.

5.2.2 Powers of the Still Life de Bruijn Matrix

The matrix elements of powers of the first level matrix tell how many paths there are from the row index node to the column index node. Diagonal elements count loops. The trace counts all possible loops, once for each node which they contain, while specific diagonal elements identify the loops through that particular node.

Not all the possible loops will participate in the second level matrix because they may not overlap correctly. The columns labelled "Initial" contain the raw data, including transients as well as the ergodic set. Dropping all those nodes and links which lack predecessors, successors, or both, refines the data. In the process a new set of nodes and links may be exposed, which in turn ought to be dropped for a lack of continuation. Eventually, after "Gen" cycles of iteration, zero, one, or more ergodic sets will be reached, in which there are no dead ends.

Width	(0, 0) element					Trace				
	Initial		Final		Gens	Initial		Final		Gens
	Nodes	Links	Nodes	Links		Nodes	Links	Nodes	Links	
1	1	1	1	1	1	3	3	3	3	1
2	1	1	1	1	1	11	23	7	9	3
3	1	4	1	1	2	57	156	57	156	1
4	10	24	10	16	2	139	499	127	309	3
5	48	103	40	73	3	583	2,613	583	2,233	2
6	161	455	149	341	4	2,519	12,320	2,515	10,986	3
7	579	2,224	523	1,559	3	8,046	57,235	7,864	44,159	3
8	2,170	10,708	2,064	7,818	3	30,947	279,523	30,743	229,341	3
9	7,963	51,006	7,667	37,708	5	118,830	1,329,996	118,362	1,078,689	3

The "Final" columns display the numbers of nodes and links in the second level de Bruijn diagram, a separate line for each width. To a certain extent, these numbers can be divided by the width to get the number of symmetry classes. If very many of the patterns lack reflective symmetry, twice the width is an appropriate divisor; in any event the internal symmetry of the classes has to be taken into account.

Still Life, Width 1

Still Life, Width 2

Still Life, Width 3

Connected component of 00

nodes	linked to	nodes	linked to
00	00, 40^{rot}, 50^{rot}	50	F0, A1, E1
40 D1	A2	F0	A0
A2	40^{rot}, 51	A0	00, 50^{rot}, 51
51	A2, E2^{rot}	A1	A0^R
E2 80	00, 40, 51	E1	A0^L

Connected component of C0

node	linked to	node	linked to	node	linked to
C0	D0, C1, 91	50	D0, C1, 91	42	D0, C1, 91
D0	E0, C2^L, 62^R	C1	E0^L, C2, 62^R	91	E0^R, C2^L, 62
E0	C0, 50^L, 42^R	C2	C0, 50, 42	62	C0, 50^L, 42^R

Width 3 presents certain peculiar features due to a number theoretic property of *Life*'s rule of evolution. On a 3×3 torus, the still lifes are those configurations which contain exactly four live cells. If the central cell is live, it has three live neighbors and thus survives; otherwise the four live neighbors prevent a live cell from forming and again the arrangement is stable.

For a $3 \times N$ torus, the same principle applies, but the arrangement of the live cells in a cross section can vary as one moves along the long dimension of the torus. A constant sum of 4 can be realized in the forms $3 + 1 + 0$, $2 + 2 + 0$, and $2 + 1 + 1$, and their permutations.

There is no way of breaking out of the sequence $2 + 1 + 1$, but the presence of zeroes in the other two allows them to terminate in quiescent regions. Moreover, the sequence $0 + 3 + 0 + 3 + \cdots$ can be judiciously interspersed in the latter two sequences to produce still further variation.

Freestanding Still Life, Width 4

nodes	links	nodes	links
00	00, 50	50	E0, E0^F, F0
E0 90 70	E0, A0	F0	A0
		A0	00, 50

Still Life, Width 4

nodes	links	nodes	links
00	00, 01^{rot}, 50^{rot}	56	A0, E8, A9
01 46	89, CC	A0	00, 50
89	02	17	6A
CC	CC, 99, 89, 89^{2L}	E8	C0, 94, 85
02	00, 01^{rot}	13	31
99	32, 23	A9	12, 52, 13, 16
23	03, 13, 43, 56, 07	AE 84	59, 5D
F0	A0	94	69
50	B0, B0^F, F0	5D	AA
B0	60, 31	A5	4A, 5A
60	D0, 60^L	AA	50, 41^{rot}
D0	A0, B0	4A	94, 95, D4, C5
03	17	55	AA

Freestanding Still Life, Width 5

nodes	links	nodes	links
000	000, 100, 500, 500^F	700	280^F, E00
100 640	980, D80, D80^F, CC0	280	000, 500^F
CC0	980	380	240
980 200	000, 100	680	940, C00
500	380^F, E00, E40, F00	4C0	D80
E00	840, 900	940	E00^F
840	1C0	C00	940
900	240, 700	D80	280^F
1C0	280, 380, 680	E40	8C0, 980, D80^F
240	1C0, 4C0	8C0	900
		F00	280^F

(The superscript F signifies reflection.)

5.2.3 Sample Still Life Strips

Except for very narrow strips it is not easy to display a comprehensive sample of still lifes. Since there are 28 symmetry classes for a strip of width 4, one would need a strip at least 28 lines long; but there are at least twice as many links so it takes an even longer strip to show every possibility of branching at least once. Typical cycles are even harder to portray in full generality.

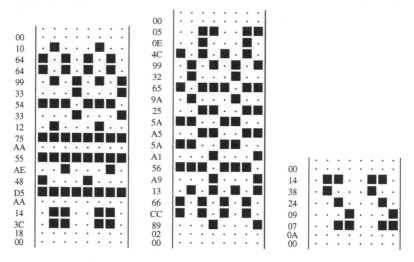

5.3 (0, 1, 1) — Longitudinal Creepers

5.3.1 First Level de Bruijn Matrix

	0	1	2	3	4	5	6	7
00	00	01	02	03	04	05	06	.
01	10	11	12	.	14	.	.	17
02	20	21	22	.	24	.	.	.
03	30	.	.	.	.	.	.	.
04	40	.	.	43	.	45	46	47
05	50	.	.	53	.	55	56	57
06	60	.	.	.	.	.	.	.
07	.	.	.	.	.	.	.	.
10	00	01	02	.	04	.	.	07
11	10	.	.	13	.	15	16	17
12	20	.	.	.	.	.	.	.
13	.	.	33	.	.	.	.	.
14	40	.	.	43	.	45	46	47
15	.	51	52	53	54	55	56	57
16	.	.	.	.	.	.	66	.
17	.	.	72	.	.	.	.	.
20	00	.	.	.	.	05	06	.
21	.	01	02	.	04	.	.	.
22	.	21	22	23	24	25	26	.
23	30	31	32	.	34	.	.	.
24	.	41	42	.	44	.	.	.
25	50	.	.	.	.	.	.	.
26	60	61	62	.	64	.	.	.
27	70	.	.	.	.	.	.	.
30	.	01	02	.	04	.	.	.
31	10	.	.	.	.	.	.	.
32	20	.	.	.	24	.	.	.
33	30	.	.	.	.	.	.	.
34	40	.	.	.	.	.	.	.
35	.	.	.	.	.	.	.	.
36	60	.	.	.	.	.	.	.
37	.	.	.	.	.	.	.	.
40	00	01	02	.	04	.	.	07
41	10	.	.	13	.	.	16	17
42	20	.	.	.	.	.	.	.
43	.	.	.	33	.	.	.	.
44	40	.	.	43	.	45	46	47
45	.	51	52	53	54	55	56	57
46	.	.	.	63	.	.	.	.
47	.	72	.	.	.	.	.	.
50	00	.	.	03	.	05	06	07
51	.	11	12	13	14	15	16	17
52	.	.	.	.	.	25	.	.
53	.	31	32	.	34	.	.	.
54	.	41	42	43	44	45	46	47
55	50	51	52	53	54	55	56	57
56	.	61	62	.	64	.	.	.
57	70	.	72	.	.	.	.	.
60	.	01	02	.	04	.	.	.
61	10	.	.	.	.	.	.	.
62	20	21	.	.	.	.	.	.
63	30	.	.	.	.	.	.	.
64	40	.	.	.	.	.	.	.
65	.	.	.	.	.	.	.	.
66	60	.	.	.	.	.	.	.
67	.	.	.	.	.	.	.	.
70	00	.	.	.	.	.	.	.
71	.	.	.	.	.	.	.	.
72	20	.	.	.	.	.	.	.
73	.	.	.	.	.	.	.	.
74	.	.	.	.	.	.	.	.
75	.	.	.	.	.	.	.	.
76	.	.	.	.	.	.	.	.
77	.	.	.	.	.	.	.	.

5.3.2 *Powers of the Longitudinal de Bruijn Matrix*

Width	(0, 0) element					Trace				
	Initial		Final		Gens	Initial		Final		Gens
	Nodes	Links	Nodes	Links		Nodes	Links	Nodes	Links	
1	1	1	1	1	1	3	3	3	3	1
2	1	1	1	1	1	3	13	3	3	2
3	1	4	1	1	2	3	51	3	3	2
4	1	13	1	1	2	35	137	3	3	4
5	8	45	1	1	3	143	533	58	93	5
6	25	148	1	1	3	349	1,909	99	135	6
7	70	524	1	1	4	1,053	6,443	283	444	5
8	184	1,827	1	1	5	3,003	23,049	415	699	8
9	488	6,419	1	1	5	8,301	80,475	1,209	2,262	7

Longitudinal, Width 1

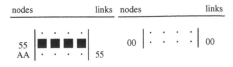

Longitudinal, Width 2

Longitudinal, Width 3

Longitudinal, Width 4

Longitudinal, Width 5

There is a fivefold symmetry, according to which the first eleven nodes shown below can be rotated by an arbitrary amount, giving 55 nodes altogether. The last three nodes are invariant to rotation, completing the total of 58 nodes shown in the table.

It is convenient to show only one member from each symmetry class, so the links between nodes need to indicate the amount of rotation or reflection implicit in the link.

nodes		links	nodes		links
5A0 A00	· · ■ ■ · / ■ ■ · · · / · · · · ·	000, 500	5E0 A80	· · ■ ■ · / ■ ■ · · · / · · · · ·	551, 540
000	· · · · · / · · · · · / · · · · ·	000, 500 (rotated)	551 AA2	■ ■ ■ ■ ■ / · · · · ·	551, 540 (rotated)
500 A50	■ ■ · · · / · · ■ ■ ·	0F0, 5E0	540 AD0	■ ■ · · · / · · ■ ■ ·	5A0, 5E0, 0B0
0F0	· · ■ ■ · / · · ■ ■ ·	5A0	0B0 070	· · ■ ■ · / · · ■ ■ ·	5E0

The simplest figure which can be constructed uses the lines of the first column above, just filling a 5×5 square. It is ubiquitous; it might be named an ant. Although not self-propelled, it can be led by any live cell placed in front of it, or by a pair of live cells flanking it. Such cells can even be the hind feet of the ant in front, or of a pair of ants straddling it. Ants can even be staggered, although there is not enough room in a strip of width 5 for them to do anything but march in parallel. Nevertheless, space allowing, they frequently trail behind other configurations, sometimes leading still smaller processions.

Longitudinal, Width 6

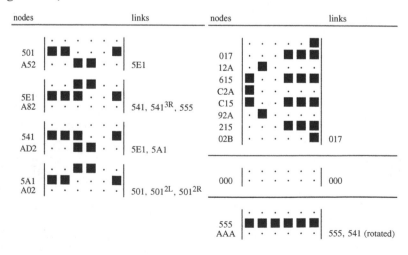

5.3.3 Sample Strips with Longitudinal Movement

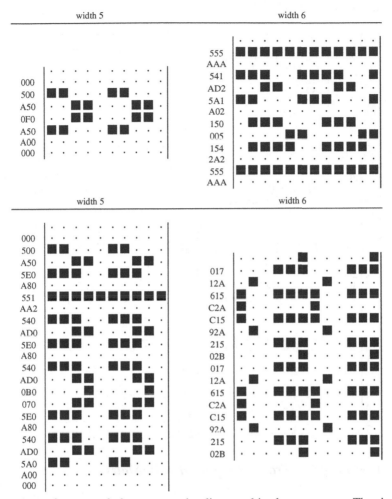

A variety of structural elements can be discerned in these patterns. The simplest is the ant, encased in its 5 × 5 square, which is almost freestanding. Except for the guiding bit or bits which is required to lead it, it can exist in isolation, or it can be stacked with other ants either in a single file or staggered in various ways. Wider strips are required to display the full variability possible.

Figures flow along channels according to the next most restrictive format. The channel boundaries can be stabilized with castellations, supporting blocks, and various other ways which are not evident when translational symmetry is restricted to a single cell of displacement. Some figures can be packed side by side within their channels, and sometimes they can be packed with relative displacements.

Finally there are figures which cannot be interrupted at all, preserving their integrity only when they fill the entire infinite strip or form a periodic pattern on a torus of appropriate length.

5.4 (1, 0, 1) — Transversal Creepers

5.4.1 First Level de Bruijn Matrix

	0	1	2	3	4	5	6	7
00	00	01	02	03	04	05	06	.
01	.	.	13	.	15	16	.	
02	20	21	22	.	24	.	.	.
03	.	31	32	33	34	.	36	.
04	40	41	42	.	44	.	.	47
05	.	51	52	54	.	.	.	.
06	60	.	.	.	.	.	.	.
07	70	.	72	74	.	.	.	.
10	00	01	02	.	04	.	.	07
11	.	11	12	.	14	.	.	.
12	20	.	.	.	.	.	.	27
13	30	31	32	.	34	.	.	.
14	40	.	.	43	.	45	46	47
15	50	.	.	.	.	.	.	.
16	.	.	.	63	.	.	66	.
17	.	.	72	.	.	.	.	.
20	00	01	02	.	04	.	.	07
21	11	.	12	.	14	.	.	.
22	20	.	.	.	.	.	.	.
23	30	31	.	.	34	.	.	.
24	40	.	.	43	.	45	46	47
25	50	.	.	.	.	.	.	.
26	.	.	.	.	.	.	.	.
27	70	.	.	.	.	.	.	.
30	00	.	.	03	.	05	06	07
31	10	.	.	.	.	.	.	.
32	.	.	.	23	.	25	.	.
33	30	.	.	.	.	.	.	.
34	.	41	42	43	44	45	46	47
35	.	.	.	.	.	.	.	.
36	.	61	.	63	64	.	.	.
37	.	.	.	.	.	.	.	.

	0	1	2	3	4	5	6	7
40	00	01	02	.	04	.	.	07
41	.	11	12	.	14	.	.	.
42	20	.	.	.	.	.	.	.
43	30	31	32	33	.	.	.	.
44	40	.	.	43	.	.	.	.
45	50	.	.	.	.	.	.	.
46	.	.	.	63	.	.	.	.
47	70	.	.	.	.	.	.	.
50	00	.	.	03	.	05	06	07
51	10	.	.	.	.	.	.	.
52	.	.	.	23	.	25	.	.
53	.	.	.	.	.	.	.	.
54	.	41	42	43	.	.	.	.
55	.	.	.	.	.	.	.	.
56	.	.	.	.	.	.	.	.
57	.	.	.	.	.	.	.	.
60	00	.	.	03	.	05	06	07
61	10	.	.	.	.	.	.	.
62	.	.	.	.	.	.	.	.
63	30	.	.	.	.	.	.	.
64	.	41	42	43	.	.	46	47
65	.	.	.	.	.	.	.	.
66	.	61	.	.	.	.	.	.
67	.	.	.	.	.	.	.	.
70	.	01	02	03	04	05	06	07
71	.	.	.	.	.	.	.	.
72	.	21	.	.	.	.	.	.
73	.	.	.	.	.	.	.	.
74	40	41	42	43	.	.	46	47
75	.	.	.	.	.	.	.	.
76	.	.	.	.	.	.	.	.
77	.	.	.	.	.	.	.	.

5.4.2 Powers of the Transversal de Bruijn Matrix

Width	(0, 0) element					Trace				
	Initial		Final		Gens	Initial		Final		Gens
	Nodes	Links	Nodes	Links		Nodes	Links	Nodes	Links	
1	1	1	1	1	1	1	3	1	1	2
2	1	1	1	1	1	6	15	5	7	3
3	2	4	1	1	3	10	42	1	1	4
4	5	16	1	1	5	80	227	37	59	6
5	25	57	1	1	11	241	633	41	66	10
6	64	187	11	15	10	617	2,367	133	243	8
7	174	682	17	27	16	1,730	8,165	302	533	9
8	510	2,496	27	45	19	5,656	30,899	757	1,571	11
9	1,570	9,144	96	172	18	16,561	110,094	1,354	2,803	14

Transversal, Width 1

Transversal, Width 2

Transversal, Width 3

Transversal, Width 4

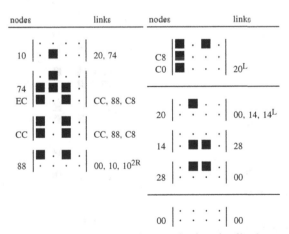

The transversal patterns are quite similar to the longitudinal patterns, except for the fact that the width 4 pattern can be stretched out to an arbitrary length and can acquire tail fins, which are capable of supporting a trailing plume. The longer of two tail fins must project an additional two cells, but a single cell projection is capable of generating interesting structures of period two which are outside the present analysis.

Transversal, Width 5

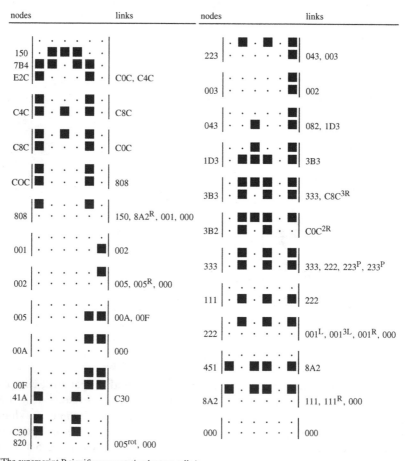

By turning the ant sideways, a strip of width 5 supports a trailing spark which in turn can lead another ant or a small plume consisting of a pair of cells.

Transversal, Width 6

(The superscript P signifies any rotation by two cells.)

5.4.3 Sample Strips with Transversal Movement

A width of 6 allows the width 4 patterns to be separated by a channel and to stretch to arbitrary lengths. Certain interruptions allow channel switching, as indicated by the alternative (permutation) at the node 333. Any permutation of the node 223 is allowed in continuation, which is to say, any rotation by two cells.

With a width of 6, ants have still more freedom of movement; even a wide bodied ant is possible, which can lead ordinary ants in its wake. Unlike an ordinary ant, the wide one requires lateral support for strips of width greater than 6.

Thus there are four basic configurations which may lead from one to another, finally trailing off into a shower of small sparks. The ants themselves can be free-standing, in the sense that parallel columns of marching ants may be separated by arbitrary quiescent regions, even though each individual column extends upward to infinity.

width 5	width 6
050	150
0F0	7B4
1A1	E2C
303	C0C
202	808
050	150
0F0	7B4
1A1	E2C
303	C4C
202	C8C
400	C0C
800	808
500	001
F00	002
A41	005
0C3	00F
082	41A
010	C30
020	820
050	140
0A0	280
000	000

5.5 (1, 1, 1) — Diagonal Creepers

5.5.1 First Level de Bruijn Matrix

	0	1	2	3	4	5	6	7		0	1	2	3	4	5	6	7
00	00	01	02	03	04	05	06	.	40	00	01	02	.	04	.	.	07
01	10	11	12	.	14	.	.	17	41	10	.	.	13	.	15	16	17
02	20	21	22	.	24	.	.	.	42	20	.	.	.	.	.	.	27
03	30	.	.	.	.	.	.	.	43	.	.	.	33	.	.	36	.
04	40	41	42	.	44	.	.	47	44	40	.	.	43	.	45	46	47
05	50	.	.	53	.	.	56	.	45	.	51	52	53	54	.	56	.
06	.	.	.	.	.	.	60	67	46	.	.	.	63	.	65	66	67
07	.	.	.	.	.	.	.	.	47	.	.	72	.	.	.	.	.
10	.	.	.	03	.	.	06	.	50	.	01	02	.	04	.	.	.
11	.	11	12	.	14	.	.	.	51	10	.	.	.	.	.	.	.
12	.	21	22	23	24	25	26	.	52	20	21	22	.	24	.	.	.
13	30	31	32	.	34	.	.	.	53	30	.	.	.	.	.	.	.
14	.	41	42	.	44	.	.	.	54	40	.	.	.	.	.	.	.
15	50	.	.	.	.	.	.	.	55	.	.	.	.	.	.	.	.
16	60	61	62	.	64	.	.	.	56	60	.	.	.	.	.	.	.
17	70	.	.	.	.	.	.	.	57	.	.	.	.	.	.	.	.
20	00	01	02	04	.	.	.	07	60	00	.	.	03	.	05	.	.
21	10	.	.	13	.	15	16	17	61	.	11	12	13	14	15	.	.
22	20	.	.	.	.	.	.	.	62	.	.	.	.	.	.	.	.
23	.	.	.	33	.	.	.	.	63	.	31	.	.	.	.	.	.
24	40	.	.	43	.	45	46	47	64	.	41	42	43	44	45	.	.
25	.	51	52	53	54	.	56	.	65	50	51	52	53	54	.	.	.
26	.	.	.	63	.	65	.	.	66	.	61	.	.	.	.	.	.
27	.	.	72	.	.	.	.	.	67	70	.	.	.	.	.	.	.
30	.	01	02	.	04	.	.	.	70	00	.	.	.	.	.	.	.
31	10	.	.	.	.	.	.	.	71	.	.	.	.	.	.	.	.
32	20	21	22	.	24	.	.	.	72	20	.	.	.	.	.	.	.
33	30	.	.	.	.	.	.	.	73	.	.	.	.	.	.	.	.
34	40	.	.	.	.	.	.	.	74	.	.	.	.	.	.	.	.
35	.	.	.	.	.	.	.	.	75	.	.	.	.	.	.	.	.
36	60	.	.	.	.	.	.	.	76	.	.	.	.	.	.	.	.
37	.	.	.	.	.	.	.	.	77	.	.	.	.	.	.	.	.

5.5.2 Powers of the Diagonal de Bruijn Matrix

Width	(0, 0) element					Trace				
	Initial		Final		Gens	Initial		Final		Gens
	Nodes	Links	Nodes	Links		Nodes	Links	Nodes	Links	
1	1	1	1	1	1	1	3	1	1	1
2	1	1	1	1	1	3	11	1	1	3
3	2	4	1	1	3	10	42	1	1	4
4	6	14	1	1	5	37	131	25	41	3
5	14	44	1	1	6	141	463	56	96	8
6	39	151	1	1	7	350	1,556	97	163	10
7	97	505	1	1	8	939	5,239	92	155	11
8	268	1,715	1	1	9	2,501	17,875	473	1,057	9
9	709	5,831	1	1	10	6,976	60,720	976	2,290	16

Diagonal, Width 1

nodes		links
0	$\vdots$	0

Diagonal, Width 2

nodes		links
0	$\vdots \quad \vdots$	0

Diagonal, Width 3

nodes		links
00	$\vdots \quad \vdots \quad \vdots$	00

Diagonal, Width 4

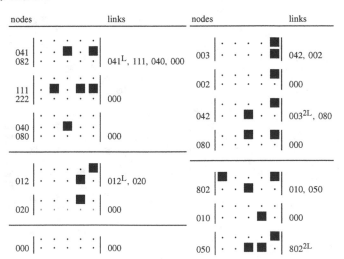

Diagonal, Width 5

Diagonal, Width 6

nodes	links	nodes	links
011	032, 202^{2L}	003	012, 002
032 / 121	202	012	003^{2L}, 002^{2L}, 130
202	011, 514, 000	130	981^{R}
514 / A28	000	00A	050, 001^{2L}, 000
001	002	050	00A^{2L}, 1A0
002	000	1A0 / 640	981
006	006^{L}, 002^{L}	981	981^{2L}
000	000		

5.5.3 Sample Strips with Diagonal Movement

For clarity, only one strand of the three diagonal strings is shown, as though they were embedded in a strip of width 12 rather than 6.

width 6

003
012
030
120
301
505
819
055
198
660

006
018
060
180
600
801
006
018
060
180
600

width 6

00A
050
0A0
500
A01
406
198
660
198
660

011
032
101
202
011
036
121
202
011
032
121

Part II
Classical Topics

Chapter 6
Growth and Decay in Life-Like Cellular Automata

David Eppstein

6.1 Introduction

Since the study of life began, many have asked: is it unique in the universe, or are there other interesting forms of life elsewhere? Before we can answer that question, we should ask others: What makes life special? If we happen across another system with life-like behavior, how would we be able to recognize it? We are speaking, of course, of the mathematical systems of cellular automata, of the fascinating patterns that have been discovered and engineered in Conway's Game of Life [5, 13], and of the possible existence of other cellular automaton rules with equally complex behavior to that of Life.

In an influential early paper [28], Stephen Wolfram proposed an answer to this question of what makes Life special. He categorized cellular automaton rules into four types, according to their behavior when started with random initial conditions:

- In Class I automata, all cells eventually become the same.
- In Class II, the field eventually degenerates into scattered stable or oscillating patterns.
- In Class III, the chaos of the initial random pattern persists indefinitely.
- Class IV contains the remaining rules, in which patterns such as those in Life exhibit complex behavior.

Wolfram initially intended this classification for one-dimensional cellular automata, but in a later paper [29] he and Packard extended it to the same set of two-dimensional automata that we consider here; Adamatzky et al. [1] performed a more detailed classification of some of these automata based on the same principle of studying their behavior on random initial conditions.

However, Wolfram's classification is problematic in more than one way. Although there is some evidence for phase transitions between regions of rule space where one class is more frequent than others [6, 19, 20, 30], the classification depends strongly on the specific behavior of an individual rule, so that one cannot use it to predict which rules are likely to have interesting behavior, but only to describe that behavior after having already observed it. The boundary between classes is less

A. Adamatzky (ed.), *Game of Life Cellular Automata*,
DOI 10.1007/978-1-84996-217-9_6, © Springer-Verlag London Limited 2010

clear-cut and more subjective than one would like, and may for some automata be impossible to decide [7, 22]. Some automata may have a constant probability of behaving in more than one of these ways [2], making them difficult to classify. It is not obvious, even, which class Life properly belongs to, and it is conventionally classified as Class IV less because the description of that class best fits our observations of Life and more because Life is the archetypical rule whose behavior Class IV was intended to capture. When the boundary is clear, it is not where we might like it to be: for instance, the rule B35/S236 (discussed below) appears to be in Class III, but can support many complex patterns similar to those in Life. Wolfram's classification defines the interesting rules negatively rather than positively: they are the rules where some known type of uninteresting behavior doesn't happen. It is predicated on the assumption that interesting behavior should emerge from uninformative initial conditions, but many of the most interesting patterns in Life could not have been found in this way.

We propose a four-way classification of two-dimensional semi-totalistic cellular automata that is different than Wolfram's, based on two questions with yes-or-no answers: do there exist patterns that eventually escape any finite bounding box placed around them? And do there exist patterns that die out completely? If both of these conditions are true, then a cellular automaton rule is likely to support spaceships, small patterns that move and that form the building blocks of many of the more complex patterns that are known for Life. If one or both of these conditions is not true, then there may still be phenomena of interest supported by the given cellular automaton rule, but we will have to look harder for them. Although our classification is very crude, we argue that it is more objective than Wolfram's (due to the greater ease of determining a rigorous answer to these questions), more predictive (as we can classify large groups of rules without observing them individually), and more accurate in focusing attention on rules likely to support patterns with complex behavior. We support these assertions by surveying a number of known cellular automaton rules.

6.2 Life-Like Rules

The space of possible cellular automaton rules is infinite and highly varied. One may define cellular automata on grids of high dimensions or on neighborhood structures more general than grids. The set of neighbors of a cell may be only those other cells nearest to it in the grid or may fall within a neighborhood of larger than unit radius. The state of a cell may depend only on the states of neighboring cells in the previous time step, or it may depend on the states of neighbors over several previous time steps. The number of states of each cell may be any finite number or even a continuously variable value, and researchers have considered update rules that are asynchronous, randomized, or quantum mechanical.

In order to impose some order on this vast wilderness, we restrict our attention to a more circumscribed set of rules that are very similar in structure to Conway's Game of Life. Specifically, we consider semi-totalistic (or outer totalistic) binary cellular automata on a two-dimensional Moore neighborhood. These are the cellular automata in which:

- The cells of the automaton form a two-dimensional square lattice.
- The neighbors of each cell are the eight lattice squares that are orthogonally or diagonally adjacent to it.
- Each cell may be in one of two states, alive or dead.
- All cells are updated simultaneously in a sequence of time steps.
- In time step i, the state of any given cell is a function of the state of the same cell in time step $i - 1$ and of the number of live neighbors it had in time step $i - 1$.

In a cellular automaton of this type, a single cell may do one of four things within a single time step: If it was dead but becomes alive, we say that it is *born*. If it was alive and remains alive, we say that it *survives*. If it was alive and becomes dead, we say that it *dies*. And if it was dead and remains dead, we say that it is *quiescent*.

We follow a standard convention for naming these cellular automata in which the update rule of the automaton is represented by a *rule string*, a sequence of characters in the form "B*xxx*/S*yyy*". The *xxx* part of the rule string is a subset of the digits from 0 to 8, representing numbers of neighbors such that a dead cell with that many neighbors would become alive in the next time step, causing a birth event: the B stands for birth. The *yyy* part of the rule string is another subset of digits, representing numbers of neighbors such that a live cell with that many neighbors would remain alive in the next time step, causing a survival event: the S stands for survival. For instance, Conway's Game of Life itself is represented by the rule string "B3/S23": a dead cell with three live neighbors leads to a birth event, and a live cell with two or three live neighbors leads to a survival event. All other combinations of cell state and number of neighbors lead to death or quiescence; we do not need to list these combinations separately as they can be inferred from the birth and survival parts of the rule string.

Each digit from 0 to 8 may be present or absent in the birth part of the rule string, and may independently be present or absent in the survival part. Therefore, there are 2^{18} different Life-like rules, too many to study in detail individually: we need a roadmap to help guide us to the interesting rules.

6.3 Natural Evolution or Intelligent Design?

Wolfram's classification takes the point of view of statistical mechanics, a field that uses probability to study large systems of interacting objects such as the molecules in an ideal gas. Wolfram posits that a typical initial state of the automaton has a 50/50 chance that any given cell is alive or dead, independently of all other cells; he then asks how the automaton behaves when started from this typical state.

From the mathematical point of view, random initial conditions are completely general: any finite configuration of an automaton, such as a glider gun centered within an otherwise quiescent $10^6 \times 10^6$ square of empty space, occurs infinitely

Fig. 6.1 Greene's period-416
$2c/5$ spaceship gun

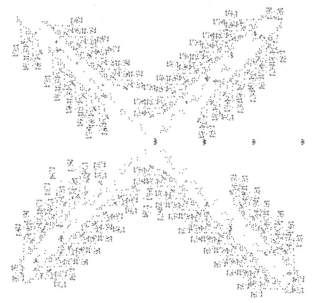

often within a random initial state, despite the tiny probability $(2^{-10^{12}})$ that this pattern occurs at any particular location. However, from the practical point of view, the cellular automaton patterns that can be seen to occur randomly and the patterns that can be constructed by human engineering are very different from each other. The patterns arising from random fields in Life, as seen in practice, consist overwhelmingly of small still lifes and small period-two oscillators, with higher-period oscillators such as the period-3 pulsar, period-15 pentadecathlon, and period-8 figure-8 arising much less frequently (once per 20,000, 1.6 million, or 33 million distinct patterns generated in this way, respectively) in experiments performed by Achim Flammenkamp [12]. In contrast, many of the most interesting patterns in Life were formed by human engineering, and may combine hundreds of individual patterns that themselves have much higher complexity than the patterns that could ever be seen in random experiments by any human observer. If we seek rules that can support similar patterns, we should use a classification scheme that classifies those rules similarly to Life, regardless of whether they behave similarly on random initial conditions. Consider, for instance, the following Life patterns:

- The pattern library included with the Golly life simulation software [27] includes a period-416 $2c/5$ gun, designed in 2003 by Dave Greene based on a reaction due to Noam Elkies (Fig. 6.1). It fits within a square of 1 million cells; every 416 time steps it produces a spaceship that moves at speed $2c/5$. It consists of 64 period-416 glider guns; each of these glider guns in turn is formed by five small still life patterns, four small low-period oscillators, and a larger high-period pattern that is contained and controlled by its interactions with the still lifes and oscillators. 60 separate patterns, each of which is again formed from a cluster of still lifes and low-period oscillators, convert the gliders from these guns into

Herschel patterns (highly reactive patterns formed from seven live cells) and then back into gliders in order to insert them into position within four diagonal glider streams that converge in the center of the pattern, where 64 gliders repeatedly react with each other to build up complex still life patterns and then trigger them to become the $2c/5$ spaceships that are emitted by the gun. Overall, the pattern has over 26,000 live cells.

- Brice Due's "Outer Totalistic Cellular Automata Meta-Pixel" [8] allows any Life-like cellular automaton to be simulated within Life, by representing each cell of the other automaton as a "meta-pixel" of approximately 4 million Life cells; it takes 35,328 steps of Life for the meta-pixel to simulate a single step of the other automaton. The interior of each meta-pixel is either quiescent (when it represents a dead cell of the simulated automaton) or covered by lightweight spaceships when it represents a live cell. The boundary area of each meta-pixel is filled with period-46 oscillators and other components which control whether the spaceships are generated or not, generate the spaceships, and communicate via gliders to the corresponding components of adjacent meta-pixels in order to compute the next state at each time step of the automaton. Many copies of the meta-pixel pattern must be combined to simulate any nontrivial pattern of another automaton.

- The Caterpillar [23] is a large spaceship that moves at speed $17c/45$, constructed in 2004 by David Bell, Gabriel Nivasch, and Jason Summers. The pattern repeats its configuration every 270 steps, advancing by 102 cells in an axis-parallel direction within that time period. It is based on a reaction in which the π *heptomino* pattern moves at this speed along a track of properly spaced blinkers without destroying the track. If two π heptominos move in tandem along parallel tracks, they can be made to emit gliders; interactions between the gliders emitted by multiple heptominos sharing a system of tracks can be used to tear down the tracks behind the heptominos and to send streams of faster spaceships toward the front of the moving system of heptominos. These spaceships interact with additional gliders to extend the tracks on which the heptominos follow. The whole pattern uses large numbers of these components; it fits within a bounding box with an area of 1.4 billion cells, of which approximately 12 million are alive at any time.

Beyond the practical unlikelihood of discovering patterns such as these from random initial conditions, there is another reason that using random initial states to classify automata is problematic: with this assumption, it is very difficult to say anything that can be backed up by rigorous mathematics. For the Game of Life, what we know rigorously is limited to random states in which the probability of a cell being live is a number ε that is very close to zero, on time and distance scales that are bounded by polynomial functions of $1/\varepsilon$ [14]. For random states with greater numbers of live cells, or over a greater number of time steps, much remains unproven. For instance, in a random Life universe, does every cell eventually becomes periodic with probability one? If so, Life should probably be classified as Class II rather than Class IV. There exist finite initial configurations for Life that do not ever become periodic, but we do not know whether it is possible for nonperiodic configurations to survive when surrounded by the ashes of a random starting state.

It is perhaps worth emphasizing that this dichotomy between nature and design in Conway's Life does not tell us much about the proper study of nature itself. The physical universe is enormously larger and has been running on far longer time scales than any cellular automaton simulation, and we have plenty of evidence in front of and behind our eyes that complex systems can evolve from simpler ones. For all we know, it may be possible that a Life simulation over similarly large time and space scales, from an initially random state, could eventually develop something recognizable as an ecology [5] rather than becoming periodic everywhere. In this respect, the frequently used term "evolution" for the behavior of a cellular automaton pattern over a sequence of time steps is unfortunate, because at observable scales this behavior is much better modeled as physics than as biology.

6.4 Growth

Because our primary interest is in engineered patterns rather than random fields, we restrict our attention to patterns that have a finite number of live cells. This eliminates the possibility of rules with births on zero live neighbors, because in such rules any finite pattern would immediately become infinite. Without B0, we need consider only 2^{17} possible rules, half as many as before.

A *bounding box* for a pattern is a rectangle with axis-parallel sides that contains all the live cells of the pattern; the *minimum bounding box* is the smallest possible such rectangle, but we also allow larger rectangles to count as bounding boxes. Every pattern with finitely many live cells can be placed within a bounding box.

We define a cellular automaton rule to be *fertile* if it has a finite pattern that eventually escapes any of its bounding boxes. That is, the rule is fertile if there exists a *growth pattern* P such that, for every bounding box $B \supset P$, after some number of time steps starting from P there will be a live cell outside B. The growth patterns in Life include many types of pattern that Life enthusiasts have found interesting, including gliders and spaceships, puffer trains, guns, rakes, breeders, and other spacefillers.

If a rule is not fertile, then every finite pattern must eventually become periodic: for every pattern P there is a bounding box B that it cannot escape, and it can only progress through $2^{|B|}$ possible states before returning to a state that it has already been in. On the other hand, if every finite pattern eventually becomes periodic, then there is no pattern that can escape all of its bounding boxes. That is, instead of defining fertility in terms of escape from bounding boxes, it would be equivalent to define an *infertile* rule as one in which every finite pattern eventually becomes periodic. However, the definition in terms of bounding boxes and growth patterns is more convenient when attempting to determine which rules are fertile and which are infertile.

Some simple case analysis allows us to determine whether a rule is fertile for most of the 2^{17} rules that support finite patterns:

- If a rule includes B1, it is fertile. In this case, the pattern consisting of a single live cell is a growth pattern. After k steps starting from this pattern, the minimum

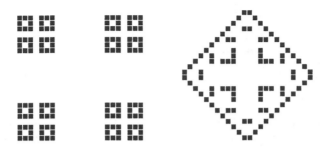

Fig. 6.2 Growth patterns in B1 and B2 rules. *Left*: 11 steps after starting from a single live cell in B1/S. As is true for all steps of this pattern in all B1 rules, the four corner cells of the minimal bounding box are alive. *Right*: 11 steps after starting from a 2 × 2 block of live cells in B2/S. As is true for all steps of this pattern in all B2 rules, in each edge of the minimal bounding box only the two middle cells are alive

bounding box will contain $(2k + 1) \times (2k + 1)$ cells, and will have a single live cell at each of its four corners (Fig. 6.2, left).

- If a rule includes B2, it is fertile. If the rule also includes B1, this follows from the previous case; otherwise, the pattern consisting of a 2 × 2 block of live cells is a growth pattern. After k steps starting from this pattern, the minimum bounding box will contain $(2k + 2) \times (2k + 2)$ cells, and will have two adjacent live cells at the center of each of its edges; the remaining cells on each edge of the bounding box will be non-live, leading the same pattern to propagate outwards by one more unit in the following time step (Fig. 6.2, right).
- If a rule does not include B1, B2, or B3, it is not fertile. The dead cells outside of a bounding box B of any pattern can have at most three live neighbors, and therefore can never become live themselves.

The remaining cases are those with rule strings that begin "B3...", as Life's rule string B3/S23 does. For these rules, often the simplest way to show that they are fertile is to exhibit a growth pattern such as Life's glider. We have used our search software [10], together with searches using small random seeds, to search for space-ships in these B3 rules; so far, we have found that 10,736 out of the 16,384 possible B3 rules have spaceships and therefore are fertile [9].

In some of the remaining cases, it is possible to prove using a more detailed analysis that a rule is not fertile. For instance, suppose that a B3 rule does not allow births with 1, 2, 4, or 5 live neighbors, nor does it allow survivals with five or fewer live neighbors. In this case, no growth pattern can exist. For, let P be any pattern, and let D be a bounding diamond of P (that is, a shape containing it bounded by lines of slope ± 1). We prove by contradiction that no cell that has at most one neighbor in D can ever become live in any future state of P. For, otherwise, suppose that (x, y) are the Cartesian coordinates of the first such cell to become live, at time step i; in Fig. 6.3, right, (x, y) is the dark marked cell, and D is the dotted polygon. We assume without loss of generality that D lies below and to the left of (x, y), as shown in the figure, as the other three cases are symmetric. Then in order to become born, (x, y) must have three live neighbors at time step $i - 1$: two

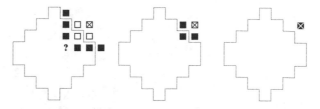

Fig. 6.3 Infertility in rules such as B36/S678 without births on 2, 4, or 5 neighbors and without survival on fewer than six neighbors. In order for a point (x, y) that has only one neighbor in an initial bounding diamond to become live in step i (*right*), three neighbors of (x, y) closer to the bounding diamond must be live in step $i - 1$ (*center*), but there is no way to configure the cells in step $i - 2$ to cause these three neighbors to be born (*left*)

neighbors $(x - 1, y)$ and $(x, y - 1)$ of x are live and outside D, and a third neighbor $(x - 1, y - 1)$ is live and inside D. These four cells (x, y), $(x - 1, y)$, $(x, y - 1)$, and $(x - 1, y - 1)$ form a 2×2 block, as that is the only way to form three or more live neighbors of (x, y) that do not themselves have at most one neighbor in D, contradicting the assumption that (x, y) is the first such live cell — see Fig. 6.3, center. Before step i, every cell outside D has too few live neighbors to survive, so $(x - 1, y)$ and $(x, y - 1)$ must have been newly born at step $i - 1$, and were not live at step $i - 2$. At step $i - 2$, $(x - 1, y - 1)$ has five or fewer live neighbors, for its three neighbors (x, y), $(x - 1, y)$, and $(x, y - 1)$ are not live; therefore, if it were live at that step it would have died, contradicting its living state in step $i - 1$. Therefore, at step $i - 1$, $(x - 1, y - 1)$ is also newly born. However, it is not possible for $(x - 1, y)$, $(x, y - 1)$, and $(x - 1, y - 1)$ to all be newly born in the same step: the three live neighbors needed for $(x - 1, y)$ to be newly born and the three live neighbors needed for $(x, y - 1)$ to be newly born (Fig. 6.3, left) cause $(x - 1, y - 1)$ to have four or five live neighbors and to remain quiescent. This contradiction implies that P cannot escape an expanded bounded diamond surrounding D and therefore that the rule is not fertile. There are 64 B3 rules without any of B1245/S012345, all of which can be proven to be infertile using the above analysis.

There remain only 5,584 rules for which we neither have a known growth pattern nor a proof of infertility. Therefore, this analysis allows us to determine in 96% of the cases whether a rule is fertile or infertile. Additionally, this analysis ignores the possibility of growth patterns other than spaceships, such as the ladders that are known in Life without Death (B3/S012345678), Maze (B3/S12345), and related rules.

A related attempt to classify cellular automata by their growth properties was made by Gravner [15]. Gravner considered initial patterns formed by setting the cells within an $n \times n$ bounding box to be alive or dead randomly, and by setting the cells outside the bounding box to be dead. A pattern of this type exhibits *quadratic growth* (or, in Gravner's terminology, *linear expansion*) if, after t steps of the automaton, it has $\Omega(t^2)$ live cells, so that it eventually grows to fill a large fraction of the plane. Gravner defined a cellular automaton rule to be *expansive* if, with probability approaching one in the limit as n goes to infinity, random patterns with $n \times n$ bounding boxes exhibit linear expansion. However, being able to cover the plane is a much stricter requirement on a pattern than being able to escape a bounding box, so it is often difficult to determine whether a pattern exhibits linear expansion, and even more difficult to determine whether a rule is expansive. More, this classification shares with Wolfram's classification the property that it is based on random initial conditions, so (as we have argued above) it does not address well the ability of a cellular automaton to support non-random structures.

6.5 Decay

In order for its patterns to exhibit the complex behavior that they do, it is important in Life that some patterns shrink as well as that others grow. In the extreme, some patterns may eventually lead to a state in which every cell is dead and quiescent; in the terminology of *Winning Ways* [5], a pattern of this type is said to *fade*. For instance, the proof that determining the eventual fate of a Life pattern is undecidable depends on patterns that fade: it is undecidable to determine, for a given Life pattern, whether it fades or whether it has some living cells in every future state [5].

If a rule supports a pattern P with a finite number of live cells, such that the state following P has no live cells, we say that the rule is *mortal*, and otherwise that it is *immortal*. It is equivalent to ask whether the rule has a finite pattern that fades, because if pattern Q fades, the pattern P formed from Q on the penultimate step before it fades has the desired property that the state following P is empty. However, requiring that P fade in the next step rather than at some future time simplifies our analysis.

As Dean Hickerson observed, a pattern fades in rule r if and only if the same pattern forms a still life in the rule $\bar{r}$ with the same birth conditions and complementary survival conditions. A 2001 analysis by Hickerson, aided by Matthew Cook, Jason Summers, and the author, determined the existence or nonexistence of a still life in most of the 2^{17} possible rules without B0, and the same analysis can also be used to determine which of these rules are mortal. In particular, in the rules that are known to be mortal, one of the patterns shown in Fig. 6.4 must fade. We provide some flavor of the analysis of immortal rules, although not the complete analysis, below.

- If a rule allows births with exactly one live neighbor, then it is immortal. For, if (x, y) are the Cartesian coordinates of a live cell in a pattern that maximizes y among all live cells in a given finite pattern, and that maximizes x among all live cells with the same y-coordinate, then in the next step cell $(x + 1, y + 1)$ will also be live — see Fig. 6.5, far left.
- If a rule causes all live cells with fewer than five live neighbors to survive, then it is immortal. For, if (x, y) is a live cell that maximizes y among all live cells in a given finite pattern, and that maximizes x among all live cells with the same y-coordinate, then (x, y) has fewer than five live neighbors and must survive into the next step — see Fig. 6.5, far left again.

Fig. 6.4 Patterns that fade in any rule that includes none of the following birth and survival possibilities. *Row 1*: B1/S0, B12/S1, B123/S2, B12/S3, B124/S2, B124/S2, B13/S34, B125/S2, B13/S14, B13/S35, B135/S23, B134/S2, B123/S47. *Row 2*: B1238/S46, B1235/S45, B1237/S456, B135/S245, B1234/S45, B134/S1578, B1236/S45, B13/S13. *Row 3*: B12347/S46, B134/S156, B1235/S46, B12346/S46, B1345/S157, B12367/S46, B1357/S256. *Row 4*: B1356/S246, B135/S2567, B135/S2578, B1357/S246, B1356/S24. *Row 5*: B135/S12, B1356/S25

- If a rule allows births with both two and three live neighbors, and survival with zero live neighbors, then it is immortal. In this case the analysis showing immortality is somewhat more intricate. Suppose that, in a given pattern P with finitely many live cells, (x, y) is the live cell maximizing $x + y$, and that among all live cells with that value of $x + y$ it is the one maximizing y. Further, suppose for a contradiction that the next state from P according to the given rule has all cells dead. Then cell $(x - 1, y)$ must be dead in P, for otherwise there would be a birth at $(x, y + 1)$. Cells $(x, y - 1)$ and $(x + 1, y - 1)$ must also be dead, for otherwise there would be a birth at $(x + 1, y)$. This eliminates all possible live neighbors of (x, y) except for $(x - 1, y - 1)$, which must be live in order for (x, y) to die at the next step. Additionally, $(x - 2, y)$ and $(x - 2, y + 1)$ must be dead, for otherwise there would be a birth at $(x - 1, y + 1)$. All of these conditions together imply that cell $(x - 1, y)$ has either two or three live neighbors: (x, y) and $(x - 1, y - 1)$ are live, and all other neighbors with the possible exception of $(x - 2, y - 1)$ are dead. But then, $(x - 1, y)$ would have a birth in the next state, contradicting our assumption. This case is shown in Fig. 6.5, center left.
- If a rule allows births with either two or three live neighbors, and it causes any live cell with fewer than four live neighbors to survive, then is immortal. For, if (x, y) is a live cell that maximizes $x + y$ among all live cells in a given finite pattern, and that maximizes y among live cells with the same value of $x + y$, then (x, y) can have at most four live neighbors, at $(x - 1, y)$, $(x - 1, y - 1)$, $(x, y - 1)$, and

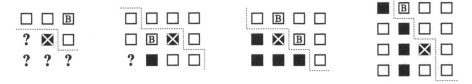

Fig. 6.5 Case analysis for immortality of certain rules. In each case the *dark squares* show live cells, the *open boxes* show dead cells, and the *dark square* marked with a *white X* is the cell (x, y) selected at the start of the case. Due to the way (x, y) was selected, all cells *above* and to the *right* of the *dashed line* must be dead. An *open box* with the *letter B* in it marks a location of cell birth. *Far left*: rules with B1 are immortal due to a birth at $(x + 1, y + 1)$, while rules with S01234 are immortal because no matter how the cells marked by *question marks* are set, cell (x, y) survives. *Center left*: Rules with B23/S0 lead to a birth at the marked *square*. *Center right*: Rules with B2/S0123 or B3/S0123 lead to a birth at one of the two marked *squares*. *Far right*: Rules with B2/S01245 lead to a birth at the marked *square*

$(x + 1, y - 1)$. If fewer than four of these neighbors are live then (x, y) survives, and if all four are live then there is a birth at $(x, y + 1)$ (for rules with B2) or $(x + 1, y)$ (for rules with B3) — see Fig. 6.5, center right.

- If a rule includes all of B2/S01245, then it is immortal. For, let (x, y) be a live cell that maximizes $x + y$ among all live cells in a given finite pattern P, and that maximizes x among all live cells with the same value of $2x + y$. Then, if P is to fade on the next step, $(x, y - 1)$ must be dead, for otherwise there would be a birth at $(x + 1, y)$. Thus (x, y) has at most three live neighbors, and if it is to die on the next step it must have exactly three, at each of the remaining neighboring locations $(x - 1, y + 1)$, $(x - 1, y)$, and $(x - 1, y - 1)$. This also provides $(x - 1, y)$ with three live neighbors; if it is to die, its neighbor count must be either three or six, and if it had six live neighbors then $(x - 1, y + 1)$ would have four or five and would survive. We can therefore conclude that the three locations $(x - 2, y + 1)$, $(x - 2, y)$, and $(x - 2, y - 1)$ are all dead. In order for $(x - 1, y + 1)$ to avoid surviving, it needs a third live neighbor at $(x - 2, y + 2)$; but then, cell $(x - 1, y + 2)$ would have two live neighbors and would lead to a birth in the next step — see Fig. 6.5, far right. We conclude that P cannot fade and therefore that the rule is immortal.

- If a rule includes all of B345/S013, then it is immortal. For, let (x, y) be a live cell that maximizes $x + y$ among all live cells in a given finite pattern, and that maximizes y among live cells with the same value of $x + y$. Then, in order for (x, y) to die in the next step, it must have two or four live neighbors among the four cells $(x - 1, y)$, $(x - 1, y - 1)$, $(x, y - 1)$, and $(x + 1, y - 1)$. However, four live neighbors would lead to a birth at $(x + 1, y)$, so the only possibility for the entire pattern to fade in the next step is to have two live neighbors. We may now consider sub-cases, according to how those live neighbors of (x, y) are arranged, to show that in each case the pattern has a birth or survival into the next step:

  – If $(x, y - 1)$ is not live, then it has as live neighbors (x, y) and the other two live neighbors of (x, y). In order to prevent a birth from occurring at $(x, y - 1)$ in the next step of the pattern, the three cells $(x - 1, y - 2)$, $(x, y - 2)$, and

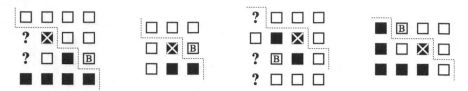

Fig. 6.6 Case analysis for immortality of rules with B345/S013. *Far left*: $(x, y-1)$ is not live. *Center left*: $(x, y-1)$ and $(x+1, y-1)$ are live. *Center right*: $(x, y-1)$ and $(x-1, y)$ are live. *Far right*: $(x, y-1)$ and $(x-1, y-1)$ are live

$(x+1, y-2)$ must also all be live. But with these choices fixed, the cell $(x+1, y-1)$ has either three or four live neighbors, and would, if dead, be the location of a birth in the next step. Thus, $(x+1, y-1)$ must be live, and in order for it to die in the next step $(x+2, y-2)$ must also be live — see Fig. 6.6, far left.

- If $(x, y-1)$ is live, and the other live neighbor of (x, y) is at $(x+1, y-1)$, then there is a birth at $(x+1, y)$ — see Fig. 6.6, center left.
- If $(x, y-1)$ is live, and the other live neighbor of (x, y) is at $(x-1, y)$, then $(x-2, y)$ must be dead, for otherwise there would be a birth in the next step at $(x-1, y+1)$. $(x, y-2)$ and $(x+1, y-2)$ must be dead, for otherwise there would be a birth at $(x+1, y-1)$. And $(x-1, y-2)$ must be dead, for otherwise the cell at $(x, y-1)$ would survive. But these choices together imply that $(x-1, y-1)$ has from three to five live neighbors, leading to a birth there in the next step — see Fig. 6.6, center right.
- If $(x, y-1)$ is live, and the other live neighbor of (x, y) is at $(x-1, y-1)$, then the three cells $(x-2, y-1)$, $(x-2, y)$, and $(x-2, y+1)$ must all be live to prevent a birth at $(x-1, y)$. But then there would be a birth at $(x-1, y+1)$ — see Fig. 6.6, far right.
- A similar analysis shows that the rules B/S0123567, B2/02345, B2/S023467, B24/S0234, B24/S01345, B245/S01356, B246/S013467, B256/S023468, B257/S023468, B2456/S013, B2457/S0135, B2467/S013468, B24678/S013478, B456/S012367, B4568/S012357, B45678/S01237, B4578/S012357, and B5678/S012357 are all immortal, as are any rules that include births or survivals on a superset of the numbers of neighbors of these rules.

These cases together show that, of the 131,072 possible life-like rules without B0, 77,563 of them are immortal and 53,214 of them are mortal. There remain 295 rules that are not classified by this analysis, so we may determine whether a rule is mortal or not in approximately 99.8% of the cases. In particular this classification covers all rules that include B3.

It does not follow from the definition of mortality that an immortal rule can have no spaceships. For instance, the immortal rule B3456/S013 has spaceships, shown in Fig. 6.7. Similarly, immortality does not provide any theoretical barrier to the existence of more complex engineered patterns such as spaceship guns, although we do not know of any such patterns for this rule. However, in many of the cases for which we can prove that a rule is immortal, the proof shows something stronger,

Fig. 6.7   A $c/4$ orthogonal spaceship (*left*) and a $c/3$ orthogonal spaceship (*right*) in the immortal rule B3456/S013

that the minimal bounding box or minimal bounding diamond of a pattern can never shrink. When this is true, it is impossible for a spaceship to exist.

6.6 The Life-Like Menagerie

It is our hypothesis that the rules most likely to support interesting patterns are the ones that are both fertile and mortal. Figure 6.8 depicts in rough terms this region of the rule space. In this section, we examine some specific rules where complex engineered patterns have already been discovered, in support of this hypothesis. Not coincidentally, these rules are both fertile and mortal, but they have differing Wolfram classes, demonstrating that Wolfram's classification does not accurately describe the existence of this sort of pattern.

HighLife (rule B36/S23) was investigated extensively by Bell [4]. Many of its patterns and behaviors are similar to those in Life, because it differs from Life only in the comparatively rare case of a dead cell with six live neighbors. However, unlike Life, it features a small pattern known as a *replicator* that (in the absence of obstacles) replaces itself with two separated copies of the same pattern every twelve steps. Many nontrivial combinations of replicators with other patterns are known:

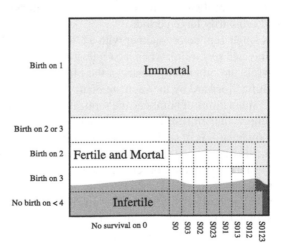

Fig. 6.8 A map of the possible life-like rules, depicting regions of fertile, infertile, mortal, and immortal rules. The area of a region represents approximately the number of different rules within it

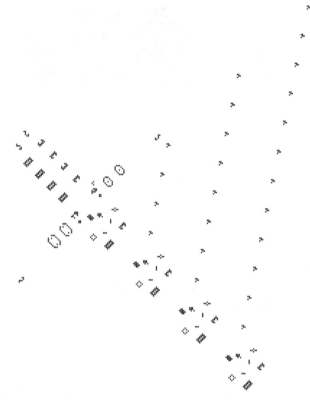

Fig. 6.9 A rake gun in HighLife. Four rows of replicators capped by eaters (*left*) interact to form pairs of bombers that move downwards and to the right. The sparks from each pair of bombers interact to form a trail of gliders that move upwards and to the right, filling a quarter of the plane with a quadratically growing number of live cells

- Rows of replicators can be capped by oscillators or blocks, producing oscillators of arbitrarily large periods.
- A single replicator together with a blinker oscillator produces the *bomber*, a $c/6$ diagonal spaceship: when the replicator copies itself, one copy is shifted forward while the other copy destroys the blinker and replaces it with another blinker shifted forward by the same amount.
- Combinations of bombers and replicators can produce puffers and rakes, moving objects that emit still lifes, gliders, or even rows of replicators.
- Combinations of replicator-based oscillators can be used to make guns of arbitrarily high period that emit gliders, bombers, rakes, or other patterns (Fig. 6.9).
- Dean Hickerson has observed that a carefully timed sequence of replicators, interacting with a blinker, can be used to push it forwards by eight steps. Together with the bomber reaction that pulls a blinker towards a group of oscillators, it should be possible to use these reactions to construct very large spaceships that move at arbitrarily slow speeds.

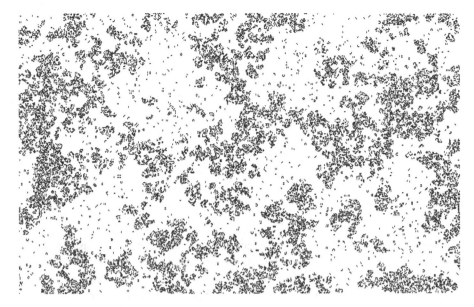

Fig. 6.10 B368/S12578, after 4,000 steps from random initial conditions

As in Life, random fields in HighLife seem to eventually settle down to still lifes and small oscillators: the replicators cannot make progress through the other patterns that surround them. Thus, HighLife should probably be assigned the same Wolfram class as Life, either Class II or Class IV.

B368/S12578 like HighLife supports a simple replicator, in the form of a 1×5 block of cells; it copies itself every 13 time steps. This replicator may be used to form oscillators of arbitrarily high periods; combinations of these oscillators can form guns for a small $c/8$ diagonal spaceship supported by this rule. As with HighLife, it seems likely that push-pull reactions can be used to form spaceships with arbitrarily slow speeds. However, with random initial conditions, this rule forms regions of chaotic activity interspersed with other regions of small oscillators and still lifes (Fig. 6.10), indicating a mixture of Class II and Class III behavior.

B36/S245. Soon after the discovery of a replicator in HighLife, Mark Niemiec found another rule with a replicator: B36/S245. The initial pattern for this replicator consists of a pair of "shuttles", sets of twelve live cells with a 3×6 bounding box, in the shape of a capital letter D. Each shuttle, if sufficiently far from other patterns, repeats its shape in 102 generations, flipped 180 degrees, after laying a pair of "eggs" (period 4 oscillators). But, if an egg is already present, it produces a collection of sparks, which in the presence of the symmetrically placed shuttle end up hatching another replicator after 96 generations. The first few generations at which a copy of the replicator reappears are 102, 204, 300 (the first hatched egg), 306, 402, 504, 606, 702, and 708. Rule B36/S245 is also interesting because, like Life, it has many small spaceships, including a 3×3 period-7 diagonal glider, and a $4c/23$ orthogonal spaceship. As in HighLife, rows of the replicators in this rule can be capped in various ways to form high-period oscillators and guns. Dean

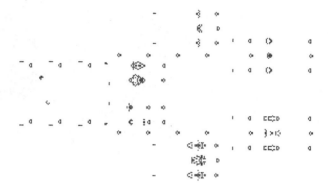

Fig. 6.11 Hickerson's 28c/1,200 spaceship in B36/S245

Hickerson also showed that the replicators in this rule can also be used to form high-period puffers that move 28 units of distance in either 600 or 1,200 time steps. By combining groups of these puffers together, he found large spaceships that move at speed $14c/300$ and $28c/1,200$ (Fig. 6.11). Like life, random states eventually but slowly settle down to scattered oscillators and still lifes, so it should be classified either as Class II or Class IV.

Morley (B368/S245) is named after Stephen Morley, who found many of its patterns; it has also been called Move. It contains small naturally-occurring spaceships of several periods, as well as a slow puffer with speed $13c/170$. Groups of puffers can be combined to form spaceships that move at the same speed, as well as breeders that emit streams of puffers moving sideways to the path of the breeder. This rule also supports several guns. Like life, random states eventually but slowly settle down to scattered oscillators and still lifes, so it should be classified either as Class II or Class IV.

B37/S23 is superficially similar to Life, but also supports a pattern unlike anything in Life in which a pair of R pentominos form a puffer that moves at speed $9c/28$, leaving a trail of pairs of pond still lifes behind them (Fig. 6.12, center). Five of these puffers can be combined, with three in a front row and two in a back row, so that all of the ponds are destroyed by sparks from other puffers, resulting in a large $9c/28$ spaceship, the *puddlejumper* [10] (Fig. 6.12, right). Along with its high-period spaceship, B37/S23 supports a glider gun found by Jason Summers (Fig. 6.12, left). Like B368/S12578, random initial conditions cause this rule to

Fig. 6.12 A gun in B37/S23, the $9c/28$ puffer formed from two R pentominos, and the $9c/28$ puddlejumper, a spaceship formed from five puffers

Fig. 6.13 A gun and an
antigun interact in Day &
Night. Pattern by David Bell,
based on a reaction by Dean
Hickerson, from the Golly
pattern collection

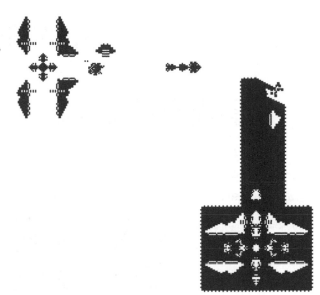

develop regions of chaotic activity interspersed with regions of scattered still life
and oscillator patterns, indicating a mix of Class II and Class III activity.

Several other rules also have large spaceships in which puffers interact to destroy
all of the debris they would otherwise leave behind. These rules include B356/S23,
B356/S238, B3678/S0345, and B38/S02456 [9]. In Life, Dean Hickerson's $c/12$
diagonal Cordership, based on Charles Corderman's switch engine puffer, also has
the same structure.

Day & Night (B3678/S34678) is another rule investigated by Bell [3]. It has the
curious property that, if one exchanges the roles of live and dead cells, the rule's
behavior is unchanged, because the numbers of dead neighbors that lead to a cell
death are exactly the same as the numbers of live neighbors that lead to a birth,
and the numbers of dead neighbors that lead to quiescence are exactly the same
as the numbers of live neighbors that lead to survival. Day & Night supports a
small diagonal spaceship and many orthogonal ones, high-period oscillators, and
patterns that mediate the boundaries between regions of live cells on a dead background and regions of dead cells on a live background (Fig. 6.13). When its initial
state is a random field of cells with a probability $p < \frac{1}{2}$ of being live, the live
cells form clusters that gradually shrink, eventually forming a collection of scattered oscillators. When the probability of being live is greater than $\frac{1}{2}$, the dead
cells form shrinking clusters and one again gets a collection of scattered oscillators in the complementary dead-on-live world. When the probability of being live
is exactly $\frac{1}{2}$, the clusters and live and dead cells gradually grow and merge, so
that the typical size of a cluster and the typical curvature of the cluster boundaries (ignoring small isolated oscillators) are monotonic functions of the number
of steps of the rule; see Fig. 6.15 for an example of similar behavior in a different self-complementary rule. The boundaries between clusters are not a type of
pattern that is directly addressed by Wolfram's classification, but if one interprets

Fig. 6.14 A $c/7$ spaceship (*top*) and David Bell's spacefiller (*bottom*) in Diamoeba (B35678/S5678)

Fig. 6.15 After 500 steps starting from a random field in Anneal (B4678/S35678)

these boundaries as being largely stable or oscillatory, one can view the behavior of Day & Night as belonging to Wolfram's Class II.

Diamoeba (B35678/S5678) was first investigated by Dean Hickerson. In this rule, patterns tend to form large diamond shapes with an irregular and difficult-to-predict growth rate. It supports a small $c/7$ spaceship (Fig. 6.14, top) [10] with some internal structure: between the head and the tail of the spaceship there is an extensible middle section, the boundary of which is lined by a sequence of one-cell and two-cell protrusions, and the time–space patterns formed by these protrusions simulate a simple one-dimensional cellular automaton. Hickerson offered a $50 prize in 1993 for finding a quadratic growth pattern and in 1998 Gravner and Griffeath [16] asked more specifically whether there exists a pattern that eventually fills the entire plane with live cells. Both problems were solved affirmatively in 1999 by David Bell, whose solution combined two oppositely-oriented copies of the $c/7$ spaceship (Fig. 6.14, bottom). Bell and Hickerson subsequently also found patterns for which the number of live cells grows linearly rather than quadratically, based on the same $c/7$ spaceship head. Under Wolfram's classification, Diamoeba appears either as Class I or Class II: random initial conditions lead to a state in which almost all cells are live, but in which there are very sparse clusters of oscillating dead cells.

B35/S236 [11] is near Life in rule space, but most patterns behave differently in the two rules. It has a small spaceship analogous to Life's glider, and as with the glider this spaceship frequently arises from small random seeds, but it moves at speed $2c/5$ orthogonally. This rule also supports small period-24 and period-68 oscillators; by combining multiple copies of either of these oscillators it is pos-

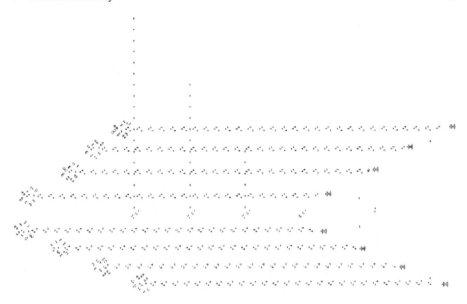

Fig. 6.16 A quadratic growth pattern in B35/S236

sible to form guns. Figure 6.16 shows a pattern in which eight banks of three period-68 guns (with two additional period-68 spaceship reflectors per bank) send out streams of small $2c/5$ spaceships behind eight slower $c/3$ spaceships; when the $2c/5$ stream collides with the rear end of the $c/3$ spaceship, it sends a single spaceship towards the middle of the pattern. These spaceships collide to produce additional guns, leading to a pattern like Life's breeder in which the number of live cells grows quadratically with the number of steps. When started from a random initial condition, this rule remains chaotic, falling into Wolfram's Class III.

B27/S0 supports a large $c/2$ spaceship (Fig. 6.17) which, despite being found by an automated search [10], contains a large amount of structure similar to that of an engineered pattern. As oriented in the figure, it moves downwards; although the

Fig. 6.17 A $c/2$ spaceship in
B27/S0

Fig. 6.18 *Left*: A spaceship sandwiched between two replicators in B25/S4 leads to pseudorandom behavior. *Right*: A glider gun in B24/S

overall pattern has bilateral symmetry with even width, it has two head components that are themselves symmetric with odd width. Behind them two repeating patterns stretch out on either side, with 11 units of repetition per side, spreading out the pattern from the head components to a wider tail that includes two more of the same head components and six smaller equal tail components. However, when started in a random configuration, patterns in this rule remain highly and uniformly chaotic, placing it clearly in Class III of Wolfram's classification.

B25/S4 contains a small period-3 replicator, and (like most B2 rules without B3/S1) supports small *photons*, spaceships that move at speed c. If a photon and a replicator are aligned on the same axis, their interaction will destroy that copy of the replicator and emit an oppositely-oriented photon. Placing a photon between two appropriately spaced replicators (Fig. 6.18, left) leads to a pattern in which the photon repeatedly bounces back and forth between copies of the replicators on both sides, following a pseudorandom walk in which the typical distance of the photon from its starting point at step n appears to be proportional to $\sqrt{n}$. After our 2001 discovery of this system, Tomas Rokicki found very efficient algorithms allowing its behavior to be simulated for billions of steps [24]. As Dean Hickerson observed, a similar system can be set up in HighLife: the bomber reaction allows two sets of replicators to play tug of war with a blinker. Like most B2 rules, B25/S4 falls into Wolfram's Class III.

B24/S supports a low-period photon gun (Fig. 6.18, right). Guns of this size and period are small enough to be found by automated search software; another, similar gun exists in B25/S45. Again, these rules belong to Class III.

B2/S7 was studied by Martínez et al. [21]. This rule supports spaceships, puffers, rakes, and a novel structure that Martínez et al. termed *avalanches* in which a diagonal chain of cells grows wider as it moves across the plane leaving a diamond-shaped trail of chaos behind it. They described methods of simulating Boolean circuits using spaceship collisions and implementing a finite-state memory with oscillators.

In contrast with these fertile and mortal rules, we describe a few rules that are infertile or immortal. Because of these properties, fewer structured patterns are known in these rules, but they may still exhibit other interesting behaviors.

Anneal (B4678/S35678) is mortal but infertile. Similarly to Day & Night, any pattern in this rule has the same behavior if all live cells are replaced by dead cells and vice versa. And as with Day & Night, its behavior on random initial configurations is to form growing clusters of live and dead cells, with scattered oscillators. The live cells predominate for initial configurations with probability greater than 0.5 of being live, the dead cells predominate when this probability is less than 0.5, and when the probability is exactly $\frac{1}{2}$ (Fig. 6.15) both types of clusters coexist with the cluster size and radius of curvature increasing over time. Thus, Wolfram's classification is incapable of distinguishing this rule from Day & Night, although its nonrandom behavior is much more constrained.

B1357/S1357 is a fertile but immortal rule investigated by Edward Fredkin in which every pattern is a replicator. Cell (x, y) is alive at step i if and only if the number of ways that a chess king could take i steps to walk from an initially-live cell to (x, y) is odd. When i is divisible by a number 2^k that is larger than the size of the bounding box of the initial pattern P, the pattern at step i consists of several disjoint copies of P, spaced at multiples of 2^k units apart, with the overall arrangement of these copies being identical to the arrangement of live cells that one would get at step $i/2^k$ starting from a single live cell. Thus, although this rule supports replicators, which in many other rules lead to other sorts of complex behavior, in this rule there is nothing but replicators. If a starting state has all cells set to live or dead uniformly and independently at random, then the same is true at each subsequent step, so this rule displays no structure whatsoever when run under random initial conditions and falls into Wolfram's Class III.

B1/S012345678 is another fertile and immortal rule, introduced as a model of snowflake formation by Packard [25] and one of several Life-like rules later mentioned by Wolfram and Packard [29]. In this rule a starting state consisting of a single live cell leads to a pattern that fills the plane with a fractal tree structure — see Fig. 6.19. The ratio of live cells to dead cells in the eventual stable pattern is exactly $\frac{4}{9}$ [16]; however, Dean Hickerson has found other starting patterns for this rule that lead to different densities [17]. See [17] for similar results in all the related *Packard snowflake* Life-like rules with rule strings of the form B1*xxx*/S012345678. When run from a random initial state, B1/S012345678 very quickly stabilizes, putting it in Class II.

B4/S01234 is both infertile and immortal. When run from a random initial state it crystallizes into regions filled with horizontal stripes of live and dead cells, mixed with similar regions with vertical stripes (Fig. 6.20); some cells on the borders between regions oscillate with low periods. Similar striped patterns also form in B35/S234578, a mortal rule that seems extremely likely to be fertile: although we do not know of a spaceship nor a proof that any other pattern is a growth pattern, most finite starting patterns in B35/S234578 form round regions with a linearly growing radius, within which these same patches of horizontal and vertical stripes predominate.

Life Without Death (B3/S012345678) (also known as Inkspot [26]) has the same birth rules as Life, but disallows any death of a live cell, causing it to be immortal. It is fertile, despite having no spaceships: its patterns frequently develop *ladders* consisting of a growing tip that moves in a straight line leaving an immortal trail of live cells behind it (Fig. 6.21). Ladders can be arranged to simulate any Boolean

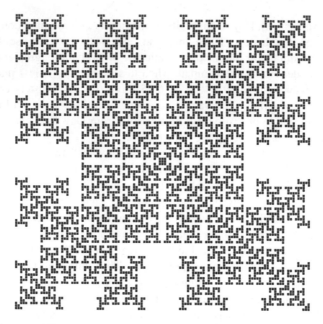

Fig. 6.19 The space-filling tree structure formed by B1/S012345678 after 57 steps starting from a single live cell

Fig. 6.20 Regions of horizontal and vertical stripes created by B4/S01234

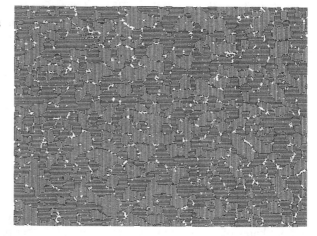

circuit, showing that it is P-complete to determine the value of a cell at a future state of the automaton [18]; this implies that it is unlikely for there to exist an algorithm that solves this cell value problem significantly more quickly than the naive algorithm that simply simulates the rule for the desired number of steps. In some sense this circuit simulation result tests the limits of our classification by showing that complex engineered patterns with a modular structure may exist even in immortal rules.

Fig. 6.21 A pattern resulting
from approximately 1,000
steps of Life Without Death
starting from a small random
seed. The long straight
patterns extending from the
central mass are ladders

6.7 Beyond Growth and Decay

Our definitions of growth and decay make the assumption that the patterns to be
analyzed have finitely many live cells, set on a background of quiescent dead cells.
However, even for the restricted family of two-dimensional semitotalistic automata
that we study here, these are not the only possibilities. Many automata, even those
that may be wildly chaotic when started randomly, may support background patterns
that are periodic in both time and space, as well as finite perturbations to these periodic background patterns that move and interact similarly to the way Life's gliders
move and interact.

   As one particular case of a periodic background, we have made some preliminary
investigations of rules in which a birth occurs with zero live neighbors and a death
occurs with eight live neighbors. In these rules, if a pattern starts with finitely many
live cells on a background of dead cells, it will continue to have finitely many live
cells on a background of dead cells in every even step, but in the odd steps the
pattern is reversed: there are finitely many dead cells on a background of infinitely
many live cells. A few of these rules have already shown interesting behavior:

B01245/S0125 is a strobing version of Day & Night. Any pattern in Day & Night
   will behave identically in this rule, except that in odd steps the live and dead cells
   will reverse roles. Similar strobing versions exist for any rule that is symmetric
   under live-dead reversal, such as Anneal.
B017/S1 supports two different replicators, with periods 8 and 14. Initial states
   in which cells are set to be alive independently at random with either a low or

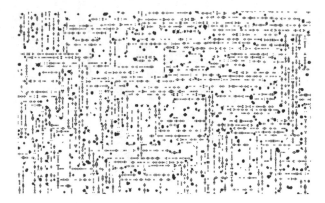

Fig. 6.22 Replicator chaos in B017/S1

high density of live cells eventually become dominated by oscillators formed by rows of these replicators, bounded at each end by stable blobs of cells (Fig. 6.22); however, for intermediate densities of initial live cells (around 25%), the pattern instead coagulates into larger stable blobs of cells that are all alive or all dead in alternating phases.

B013468/S02 (a rule investigated by the author in 2002) has a small spaceship, and a very small period-36 double-barreled gun that in some phases fits into a 4×6 bounding box. As in Day & Night, it also has several larger high-period $c/2$ puffers and spaceships, as well as ladders like those in Life Without Death. Figure 6.23 shows a pattern formed from eight puffers that would, if by themselves, leave trails of domino oscillators. Pairs of puffers combine to form high-period spaceships with large trailing sparks, one pair of spaceships combines to form a much messier puffer, and the other pair of spaceships combine to form a rake, whose output spaceships crash into the puffer trail to eliminate some unwanted debris, leaving behind a sequence of ladders with a quadratic growth rate in the number of non-background cells. Random initial conditions tend to settle down to large regions of still life, oscillator, and spaceship patterns, like those in Life, but as in Day & Night some of these regions have live cells on dead backgrounds in even steps and dead cells on live backgrounds in odd steps, while for other regions this pattern is reversed. The boundaries of these regions are chaotic and send spaceships into the interiors of the regions, so patterns tend to stabilize much more slowly than in Life. As in Day & Night, sparse random conditions eventually settle down to live cells on dead backgrounds on even steps and dense random conditions eventually settle down to the opposite parity, with the transition between these two phases occurring at roughly a 46% ratio of live to total cells.

B01367/S0124 has an unusual replicator in the shape of a W pentomino with the property that the two copies formed from this replicator are turned at right angles from its original orientation.

B01367/S012 has replicators of periods 20 and 22 that can interact to form space-ship guns.

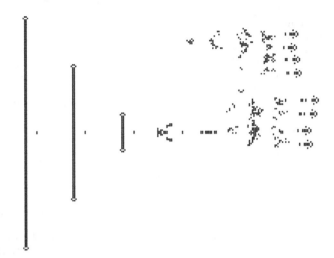

Fig. 6.23 A quadratic growth pattern in B013468/S02

B02346/S023 has very tiny replicators consisting of two live cells a knight's move apart; the copies this pattern makes of itself spread themselves out along a line of slope ±2 or ±1/2.

One may generalize our definitions of fertility and mortality for such rules by replacing the finite sets of live cells in the definition by a finite set of cells that is different from the background: a fertile combination of a rule and a background is one for which some finite perturbation to the background escapes any bounding box, and a mortal combination of rule and background is one in which some finite perturbation to the background eventually stabilizes so that only the background remains. We know little about which combinations of rules and backgrounds are likely to be both fertile and mortal, but such knowledge would be very helpful as a guide in exploring the limitless possibilities these combinations have to offer.

Acknowledgements   We thank Calcyman, David Griffeath, Dean Hickerson, Nathaniel Johnston, Harold McIntosh, and Andrew Trevorrow for helpful comments on a draft of this chapter. The author is supported in part by NSF grant 0830403 and by the Office of Naval Research under grant N00014-08-1-1015.

References

1. Adamatzky, A., Martínez, G.J., Mora, J.C.S.T.: Phenomenology of reaction–diffusion binary-state cellular automata. Int. J. Bifurc. Chaos Appl. Sci. Eng. **16**(10), 2985–3006 (2006). http://uncomp.uwe.ac.uk/genaro/Papers/Papers_on_CA_files/rdca.pdf
2. Baldwin, J.T., Shelah, S.: On the classifiability of cellular automata. Theor. Comput. Sci. **230**(12), 117–129 (2000). doi:10.1016/S0304-3975(99)00042-0. arXiv:math.LO/9801152
3. Bell, D.I.: Day & Night — An interesting variant of Life. http://www.tip.net.au/~dbell/articles/HighLife.zip (1994). Unpublished article

4. Bell, D.I.: HighLife — An interesting variant of Life. http://www.tip.net.au/~dbell/articles/ HighLife.zip (1994). Unpublished article
5. Berlekamp, E.R., Conway, J.H., Guy, R.K.: Winning Ways for Your Mathematical Plays, vol. 4, 2nd edn. AK Peters, Wellesley (2004)
6. Chaté, H., Manneville, P.: Criticality in cellular automata. Physica D **45**, 122–135 (1990). doi:10.1016/0167-2789(90)90178-R. Special issue of Physica D, reprinted as: Gutowitz, H. (ed.) Cellular Automata: Theory and Experiment. MIT/North-Holland, Cambridge/Amsterdam (1991)
7. Culik, K. II, Yu S.: Undecidability of CA classification schemes. Complex Syst. **2**, 177–190 (1988)
8. Due, B.: Outer totalistic cellular automata meta-pixel. http://otcametapixel.blogspot.com/ (2006). See also Dave Greene's "metafier" script, and several examples of metafied patterns, included as part of the Golly cellular automaton software. Retrieved November 1, 2009. Unpublished web page
9. Eppstein, D.: Gliders in life-like cellular automata. http://fano.ics.uci.edu/ca/. Web database of spaceships in life-like cellular automaton rules
10. Eppstein, D.: Searching for spaceships. In: More Games of No Chance. MSRI Publications, vol. 42, pp. 433–453. Cambridge University Press, Cambridge (2002). arXiv:cs.AI/0004003. http://www.msri.org/publications/books/Book42/files/eppstein.pdf
11. Eppstein, D.: B35/S236. http://www.ics.uci.edu/~eppstein/ca/b35s236/ (2003). Unpublished web site
12. Flammenkamp, A.: Most seen natural occurring ash objects in Game of Life. http://wwwhomes.uni-bielefeld.de/achim/freq_top_life.html (2004). Retrieved November 1, 2009. Unpublished web page
13. Gardner, M.: Mathematical Games: The fantastic combinations of John Conway's new solitaire game "Life". Sci. Am. **223**, 120–123 (1970)
14. Gotts, N.M.: Self-organized construction in sparse random arrays of Conway's Game of Life. In: New Constructions in Cellular Automata, pp. 1–53. Oxford University Press, London (2003)
15. Gravner, J.: Growth phenomena in cellular automata. In: New Constructions in Cellular Automata, pp. 161–181. Oxford University Press, London (2003)
16. Gravner, J., Griffeath, D.: Cellular automaton growth on Z^2: theorems, examples, and problems. Adv. Appl. Math. **21**(2), 241–304 (1998). doi:10.1006/aama.1998.0599. http://psoup.math.wisc.edu/extras/r1shapes/r1shapes.html
17. Gravner, J., Griffeath, D.: Asymptotic densities for Packard *Box* rules. Nonlinearity **22**, 1817–1846 (2009). doi:10.1088/0951-7715/22/8/003. http://psoup.math.wisc.edu/papers/box.pdf
18. Griffeath, D., Moore, C.: Life without Death is P-complete. Complex Syst. **10**, 437–447 (1996). http://psoup.math.wisc.edu/java/lwodpc/lwodpc.html
19. Lafusa, A., Bossomaier, T.: Localisation of critical transition phenomena in cellular automata rule-space. In: Recent Advances in Artificial Life. World Scientific, Singapore (2005). doi:10.1142/9789812701497_0010
20. Li, W., Packard, N.H., Langton, C.G.: Transition phenomena in cellular automata rule space. Physica D **45**, 77–94 (1990). doi:10.1016/0167-2789(90)90175-O. Special issue of Physica D, reprinted as: Gutowitz, H. (ed.) Cellular Automata: Theory and Experiment. MIT/North-Holland, Cambridge/Amsterdam (1991)
21. Martínez, G.J., Adamatzky, A., McIntosh, H.V.: Localization dynamics in a binary two-dimensional cellular automaton: the Diffusion Rule. arXiv:0908.0828. J. Cell. Autom. (2008, in press)
22. McIntosh, H.V.: Wolfram's class IV automata and a good life. Physica D **45**, 105–121 (1990). doi:10.1016/0167-2789(90)90177-Q. Special issue of Physica D, reprinted as: Gutowitz, H. (ed.) Cellular Automata: Theory and Experiment. MIT/North-Holland, Cambridge/Amsterdam (1991)
23. Nivasch, G.: The 17c/45 caterpillar spaceship. http://www.yucs.org/~gnivasch/life/article_cat/ (2005). Retrieved November 8, 2009. Unpublished web page

24. Nivasch, G.: The photon/XOR system. http://yucs.org/~gnivasch/life/photonXOR/ (2007). Retrieved November 8, 2009. Unpublished web page
25. Packard, N.H.: Lattice models for solidification and aggregation. Inst. for Advanced Study preprint (1984). Reprinted in: Wolfram, S. (ed.) Theory and Applications of Cellular Automata, pp. 305–310. World Scientific, Singapore (1986)
26. Toffoli, T., Margolus, N.: Cellular Automata Machines: A New Environment for Modeling, pp. 6–7. MIT Press, Cambridge (1987)
27. Trevorrow, A., Rokicki, T.: Golly. http://golly.sourceforge.net/ (2009). Multiplatform open-source software, version 2.1
28. Wolfram, S.: University and complexity in cellular automata. Physica D **10**, 1–35 (1984). doi:10.1016/0167-2789(84)90245-8. Reprinted in: Cellular Automata and Complexity, pp. 115–157. Addison–Wesley, Reading (1994)
29. Wolfram, S., Packard, N.H.: Two-dimensional cellular automata. J. Stat. Phys. **38**, 901–946 (1985). doi:10.1007/BF01010423. Reprinted in: Cellular Automata and Complexity, pp. 211–249. Addison–Wesley, Reading (1994)
30. Wootters, W.K., Langton, C.G.: Is there a sharp phase transition for deterministic cellular automata? Physica D **45**, 75–104 (1990). doi:10.1016/0167-2789(90)90176-P. Special issue of Physica D, reprinted as: Gutowitz, H. (ed.) Cellular Automata: Theory and Experiment. MIT/North-Holland, Cambridge/Amsterdam (1991)

Chapter 7
The B36/S125 "2x2" Life-Like Cellular Automaton

Nathaniel Johnston

The B36/S125 (or "2x2") cellular automaton is one that takes place on a 2D square lattice much like Conway's Game of Life. Although it exhibits high-level behaviour that is similar to Life, such as chaotic but eventually stable evolution and the existence of a natural diagonal glider, the individual objects that the rule contains generally look very different from their Life counterparts. In this article, a history of notable discoveries in the 2x2 rule is provided, and the fundamental patterns of the automaton are described. Some theoretical results are derived along the way, including a proof that the speed limits for diagonal and orthogonal spaceships in this rule are $c/3$ and $c/2$, respectively. A Margolus block cellular automaton that 2x2 emulates is investigated, and in particular a family of oscillators made up entirely of 2×2 blocks are analyzed and used to show that there exist oscillators with period $2^{\ell}(2^k - 1)$ for any integers $k, \ell \geq 1$.

7.1 Introduction

One cellular automaton that has drawn a fair amount of interest recently is the one that takes place on a grid like Conway's Game of Life, except that dead cells are born if they have 3 or 6 live neighbours, and alive cells survive if they have 1, 2, or 5 live neighbours. Its birth and survival information is conveyed by the rulestring "B36/S125". This rule exhibits many qualities that are similar to those of Life — for example, evolution seems unpredictable and random patterns tend to evolve into "ash fields" consisting of several small stable patterns (known as *still lifes*) and periodic patterns (known as *oscillators*).

The B36/S125 automaton has become known as "2x2" because of the fact that it emulates a simpler cellular automaton that acts on 2×2 blocks of cells. In particular, this means that patterns that are initially made up of 2×2 blocks will forever be made up of 2×2 blocks under this evolution rule. Because of the simplicity of the emulated block cellular automaton, it has many properties in common with elementary cellular automata [3, 4] and in particular emulates Wolfram's rule 90 [2].

Although the rough behaviour and statistics of 2x2 are similar to those of Life, the patterns of 2x2 have completely different structure and thus are interesting in their

Fig. 7.1 A large still life

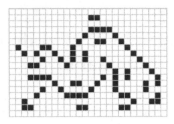

Fig. 7.2 The wickstretcher

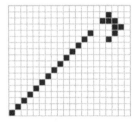

own right. Furthermore, many questions that have been answered about Life remain open in 2x2, such as whether or not it contains guns or replicators. Interestingly, both of these types of patterns are known to exist in the nearby rule B368/S12578. However, this rule lacks the block cellular automaton simulation properties of B36/S125 that will be discussed later.

2x2 has a basic $c/8$ diagonal glider that occurs fairly commonly, though it is larger than the standard Life glider and is thus more difficult to construct. The first infinitely-growing pattern to be discovered was a wickstretcher based on the glider. This wickstretcher is displayed in Fig. 7.2 with alive cells in black and dead cells in white.

Other spaceships were found via computerized searches carried out by Alan Hensel, Dean Hickerson, David Bell, and David Eppstein in the 1990s. One of the most important spaceship discoveries was a $c/3$ diagonal spaceship, which showed that it is possible for spaceships to travel faster in the 2x2 universe than in the standard Life universe, despite most of the easy-to-find spaceships being quite slow. We will see that $c/3$ is the diagonal speed limit in 2x2, much like $c/4$ is the diagonal speed limit in Life. Several derived results will also apply to other Life-like cellular automata, and we will note when this is the case, although our focus and motivation will be the 2x2 rule.

7.2 Ash and Common Patterns

One of those most interesting aspects of 2x2 is the large variety of still lifes and oscillators that appear naturally as a result of evolving randomly-generated starting patterns (known as *soup*). Many simple still lifes are familiar from the standard Life rules, such as the tub, beehive, aircraft carrier, loaf, and pond. More commonly-occurring, however, are simple "sparse" still lifes that are not stable in Life, such

Table 7.1 The 20 most common naturally-occurring still lifes in the 2x2 rule and their approximate frequency (out of 1.000) relative to all still lifes

#	Pattern	Rel. frequency	#	Pattern	Rel. frequency
1		6.076×10^{-1}	11		8.895×10^{-4}
2		2.130×10^{-1}			
3		5.038×10^{-2}	12		7.738×10^{-4}
4		4.898×10^{-2}	13		6.618×10^{-4}
5		2.964×10^{-2}	14		5.604×10^{-4}
6		2.575×10^{-2}	15		3.640×10^{-4}
7		9.784×10^{-3}	16		2.831×10^{-4}
8		4.232×10^{-3}	17		2.487×10^{-4}
9		4.077×10^{-3}	18		1.385×10^{-4}
10		1.949×10^{-3}	19		1.053×10^{-4}
			20		7.571×10^{-5}

as a horizontal or diagonal row of 2 cells. Table 7.1 shows the 20 most commonly-occurring still lifes in 2x2.[1]

Because cells stay alive if they have only one live neighbour there are many small still lifes, most of which are made up of islands, with each island being a chain or loop of a few cells. This leads to a simple grammar for constructing large still lifes — see Fig. 7.1. The number of distinct strict still lifes with n cells for $n - 1, 2, 3, \ldots$ is given by $0, 2, 1, 3, 4, 9, 10, 27, 48, 126, \ldots$.[2] Compare this with the corresponding sequence for Life, which is $0, 0, 0, 2, 1, 5, 4, 9, 10, 25, \ldots$.[3]

The oscillators that occur naturally in 2x2 do not occur in Life. The majority of common oscillators have period 2 or 4, but some small patterns give rise to very high-period oscillators. For example, the fourth most common oscillator is simply the stairstep hexomino in one of its phases, yet it has period 26. The thirteenth most common oscillator, which it might be appropriate to name the "decathlon," has period 10 and evolves out of a horizontal row of 5 adjacent cells, much like the period 15 "pentadecathlon" of Life evolves out of a horizontal row of 10 adjacent cells. Oscillators with periods 14 and 22 are also relatively frequent, as demonstrated by Table 7.2.[4]

---

[1]Computed by evolving 22,846,665 random patterns of size 20 × 20 and initial density 37.5%. A total of 255,689,477 (non-distinct) still lifes were catalogued as a result. Statistics compiled by the *Online Life-Like CA Soup Search*, http://www.conwaylife.com/soup.

[2]Sloane's A166476 — the still lifes with 9 or fewer cells are shown in Appendix.

[3]Sloane's A019473.

[4]Based on data from the *Online Life-Like CA Soup Search*. A total of 11,270,020 (non-distinct) oscillators were catalogued.

Table 7.2 The 20 most common naturally-occurring oscillators in the 2x2 rule and their approximate frequency (out of 1.000) relative to all oscillators

#	Pattern	Period	Rel. freq.	#	Pattern	Period	Rel. Freq.
1		2	4.824×10^{-1}	11		2	9.330×10^{-3}
2		2	2.170×10^{-1}	12		2	7.766×10^{-3}
3		2	5.741×10^{-2}	13		10	5.188×10^{-3}
4		26	5.515×10^{-2}	14		2	2.042×10^{-3}
5		4	3.718×10^{-2}	15		2	1.633×10^{-3}
6		14	3.364×10^{-2}	16		2	1.559×10^{-3}
7		4	3.104×10^{-2}	17		14	1.182×10^{-3}
8		2	1.795×10^{-2}	18		2	9.618×10^{-4}
9		4	1.766×10^{-2}	19		6	9.539×10^{-4}
10		4	1.745×10^{-2}	20		22	4.423×10^{-4}

The only other notable patterns that have been known to appear spontaneously from random soup are the fairly common $c/8$ diagonal glider and the related $c/8$ wickstretcher. Although the glider itself was known of by no later than 1993, the wickstretcher, which works simply by placing the glider next to a diagonal wick, was not found until June 2009. In fact, the $c/8$ wickstretcher and its slight modifications are currently the only known infinitely-growing patterns in 2x2.

7.3 Oscillators and Spaceships

Beyond the standard oscillators that appear naturally, many oscillators of period 2 through 4 have been constructed by hand and computer search by Alan Hensel, Dean Hickerson, and Lewis Patterson over the years. In 1993, Hensel discovered the first known oscillator with odd period, the small period 5 pattern shown in Fig. 7.3. David Bell soon thereafter noticed that it can be combined with itself and extended in a variety of different ways, creating the first known extensible oscillator. Hensel discovered the first known period 3 oscillator in 1994, and Hickerson found the first known period 11 and 17 oscillators later that same year. To date, the only odd periods for which there are known oscillators are 3, 5, 11, and 17, which shows that odd-period oscillators seem to be much more difficult to construct in this rule than even-period oscillators. We will see later that there is an infinite family of even periods that are easily realized by simple block oscillators. The least period for which there is no known oscillator is 7, while the least even period for which there is no known oscillator is 18.

The $c/8$ diagonal glider is the only spaceship that has ever been seen to occur naturally in 2x2, though several others have been found via computer search. In

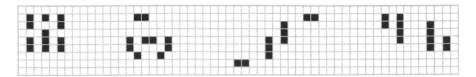

Fig. 7.3 The first known oscillators of period 3, 5, 11, and 17, respectively

Fig. 7.4 The first discovered
spaceship other than the $c/8$
glider

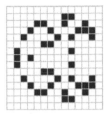

Fig. 7.5 The $c/3$ diagonal
spaceship

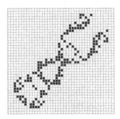

February 1994, Hensel found the first such spaceship — the $c/3$ orthogonal pattern shown in Fig. 7.4. He also found the next three spaceships, which were orthogonal with speed $c/3$, $c/3$, and $c/4$. As a result of the relative ease of finding slow spaceships in this rule, it was initially suspected that it does not contain spaceships that travel as fast as their Life counterparts ($c/2$ orthogonally and $c/4$ diagonally). However, David Eppstein found a $c/2$ orthogonal spaceship in October 1998 using his gfind program [1], and several others have been found since then (see Appendix). Hickerson found the first $c/4$ diagonal spaceship in 1999. Eppstein has since found a $c/3$ diagonal spaceship (see Fig. 7.5), which shows that it is possible for spaceships in 2x2 to travel *faster* than spaceships in Life.

With the discovery of the $c/3$ diagonal spaceship comes the natural question of what the speed limits in 2x2 are for spaceships. It was proved by Conway in the early 1970s that spaceships in Life can not travel faster than $c/2$ orthogonally or $c/4$ diagonally. Using similar methods it is possible to prove that spaceships in 2x2 (and many other rules) can not travel faster than $c/2$ orthogonally or $c/3$ diagonally.

Theorem 1 *In a Life-like cellular automaton in which births occurs for three live neighbours but not for two or fewer live neighbours, spaceships can not travel faster than $c/3$ diagonally or $c/2$ orthogonally.*

Fig. 7.6 If a spaceship is on or below the line defined by A, B, C, D, E, and F in generation 0, supposing X or Y is alive in generation 2 gives a contradiction

Proof Assume that the spaceship is on or below the line of slope $-1/2$ defined by cells A, B, C, D, E, and F in Fig. 7.6 in generation 0. To see the result, we will assume that either X or Y can be alive in generation 2 and derive a contradiction.

If Y is alive in generation 2, then it must have at least three live neighbours in generation 1. However, the only one of Y's neighbours that can possibly be alive in generation 1 is N, so Y can not be alive in generation 2. Similarly, the only of X's neighbours that can be alive in generation 1 are B and N, so X can not possibly be alive in generation 2. This shows that neither X nor Y can be alive in generation 2.

In particular, this shows that orthogonal spaceships can not travel faster than $c/2$ and diagonal spaceships can not travel faster than $c/3$ (by applying the above semi-diagonal line logic twice).[5]   □

Because a $c/3$ diagonal spaceship is already known in 2x2, as are several $c/2$ orthogonal spaceships, the speed limit question for 2x2 has been answered. Additionally, it is worth noting that the proof of Theorem 1 implies not only the $c/2$ and $c/3$ speed limit results for orthogonal and diagonal spaceships, but also that any spaceship that travels a cells vertically for every b cells horizontally (if such a spaceship exists) can not travel faster than $\max\{a, b\}c/(2a + b)$.[6] In particular, this says that knightships (which travel two cells vertically for every one cell horizontally), if they exist, can not travel faster than $2c/5$.

7.4 As a Block Cellular Automaton

One of the most interesting properties of 2x2 is that any patterns that is initially made up of 2×2 blocks of the same state will forever be made up of 2×2 blocks of the same state. A bit more specifically, it emulates the block cellular automaton that makes use of the Margolus neighbourhood and evolves according to the six rules given by Fig. 7.7.

By saying that 2x2 emulates a Margolus block cellular automaton, we mean that the resulting block appears at the center of the original four blocks. Thus, patterns that are originally made up of 2×2 blocks will forever be made up of 2×2 blocks,

---

[5]In 1994, Dean Hickerson used alternate means to show that spaceships in 2x2 can not travel faster than $c/3$ diagonally or $2c/3$ orthogonally.

[6]David Eppstein was aware of this speed upper bound for rules in which birth occurs when a cell has three live neighbours back in 1999 and provided the first proof that the orthogonal speed limit is $c/2$.

Fig. 7.7 The block cellular
automaton emulated by 2x2

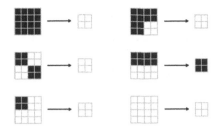

but the block partition will be offset diagonally by one cell in the odd generations from the even generations. Of course, 2x2 is not the only Life-like cellular automaton that emulates a Margolus block cellular automaton; such a cellular automaton is emulated if and only if the rule satisfies the following four conditions:

- Birth occurs for 3 neighbours if and only if survival occurs for 5 neighbours.
- Birth occurs for 4 neighbours if and only if survival occurs for 4 neighbours.
- Birth occurs for 5 neighbours if and only if survival occurs for 6 neighbours if and only if survival occurs for 7 neighbours.
- Birth occurs for 1 neighbour if and only if birth occurs for 2 neighbours if and only if survival occurs for 3 neighbours.

More succinctly, a Life-like cellular automaton emulates a Margolus block cellular automaton if and only if, in its rulestring, B3 = S5, B4 = S4, B5 = S6 = S7, and B1 = B2 = S3. 2x2 can be seen to satisfy these conditions because 4 is neither a birth condition nor a survival condition, 5 is not a birth condition and 6 and 7 are not survival conditions, 3 is a birth condition and 5 is a survival condition, and 3 is not a survival condition and 1 and 2 are not birth conditions. There are $2^{12} = 4,096$ Life-like cellular automata that emulate $2^6 = 64$ different Margolus block cellular automata. The 64 Life-like cellular automata from B3/S5 to B3678/S0125 all emulate the same Margolus block cellular automaton given by Fig. 7.7.

In fact, it was noticed by David Eppstein in 1998 that the Margolus block cellular automaton that 2x2 emulates also emulates itself via 2×2 blocks *of* 2×2 blocks, but with a slowdown of one simulated generation per real generation. That is, each 2×2 block in Fig. 7.7 can be replaced by the corresponding 4×4 block as long as it is understood that the transition time indicated by the arrows is two generations instead of one (see Fig. 7.8).

It follows naturally that 8×8 blocks can be used to simulate 4×4 blocks, again doubling the number of real generations per simulated generation. In general, $2^k \times 2^k$ blocks can be used to simulate 2×2 blocks, with each simulated generation requiring 2^{k-1} real generations. Because the 2×4 rectangle is an oscillator with period 2, it follows that the 4×8 rectangle is an oscillator with period 4, the 8×16 rectangle is an oscillator with period 8, and in general the $2^k \times 2^{k+1}$ rectangle is an oscillator with period 2^k. This was the first known proof that 2x2 contains oscillators of arbitrarily large period — we will see another proof (also related to 2×2 block oscillators) in the next section.

One might wonder what types of patterns exist in this block cellular automaton – after all, it is a simpler rule so perhaps we can prove the existence of certain types of

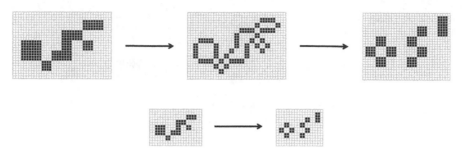

Fig. 7.8 An example of 4 × 4 blocks (*top*) taking two generations to emulate one generation of a 2 × 2 block pattern (*bottom*)

patterns in 2x2 by simply trying to find interesting block patterns. However, no block pattern can ever escape its initial "bounding diamond," so we can not hope to find spaceships or infinitely-growing patterns in this manner. Additionally, because the block partition in even generations is offset by one cell from the block partition in odd generations, we can't hope to find odd-period oscillators. Thus, we investigate an infinite family of even-period block oscillators.

7.5 Block Oscillators

One particularly interesting family of oscillators in 2x2 are those that in one phase are a $2 \times 4n$ rectangle of alive cells for some integer $n \geq 1$. For $n = 1$, the oscillator has period 2 and simply rotates by 90 degrees every generation. That is, it starts as a 2×4 rectangle, evolves into a 4×2 rectangle after one generation, and then evolves back into a 2×4 rectangle after the next generation. As n increases, these oscillators behave more and more unpredictably. For $n = 2$, the oscillator has period 6 as shown in Fig. 7.9.

By simply playing around with Life simulation software, it is not difficult to compute the period of the $2 \times 4n$ block oscillator for $n = 1, 2, 3, \ldots$ to be

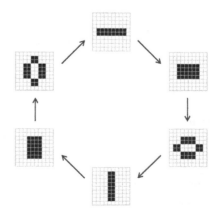

Fig. 7.9 The 2 × 8 period 6 block oscillator

Fig. 7.10 Generation 6 of
the $n = 3$ block oscillator,
depicted as the XOR of a
2×12 rectangle, a 6×8
rectangle, and a 10×4
rectangle

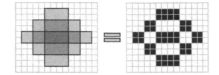

$2, 6, 14, 14, 62, 126, 30, 30, 1{,}022, 126, \dots$ [7] Other than the fact that each of these periods is two less than a power of two, this sequence does not have any obvious pattern. The following theorem shows that the sequence is in fact related to a well-studied mathematical phenomenon, albeit one that does not admit what most people would consider a closed form solution. Before presenting the theorem, it is perhaps worth noting that B36/S125 is not the only rule in which these oscillators work; these rectangular oscillators behave the same in the $2^9 = 512$ rules from B3/S5 to B35678/S012567.

Theorem 2 *The period of the $2 \times 4n$ block oscillator is $2(2^k - 1)$, where k is the multiplicative suborder of $2 \pmod{(2n + 1)}$.* [8]

The multiplicative suborder of $a \pmod{b}$, denoted $\mathrm{sord}_b(a)$, is defined as the least natural number k such that $a^k \equiv \pm 1 \pmod{b}$. There exists a k satisfying this condition if and only if $GCD(a, b) = 1$. Since $2n + 1$ is odd we know that the k mentioned by Theorem 2 is well-defined. We now sketch a proof of the theorem.

Proof The key step in the proof is to notice that each phase of these oscillators can be described as an XOR of rectangles — that is, an intersection of rectangles where we keep the sections that consist of an odd number of overlapping rectangles and we discard the sections that consist of an even number of overlapping rectangles. For example, Fig. 7.10 shows a phase of the $n = 3$ block oscillator represented as the XOR of a 2×12 rectangle, a 6×8 rectangle, and a 10×4 rectangle.

Different phases may require a different number of rectangles to be XORed, but every phase can always be represented in this way. More important, however, is the fact that the evolution of the oscillators occurs in a predictable way when modeled like this. Consider the grid given in Fig. 7.11, which is helpful in analyzing the evolution of the block oscillators. The fact that it resembles the Sierpinski triangle is no coincidence; this block automaton is, in a sense, emulating the Rule 90 elementary cellular automaton.

The way to read Fig. 7.11 is that the row represents the generation number and the column represents the size of the rectangle that is being XORed. The first column represents a $2 \times 4n$ rectangle, the second column represents a $4 \times (4n - 2)$ rectangle, the third column represents a $6 \times (4n - 4)$ rectangle, and so on. Thus, you "start" in the top-left cell, and that represents the $2 \times 4n$ rectangle of generation 0. To go

---

[7] Sloane's A160657.

[8] The values of k given by Theorem 2 for $n = 1, 2, 3, \dots$ are $1, 2, 3, 3, 5, 6, 4, 4, 9, 6, \dots$ (Sloane's A003558).

Fig. 7.11 A grid that can be used to determine future phases of the block oscillators

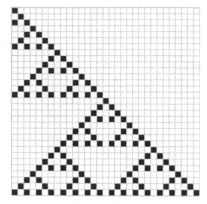

Fig. 7.12 For $1 \le k \le 6$, the columns marked by the number k correspond to a $2k \times (4n - 2(k - 1))$ rectangle — columns marked 0 or 7 are ignored

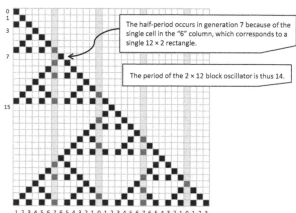

The half-period occurs in generation 7 because of the single cell in the "6" column, which corresponds to a single 12 × 2 rectangle.

The period of the 2 × 12 block oscillator is thus 14.

to generation 1, go to the next row, where we see that the only filled in cell is in the second column, which is the $4 \times (4n - 2)$ column. Thus, generation 1 will just be a filled-in $4 \times (4n - 2)$ rectangle. To see what generation 2 will look like, go to the next row, where we see that two cells are filled in, corresponding to $2 \times 4n$ and $6 \times (4n - 4)$ rectangles. Thus, XOR together two rectangles of those sizes (in the sense described earlier) to get what generation 2 looks like.

In order to determine the oscillators' periods, observe that if we continue to label the columns in the way described, eventually we hit zero-length rectangles, which does not make a whole lot of sense. Thus, we simply ignore any rectangles that are of length zero. But what about rectangles of negative length? Instead of counting down into negative numbers, start counting back up. Thus, the columns corresponding to a $2k \times (4n - 2(k - 1))$ rectangle are given by $2\ell(2n + 1) + k$ for $\ell \in \{0, 1, 2, \ldots\}$ and $2\ell(2n + 1) - k$ for $\ell \in \{1, 2, 3, \ldots\}$. To help illustrate this idea, consider the grid given in Fig. 7.12, which shows the lengths associated with each column in the $n = 3$ case.

For example, in generation 6, there is a live cell in columns marked "1", "3", "5", and "7". The column marked "7" is ignored, and the columns marked "1", "3", and "5" correspond to rectangles of size 2×12, 6×8, and 10×4, respectively. Generation 6 of the $n = 3$ oscillator should thus be the XOR of these three rectangles, which we saw in Fig. 7.10 is indeed the case.

Now, notice that, because the multiplicative suborder of 2 (mod $(2n + 1)$) is well-defined, there must exist some ℓ, m such that $2^\ell = m(2n + 1) \pm 1$. Because 2^ℓ is even, it must be the case that m is odd. In terms of Fig. 7.12 this means that there is some row containing only one cell (i.e., one of the rows labelled $3, 7, 15, 31, \ldots, 2^\ell - 1, \ldots$) such that its cell is in a "$2n$" column (i.e., a "6" column in the case of the example provided). This means that, at some point during its evolution, a $2 \times 4n$ rectangle evolves into a $4n \times 2$ rectangle. A simple symmetry argument shows that this must occur at exactly half of its period.

It follows that the oscillator returns to its original $2 \times 4n$ form in generation $2(2^k - 1)$, where k is the multiplicative suborder of 2 (mod $(2n + 1)$). Thus, the period of the $2 \times 4n$ block oscillator must be a factor of $2(2^k - 1)$. It is not difficult to see that the period must actually be of the form $2(2^\ell - 1)$ for some integer ℓ, so the result follows.   □

Even though there is no known closed form formula for the multiplicative suborder of a (mod b), it has several simple properties that can be verified without much difficulty. Importantly, it can be bounded as follows:

$$\log_a(b - 1) \le \text{sord}_b(a) \le \frac{b - 1}{2}. \tag{7.1}$$

The lower bound follows simply by noting that if $k = \text{sord}_b(a)$, then $a^k \ge b - 1$. To see that the upper bound holds, recall that Euler's theorem says that $a^{\phi(b)} \equiv 1$ (mod b), where $\phi(b)$ is the totient of b. Since $\phi(b) \le b - 1$, the inequality follows. Also, since $\phi(b) = b - 1$ exactly when b is prime, we see that equality is attained on the right in inequality (7.1) only if b is prime.

The following corollary of Theorem 2 and inequality (7.1) follows immediately.

Corollary 1 *Let p be the period of the $2 \times 4n$ block oscillator. Then*

$$2(2n - 1) \le p \le 2^{n+1} - 2.$$

Proof By simply letting $a = 2$, $b = 2n + 1$, and $k = \text{sord}_b(a)$, inequality (7.1) says

$$\log_2(2n) \le k \le n.$$

Rearranging the inequality gives

$$2(2n - 1) \le 2^{k+1} - 2 \le 2^{n+1} - 2.$$

The result follows from Theorem 2.   □

This result provides, for example, a second proof that 2x2 contains oscillators with arbitrarily large period. In fact, any period of the form $2(2^k - 1)$ for an integer $k \ge 1$ is attainable as a $2 \times 4n$ block oscillator by simply choosing $n = 2^{k-1}$. Combining this with the period-doubling method of Sect. 7.4, we can construct a block oscillator with period $2^\ell(2^k - 1)$ for any integers $k, \ell \ge 1$ — one such oscillator is a solid rectangle of alive cells of size $2^\ell \times 2^{k+\ell}$. It is still an open question whether or

Table 7.3 The period of the $2 \times 4n$ block oscillator, as well as the bounds of Corollary 1, for small values of n

n	$2(2n-1)$	Period	$2^{n+1}-2$	n	$2(2n-1)$	Period	$2^{n+1}-2$
1	2	2	2	13	50	1,022	16,382
2	6	6	6	14	54	32,766	32,766
3	10	14	14	15	58	62	65,534
4	14	14	30	16	62	62	131,070
5	18	62	62	17	66	8,190	262,142
6	22	126	126	18	70	524,286	524,286
7	26	30	254	19	74	8,190	1,048,574
8	30	30	510	20	78	2,046	2,097,150
9	34	1,022	1,022	21	82	254	4,194,302
10	38	126	2,046	22	86	8,190	8,388,606
11	42	4,094	4,094	23	90	16,777,214	16,777,214
12	46	2,046	8,190	24	94	4,194,302	33,554,430

not 2x2 is omniperiodic; that is, whether or not it contains an oscillator of any given period.

Table 7.3 shows the period of the $2 \times 4n$ block oscillator for several values of n, as well as the bounds given by Corollary 1. Values for which either of the bounds are attained have been highlighted.

It is clear via Fig. 7.12 that a $2 \times 4n$ oscillator is a solid rectangle in any generation of the form $2^\ell - 1$, where ℓ is an integer. In fact, the number of distinct solid rectangles that the oscillator will produce is exactly $2k$ (or simply k if you don't double-count the 90 degree rotations of rectangles that appear during the second half of the oscillator's period), where k is the multiplicative suborder of 2 (mod $(2n+1)$) (as in Theorem 2). Thus, if the upper bound of Theorem 1 is not attained, then there are rectangular blocks of size $2\ell \times (4n - 2(\ell - 1))$ for some integer ℓ that are not part of the evolution of the $2 \times 4n$ block oscillator. Table 7.3 shows us that the smallest example of this happening is in the $n = 4$ case because the 2×16 oscillator has period 14, which is less than the upper bound of 30. Indeed, the $\ell = 3$ case of a 6×12 rectangle is another oscillator of period 14. This tells us what happens for 28 of the $2(2^4 - 1) = 30$ possible nonempty combinations of four rectangles being XORed together (with an overall 90 degree rotation being allowed). So what happens to the two missing combinations? Well, if we XOR a 2×16 rectangle with a 10×8 rectangle and a 14×4 rectangle, we get a seemingly atypical period 2 block oscillator that simply rotates 90 degree every generation (see Fig. 7.13).[9,10]

While solid rectangles of size $2 \times 4n$ are always an oscillator, this is not the case for rectangles of size $2 \times (4n - 2)$ — they sometimes evolve into an oscillator

---

[9]In 1993, Dean Hickerson observed that a similar phenomenon occurs in the $n = 7$ case of a 6×24 rectangle XORed with an 18×12 rectangle, although it was unknown whether or not there was a smaller block oscillator that did not turn into a single rectangle in one of its phases.

[10]The oscillator of Fig. 7.13 is actually a member of an infinite family of period 2 block oscillators that can be constructed by tiling 2×4 rectangles. Thus a period 2 oscillator like this appears for any $n = 4 + 3m$ with m a positive integer.

Fig. 7.13 The XOR of a 2 × 16 rectangle, a 10 × 8 rectangle and a 14 × 4 rectangle has period 2 and rotates by 90 degrees every generation

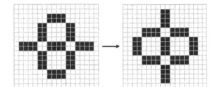

and they sometimes vanish completely. It was shown by Dean Hickerson in 1993 they eventually vanish if and only if $n = 2^\ell$ for some integer $\ell \geq 0$. Furthermore, he showed that they must vanish by no later than generation $2^{\ell+1}$. To find out what happens when n is not a power of 2, we will make use of all of the ideas presented so far in this and the previous section.

If n is odd, say $n = 2m + 1$ for some integer $m \geq 1$, then generation 1 of the oscillator will be a single rectangle of the size $4 \times 8m$. Since this rectangle is made up of 4×4 blocks, we know from Sect. 7.4 that it is an oscillator with period that is double the period of the $2 \times 4m$ block oscillator. It follows from Theorem 2 that it has period $4(2^k - 1)$, where k is the multiplicative suborder of 2 (mod $(2m + 1)$).

Now let's suppose n is divisible by 2 but not by 4 — that is, $n = 2(2m + 1)$ for some integer m. Then generation 3 of the oscillator will be a single rectangle of size $8 \times 16m$. Since this rectangle is made up of 8×8 blocks, we can use the same logic as earlier to see that this pattern is an oscillator with period that is quadruple the period of the $2 \times 4m$ block oscillator. That is, it has period $8(2^k - 1)$, where k is the multiplicative suborder of 2 (mod $(2m + 1)$).

Carrying on in this way, one can easily prove the following result.

Theorem 3 *Let $n = 2^\ell(2m + 1)$ for some $\ell, m \geq 0$. If $m = 0$ then the $2 \times (4n - 2)$ rectangle vanishes in the $(2^{\ell+1} - 1)$th generation. Otherwise, the $2 \times (4n - 2)$ rectangle evolves in the $(2^{\ell+1} - 1)$th generation into a $2^{\ell+2} \times m2^{\ell+3}$ block oscillator with period $2^{\ell+2}(2^k - 1)$, where k is the multiplicative suborder of 2 (mod $(2m + 1)$).*

Acknowledgements Thanks are extended to Alan Hensel and David Eppstein for helpful e-mails and comments, as well as Lewis Patterson and the rest of the ConwayLife.com community for helping dig up information about this rule. Thanks also to Dean Hickerson, David Bell, and the other Life enthusiasts who have investigated B36/S125 over the years. The author was supported by an NSERC Canada Graduate Scholarship and the University of Guelph Brock Scholarship.

Appendix: Pattern Collection

An appendix is provided that displays most of the oscillators (Fig. 7.14), still lifes (Fig. 7.15), and spaceships (Figs. 7.16–7.18) that have been found or constructed in 2x2.

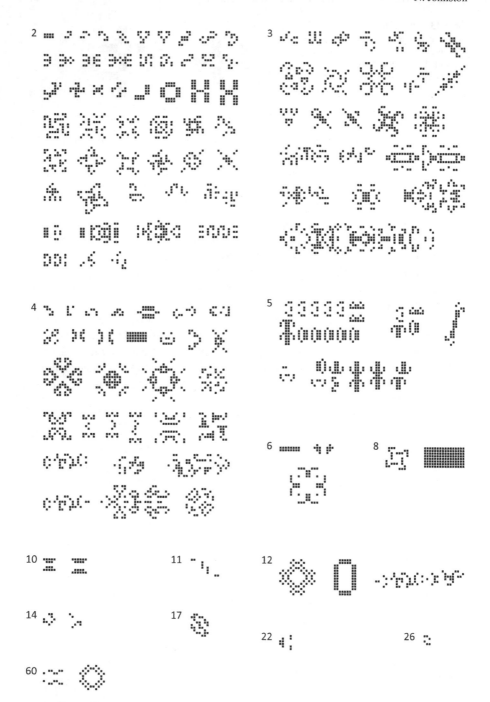

Fig. 7.14 Many oscillators of periods 2 through 60, most of which were found by Alan Hensel, Dean Hickerson, and Lewis Patterson, with contributions by David Bell

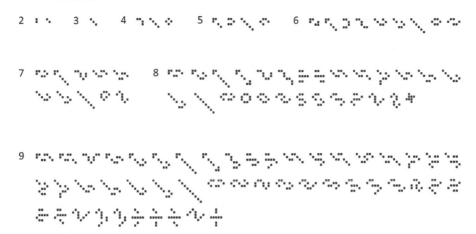

Fig. 7.15 The 104 strict still lifes with 9 or fewer cells, organized by their cell count

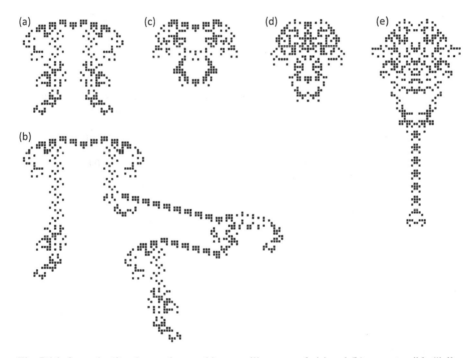

Fig. 7.16 Several $c/2$ orthogonal spaceships travelling upward: (**a**) and (**b**) are extensible "jelly-fish" found by David Bell in December 1999, (**c**), (**d**) and (**e**) were found by David Eppstein, with (**d**) being the first discovered $c/2$ in October 1998

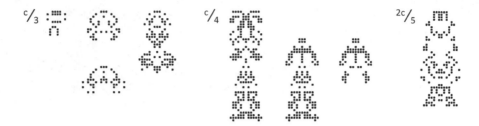

Fig. 7.17 Several orthogonal spaceships of various speeds travelling upward found by David Eppstein, Alan Hensel, and Dean Hickerson

Fig. 7.18 Diagonal spaceships of various speeds travelling up and to the right: the $c/8$ spaceship occurs naturally, the $c/4$ spaceship was found by Dean Hickerson in January 1999, and the $c/3$ spaceship was found by David Eppstein

References

1. Eppstein, D.: Searching for spaceships. In: More Games of No Chance. MSRI Publications, vol. 42, pp. 433–453. Cambridge University Press, Cambridge (2002). arXiv:cs/0004003v2 [cs.AI]
2. Weisstein, E.W.: rule 90, from MathWorld — A Wolfram web resource. http://mathworld.wolfram.com/Rule90.html
3. Wolfram, S.: Statistical mechanics of cellular automata. Rev. Mod. Phys. **55**, 601–644 (1983)
4. Wolfram, S.: A New Kind of Science. Wolfram Media, Champaign (2002)

Chapter 8
Object Synthesis in Conway's Game of Life and Other Cellular Automata

Mark D. Niemiec

8.1 Introduction

Of the very large number of cellular automata rules in existence, a relatively small number of rules may be considered *interesting*. Some of the features that make such rules interesting permit patterns to expand, contract, separate into multiple sub-patterns, or combine with other patterns. Such rules generally include still-lifes, oscillators, spaceships, spaceship guns, and puffer trains. Such structures can often be used to construct more complicated computational circuitry, and rules that contain them can often be shown to be computationally universal. Conway's Game of Life is one rule that has been well-studied for several decades, and has been shown to be very fruitful in this regard.

One of the things that are necessary for construction of virtually all regular growing patterns is the ability to dynamically construct component objects out of easily-produced spaceships or other objects. For example, a breeder, which demonstrates quadratic growth, consists of three parts: the first part is a group of guns or puffers which periodically produce spaceships or possibly other stationary objects which collide to form a different set of guns or puffers, each of which in turn periodically produces a wave of a third kind — spaceships or stationary objects. In order for such patterns to be viable, it is necessary to be able to construct complicated machinery like guns and puffers from simpler objects — namely objects that can themselves be produced by other guns and puffers. Spaceships are usually the most convenient, since multiple spaceships can be brought to bear, usually from several different directions.

In order to be able to construct interesting patterns, a technology must be developed to synthesize vital component objects from simple spaceships. In addition, it can also prove useful to be able to synthesize non-vital objects as well, as the technologies required can often prove useful in other areas.

This chapter will deal with the techniques and problems of such object synthesis. It is an expansion of ideas I presented in a paper on a similar topic for the CA 98 conference [1]. While most of the examples will be from Life, most of the issues discussed also apply to most other interesting rules as well.

A. Adamatzky (ed.), *Game of Life Cellular Automata*,
DOI 10.1007/978-1-84996-217-9_8, © Springer-Verlag London Limited 2010

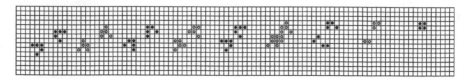

Fig. 8.1 Synthesis of a block. Generations 0–8: 2 gliders approach each other, and collide to form a block

   David Buckingham pioneered much of the work in this area. Between the 1970s and the 1990s, he created systematic syntheses for all stable, oscillating, and moving Life objects up to 14 cells, and larger ones — including many synthesis tools that were instrumental in these syntheses. Before that point, most of the object syntheses resulted from observing the results of random glider collisions, so finding syntheses for specific objects was largely a matter of luck. While many of the syntheses herein are new, most are based on refinements of tools and mechanisms originally discovered by Buckingham [2].

8.2 Simple Object Synthesis

One Life pattern discovered very early on was the *glider* — a simple five-cell spaceship. Naturally, one of the first things to try was to smash gliders together in every possible way and observe the results — which resulted in seven new simple unique objects, as well as some more complicated constellations of multiple objects. This could be repeated by smashing gliders into each of the resulting objects, sometimes yielding more objects. This process could be repeated as often as desired to produce a set of "naturally synthesizable objects". More complicated results could be produced by smashing together three or even more gliders together. Many interesting objects can also arise when smashing multiple gliders into objects along axes of symmetry. Figure 8.1 shows a collision of two gliders to form the block, the simplest and most abundant still-life. Figure 8.2 shows all objects and pseudo-objects that can be formed by colliding 2 or 3 gliders together.

8.3 Art Imitates Life

Sometimes, when one observes the natural evolution of a Life pattern, one may note that a random mass of cells will spontaneously emit a sophisticated object, in a way evocative of Botticelli's *The Birth of Venus*, only to be later consumed by other nearby masses of cells running out of control. In such cases, it is often possible to save the fragile object by smashing gliders into one of the errant masses, with the hopes that some glider or combination of gliders will sufficiently perturb it to cause it to die, or at least avoid the desired object.

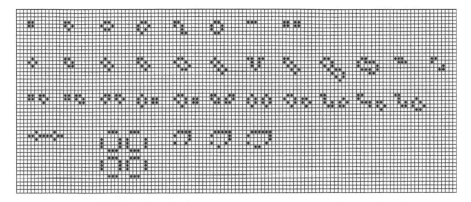

Fig. 8.2 Simple objects resulting from 2–3 gliders

- *First row*: Objects and pseudo-objects resulting from collisions of 2 gliders
- *Remaining rows*: Objects and pseudo-objects resulting from collisions of 3 gliders

One reason Life works so well in this regard is that it exhibits a very simple and profound property: *almost everything dies*. Because of this, it is usually quite easy to perturb an object to cause all or part of it to die, while not affecting nearby objects.

One particularly striking example of such a synthesis is Buckingham's synthesis of the *Hertz Oscillator*, a period-8 billiard table oscillator. Before this time it was generally believed that billiard tables were so complicated that it would never be possible to construct any of them from gliders. However, Buckingham observed that a very simple (and fairly common) interaction produced a mass of cells which at one point in their history resemble a predecessor to the interior of the *Hertz Oscillator*. In fact, this interaction appears in a simple two-glider collision called the *Die Hard*. In order to save this, all that was required was to add two additional gliders from each side to add the required external stabilizers. Even though the *Die Hard* collision to generate the interior is not suitable for this purpose, the same mechanism can be generated easily from a collision of three gliders. This permits the *Hertz Oscillator* to be synthesized from a mere 11 gliders, far cheaper than many less complicated objects. Figure 8.3 shows the synthesis of the *Hertz Oscillator*.

8.4 Life Imitates Art

8.4.1 Wishing for the Impossible

Another thing that often happens is that a natural evolution of a Life pattern may pass through a phase which very closely resembles a desirable object, or a known predecessor for such an object. One is left with thoughts like "If only that one cell didn't die" or "If only there was an extra birth there". Unfortunately, such thoughts

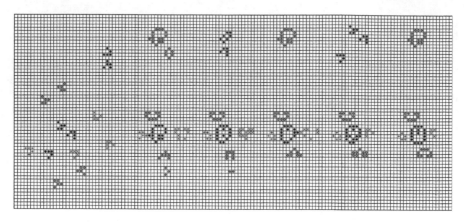

Fig. 8.3 Synthesis of the *Hertz Oscillator*

Top row:

- Generations 0+44 of 2-glider *Die Hard* collision, forming *Hertz Oscillator* core and beehive
- Generations 0+9 of glider/long-boat collision, forming *Hertz Oscillator* core
- Generations 0+19 of three-glider collision, forming *Hertz Oscillator* core

Bottom row:

- Generation 0: Five sets of gliders (shown in groups in different colors) approach from all directions
- Generation 19: Middle three gliders (*solid*) make expanding oscillator core with fully formed top and left sides, and one extra cell protruding on the left. Glider on the left (+) simultaneously hits that cell. Meanwhile, two top gliders (o) form a pi-heptomino that turns into a house that stabilizes the top side of the oscillator
- Generation 20: Leftmost glider slides into place as a boat stabilizing the oscillator from the left side
- Generation 21: Extra cell protrudes from right side of core. Three gliders on the right (x) form a 4-cell spark that touches the protrusion
- Generation 22: Sparks on right slide into place to temporarily stabilize length-3 wall on right side. Meanwhile, bottom right side of core prepares to expand into place
- Generation 23: Bottom and right sides expand into place, and right side stabilizer retracts. into a block predecessor (which will become a stable block two generations later). Meanwhile, two bottom gliders (o) form a pi-heptomino that turns into a house that stabilizes the bottom side of the oscillator

would rely on locally and briefly altering the laws of physics (i.e. the rules of the cellular automata), which is, unfortunately, impossible.

> Data: *"Can you recommend a way to counter the effect?"*
> Q: *"Simple. Change the gravitational constant of the universe." [3]*

However, all is not lost. One of the fundamental concepts of cellular automata is that each individual cell is only aware of its own state, and that of its immediate neighbors. So, for the purposes of being able to construct a desirable object, only the cells whose rules we wish to alter (and possibly their immediate neighbors) need to be aware of any such change.

8.4.2 Altering the Laws of Physics

Furthermore, a cell is not "consciously" aware of the rules that govern its life. For example, an empty cell with two neighbors and a "birth on two" rule behaves exactly the same way as an empty cell with three neighbors and a "birth on three" rule. One can simulate altering the rule for a single key cell by temporarily altering its neighbors. This can usually be accomplished most easily by smashing several gliders together to produce an explosion that dies out, leaving nothing but a dying spark of one or two cells on an edge or corner. Such a spark can have a minimal influence on a small number of key cells, but leave the rest of the pattern unaffected. In some cases, a spark cannot help but cause undesirable side-effects, so greater perturbations are necessary to counteract those. (For example, if a spark increases the neighborhood of an adjacent dead cell from 2 to 3 (causing an undesired birth), the spark's effect must be expanded to increase the neighborhood to at last 4 to suppress that same birth that would otherwise occur.)

There are many mechanisms using such sparks that can change certain simple objects or object parts into other objects or parts. While one rarely needs to change one simple object into another, many cases arise where one simple object is used as part of a larger object (for example, the block and boat stabilizers on the *Hertz Oscillator*), and the larger object can only be synthesized with certain simple stabilizing components. Tools that can reshape these components are useful for creating variants of the original object.

Figure 8.4 shows two trivial examples. In the first case, a boat can be changed into a ship by adding one extra cell. The cell can be added directly by one other cell, or indirectly by having a two-cell spark add an extra cell that flips in. In the second case, a block can be changed into a boat by adding an extra cell onto one corner. Unfortunately, there is no way to cause such a cell to be born without also affecting other nearby cells as well. The solution to this is to create a large spark that encourages the birth of the cell in question, while simultaneously overwhelming and suppressing all other nearby births.

8.4.3 The Butterfly Effect

In some cases, the cell whose state needs to be altered may not be easily accessible to a spark. In such cases, one may sometimes work backwards in time to try to find a more accessible spot at an earlier time. For example: to cause a birth at point x at generation n, this requires that cell y be alive at generation $n - 1$, which requires that one of y's neighbors must be suppressed at generation $n - 2$, and so on. If such a sequence can be found, it leaves one with a chain of events where a small spark that occurs at one point in time will cause a specific change of one cell elsewhere at a later point in time. This is similar to the Butterfly Effect of chaos theory.

One good example of this is Buckingham's synthesis of still-life 14.35, which he called *The Still Life From Hell*, since at the time it was the most difficult to

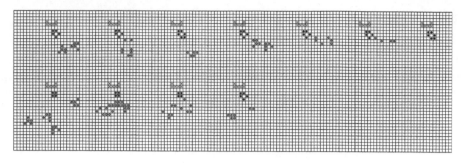

Fig. 8.4 Altering simple objects

Top row: Changing a boat into a ship

- Generation 0: 2 gliders collide to create a diagonally-accessible one-cell spark
- Generation 7: One-cell spark encourages birth at end of boat
- Generation 8: End cell is born, turning boat into ship; remaining sparks soon die
- Generation 0: 2 gliders collide to create an orthogonally-accessible domino spark
- Generation 4: Domino spark encourages birth diagonally from corner of boat
- Generation 5: Unstable cell on corner of boat encourages birth at end of boat
- Generation 8: Unstable cell dies, end cell is born, turning boat into ship

Bottom row: Changing a block into a boat

- Generation 0: 4 gliders collide to create a spark
- Generation 13: Spark wall encourages birth on corner of block, while suppressing other births
- Generation 14: New cell on corner of block forms a boat predecessor
- Generation 15: Block changes into boat; remaining sparks soon die

synthesize of all the objects in his collection. His original synthesis (shown in Sect. 8.5) took more than 30 gliders to synthesize the object incrementally, by brute force. However, he later found a natural object that could be slightly altered to form 14.35. Figure 8.5 shows this synthesis, which involves hitting the forming mess with three different sets of sparks, none of which has an immediate effect, but each of which eventually causes a very small alteration in the resulting object.

8.5 Incremental Synthesis

8.5.1 *Better Living Through Chemistry*

Most complicated objects can be synthesized in many stages by applying certain standard synthesis tools (for example "lengthen snake by one cell") to get to a desired complicated object from a known simpler object via a series of one or more intermediate steps. Most of Buckingham's more complicated syntheses are constructed in this way. This process resembles the synthesis of complicated chemicals from simpler ones in multiple stages.

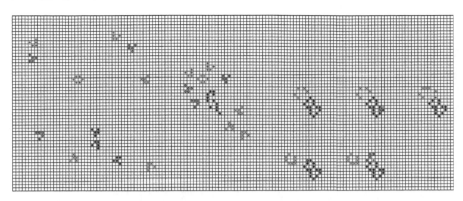

Fig. 8.5 Optimal synthesis of still-life 14.35

Top row:

- Generation 0: Five groups of gliders approach a beehive:
  - Middle 2 gliders (*solid*) form a B-heptomino, and bottom glider suppresses debris block
- Generation 40:
  - Leftmost two gliders (o) transform evolving B-heptomino into mess resembling 14.35
  - Top left glider (o) eventually rescues top cell of result
  - Top right 2 gliders and beehive (+) eventually suppress extra cell on top
  - Bottom left 3 gliders eventually suppress extra cell on bottom
- Generation 49: Completed 14.35, with 5 sparks (x) that die in one generation
- Generation 49: With bottom left group removed; one cell (o) is spurious, and one cell (+) is missing
- Generation 49: With top left glider removed; top cell (+) is missing

Bottom row:

- Generation 49: With top gliders and beehive removed; one cell on top (o) is spurious
- Generation 49: With both above sets removed; top has one missing (+) and many spurious (o) cells

Figure 8.6 shows a typical example: Buckingham's original synthesis of still-life 14.35.

8.5.2 *The Joy of Cooking*

Several years ago, I wrote a computer program whose purpose is to attempt syntheses of all objects in a list from smaller objects, given a list of recipes or templates for various synthesis tools. The current recipe database includes around 750 distinct recipes, representing slightly over 400 distinct object component transformations, some of which require multiple slightly-different versions in order to handle different object geometries that place different constraints on how transformations can take place, or where incoming gliders can come from.

This method has eliminated large numbers of objects as being "trivially synthesizable" from simpler objects using known recipes. Of the 1,353 15-cell still-lifes, only three still elude synthesis, while for larger sizes (16–24 cells), around 96% can

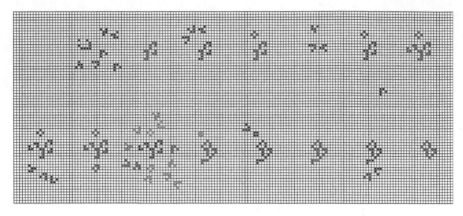

Fig. 8.6 Incremental synthesis of still-life 14.35

Top row:

- Generations 0+13: Lightweight spaceship and 6 gliders make 14.95
- Generations 0+4: 3 gliders make nearby tub
- Generations 0+52: 4 gliders attach a bridged carrier

Bottom row:

- Generations 0+10: 3 gliders create a nearby boat
- Generations 0+18 (key step): 3 gliders and tub on top (+) make a wide spark that hits the top in two places and welds them closed and leave a spurious block; 1 glider on right (o) flips open right side, and 3 gliders on bottom right (x) make a spark to suppress one birth; 1 glider on left (x) deletes back of carrier, and makes spurious pond; glider and boat on bottom (+) and 2 gliders on bottom left (o) create two sparks to turn table leg into tail, and make spurious glider that deletes above pond
- Generations 0+5: 1 glider removes spurious block
- Generations 0+6: 2 gliders remove spurious tail, yielding 14.35

be synthesized in this way. Of the 13,494 pseudo-still-lifes up to 17 cells, only one eludes synthesis, while for large sizes (18–24 cells), around 99.6% can be synthesized this way. For period-2 oscillators, the rate varies from 79–95% for large sizes (15–21 cells). For period-2 pseudo-oscillators, all up to 21 cells can be made except for one 20-cell and 12 21-cells (plus 101 from 19–21 cells whose synthesis requires smaller as-yet-unbuildable oscillators), for a rate of 99.7%, 97% of higher-period oscillators (and all pseudo-oscillators) up to 20 cells can also be built.

Figure 8.7 shows several pseudo-objects being synthesized in a number of steps. The first is one of the two 20-cell period-2 pseudo-objects that cannot be synthesized incrementally, but fortunately, it can be synthesize by forming both parts simultaneously, as is shown here.

8.5.3 Fireworks

The synthesis of oscillators presents special problems. Unlike still-lifes, which can be built up at one's leisure, oscillators are constantly in motion. One model of

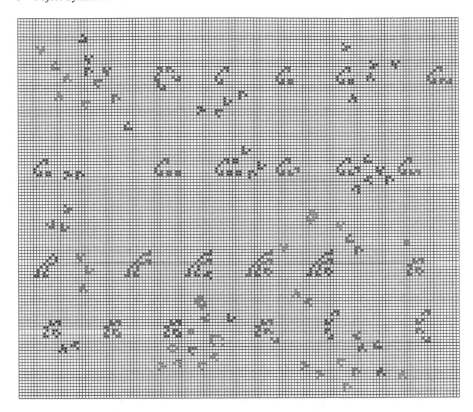

Fig. 8.7 Synthesis of several pseudo-objects

First row: Simultaneous bipole on 12.91

- Generations 0+24: 2 bottom left gliders make eater; Inducting snake on 10.20 (part 1) 3 top left gliders make piece that attaches long snake to eater forming 12.91; 3 top right gliders make long boat; 3 bottom right gliders make half traffic lights that interact with forming long boat to form bipole
- Generations 0+23: 4 gliders add inducting block (hard method)
- Generations 0+41: 5 gliders turn block into down snake (easy)

Second row: Inducting snake on 10.20 (part 2)

- Generations 0+90: 3 gliders add another inducting block (standard method). (Not shown: 4 gliders add inducting block on top (hard method))
- Generations 0+7: 3 gliders turn block and block pair into carrier
- Generations 0+10: 4 gliders turn carrier into up snake (hard)

Third row: Inducting pair of snakes. (Not shown: 5 gliders peel bottom side of half-bakery, similar to following step)

- Generations 0+36: 3 top gliders (o) make spark above and 2 right gliders (x) make domino spark, peeling open right side of half-bakery
- Generations 0+3: 1 glider adds inducting boat
- Generations 0+28: 2 pairs of gliders (o and x) make blinkers, while 2 gliders and lightweight spaceship hit object from behind, ripping away top and left sides. (Not shown: 1 glider eliminates spurious block)

Fourth row: Inducting bipole

- Generations 0+13: 2 gliders turn boat into ship. (Not shown: 2 gliders make nearby block, 2 gliders make nearby beehive)
- Generations 0+32: 1 glider (o) expands ship into bipole, with the assistance of a block, 2 gliders, and 2 lightweight spaceships make (x) complex right spark, and 3 gliders and beehive (+) make complex bottom spark. 1 glider (*solid*) cleans up debris
- Generations 0+38: 2 sets of 4 gliders (o and x) make complex sparks turning carrier into bipole; 2 more gliders (+) clean up debris

Fig. 8.7 (continued)

oscilator resembles creating a fireworks display — that is, several stationary components are first created, and then either one final moving component is added, or one of the stationary components is ignited. Figure 8.8 shows a synthesis of one such oscillator.

8.5.4 Open Heart Surgery

Unfortunately, for some kinds of oscillators, one cannot synthesize the oscillator from scratch; rather, a simpler oscillator must be made first, and then altered into a different form. The problem with this is that any alterations must be made while the

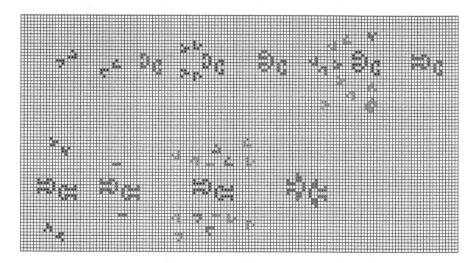

Fig. 8.8 Synthesis of *Skewed Jack* (period 4 oscillator)

Top row: Create pairs of houses with attached tables

- Generation 0+12: Two pairs of gliders make pi-heptominos that form inducting houses
- Generation 0+10: Two pairs of gliders turn house corners into fly-like wings
- Generation 0+28: Two gliders on top left (x) form a one-cell spark, and one glider on top right (o) hits a wing, turning it into a table. A similar mechanism is used on the bottom, but is more complicated to avoid the protrusion on the right: 3 gliders from left (o) form one-cell spark, 3 gliders plus lightweight spaceship from bottom right (x) do same job as top right (o) glider. One extra bottom left glider (+) removes debris. (Not shown: repeat above two steps on right side)

Bottom row: Changing houses with inducting tables into *Skewed Jack* (period 4)

- Generation 0+26: Two pairs of gliders create nearby blinkers
- Generation 0+28: Two top gliders (o) make forming traffic lights, and glider and blinker (x) also make forming traffic lights; these interact to form a skewed pulsar. One glider on the left and two on the right (+) hit the pulsar pieces to create held one-cell sparks that activate the top two quadrants of the *Jack*. (A similar set of gliders does exactly the same thing on the bottom.) (Note: this synthesis is a slight variation of Buckingham's original synthesis of the symmetrical version of *Jack*, where both sides are in-line rather than skewed. Step 3 is simplified, using the simple 3-glider top mechanism on the bottom as well. Step 4 is eliminated. Step 5 requires only 10 gliders: each side needs 3 gliders for a pulsar and one glider each from left and right)

Fig. 8.8 (continued)

oscillator is in motion. This raises certain complicated technical issues; rather than merely resembling a simple process of chemical synthesis, the process involved here is much more akin to performing open heart surgery on a beating heart — it must be done very carefully. Figure 8.9 shows synthesis of three such oscillators.

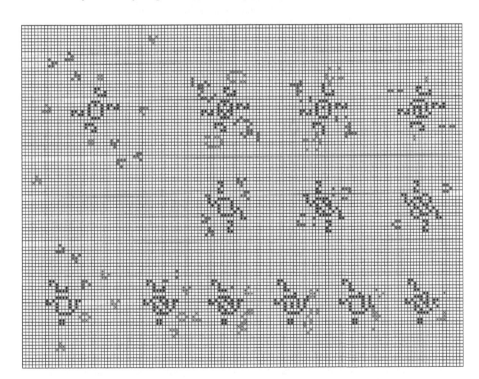

Fig. 8.9 Synthesis of *Hustler*, *Hustler II*, and *$Rats*

Top row: Changing *Hertz Oscillator* (period 8) into *Hustler* (period 3)

- Generation 0: Four groups of gliders approach *Hertz Oscillator*
- Generation 41: Two gliders plus block create C-shaped sparks to change bookends into long bookends

- Generation 43: Three gliders create domino sparks that attack corners of *Hertz Oscillator* core
- Generation 45: Corners of the oscillator core puff out, changing it into a *Hustler*. Meanwhile, the stabilizing bookends are correspondingly lengthened

Middle row: Changing *Hustler* (period 3) into *Hustler II* (period 4)

- Generation 0: Four groups of gliders approach *Hustler*
- Generation 7: Two gliders create one-cell sparks that attack corners of *Hustler* oscillator core
- Generation 10: Corners of the oscillator core puff out, changing it into a *Hustler II*

Bottom row: Changing *Hustler* (period 3) into *$Rats* (period 6)

- Generation 0: Three groups of gliders approach *Hustler*
- Generation 19: Top three gliders create domino spark to decimate top table stabilizer
- Generation 22: Top table starts to collapse, allowing oscillator corner to escape up
- Generation 23: Top stabilizer becomes an eater and oscillator corner starts to pop up. Meanwhile, glider attacks top of right tail stabilizer, and three gliders form a spark on bottom right of same tail stabilizer
- Generation 24: Oscillator left side lengthens by one. Meanwhile, left tail stabilizer momentarily grows correspondingly, while bottom left spark attacks it
- Generation 25: Oscillator fully pops out and becomes *$Rats*. Meanwhile, left side tail returns to its original shape, and fully reforms one generation later. Bottom left spark leaves a spurious loaf which can be removed later. (Note: many other stages in these syntheses involve making cosmetic alterations to the oscillator stators (i.e. the stationary external stabilizers), but these are omitted for brevity)

Fig. 8.9 (continued)

8.6 Synthesis of Moving Objects

8.6.1 Spaceships and Flotillas

Synthesis of spaceship flotillas is even more complicated than synthesis of oscillators, since spaceships are like oscillators that move, and usually at a high rate of speed. While basic spaceships can be constructed simply, flotillas of two or more spaceships require very careful timing and access considerations. For example, adding an additional spaceship onto the side of an existing flotilla is somewhat akin to adding a new seat to a car — while it is driving by at full speed. Side spaceship escorts must be formed in place fully functional, ready to take off the moment the other spaceships arrive from behind. Fortunately, of the 42 different ways in which two simple spaceships can travel in tandem, in all but one case, either of the two ships can be added to pre-existing first ship.

Unfortunately, most Life spaceships (at least the synthesizable ones) travel at a speed of $c/2$ orthogonally, and any kind of sustained object growth cannot exceed a speed of $c/2$ orthogonally, so it is virtually impossible to add anything from behind. If a flotilla has any kind of tag-along, the tag-along must usually be built from behind at the same time the escorting spaceship is built from the front. Figure 8.10 shows syntheses for several spaceship flotillas.

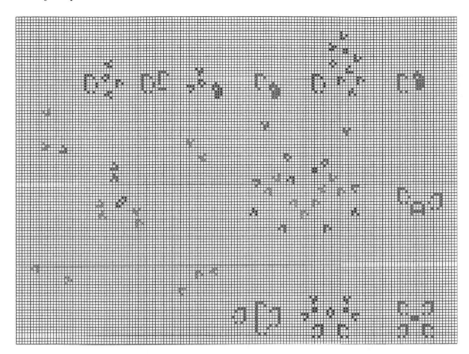

Fig. 8.10 Spaceship flotilla syntheses

First row: Unusual escorts

- Generations 0+12: boat and 2 gliders make middleweight spaceship that is damaged, and
  later turned into a sidecar by rear glider
- Generations 0+4: boat and 2 gliders add forward escort middleweight spaceship (the only
  standard escort spaceship configuration that cannot be made from both sides)
- Generations 0+16: block, tub, and 7 gliders make hard side escort heavyweight spaceship

Second row: 11-long heavyweight spaceship flotilla, and flotilla dragging 'A'

- Generations 0+88: 3 top gliders (o) make beehive predecessor that ignites long ship into
  expanding double-headed spaceship (and top left glider (+) deletes dying debris before
  spaceships hits it); 2 bottom gliders (+) truncate it to length 11; 2 left gliders (x) make
  boat and 2 more gliders turn it into left middleweight spaceship escort; 2 right gliders (X)
  make pond and 4 more gliders turn it into right heavyweight spaceship escort
- Generations 0+56: boat + 2 gliders (o) make escort middleweight spaceships on left and
  right sides on top; 2 pairs of gliders (x) make blocks on left and right sides; block and
  glider (*solid*) make honeyfarm; 4 gliders (+) make 2 inducting traffic lights that 2 bottom
  gliders (o) turn into B-heptominos; B-heptominos hit blocks and turn forming honeyfarm
  into 'A' (Jason Summers's method); 2 gliders (*solid*) clean up debris

Third row: Beehive-dragging flotilla

- Generations 0+8: 2 rear escort spaceships approach beehive from behind, while two front
  escort spaceships are constructed in place

8.6.2 Puffer Trains

Many spaceship flotillas can produce plumes of debris in their wake. Sometimes
such plumes die cleanly, sometimes they produce simple periodic output, and

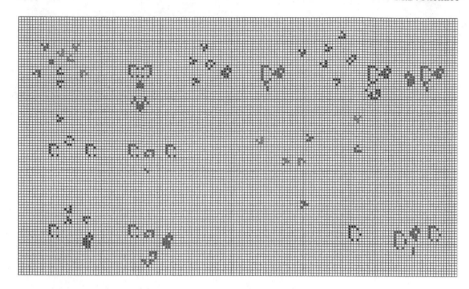

Fig. 8.11 Puffer train syntheses

Top row: Clean puffers: Schick Ship (period 12), Coe Ships (period 16+12)

- Generations 0+84: 2 gliders (*solid*) make traffic lights; block and 2 gliders (o) make left escort lightweight spaceship, while 4 gliders (x) make right escort spaceship, resulting in Schick Ship
- Generations 0+16: 3 gliders turn loaf into Coe ship. (Note: combining Schick ship with Coe escort produces a dirty puffer)
- Generations 0+48: 5 gliders turn long boat into middleweight spaceship escort, changing period of Coe ship from 16 to 12

Middle row: Dirty Puffers

- Generations 0+20: Glider turns toad into B-heptomino engine for period-20 Buckingham puffer
- Generations 0+64: 3 gliders (o) make unstable engine while 4 gliders (x) make heavy-weight spaceship escort for period-8 Wainwright blinker puffer

Bottom row: Dirty Puffers

- Generations 0+24: 3 gliders make B-heptomino engine for period-24 Gosper puffer

sometimes they produce patterns that interact with each other in very complex and messy ways. Many such plumes can be tamed to produce useful results by using sparks from escort spaceships. Generally, the hardest part about synthesizing such puffers is the puffer engine. Figure 8.11 shows syntheses of several puffer engines originally discovered by Paul Schick, Tim Coe, Robert T. Wainwright, David Buckingham, and R.W. Gosper.

8.6.3 Glider Guns

The stable analog to the puffer train is the glider gun, which is an oscillator that produces moving debris (usually a glider) that flies away. Guns for other simple

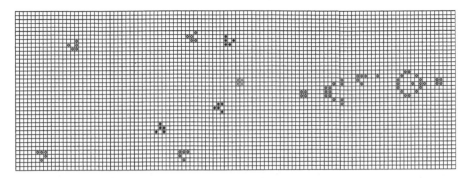

Fig. 8.12   Glider gun synthesis
- Generations 0+90: 3 gliders (*solid*) make rightmost shuttle, stabilized by block (x); 4 gliders (o) make leftmost shuttle and a stabilizing block; shuttles interact to produce gliders every 30 generations

spaceships and flotillas can also be constructed; however, as gliders frequently occur spontaneously from random patterns, while other spaceships occur only rarely, guns for anything other than gliders usually form such spaceships by colliding glider streams. True-period guns (i.e. guns which are oscillators of the same period as the resulting glider stream) are possible for all periods 62 and up by using Herschel tracks. Some smaller periods are also possible, usually by smashing the edges of oscillators together to produce debris that turns into a glider without damaging the oscillators themselves. Pseudo-period guns are possible for all periods 14 and up. These are guns which are oscillators of larger periods thinned out to *lcm(oscillator period, glider stream period)* and then multiplexed along the same stream to produce gliders of any desired period. Figure 8.12 shows synthesis of the original period-30 glider gun. Most other guns are too large to show their syntheses here, although many use small components that can be easily synthesized.

8.6.4 Breeders and Other Large Patterns

Breeders (patterns whose population expands quadratically with time), or other patterns whose populations exceed linear growth are generally not suitable for glider synthesis, since a small fixed number will eventually fill the entire universe. Patterns whose population growth is linear or slower may themselves be constructed by glider streams coming from other guns or puffer trains. Since most gun and puffer components are fairly simple, the main difficulties are not with synthesizing the components, but rather in delivering all the gliders at the right time and place, especially if many glider streams are crossing each other, attempting to construct many copies of the same object in rapid succession, in assembly-line fashion.

Life patterns that act as computational circuitry (signal sources, logic gates, memory cells, etc.) typically use simple components — gliders and simple spaceships, still-lifes, guns, puffers, etc. Despite their organizational complexity, synthesis of their components is generally quite straightforward, subject to spacing and timing constraints similar to those found in other large patterns such as breeders.

8.7 Other Rules

8.7.1 B34/S34 Life

One rule that was investigated in the early 1970s was called "3–4 Life". This rule is characterized by the rule B34/S34 — that is, both birth and survival on 3 or 4 neighbors. One thing notable about such rules where the birth and survival conditions are the same is that cells are memoryless — a cell's state depends only on its neighbors, but *not* on its own state. As such, information can never be preserved by a

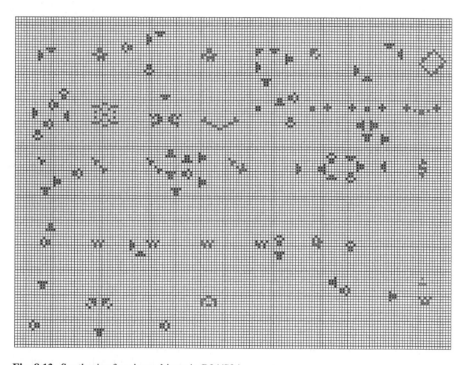

Fig. 8.13 Synthesis of various objects in B34/S34

First row: Two gliders, and an object

- Generations 0+18: 2 glider collision makes object that is almost asymmetric glider (missing one cell), but dies in 13 generations
- Generations 0+18: as above, but 2 more gliders provide 1-cell spark to successfully make asymmetric glider
- Generations 0+12: 2 pairs of 2 gliders add tail sparks to period-12 oscillator turning into a diagonal glider
- Generations 0+42: 4 gliders make a period-6 oscillator resembling Life's *Bakery*

Second row: Two objects, and an ignited oscillator

- Generations 0+14: 2 pairs of 2 gliders make two infinitely-expanding M1 methuselahs, which collide and form a 26-cell flip-flop
- Generations 0+26: glider plus 2 asymmetric gliders make necklace-like 16.1573
- Generations 0+24: 3 gliders add period-3 star near a block. (Another 3 will do the same to the other side)
- Generations 0+9: 4 gliders add a domino spark, igniting a block between two stars into a block hassled by two stars

Third row: Two tools, and an object

- Generations 0+34: 3 gliders attach a diagonal *y*
- Generations 0+36: 3 gliders change *y* into clock, and 4 more clean up resulting mess
- Generations 0+40: 6 gliders make 11.2 (small dollar sign) and 2 more clean up resulting mess

Fourth row: Buckingham's *Roteightor*

- Generations 0+88: 3 gliders slowly make 7.2
- Generations 0+6: 2 gliders add diagonal spark, turning 7.2 into 8.1
- Generations 0+16: 2 gliders add domino spark, turning 8.1 into *Roteightor*

Fifth row: Rescuing natural objects

- Generations 0+72: glider and 2 diagonal gliders make 12.7 and mess, which is eliminated by 2 more gliders
- Generations 0+60: 2 leftmost gliders make mess almost yielding 8.2; rightmost glider rescues it; top glider tames mess into a spurious 4.3

Fig. 8.13   (continued)

cell, but only propagated to nearby cells. This makes period-2 oscillators extremely common, but still-lifes very rare. (The smallest six still-lifes have populations of 4, 36, 44, 50, 50, 51 respectively.) The rule also supports three small spaceships, all which move at speeds of $c/3$ and periods of 3. These will be referred to respectively as the orthogonal, asymmetric, and diagonal spaceships.

Only the symmetric orthogonal spaceship (also commonly called a glider) is common, although the other two do spontaneously occur on very rare occasions. One of the collisions of two gliders dies out, but before doing so, produces an unstable object that is identical to the asymmetric spaceship with one cell missing. It is a simple matter of adding two more gliders to create a one-cell spark to supply that missing cell, making the asymmetric spaceship constructible from four gliders.

Similarly, the diagonal spaceship resembles a simple period-12 oscillator with two extra cells, so it can be synthesized by creating the oscillator from three gliders, and two more pairs of gliders add sparks that attach the two additional cells to the back.

One challenge to object synthesis in this rule is that, unlike in Life, most things do *not* die, but rather most unstable masses tend to foam and grow without limit. This makes creation of sparks more difficult. Also, because birth is allowed on 4, births cannot be suppressed by just adding a small spark to the outside of an object as can be done in Life. For the same reason, pseudo-objects are impossible in this rule. Figure 8.13 shows syntheses of several objects in the B34/S34 rule.

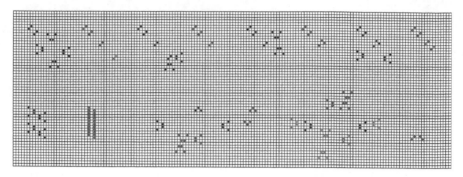

Fig. 8.14 Synthesis of various objects in B2/S2

Top row:

- Generations 0+8: 4 gliders create two 2-cell flip-flops, one of them spurious
- Generations 0+8: 2 gliders eliminate spurious 2-cell flip-flop
- Generations 0+8: 2 gliders create one-cell spark turning 2-cell flip-flop into 3-cell flip-flop
- Generations 0+8: 2 gliders create one-cell spark turning 3-cell flip-flop into 4-cell oscillator. (If these are done near another 4-cell oscillator, they produce an extensible chain that interact trivially)

Bottom row:

- Generations 0+2: 2 pairs of gliders generate inducting walls (*n* pairs generate walls of any length $10 + 12n$)
- Generations 0+16: 3 gliders (o) generate starburst spark; 2 gliders interact with starburst to make 3-cell period-4 oscillator. (If this is done near another similar oscillator, it produces an extensible chain that interacts trivially)
- Generations 0+12: 2 sets of 3 gliders (o and x) generate starburst sparks; 2 gliders (*solid*) interact with starbursts to make crown; 2 gliders (+) eliminate toxic debris

8.7.2 B2/S2 Life

One mildly interesting rule is characterized by the rule B2/S2 — that is, both birth and survival on exactly 2 neighbors. Like B34/S34, it is memoryless, but as with any rule that allows birth on 2, any object that has 2 adjacent cells on one side will expand indefinitely at speed *c*, making it unstoppable. Such rules exhibit explosive behavior, and it is quite difficult to create patterns that do *not* explode. One interesting property is that long straight lines act as immovable objects except at their ends, so they can be used to form arbitrarily long walls. Also, they can be formed in the shape of a box, and such boxes can be used to contain arbitrarily complex chaotic "plasma", which cannot escape; this results in billiard table configurations of arbitrarily large periods.

Object synthesis in this rule is especially problematic, as almost everything explodes, much more so than even B34/S34. Figure 8.14 shows syntheses of several objects in the B2/S2 rule.

8.8 Remaining Issues

8.8.1 General Problems

One of the main problems with object synthesis is the relatively small number of available tools. Despite the fact that hundreds of mechanisms to construct various object pieces or transform one kind of piece into another are known, these likely only scratch the surface of what is possible. In particular, tools that simultaneously affect multiple sites, or produce several sequential effects, are sorely lacking.

Another problem is one of entropy — most tools require several gliders and a fair amount of space to produce a very small effect (such as a one or two cell spark). If multiple effects are required in a small area, there isn't enough room to bring enough gliders into the desired area. This can sometimes be modified by using temporary still-lifes, or bringing in large numbers of extra gliders or spaceships from one side, but this can only be done so far.

8.8.2 Problems with Still-Lifes

While most still-lifes and pseudo-still-lifes can be synthesized using incremental methods, there remain a large number whose geometries can best be described as "bizarre", and cannot easily be broken down into anything simpler. More advanced techniques will be needed to construct such objects. (Or, alternatively, many more random broth experiments will need to be done, in the hopes that such objects just might randomly occur — and that their method of natural evolution may suggest synthesis methods.)

8.8.3 Problems with Pseudo-objects

The vast majority of pseudo-still-lifes (and also pseudo-objects containing oscillators) are formed by two objects placed against each other with lines of 2 cells inducting each other. Most such pseudo-objects can be synthesized by creating one of the objects, then attaching a small simple seed (such as a block, boat, ship, etc.) and then growing the seed into the second object. There are still a small number of such objects or object geometries that cannot be synthesized in this way. The 4-glider block mechanism used in Fig. 8.7 is an example of one such geometry that initially proved problematic, but was later solved.

Most other induction geometries used in pseudo-still-lifes (for example, two side-by-side snakes) are still very difficult to synthesize incrementally; pseudo-objects of this kind must currently be synthesized by simultaneously building seeds of all objects involved, and then growing each one separately. Another unsolved problem exists where two objects touch in more than one place.

8.8.4 Problems with Oscillators

Many oscillators, especially period-2 oscillators, resemble clouds of "space dust" —
that is, apparently random arrangements of disconnected cells. These are usually
very hard to synthesize, since it requires a collision of gliders to not only produce
a specific two-dimensional pattern of disconnected cells, but these cells must occur
at the very edge of the pattern so they can interact with other pieces being formed.
While mechanisms exist for creating small pieces of this type, technology to do this
for larger patterns is largely unexplored territory.

8.8.5 Problems with Spaceships and Puffer Trains

While mechanisms exist for creating small overweight spaceship flotillas, the larger
the size of the engine, the fewer ways there are to synthesize them. There are currently no known syntheses of spaceships that can safely escort engines from 12–14
cells in length, even though simple three-spaceship flotillas containing them are
possible. Also, since many spaceships (other than the ones related to the standard
simple $c/2$ spaceships) have mechanisms that contain space dust, no syntheses are
known for any spaceships other than the "natural" ones (with one exception).

8.8.6 Problems with Glider Guns

While many glider guns rely on simple pieces, some use complicated sparking oscillators (such as fountains) which contain much space dust, and are not yet synthesizable. If guns are ever found with low periods (down to 14), gliders will be emerging
so rapidly that parts of such guns will necessarily be very tightly coupled. It will be
very difficult to build part of the gun without the rest of the gun already operating.
Similar problems occur with trying to synthesize spaceship streams closer than 22
generations apart — as one spaceship is being constructed, the previous one's sparks
interfere with it. To synthesize denser streams, much more sophisticated techniques
will be required.

References

1. Niemiec, Mark D.: Synthesis of complex life objects from gliders. In: New Constructions in
   Cellular Automata, pp. 55–78. Oxford University Press, London (2003)
2. Unpublished correspondence; some portions may appear in various web archives
3. Trek, Star: The Next Generation, Season 3 Episode 13 "Deja Q". Paramount Television (1990)

Chapter 9
Gliders and Glider Guns Discovery in Cellular Automata

Emmanuel Sapin

The theories of complexity are the understanding of how independent agents are interacting in a system to influence each other and the whole system [32]. Surprising computational tasks could result from interactions of independent agents in complex systems as emergence of computation is a hot topic in the science of complexity [34]. A promising environment to study emergent computation is cellular automata [31] which are the simplest mathematical representation of complex systems [8]. They are uniform frameworks in which the simple agents are cells evolving through time on the basis of a local function, called the transition rules [33].

Emerging computation in cellular automata has different forms. Some have studied specific computation like density and synchronization tasks [7, 14, 15, 18, 28] and pattern recognition [35]. While others have considered *Turing-universal automata* [1, 2, 10, 12, 16, 19], i.e. automata encompassing the whole computational power of the class of Turing machines [17]. The well-established problems of universality in cellular automata remains an area where amazing phenomena at the edge of theoretical computer science and non-linear science can be discovered.

One of the demonstrations of universality is based on mobile self-localized patterns of non-resting states [19], called *gliders* and their generators called *glider guns* which, when evolving alone, periodically recover their original shape after emitting a number of gliders. As an example, the famous Game of Life of Conway et al. [4] has been shown universal thanks to a gun called *gosper gun*. This demonstration uses gliders to carry information and glider guns as a clock.

We aim to construct an automatic system for the discovery of Turing-universal cellular automata. A first step presented here is the discovery of gliders and glider guns. The search for gliders was notably explored by Adamatzky et al. with a phenomenological search [13], Wuensche who used his Z-parameter and entropy [36] and Eppstein [6]. Lohn et al. [11] and Ventrella [30] have searched for gliders using stochastic algorithms.

The chapter is arranged as follows: Sect. 9.2 presents some formalisations while Sect. 9.3 gives definitions of glider and glider gun. The relation between automata and these patterns are studied in Sect. 9.4. The search for gliders is described in Sect. 9.5 while Sect. 9.6 sets out the search for glider guns. Sect. 9.7 considers other

A. Adamatzky (ed.), *Game of Life Cellular Automata*,
DOI 10.1007/978-1-84996-217-9_9, © Springer-Verlag London Limited 2010

guns then the last section summarizes the presented results and discusses directions for future research.

9.1 Formalisations of Cellular Automata

9.1.1 Set of Cellular Automata

A cellular automaton of dimension d is a 4-uplet $(\mathbb{Z}^d, S, V, \delta : S^{n+1} \mapsto S)$ where:

- S is a finite set of states.
- V is a finite ordered subset of $\mathbb{Z}^d$ of cardinal n called a neighbourhood.
- $\delta : S^{n+1} \mapsto S$ is the *local transition rule* of the cellular automaton.

In the following, we will consider only the automata with the following characteristics:

- $d = 2$.
- $S = \{0, 1\}$.
- V will be the Moore neighbourhood, i.e. the 8 direct neighbours of the element (x, y) called *cell*, so $n = 8$.

The space $(\mathbb{Z}^2, \{0, 1\}$, Moore neighbourhood, $\delta : S^9 \mapsto S)$ of automata is called the space $\mathscr{E}$. The next formalisations will be given for automata of $\mathscr{E}$ but could be generalized for automata with higher dimensions and more states.

9.1.2 Evolution of Cellular Automata

A configuration at time t, or generation t, of an automaton A is an application $c_t^A : (\mathbb{Z}^2) \mapsto \{0, 1\}$. Cells in states 0 and 1 are respectively called dead and alive. The sequence $(c_t^A)_{t \geq 0}$ is said the evolution of the cellular automaton A from the configuration c_0^A if and only if:

$$c_{t+1}^A(x, y) = \delta\big(c_t^A(x + 1, y - 1), c_t^A(x + 1, y), c_t^A(x + 1, y + 1), c_t^A(x, y - 1),$$
$$c_t^A(x, y), c_t^A(x, y + 1), c_t^A(x - 1, y - 1), c_t^A(x - 1, y),$$
$$c_t^A(x - 1, y + 1)\big).$$

For visibility, in the following, the arguments of the function δ will appear in this disposition:

$$c_{t+1}^A(x, y) = \delta\big(c_t^A(x + 1, y - 1), c_t^A(x + 1, y), c_t^A(x + 1, y + 1),$$
$$c_t^A(x, y - 1), c_t^A(x, y), c_t^A(x, y + 1),$$
$$c_t^A(x - 1, y - 1), c_t^A(x - 1, y), c_t^A(x - 1, y + 1)\big).$$

Fig. 9.1 Two sequences $(c_t^A)_{t \geq 0}$ for t equals 0, 1, 2 and 3. Only the sequence at the *bottom* is the evolution of a cellular automaton that is the R-pentomino of the Game of Life [4]

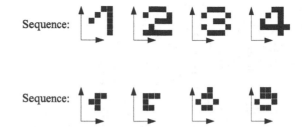

The function δ determines what will become to a cell at the generation depending on its neighbourhood.

Figure 9.1 shows the first four generations of two sequences $(c_t^A)_{t \geq 0}$, only the second one is the evolution an automaton.

9.1.3 Isotropy

An automaton ($\mathbb{Z}^2$, $\{0, 1\}$, Moore neighbourhood, $\delta : S^9 \mapsto S$) is said *isotropic* if and only if for all binary numbers $x_1, x_2, x_3, x_4, x_5, x_6, x_7, x_8$ and x_9 we get:

$$\delta \begin{pmatrix} x_1, x_2, x_3, \\ x_4, x_5, x_6, \\ x_7, x_8, x_9 \end{pmatrix} = \delta \begin{pmatrix} x_3, x_2, x_1, \\ x_6, x_5, x_4, \\ x_9, x_8, x_7 \end{pmatrix} = \delta \begin{pmatrix} x_7, x_8, x_9, \\ x_4, x_5, x_6, \\ x_1, x_2, x_3 \end{pmatrix} = \delta \begin{pmatrix} x_1, x_4, x_7, \\ x_2, x_5, x_8, \\ x_3, x_6, x_9 \end{pmatrix}$$

$$= \delta \begin{pmatrix} x_9, x_6, x_3, \\ x_8, x_5, x_2, \\ x_7, x_4, x_1 \end{pmatrix} = \delta \begin{pmatrix} x_7, x_4, x_1, \\ x_8, x_5, x_2, \\ x_9, x_6, x_3 \end{pmatrix} = \delta \begin{pmatrix} x_9, x_8, x_7, \\ x_6, x_5, x_4, \\ x_3, x_2, x_1 \end{pmatrix}$$

$$= \delta \begin{pmatrix} x_3, x_6, x_9, \\ x_2, x_5, x_8, \\ x_1, x_4, x_7 \end{pmatrix}.$$

The space of isotropic automata is called the space $\mathscr{I}$.

9.1.4 Number of Automata

The number of automata in $\mathscr{E}$ is the number of possible transition rules δ : $S^{8+1} \mapsto S$. There are $2^9 = 512$ different rectangular 9-cell neighbourhood states (including the central cell) so the space $\mathscr{E}$ contains 2^{512} automata. An automaton of $\mathscr{E}$ can be described by telling what will be the new state of a cell at the next generation depending on its neighbourhood as shown in Fig. 9.2.

The number of automata of $\mathscr{I}$ depends on in how many subsets of isotropic neighbourhood states the 512 different rectangular 9-cell neighbourhood states can be put. Let:

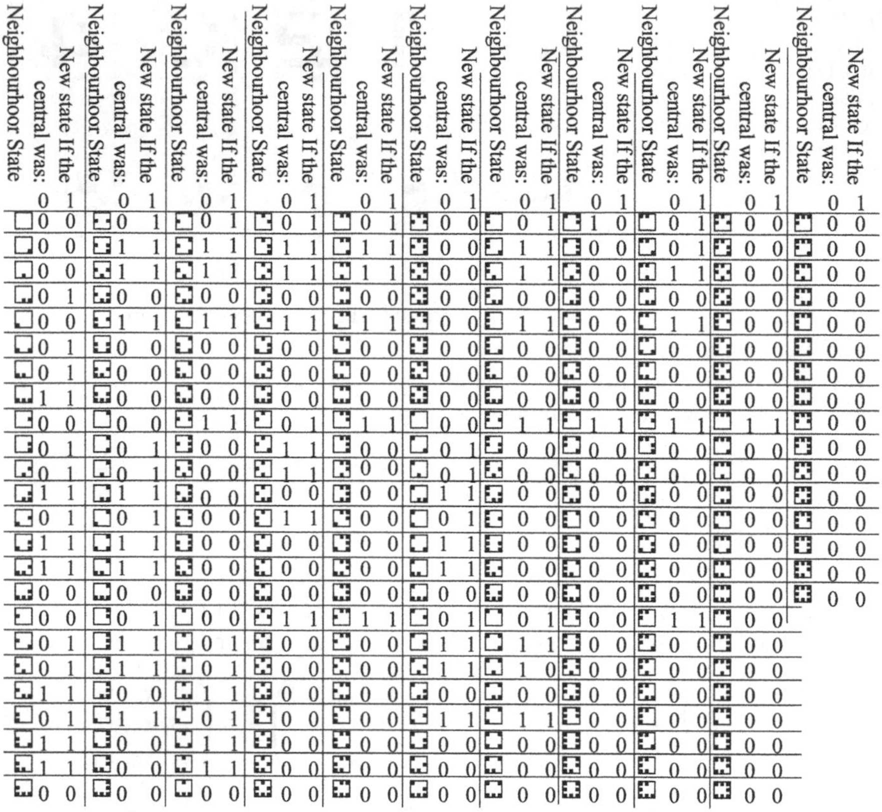

Fig. 9.2 An automaton of the space $\mathscr{E}$

$$\begin{pmatrix} x_1, x_2, x_3, \\ x_4, x_5, x_6, \\ x_7, x_8, x_9 \end{pmatrix}$$

be a rectangular 9-cell neighbourhood states. Its isotropic neighbourhood states are

$$\begin{pmatrix} x_3, x_2, x_1, \\ x_6, x_5, x_4, \\ x_9, x_8, x_7 \end{pmatrix}, \begin{pmatrix} x_7, x_8, x_9, \\ x_4, x_5, x_6, \\ x_1, x_2, x_3 \end{pmatrix}, \begin{pmatrix} x_1, x_4, x_7, \\ x_2, x_5, x_8, \\ x_3, x_6, x_9 \end{pmatrix}, \begin{pmatrix} x_9, x_6, x_3, \\ x_8, x_5, x_2, \\ x_7, x_4, x_1 \end{pmatrix}, \begin{pmatrix} x_7, x_4, x_1, \\ x_8, x_5, x_2, \\ x_9, x_6, x_3 \end{pmatrix},$$

$$\begin{pmatrix} x_9, x_8, x_7, \\ x_6, x_5, x_4, \\ x_3, x_2, x_1 \end{pmatrix}, \begin{pmatrix} x_3, x_6, x_9, \\ x_2, x_5, x_8, \\ x_1, x_4, x_7 \end{pmatrix}.$$

Depending on the values the nine binary numbers $x_1, x_2, x_3, x_4, x_5, x_6, x_7, x_8$ and x_9, the eight isotropic neighbourhood states could be:

- all different from one another or
- equal to one another or
- equal two-by-two or
- equal four-by-four.

$$\{ \boxminus \}, \{ \blacksquare \}, \{ \boxminus \}, \{ \blacksquare \}$$

Fig. 9.3 The four subsets of one isotropic neighbourhood state in which $x_5 = 1$. The *first* subset is composed with the element $(0, 0, 0, 0, 1, 0, 0, 0, 0)$ as states 0 and 1 are shown respectively by *white* and *black coloured squares*

Fig. 9.4 The four subsets of two isotropic neighbourhood states

$$\{ \blacksquare, \blacksquare \}, \{ \blacksquare, \blacksquare \}, \{ \blacksquare, \blacksquare \}, \{ \blacksquare, \blacksquare \}$$

$$\{ \blacksquare, \blacksquare, \blacksquare, \blacksquare \}, \{ \blacksquare, \blacksquare, \blacksquare, \blacksquare \}, \{ \blacksquare, \blacksquare, \blacksquare, \blacksquare \}, \{ \blacksquare, \blacksquare, \blacksquare, \blacksquare \}, \{ \blacksquare, \blacksquare, \blacksquare, \blacksquare \}, \{ \blacksquare, \blacksquare, \blacksquare, \blacksquare \},$$

$$\{ \blacksquare, \blacksquare, \blacksquare, \blacksquare \}, \{ \blacksquare, \blacksquare, \blacksquare, \blacksquare \}, \{ \blacksquare, \blacksquare, \blacksquare, \blacksquare \}, \{ \blacksquare, \blacksquare, \blacksquare, \blacksquare \}, \{ \blacksquare, \blacksquare, \blacksquare, \blacksquare \}, \{ \blacksquare, \blacksquare, \blacksquare, \blacksquare \},$$

$$\{ \blacksquare, \blacksquare, \blacksquare, \blacksquare \}, \{ \blacksquare, \blacksquare, \blacksquare, \blacksquare \}, \{ \blacksquare, \blacksquare, \blacksquare, \blacksquare \}, \{ \blacksquare, \blacksquare, \blacksquare, \blacksquare \}, \{ \blacksquare, \blacksquare, \blacksquare, \blacksquare \}, \{ \blacksquare, \blacksquare, \blacksquare, \blacksquare \},$$

$$\{ \blacksquare, \blacksquare, \blacksquare, \blacksquare \}, \{ \blacksquare, \blacksquare, \blacksquare, \blacksquare \}, \{ \blacksquare, \blacksquare, \blacksquare, \blacksquare \}, \{ \blacksquare, \blacksquare, \blacksquare, \blacksquare \}, \{ \blacksquare, \blacksquare, \blacksquare, \blacksquare \}, \{ \blacksquare, \blacksquare, \blacksquare, \blacksquare \},$$

$$\{ \blacksquare, \blacksquare, \blacksquare, \blacksquare \}$$

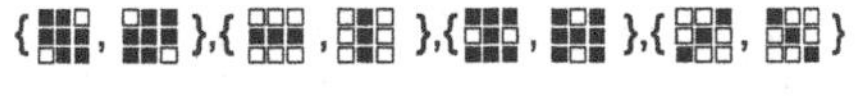

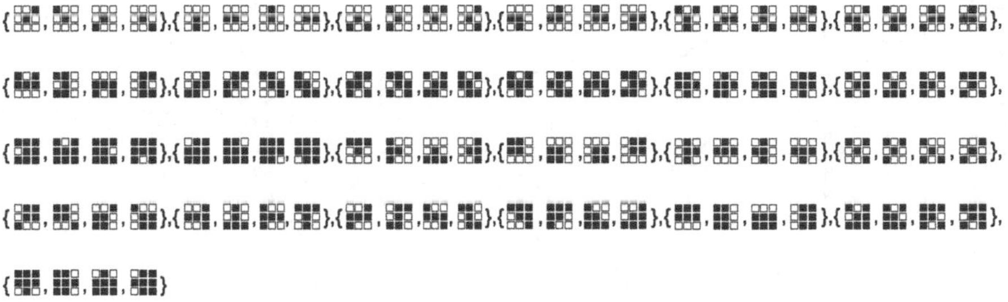

Fig. 9.5 The twenty five subsets of four isotropic neighbourhood states

$$\{ \blacksquare, \blacksquare, \blacksquare, \blacksquare, \blacksquare, \blacksquare, \blacksquare, \blacksquare \}, \{ \blacksquare, \blacksquare, \blacksquare, \blacksquare, \blacksquare, \blacksquare, \blacksquare, \blacksquare \}, \{ \blacksquare, \blacksquare, \blacksquare, \blacksquare, \blacksquare, \blacksquare, \blacksquare, \blacksquare \}$$

$$\{ \blacksquare, \blacksquare, \blacksquare, \blacksquare, \blacksquare, \blacksquare, \blacksquare, \blacksquare \}, \{ \blacksquare, \blacksquare, \blacksquare, \blacksquare, \blacksquare, \blacksquare, \blacksquare, \blacksquare \}, \{ \blacksquare, \blacksquare, \blacksquare, \blacksquare, \blacksquare, \blacksquare, \blacksquare, \blacksquare \}$$

$$\{ \blacksquare, \blacksquare, \blacksquare, \blacksquare, \blacksquare, \blacksquare, \blacksquare, \blacksquare \}, \{ \blacksquare, \blacksquare, \blacksquare, \blacksquare, \blacksquare, \blacksquare, \blacksquare, \blacksquare \}, \{ \blacksquare, \blacksquare, \blacksquare, \blacksquare, \blacksquare, \blacksquare, \blacksquare, \blacksquare \}$$

$$\{ \blacksquare, \blacksquare, \blacksquare, \blacksquare, \blacksquare, \blacksquare, \blacksquare, \blacksquare \}, \{ \blacksquare, \blacksquare, \blacksquare, \blacksquare, \blacksquare, \blacksquare, \blacksquare, \blacksquare \}, \{ \blacksquare, \blacksquare, \blacksquare, \blacksquare, \blacksquare, \blacksquare, \blacksquare, \blacksquare \}$$

$$\{ \blacksquare, \blacksquare, \blacksquare, \blacksquare, \blacksquare, \blacksquare, \blacksquare, \blacksquare \}, \{ \blacksquare, \blacksquare, \blacksquare, \blacksquare, \blacksquare, \blacksquare, \blacksquare, \blacksquare \}, \{ \blacksquare, \blacksquare, \blacksquare, \blacksquare, \blacksquare, \blacksquare, \blacksquare, \blacksquare \}$$

$$\{ \blacksquare, \blacksquare, \blacksquare, \blacksquare, \blacksquare, \blacksquare, \blacksquare, \blacksquare \}, \{ \blacksquare, \blacksquare, \blacksquare, \blacksquare, \blacksquare, \blacksquare, \blacksquare, \blacksquare \}, \{ \blacksquare, \blacksquare, \blacksquare, \blacksquare, \blacksquare, \blacksquare, \blacksquare, \blacksquare \}$$

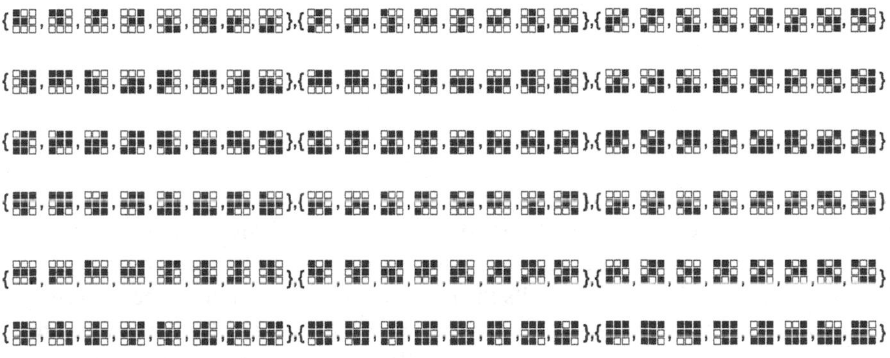

Fig. 9.6 The eighteen subsets of eight isotropic neighbourhood states

That leads to having subsets of isotropic neighbourhood states with a different number of elements. Figures 9.3–9.6 show subsets of 1, 2, 4 and 8 elements in which cells in states 0 and 1 are represented respectively by white and black coloured squares. These figures show the subsets of isotropic neighbourhood states in which $x_5 = 1$.

All the 512 different rectangular 9-cell neighbourhood states can be put in 102 subsets of isotropic neighbourhood states, meaning that there are 2^{102} different automata in $\mathscr{I}$. In order to describe an automaton of $\mathscr{I}$, one just needs to be able to tell what the state of a cell will be at the next generation, depending on which subset of isotropic neighbourhood states its neighbourhood is in. The subset of isotropic

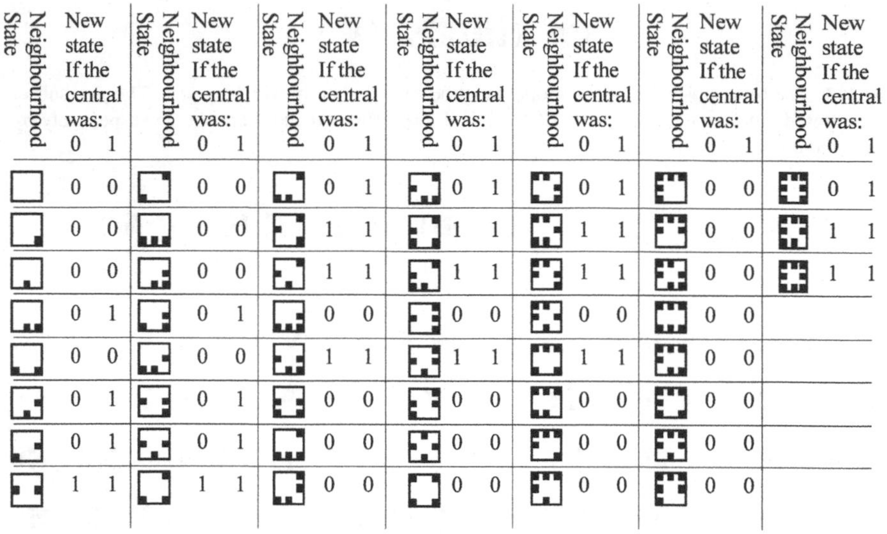

Neighbourhood State	New state If the central was: 0	1	Neighbourhood State	New state If the central was: 0	1	Neighbourhood State	New state If the central was: 0	1	Neighbourhood State	New state If the central was: 0	1	Neighbourhood State	New state If the central was: 0	1	Neighbourhood State	New state If the central was: 0	1	Neighbourhood State	New state If the central was: 0	1
(icon)	0	0	(icon)	0	0	(icon)	0	1	(icon)	0	1	(icon)	0	1	(icon)	0	0	(icon)	0	1
(icon)	0	0	(icon)	0	0	(icon)	1	1	(icon)	1	1	(icon)	1	1	(icon)	0	0	(icon)	1	1
(icon)	0	0	(icon)	0	0	(icon)	1	1	(icon)	1	1	(icon)	1	1	(icon)	0	0	(icon)	1	1
(icon)	0	1	(icon)	0	1	(icon)	0	0	(icon)	0	0	(icon)	0	0	(icon)	0	0			
(icon)	0	0	(icon)	0	0	(icon)	1	1	(icon)	1	1	(icon)	1	1	(icon)	0	0			
(icon)	0	1	(icon)	0	1	(icon)	0	0	(icon)	0	0	(icon)	0	0	(icon)	0	0			
(icon)	0	1	(icon)	0	1	(icon)	0	0	(icon)	0	0	(icon)	0	0	(icon)	0	0			
(icon)	1	1	(icon)	1	1	(icon)	0	0	(icon)	0	0	(icon)	0	0	(icon)	0	0			

Fig. 9.7 An automaton of the space $\mathscr{I}$

neighbourhood states is represented by one element on Fig. 9.7 that shows an automaton of $\mathscr{I}$.

9.1.5 Game of Life

The Game of Life is an automaton of $\mathscr{I}$ ($\mathbb{Z}^2$, $\{0, 1\}$, Moore neighbourhood, $\delta : S^9 \mapsto S$). The state of a cell of the Game of Life at the next generation depends on it's own state and the sum of cells in state 1 among its eight direct neighbours.

The behaviour of the cells of the Game of Life is inspired by the ones of real cells as a cell could die if there not enough cells surrounding it or if there too many. So if a cell in state 1 has 0 or 1 neighbours in state 1 then this cell will be in state 0 at the next generation. If a cell in state 1 has more than 3 neighbours in state 1 then the state of this cell will be 0 at the next generation. A birth happens if a cell has three neighbours in state 0. So if a cell in state 0 has 3 neighbours in state 1 then the state of this cell will be 1 at the next generation.

These rules can be formalised saying that the Game of Life is the automaton of $\mathscr{I}$ ($\mathbb{Z}^2$, $\{0, 1\}$, Moore neighbourhood, $\delta : S^9 \mapsto S$) such that for all binary numbers $x_1, x_2, x_3, x_4, x_5, x_6, x_7, x_8$ and x_9 we get:

$$\delta \begin{pmatrix} x_1, x_2, x_3, \\ x_4, x_5, x_6, \\ x_7, x_8, x_9 \end{pmatrix} = \begin{cases} 1 & \text{if } x_1 + x_2 + x_3 + x_4 + x_5 + x_6 + x_7 + x_8 + x_9 = 3 \\ 1 & \text{if } x_1 + x_2 + x_3 + x_4 + x_6 + x_7 + x_8 + x_9 = 2 \\ 0 & \text{otherwise.} \end{cases}$$

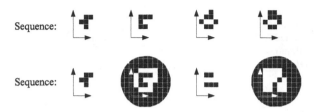

Sequence:

Sequence:

Fig. 9.8 Two sequences $(c_t^A)_{t \geq 0}$ for t equals 0, 1, 2 and 3. The *top* one is the R-pentomino of the Game of Life. The *bottom* one is the evolution of a cellular automaton without quiescent state. The transition rule of this automaton is the same of the Game of Life except that $\delta(0, 0, 0, 0, 0, 0, 0, 0, 0) = 1$. On this sequence, at generations 2 and 4, the only dead cells in existence are represented on the figure. The rest of $\mathbb{Z}^2$ is filled with living cells

9.1.6 Quiescent State

Among the states of a cellular automaton, sometime, a state s, called *quiescent state*, is such that $\delta(s, \ldots, s) = s$ [5]. In the following the state 0 will be the quiescent state. There are 2^{511} and 2^{101} automata of $\mathscr{E}$ and $\mathscr{I}$ for which 0 is the quiescent state. Figure 9.8 shows the evolution of automata with and without quiescent state.

9.2 Definitions of Patterns

9.2.1 Definition

A *pattern* P is a configuration of cells in a 2D space with relative positioning to one another. The evolution of P in the automaton A is the sequence $(P_t^A)_{t \geq 0}$ such that P_0^A is the pattern P positioned alone in a 2 dimensional lattice $\mathbb{Z}^2$.

A pattern is included in a configuration c_t^A of an automaton A of the space $\mathscr{E}$ if c_t^A is the pattern P is positioned possibly with other cells. A pattern is included in c_t^A of an automaton A of the space $\mathscr{I}$ if c_t^A is P after applying rotations and symmetries possibly with other cells around.

Figure 9.9 shows the R-pentomino of Game of Life after applying rotations and symmetries and Fig. 9.10 shows this pattern included in a configuration.

9.2.2 Periodic Pattern

A pattern P is a *periodic pattern* of the automaton A if and only if the evolution of P in the automaton A is such that there exists a non-zero natural number t such

Fig. 9.9 The R-pentomino of Game of Life after applying rotations and symmetries

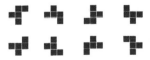

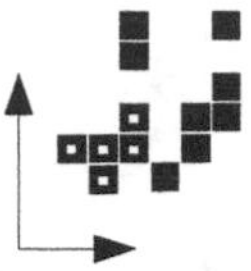

142   E. Sapin

Fig. 9.10 Configuration in
which the R-pentomino of the
Game of Life is included. The
cells of the R-pentomino is
shown with a with *dot* on the
center

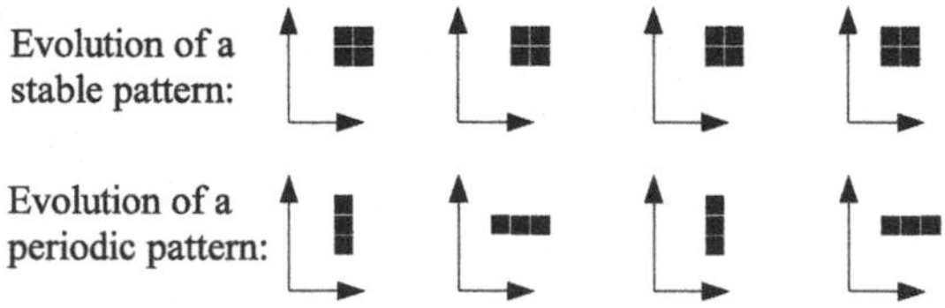

Fig. 9.11 Evolutions of
stable and periodic patterns
for generations from 0 to 3 of
the Game of Life

Evolution of a
stable pattern:

Evolution of a
periodic pattern:

that $P_0^A = P_t^A$. t is called the period of the periodic pattern. If for all t, $c_0^A = c_t^A$ the
pattern is called *stable pattern*. Figure 9.11 shows a periodic pattern and a stable
pattern of the Game of Life.

9.2.3 Glider

A pattern G is a *glider* of the automaton A if and only if the evolution of G in the
automaton A is such that there exists a non-zero natural number t and two integer
numbers D_x and D_y such that:

- $\|D_x\| + \|D_y\| > 0$.
- For all integer numbers x and y we get:

$$G_t^A(x, y) = G_0^A(x + D_x, y + D_y).$$

The number t is called the period of the glider. D_x and D_y are respectively the
displacements along X-axis and Y-axis. The movement is how much the glider
moves during a period. The definition of the movement depends on the chosen
neighbourhood. Considering Moore neighbourhood, the movement is defined by
the maximum between $\|D_x\|$ and $\|D_y\|$.

The definition of the velocity S of a glider is the movement divides by the period t.

A glider can be *orthogonal*, *diagonal* or *oblique*:

- If D_x equals 0 or D_y equals 0, G is said an orthogonal glider.
- If $\|D_x\|$ equals $\|D_y\|$, G is said a diagonal glider.
- If G is neither diagonal nor orthogonal, G is said an oblique glider.

Figure 9.12 shows orthogonal and diagonal gliders of the Game of Life.

Evolution of a diagonal glider:

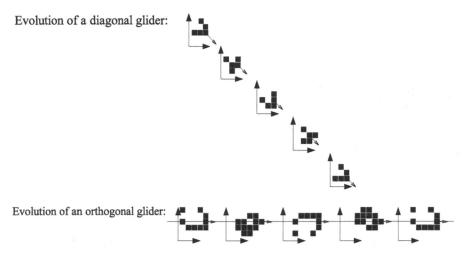

Evolution of an orthogonal glider:

Fig. 9.12 Evolutions of orthogonal and diagonal gliders for generations from 0 to 5 of the Game of Life. For the diagonal glider $t = 4$, $D_x = -1$ and $D_y = -1$ and for the orthogonal glider $t = 4$, $D_x = 0$ and $D_y = 1$

9.2.4 Glider Gun

A pattern G is a *glider gun* of the automaton A if and only if the evolution of G in the automaton A is such that there exists a non-zero natural number t such that for all natural numbers n, we get:

- For all integer numbers x and y, $G^A_{(n+1)\times t}(x, y) = 1$ if $G^A_{n\times t}(x, y) = 1$.
- Let P be a pattern for which for all integer numbers x and y we get:

$$P^A_0(x, y) = G^A_{(n+1)\times t}(x, y) - G^A_{n\times t}(x, y).$$

The pattern P is such that:
- There exists two integer numbers x and y such that $P^A_0(x, y) = 1$.
- For all integer numbers x and y such that $P^A_0(x, y) = 1$ we get (x, y) is in a glider included in P.

The number t is called the period of the glider gun. The gliders included in the pattern P is said emitted by G. Figure 9.13 shows the Gosper gun of the Game of Life.

The period of Gosper gun is 30 so for all integer numbers x and y, $G^{Life}_{30}(x, y) = 1$ if $G^{Life}_0(x, y) = 1$. Figure 9.14 shows the pattern P, in which for all integer numbers x and y, we get:

$$P^{Life}_0(x, y) = GosperGun^{Life}_{30}(x, y) - GosperGun^{Life}_0(x, y).$$

All the living cells of P^{Life}_0 are included in a glider of Game of Life.

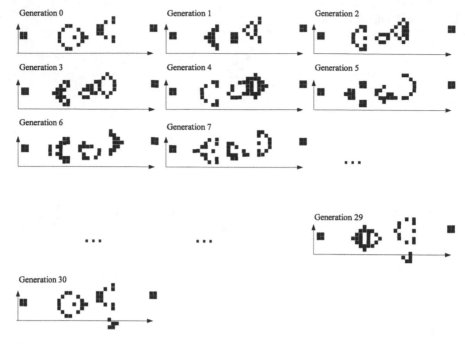

Fig. 9.13 Evolutions of the Gosper gun of the Game of Life

Fig. 9.14 The pattern P_0^{Life}

9.2.5 Symmetric Pattern

A pattern P of the automaton A is said symmetric if and only if the evolution of P in the automaton A, $(P_t^A)_{t\geq0}$, is such that for all natural numbers t we get P_t^A with a line that divides it into two symmetrical parts in such a way that it on one side is the mirror image of the other side.

This line is the axis of symmetry of the pattern P as shown in Fig. 9.15 for a symmetric pattern called the blinker of the Game of Life.

A glider G of period p of the automaton A is said half symmetric if and only if p is even and the evolution of P in the automaton A, $(P_t^A)_{t\geq0}$, is such that for all natural numbers t there exists a line which divides P_t^A into two parts such that P_t^A on each side of its line is the mirror image of $P_{t+\frac{p}{2}}^A$ on the other side of its line.

The diagonal glider of the Game of Life is half symmetric as shown in Fig. 9.16.

Fig. 9.15 The blinker of the
Game of Life and its axis of
symmetry

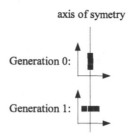

Fig. 9.16 The glider of the
Game of Life

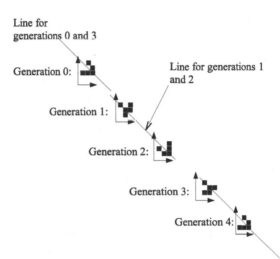

9.3 Automata and Patterns

9.3.1 Automata Accepting Patterns

An automaton B ($\mathbb{Z}^2$, $\{0, 1\}$, Moore neighbourhood, $\delta_B : S^9 \mapsto S$) is said to accept
a pattern P of the automaton A ($\mathbb{Z}^2$, $\{0, 1\}$, Moore neighbourhood, $\delta_A : S^9 \mapsto S$)
if and only if the evolution $(c_t^A)_{t \geq 0}$ of P is such that for all binary numbers
$x_1, x_2, x_3, x_4, x_5, x_6, x_7, x_8$ and x_9 we get:

$$\exists t / \begin{pmatrix} x_1, x_2, x_3, \\ x_4, x_5, x_6, \\ x_7, x_8, x_9 \end{pmatrix} \in P_t^A$$

implies:

$$\delta_B \begin{pmatrix} x_1, x_2, x_3, \\ x_4, x_5, x_6, \\ x_7, x_8, x_9 \end{pmatrix} = \delta_A \begin{pmatrix} x_1, x_2, x_3, \\ x_4, x_5, x_6, \\ x_7, x_8, x_9 \end{pmatrix} .$$

A given pattern could be accepted by one or several automata so the question
"How many automata accept a given pattern?" arises.

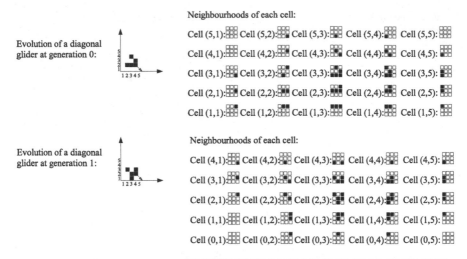

Fig. 9.17 Details of the construction of sets of neighbourhood states that are used by a glider. There are 29 neighbourhoods that have been memorized meaning that 2^{483} automata of the space $\mathscr{E}$ accept this glider in this direction

9.3.2 Automata of $\mathscr{E}$ Accepting Patterns

In order to determine the set of automata B that accept a pattern P of an automaton A, an image-filtering of P_t^A for all t is used. For all t, all neighbourhoods in P_t^A, called used neighbourhoods, are memorized. For the neighbourhoods that have been memorized, δ_B is determined by δ_A and only for the other ones the function δ_B is free. Let n be the number of neighbourhoods that have been memorized. For $512 - n$ neighbourhoods, δ_B is free so there are 2^{512-n} automata that accept the pattern P.

This process is shown on Fig. 9.17 for the glider of the Game of Life at generations 0 and 1. 29 neighbourhoods have been memorized that leads to have 2^{483} automata of the space $\mathscr{E}$ that accept this glider during the two first generations in this direction.

9.3.3 Automata of $\mathscr{I}$ Accepting Patterns

A more difficult question to answer is 'how many automata of the space $\mathscr{I}$ accept a pattern in any direction?'.

In order to answer to it, the memorized neighbourhoods are put in subsets of isotropic neighbourhoods. For the subsets of isotropic neighbourhoods in which memorized neighbourhoods have been put, δ_B is determined by δ_A and only for

The 29 Neighbourhoods that have been memorized and the corresponding subset:

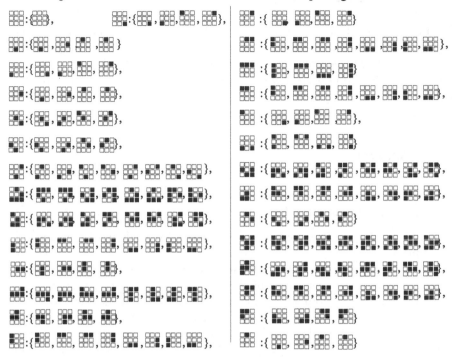

Fig. 9.18   Subset in which the 29 memorized neighbourhood can be put

the other ones the function δ_B is free. Let n be the number of subsets of isotropic neighbourhoods in which memorized neighbourhoods have been put. For $102 - n$ neighbourhoods, δ_B is free so there are 2^{102-n} automata that accept the pattern P.

If a glider G is half symmetric, the image-filtering only needs to be done on half the period. The memorized neighbourhoods of the second half of the period can be put in subsets of isotropic neighbourhoods in which a symmetric neighbourhood has already been put. In Fig. 9.17, the memorized neighbourhoods are put in subsets of isotropic neighbourhoods for the glider of the Game of Life at generations 0 and 1.

As the glider of the Game of Life is half symmetric, one needs only to consider half of the period. There are 18 different subsets of isotropic neighbourhoods in which the 29 memorized neighbourhoods can be put. This leads to having 2^{84} automata of the space $\mathscr{I}$ that accept this glider in any direction.

From the evolution of a pattern P, it is possible to know which automata accept the pattern P. So only the evolution of the gliders and glider guns will be given that is enough to know which automata accept it.

9.4 Searching for Gliders

We aim to construct an automatic system for the discovery of Turing-universal cellular automata and the first step is the discovery of gliders and glider guns. The
search for gliders is described here by setting out the rational in the first subsection and then the detail of the used algorithm. The results are described in the last
subsection.

9.4.1 Rational

Different algorithms have been tried to search for glider as Monte Carlo method,
Tabu search and evolutionary algorithms:

- The Monte Carlo method consists solely of generating random solutions and testing them.
- Evolutionary algorithms incorporate aspects of natural selection or survival of the
  fittest. It maintains a population of structures (usually initially generated at random) that evolves according to rules of selection, recombination, mutation, and
  survival referred to as genetic operators. A shared environment is used to determine the fitness or performance of each individual in the population. The fittest
  individuals are more likely to be selected for reproduction through recombination
  and mutation, in order to obtain potentially superior ones.

Throughout our experiment, no glider was found with a Monte Carlo method in
one million randomly generated cellular automata. Then, an evolutionary approach
was tried.

The next subsection presents the choice of the evolutionary algorithm.

9.4.2 Algorithm

In the algorithms we used, individuals are randomly generated. The individuals are
evaluated by a fitness function and the ones with the best scores are kept. Crossovers
and mutations are applied to these individuals in order to have new ones. The new
individuals are evaluated again by the fitness function and this cycle is iterated. More
details about this algorithm can be found in [24–27].

In order to search for gliders, the questions "what are individuals?" and "how can
a fitness function evaluate them?" have to be answered. An individual could be two
different things:

- A pattern, this pattern could evolve thanks to a mutation operator that adds or
  removes a cell. The fitness function could be based on the evolution of this pattern
  with cellular automata.

- A transition rule of an automaton. The fitness function could be based on the evolution of random configurations with this transition rule.

Cellular automata are highly sensitive to the initial configuration of cells [29] so patterns have been considered too difficult to evaluate by a fitness function. Thus we choose to consider transition rules of automata as individuals.

In this algorithm, the choice of the fitness function is perhaps the greatest difficulty. Lohn et al. [11] have chosen a multi-objective function, Ventrella [30] employed a particle swarm [9] interacting with the cellular automata. The first automata the fitness function has to evaluate are randomly generated. Most of the time, these automata do not accept any gliders as no gliders have been found using Monte Carlo method. The fitness function has to give good evaluation of the automata that do not accept gliders but could be evolved into automata accepting gliders thanks to crossover and mutation operators. In other terms, the fitness function has to evaluate how an automaton that does not accept gliders could evolve into an automaton that accepts gliders.

To achieve this task, a random configuration of cells is evolved by the tested automaton. After this evolution, different functions have been considered:

- The value of the first fitness function was the division of the size of the biggest rectangle included all cells by the total number of cells, the idea is: if the size of this rectangle increases without a great number of cells it could be thanks to the appearance of gliders or something close to being a glider.
- The value of the second fitness function was the sum of the number of apparitions of gliders by the number of apparitions of periodic patterns was the second fitness function, the idea is: the periodic pattern is the first step to the appearance of gliders.

After both fitness functions, inspired by Bays' test [3], the presence of gliders is checked. Each pattern P is isolated in an empty universe and evolves by the transition rule of the tested automaton. If a pattern P reappears alone in a place different from the one it appeared at in the beginning then the pattern P is a glider. If it reappears at the same place then it is a periodic pattern. When a glider is found the algorithm stops.

9.4.3 Gliders

The space $\mathcal{E}$ has been considered first. The first fitness function was used. The idea was: if the size of this rectangle increases without a great number of cells it could be thanks to the appearance of gliders or something close to being a glider.

Most of the gliders found by this algorithm that are accepted by automata of the space $\mathcal{E}$ were unidirectional. The scheme of demonstration of universality of the game of life uses multi-directional gliders. So we chose to restrain to gliders that are accepted only by automata of the space $\mathcal{I}$.

		Period									
		1	2	3	4	5	6	7	9	10	24
M	1	Yes	Yes	Yes	Yes	Yes	Yes	Not yet	Not yet	Yes	Not yet
o	2	No	Yes	Not yet	Yes	Not yet	Not yet	Yes	Not yet	Not yet	Yes
v	3	No	No	Yes	Not yet	Not yet	Not yet	Not yet	Yes	Not yet	Not yet
i	4	No	No	No	Yes	Not yet	Not yet	Not yet	Not yet	Not yet	Not yet
n	5	No	No	No	No	Not yet	Not yet	Not yet	Not yet	Yes	Not yet
g	6	No	No	No	No	No	Not yet	Not yet	Not yet	Not yet	Not yet
	7	No	No	No	No	No	No	Not yet	Not yet	Not yet	Not yet
	8	No	No	No	No	No	No	No	Not yet	Not yet	Not yet

Fig. 9.19 Combination of movement and period for which gliders were found. '*Yes*' means gliders of this combination were found, '*No*' means gliders of this combination can not exist and '*Not yet*' means gliders of this combination can exist but have not been found

		Period									
		1	2	3	4	5	6	7	9	10	24
V	1	Yes	Yes	Yes	Yes	Not yet	Not yet	Not yet	Not yet	Not yet	Not yet
e	$\frac{1}{2}$	No	Yes	No	Yes	No	Yes	No	No	Yes	Not yet
l	$\frac{1}{3}$	No	No	Yes	No	No	Not yet	No	Yes	No	Not yet
o	$\frac{2}{7}$	No	No	No	No	No	No	Yes	No	No	No
c	$\frac{1}{4}$	No	No	No	Yes	No	No	No	No	No	Not yet
i	$\frac{1}{5}$	No	No	No	No	Yes	No	No	No	Not yet	No
t	$\frac{1}{6}$	No	No	No	No	No	Yes	No	No	No	Not yet
t	$\frac{1}{10}$	No	No	No	No	No	No	No	No	Yes	No
y	$\frac{1}{12}$	No	No	No	No	No	No	No	No	No	Yes

Fig. 9.20 Combination of velocity and period for which gliders were found. '*Yes*' means gliders of this combination were found, '*No*' means gliders of this combination can not exist and '*Not yet*' means gliders of this combination can exist but have not been found

In order to search for isotropic gliders, the second fitness function is used because the computation of the second fitness function require less time than the computation of the first one.

Gliders of different movements, velocities, periods and directions were found. Figure 9.19 shows for which periods and movements during a period, gliders were found.

The movement during a period has to be at least 1 considering the definition of a glider, and can not be higher than the period because it is impossible for a glider to move faster than 1, this is called the speed of light [3].

Figure 9.20 shows the velocities and the periods at which gliders were found. The velocity of a glider is its movement divided by its period, so an integer number divided by its period. So for a glider of period N, the velocity can only be $\frac{D}{N}$ where D is an integer number between 1 and N.

Figures 9.21–9.31 show a sample of one hundred gliders of different periods and velocities. These gliders are numbered on these figures.

11% of the gliders found by the algorithm are diagonal and only 3 percent are oblique. The other 86% of the gliders found are orthogonal. 56% of the gliders are symmetric and 16% are half symmetric.

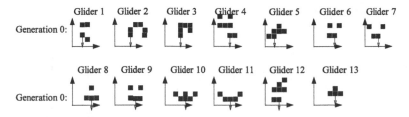

Fig. 9.21 Gliders of period 1 found by our algorithm

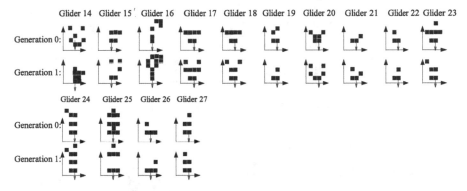

Fig. 9.22 Gliders of period 2 and velocity 1 found by our algorithm

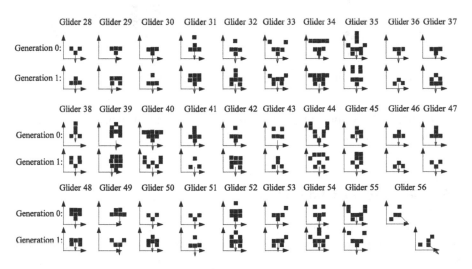

Fig. 9.23 Gliders of period 2 and velocity $\frac{1}{2}$ found by our algorithm

Fig. 9.24 Gliders of period 3 and velocity 1 found by our algorithm

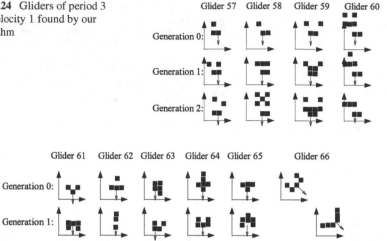

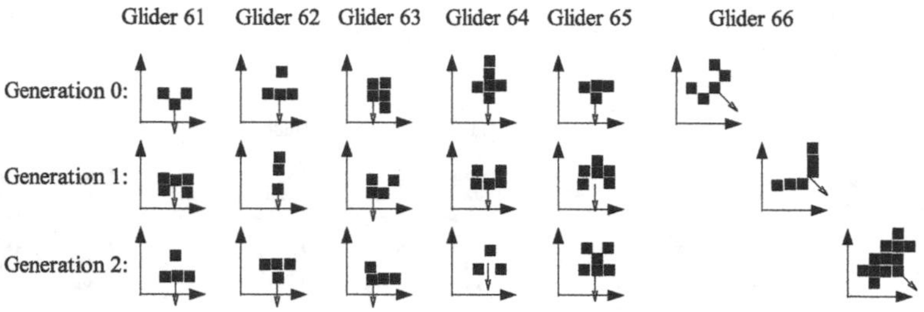

Fig. 9.25 Gliders of period 3 and velocity $\frac{1}{3}$ found by our algorithm

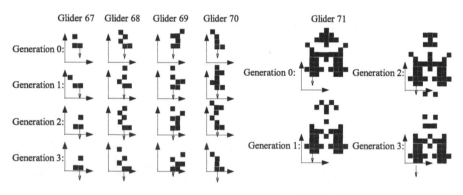

Fig. 9.26 Gliders of period 4 and velocity 1 found by our algorithm

The pattern of cells that is the most common, is the one shown in Fig. 9.32. These are gliders 28, 36, 46, 47, 51, 61 ,79 or 80.

9.5 Searching for Glider Guns

In order to demonstrate the universality of the Game of Life, Conway et al. used gliders and glider guns. Here, Turing-universal cellular automata are searching for. So, after discovering gliders, glider guns were searched for with evolutionary algorithms. All the details about this algorithm can be found in [22, 23].

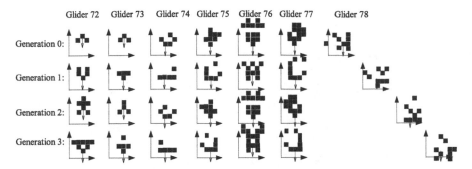

Fig. 9.27 Gliders of period 4 and velocity $\frac{1}{2}$ found by our algorithm

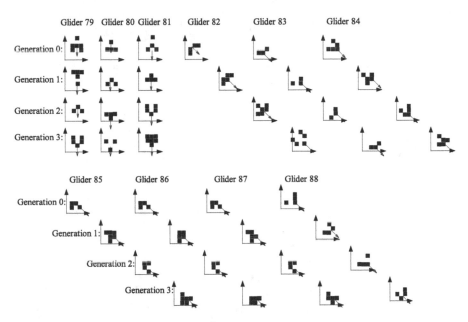

Fig. 9.28 Gliders of period 4 and velocity $\frac{1}{4}$ found by our algorithm

9.5.1 Algorithm

The idea is to focus on an existing glider like those found in the last section and to try to find guns that emit this glider.

As in the algorithm that searched for gliders, an individual could be two different things:

- a configuration of cells,
- a transition rule of an automaton.

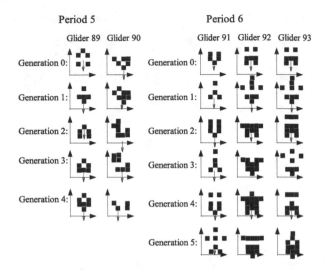

Fig. 9.29 Gliders of period 5 and velocity $\frac{1}{5}$ and gliders of period 6 and velocity $\frac{1}{2}$ found by our algorithm

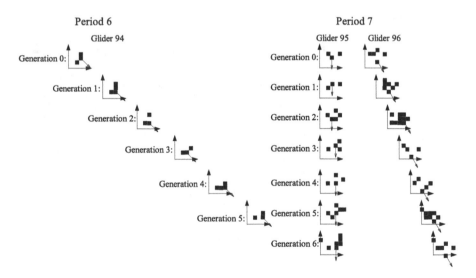

Fig. 9.30 A glider of period 6 and velocity $\frac{1}{6}$ and gliders of period 7 and velocity $\frac{2}{7}$ found by our algorithm

As explained in the last section, configurations of cells have been considered too difficult to be evaluated by a fitness function so the individuals of the algorithm that searched for glider guns are chosen to be the transition rules of automata.

The fitness function has to evaluate how an automaton that does not accept glider guns is similar to an automaton that accepts gliders. In order to do it, the optimization of the number of apparitions of gliders is attempted, as is the minimization of

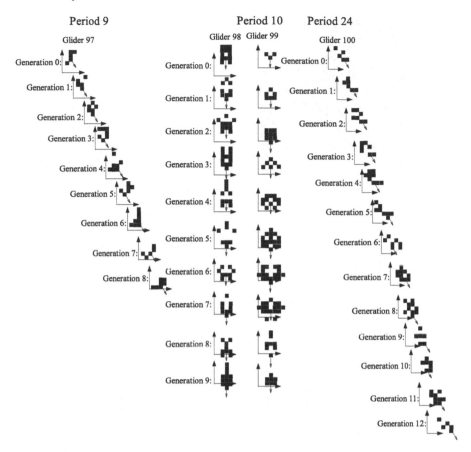

Fig. 9.31 A glider of period 9 and velocity $\frac{1}{3}$, gliders of period 10 and velocity $\frac{1}{2}$ and $\frac{1}{10}$ and half of the period of a glider of period 24 and velocity $\frac{1}{2}$ found by our algorithm. This glider is half symmetric so the second half of it, is symmetric to the first half

Fig. 9.32 Pattern the most common in the sample of one hundred gliders found by our algorithm. This pattern could be in the evolution of gliders which move up or gliders which move down

the total number of cells. Then the fitness function is the number of apparitions of gliders divide by the total number of cells plus one to avoid division by zero. The idea of this fitness function is, if a lot of gliders appear without the presence of a lot of cells it may indicate the presence of a glider gun.

An example of computation of these fitness functions is shown in Fig. 9.33.

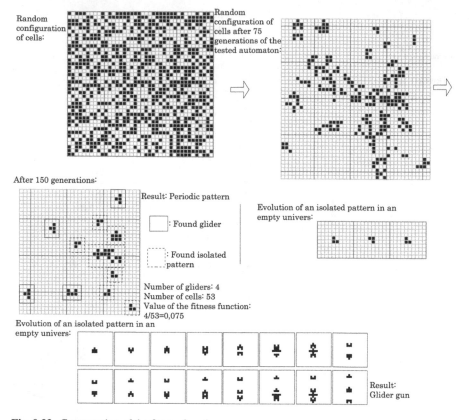

Fig. 9.33 Computation of the fitness functions

An automaton can not accept a glider gun if it does not accept the corresponding glider. Then, in order to optimize the search for glider guns the search space is the space of all the automata that accept the given glider. This space is determined following the method explained in Sect. 9.3.1.

9.5.2 Results

A gun has been searched for all the one hundred gliders on Figs. 9.21–9.31. For each glider, ten runs of the algorithm have been realized. For each run, either a gun can be found or not so the number of guns that can be found varies from zero to ten.

The number of guns the algorithm found for each glider is shown on Fig. 9.34. At least one gun has been found for 55 gliders. We can acknowledge that less guns were found for gliders of high periods and guns were found for only two diagonal gliders. For the glider of the Game of Life but no gun were found so the evolutionary

Glider	Guns	Glider	Guns	Glider	Guns	Glider	Guns	Glider	Guns
1	0	11	1	21	1	31	1	41	0
2	0	12	1	22	2	32	1	42	1
3	0	13	4	23	1	33	2	43	2
4	0	14	0	24	1	34	1	44	0
5	0	15	0	25	1	35	1	45	0
6	0	16	0	26	1	36	3	46	0
7	0	17	0	27	1	37	1	47	3
8	1	18	0	28	3	38	1	48	4
9	1	19	3	29	10	39	1	49	1
10	1	20	2	30	4	40	1	50	1
Glider	Guns	Glider	Guns	Glider	Guns	Glider	Guns	Glider	Guns
51	2	61	2	71	0	81	0	91	1
52	2	62	0	72	0	82	0	92	1
53	2	63	1	73	3	83	1	93	1
54	1	64	0	74	0	84	0	94	1
55	1	65	0	75	1	85	0	95	0
56	0	66	1	76	0	86	0	96	0
57	1	67	2	77	0	87	0	97	0
58	1	68	1	78	0	88	0	98	1
59	0	69	0	79	0	89	0	99	0
60	0	70	1	80	0	90	0	100	0

Fig. 9.34 Number of guns found for each glider. "*Glider*" is the number of the glider and "*Guns*" is the number of emitting guns

process did not find the Gosper gun obtained by Gosper through the interaction of several patterns with each other.

During its period, a gun can emit more than one glider whereas the Gosper gun emits only one glider.

Figures 9.35 and 9.36 show guns emitting four gliders. The gun in Fig. 9.35, emitted orthogonal gliders, have a period of 7. All the four gliders are emitting at the end of the period. For the gun in Fig. 9.36 the four diagonal gliders are emitted one by one at each quarter of the period.

The guns of Figs. 9.38 and 9.37 emit two gliders. The former emits orthogonal gliders and the latter emits diagonal gliders. For both, a glider is emitted at half of the period and the other one at the end of the period.

Figure 9.39 shows a gun of period 7 emitting only one orthogonal glider.

9.6 Guns of a Specific Glider

The search for glider guns leads to the discovery of a surprisingly important number of glider guns. For most of the gliders the number of glider guns discovered is

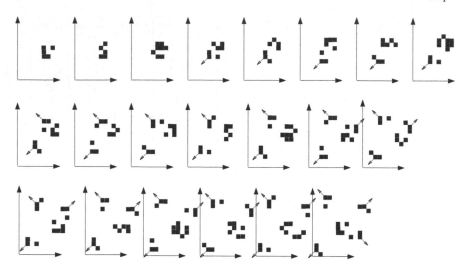

Fig. 9.35 A gun emitting four diagonal gliders

Fig. 9.36 A gun emitting
four orthogonal gliders

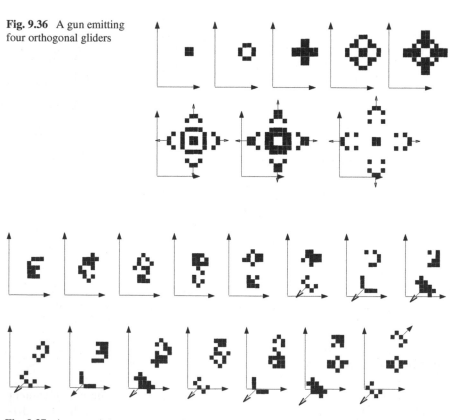

Fig. 9.37 A gun emitting two diagonal gliders

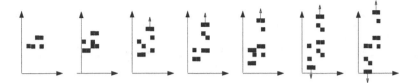

Fig. 9.38 A gun emitting two diagonal gliders

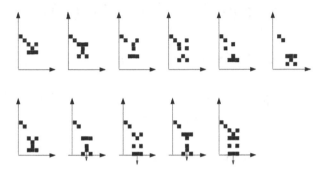

Fig. 9.39 A gun emitting two diagonal gliders

lower than 5. For the glider 29, ten different guns were found over ten runs of the algorithm. This glider has been the subject of a particular study.

Ten new runs have been done and each time a gun was found, this gun was different from the ones already discovered one. A question naturally arises: "How many glider guns emitting any particular glider exist?". In order to answer to this question, the first subsection sets out the definition of different guns while the second subsection give results of the experiments.

9.6.1 Different Guns

In order to determine how many different glider guns were found, a system that determines if a gun is new, is required. This system needs to have several properties:

- This system needs to be automatic. The number of discovered guns is more than 20 so a visual checking is not possible.
- This system needs to insure that all the discovered guns are really different. Otherwise claiming that a given number of different guns have been discovered could be false.
- This system needs to consider two guns to be the same gun if they are the same gun but discovered at two different generations.
- This system needs to consider two guns to be the same gun if one gun is the result of the other gun after some symmetries or rotations, Fig. 9.40 shows the same gun twice after some symmetries or rotations discovered at two different generations.

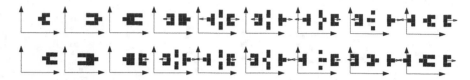

Fig. 9.40 Guns emitting glider 29, discovered at two different generations and after a symmetry

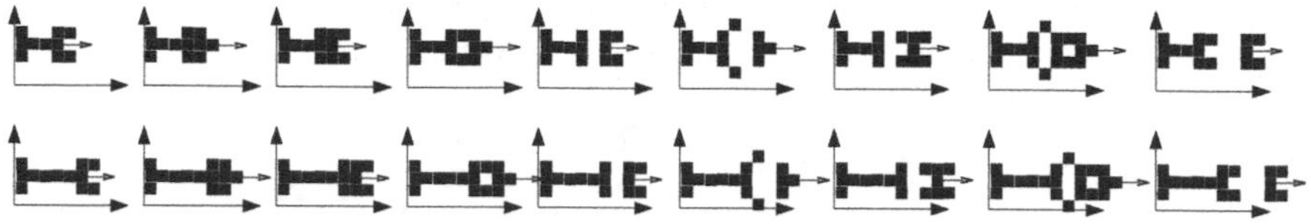

Fig. 9.41 Guns emitting glider 29. Only the right halves of the two guns vary throughout the generations and this variation is the same for the two guns. The left halves of the two guns are different from each other but remain the same throughout the generation

- This system needs to consider two guns to be the same guns if:
  - They are a part of cells that are different, and a part that is the same.
  - Only the part of the cells that is different is stable.
  - The part of the cells that die and born is the same in the two guns.
  Figure 9.41 shows two guns corresponding to this situation.
- This system need to be fast enough. Considering that there might exist more than ten thousand guns and each gun could have a period higher than 10, the number of patterns could be higher than one hundred thousand.

In order to determine if a gun is new, at least four criteria can be used:

- The period of the guns.
- The number of gliders emitted.
- The shapes of the guns at each generation of the period.
- The subsets of isotropic neighbourhoods.

The period and the number of emitted gliders of a gun have been considered as not relevant enough. The shapes of the guns at each generation of the period have been considered too long to compare. So, in order to determine if a gun is new, the subsets of isotropic neighbourhoods are considered.

The subsets of isotropic neighbourhoods in which the neighbourhood states, used by the given gun, can be put are determined by the method described in Sect. 9.4. For these subsets, all the automata that accept this gun have the same value. This value is stored and for the other subsets, the value x is stored. The list of stored values are compared to the ones of the other guns. For each guns Gun_{n+1}, Gun_{n+1} is a new gun if and only if the list of stored values is different than the list of stored values of each gun from Gun_0 to Gun_n.

This system responds to all the requirements:

- This system is automatic.

- This system insures that all the discovered guns are really different. If two patterns *A* and *B* are said to be different, at least a neighbourhood has to be different.
- This system does not take into account the generation in which the guns have been discovered.
- This system considers two guns to be the same gun if one gun is the result of the other gun after some symmetries or rotations. The subsets of isotropic neighbourhoods in which the neighbourhood states that are used by the given gun can be put, are the same after symmetries or rotations of a gun.
- This system needs to consider the two guns of Fig. 9.41 the same as the subsets of isotropic neighbourhood in which the neighbourhood states that are used by the given gun can be put, are the same for the two guns.
- This system is fast as its complexity is *n*, where *n* is the number of discovered guns.

One can notice that the definition implies that the number of different guns can not be higher than 2^{102-12}, 12 being the number of subsets of isotropic neighbourhood in which the neighbourhood states used by the glider 29 can be put.

9.6.2 Results

Thanks to this qualification of different guns, through the experiments, 26,039 different glider guns were discovered. All these guns have emerged spontaneously from random configurations of cells. The 26,039 guns can be found in [21] in Life format.

So at least 26,039 glider guns that emitted the glider 29 exist. From the probability of a discovered gun being new, it is possible to estimate the total number of guns.

Suppose each gun has the same probability to be found. Let N be the total number of guns that could be found by this algorithm. The probability of a gun found by the algorithm of being new would be $1 - \frac{26,039}{N}$. The number of new guns among the last 1,000 different found guns is 755. So the total number of guns that could be found by the algorithm could be estimated by $N = 26,039 \times \frac{1,000}{245}$ about 106,282.

Most of the discovered guns are an even period as shown in Fig. 9.42 which represents he distribution of the periods of the discovered guns.

Most of the discovered guns emit an even number of glider as shown in Fig. 9.43 which represents the distribution of the number of guns found emitting *n* gliders during a period.

Guns with even period and even number of gliders emitted recover a symmetric of their original shape after half of their period has elapsed, after having emitted one or several gliders.

The guns emitting an odd number of gliders are not common as only guns emitting one and three gliders were found.

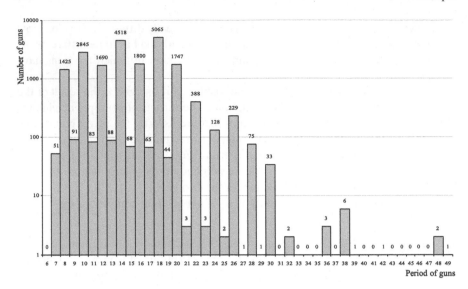

Fig. 9.42 Distribution of the period of the found guns. Three other guns of period 62, 74 and 243 are not shown on the graphic

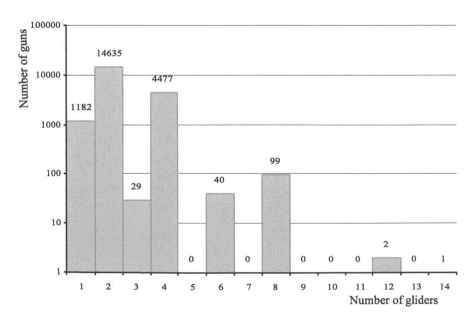

Fig. 9.43 Distribution of the number of guns found emitting *n* gliders during a period

9.7 Synthesis and Perspectives

This chapter deals with the well established problem of universality in cellular automata. A part of this problem is the discovery of gliders and glider guns that have been presented. A stochastic search method is tried to find gliders and glider guns, building on previous work shown in [25, 26]. The choice of an evolutionary algorithm led to the discovery of new gliders and a sample of one hundred new gliders have been shown. Properties of symmetry and periodicity have been studied.

Concerning the search for glider guns, the Monte Carlo method and the evolutionary algorithms are explored with different parameters. No guns were found with the Monte Carlo method but an evolutionary algorithm has led to the discovery of guns for 55 gliders of the 100 gliders sampled. These guns were not known before and a sample of them is shown.

For a particular glider, the guns seemed to be very easy to find. The study of the emergence of glider guns emitting this glider leads to the discovery of 26,039 glider guns. A study of the number of guns emitting this glider suggest that more than one hundred thousand guns exist.

All the discovered guns have emerge spontaneously from the evolution a random configuration of cells. The emergence of this number of glider guns is a very surprising result as Resnick et al. [20] claimed: "You would be shocked to see something as large and organized as a glider gun emerge spontaneously." The existence of so many different glider guns shows that they are far more easy to find than Wolfram claimed [33]: "There are however indications that the required initial configuration is quite large, and is very difficult to find." Thus the discovery of so many different glider guns represents a significant contribution to the theory of cellular automata and complex systems that considers computational theory.

Further goals can be to search for more glider guns and for more gliders with different methods. All these automata accepting a discovered glider gun may be potential candidates for being shown universal automata thanks to an automatic system for the demonstration of universal automata that can be developed. Then, another domain that seems worth exploring is how this approach could be extended to automata with more than 2 states [37].

Future work could also be to study how gliders appear from a random configuration of cells or a gun and which part of the transition rule makes them move throughout generations. This could lead to a better understanding of the link between the transition rule and the emergence of computation in cellular automata and therefore the emergence of computation in complex systems with simple components.

Acknowledgements   The work was supported by the Engineering and Physical Sciences Research Council (EPSRC) of the United Kingdom, Grant EP/E005241/1.

References

1. Adamatzky, A.: Universal dynamical computation in multi-dimensional excitable lattices. Int. J. Theor. Phys. **37**, 3069–3108 (1998)
2. Banks, E.R.: Information and transmission in cellular automata. PhD thesis, MIT (1971)

3. Bays, C.: Candidates for the game of life in three dimensions. Complex Syst. **1**, 373–400 (1987)
4. Berlekamp, E., Conway, J.H., Guy, R.: Winning Ways for Your Mathematical Plays. Academic Press, New York (1982)
5. Delorme, M.: An introduction to cellular automata. In: Delorme, M., Mazoyer, J. (eds.) Cellular Automata: A Parallel Model, pp. 5–51. Kluwer, Dordrecht (1999)
6. Eppstein, D.: http://www.ics.uci.edu/~eppstein/ca/
7. Hordijk, W., Crutchfield, J.P., Mitchell, M.: Mechanisms of emergent computation in cellular automata. In: Eiben, A.E., Bäck, T., Schoenauer, M., Schwefel, H.-P. (eds.) Parallel Problem Solving from Nature-V, pp. 344–353. Springer, Berlin (1998)
8. Ilachinski, A.: Cellular Automata. World Scientific, Singapore (1992)
9. Kennedy, J., Eberhart, C.: Particle swarm optimization. In: Proceedings of the 1995 IEEE Conference on Neural Networks (1995)
10. Lindgren, K., Nordahl, M.: Universal computation in simple one dimensional cellular automata. Complex Syst. **4**, 299–318 (1990)
11. Lohn, J.D., Reggia, J.A.: Automatic discovery of self-replicating structures in cellular automata. IEEE Trans. Evol. Comput. **1**, 165–178 (1997)
12. Margolus, N.: Physics-like models of computation. Physica D **10**, 81–95 (1984)
13. Martinez, G.J., Adamatzky, A., McIntosh, H.V.: Phenomenology of glider collisions in cellular automaton rule 54 and associated logical gates chaos. Fractals Solitons **28**, 100–111 (2006)
14. Mitchell, M., Hraber, P.T., Crutchfield, J.P.: Revisiting the edge of chaos: evolving cellular automate to perform computations. Complex Syst. **7**, 89–130 (1993)
15. Mitchell, M., Crutchfield, J.P., Hraber, P.T.: Evolving cellular automata to perform computations: mechanisms and impediments. Physica D **75**, 361–391 (1994)
16. Morita, K., Tojima, Y., Katsunobo, I., Ogiro, T.: Universal computing in reversible and number-conserving two-dimensional cellular spaces. In: Adamatzky, A. (ed.) Collision-Based Computing, pp. 161–199. Springer, Berlin (2002)
17. Ollinger, N.: Universalities in cellular automata a (short) survey. In: Durand, B. (ed.) Symposium on Cellular Automata Journees Automates Cellulaires (JAC'08), pp. 102–118 (2008)
18. Packard, N.H.: Adaptation toward the edge of chaos. In: Kelso, J.A.S., Mandell, A.J., Shlesinger, M.F. (eds.) Dynamic Patterns in Complex Systems, pp. 293–301 (1988)
19. Rendell, P.: Turing universality in the game of life. In: Adamatzky, A. (ed.) Collision-Based Computing, pp. 513–539. Springer, Berlin (2002)
20. Resnick, M., Silverman, B.: Exploring emergence. http://llk.media.mit.edu/projects/emergence/ (1996)
21. Sapin, E.: http://uncomp.uwe.ac.uk/sapin/gun.zip (2009)
22. Sapin, E., Bull, L.: A genetic algorithm approach to searching for glider guns in cellular automata. In: IEEE Congress on Evolutionary Computation, pp. 2456–2462
23. Sapin, E., Bull, L.: Searching for glider guns in cellular automata: Exploring evolutionary and other techniques. In: EA07. Lecture Notes in Computer Science, vol. 4926, pp. 255–265. Springer, Berlin (2007)
24. Sapin, E., Bailleux, O., Chabrier, J.: Research of complexity in cellular automata through evolutionary algorithms. Complex Syst. **?**, 17 (?)
25. Sapin, E., Bailleux, O., Chabrier, J.J.: Research of a cellular automaton simulating logic gates by evolutionary algorithms. In: EuroGP03. Lecture Notes in Computer Science, vol. 2610, pp. 414–423. Springer, Berlin (2003)
26. Sapin, E., Bailleux, O., Chabrier, J.J.: Research of complex forms in the cellular automata by evolutionary algorithms. In: EA03. Lecture Notes in Computer Science, vol. 2936, pp. 373–400. Springer, Berlin (2004)
27. Sapin, E., Bailleux, O., Chabrier, J.J., Collet, P.: Demonstration of the universality of a new cellular automaton. IJUC **2**(3), ? (2006)
28. Sipper, M.: Evolution of parallel cellular machines. In: Stauffer, D. (ed.) Annual Reviews of Computational Physics, V, pp. 243–285. World Scientific, Singapore (1997)

29. Urías, J., Rechtman, R., Enciso, A.: Sensitive dependence on initial conditions for cellular automata. Chaos **7**(4), 688–693 (1997)
30. Ventrella, J.J.: A particle swarm selects for evolution of gliders in non-uniform 2d cellular automata. In: Artificial Life X: Proceedings of the 10th International Conference on the Simulation and Synthesis of Living Systems, pp. 386–392 (2006)
31. von Neumann, J.: Theory of Self-reproducing Automata. University of Illinois, Urbana (1966)
32. Waldrop, M.M.: Complexity: The Emerging Science at the Edge of Chaos. Simon and Schuster, New York (1992)
33. Wolfram, S.: Universality and complexity in cellular automata. Physica D **10**, 1–35 (1984)
34. Wolfram, S.: A New Kind of Science. Wolfram Media, Champaign (2002)
35. Wolz, D., de Oliveira, P.B.: Very effective evolutionary techniques for searching cellular automata rule spaces. J. Cell. Autom. **3**(4), 289–312 (2008)
36. Wuensche, A.: Discrete dynamics lab (ddlab). www.ddlab.org (2005)
37. Wuensche, A., Adamatzky, A.: On spiral glider-guns in hexagonal cellular automata: activator–inhibitor paradigm. Int. J. Mod. Phys. C **17**(7), 1009–1026 (2008)

Chapter 10
Constraint Programming to Solve Maximal Density Still Life

**Geoffrey Chu, Karen Elizabeth Petrie,
and Neil Yorke-Smith**

The Maximum Density Still Life problem fills a finite Game of Life board with a stable pattern of cells that has as many live cells as possible. Although simple to state, this problem is computationally challenging for any but the smallest sizes of board. Especially difficult is to prove that the maximum number of live cells has been found. Various approaches have been employed. The most successful are approaches based on Constraint Programming (CP). We describe the Maximum Density Still Life problem, introduce the concept of constraint programming, give an overview on how the problem can be modelled and solved with CP, and report on best-known results for the problem.

10.1 Introduction

The Game of Life was invented by John Horton Conway in the 1960s and popularized by Martin Gardner in his *Scientific American* columns (e.g., [8]). Many variants of the game and many problems arising from them have been studied. This chapter describes one beautiful and simple problem that has taxed academics for over a decade, and recounts how it has been solved using the technique of constraint programming.

Problems often consist of choices. Making an optimal choice that is compatible with all other choices made is difficult. *Constraint programming* (CP) [12] is a branch of Artificial Intelligence, where computers help people to make these choices. CP is a multidisciplinary technology combining computer science, operations research, and mathematics. Constraints arise in design and configuration, planning and scheduling, diagnosis and testing, and in many other contexts. This means that constraints are a powerful and natural means of knowledge representation and inference in many areas of industry and academia.

The problem studied in this chapter is the *Maximum Density Still Life* problem. The Maximum Density Still Life fills a finite Game of Life board with a *stable pattern* of cells — one that does not change from generation to generation — that

A. Adamatzky (ed.), *Game of Life Cellular Automata*,
DOI 10.1007/978-1-84996-217-9_10, © Springer-Verlag London Limited 2010

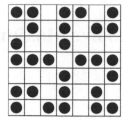 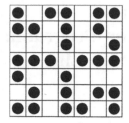

Fig. 10.1 The two optimal solutions for the 7×7 board

has as many live cells as possible. Figure 10.1 shows two optimal solutions for the 7×7 board.

We introduce more fully the Maximum Density Still Life (MDSL) problem (Sect. 10.2), give an introduction to the methodology of constraint programming (Sect. 10.3), and then describe how the MDSL problem can be modelled using constraint programming (Sect. 10.4). We report progress on the problem over a number of years, leading to the current best-known results which have essentially solved the problem (Sect. 10.5).

10.2 Maximum Density Still Life

The *Maximum Density Still Life* (MDSL) problem, or *Still Life* for short, is a variant of the Game of Life. The problem is to find a stable pattern of maximum density [1, 13, 16]. Still Life has been tackled through a variety of approaches, including constraint programming.

The Game of Life is played on a nominally infinite board, constructed of square cells where every cell has eight neighbours. The configuration of live and dead evolves over the generations. The configuration at time t leads to a new configuration at time $t + 1$ according to the three familiar rules of the game:

1. If a cell has exactly 3 living neighbours at time t, it is alive at time $t + 1$. This is the *birth condition*.
2. If a cell has exactly 2 living neighbours at time t, it remains in the same state at time $t + 1$. That is, if dead at time t, it remains dead at time $t + 1$; if live at time t, it remains live at time $t + 1$.
3. If a cell has less than 2 or more than 3 living neighbours at time t, then it is dead at time $t + 1$. These are the *death by isolation* and *death by overcrowding* conditions respectively.

A *stable pattern* (or a *still life*) is a pattern that is not changed from time t to time $t + 1$. The question addressed in this chapter is: On a $N \times N$ section of the board, with all the rest of the board dead (the finite board restriction), what is the densest possible still life pattern?

The *density* of a stable pattern is the ratio of the number of live cells in the board. For example, the optimal solutions for the 7×7 board (Fig. 10.1) have 28 live cells.

They have a density of $\frac{28}{49} = 0.571$. As a formula, the density $d(N)$ for a board of size $N \times N$ is defined by $d(N) = a(N)/N^2$, where $a(N)$ is the number of live cells.

The MDSL problem is distinguished by the finite board restriction and by the search for the maximum density stable pattern. A separate but related concept in the Game of Life is exploring stable patterns of various interesting forms [6, 10].

Especially difficult is to prove that the maximum number of live cells has been found. That is, to prove that the highest density solution found for a given board size N is in fact the *maximal density* $D(N)$ for that size. An *optimal solution* (or optimal stable pattern) of size N is one that attains the maximum density $D(N)$.

An even more difficult problem is to count the number of *non-symmetric optimal solutions* for a given board size. Two optimal solutions are *symmetric* if one can be transformed into the other by rotating the board (through 90, 180, or 270 degrees) or by reflecting it horizontally, vertically, or diagonally. For example, the two optimal solutions for size $N = 7$ (Fig. 10.1) are not unique, as the one on the right is the reflection of the one on the left in the vertical axis. Solving the MDSL problem to *uniqueness* means counting all the unique optimal solutions; it is a different and harder problem to just finding (and proving) the maximal density. For the case of $N = 7$, there is only one unique optimal solution.

10.3 Constraint Programming

Constraints are a powerful and natural means of knowledge representation and inference in many disparate fields, both in industry and academia [12, 15]. Simple illustrative examples include: adjacent countries on the map cannot be coloured the same; the orders assigned to a slab of steel must not exceed the slab's capacity; and a warehouse should only be opened if it serves at least one store. This generality underpins the success with which constraint programming has been applied to disciplines such as scheduling, industrial design, and combinatorics.

A *constraint satisfaction problem* (CSP) is a set of decision variables, each with an associated domain of potential values, and a set of constraints. For example, the problem might be to fit components (values) to circuit boards (decision variables), subject to the constraint that no two components can be overlapping. An assignment maps a variable to a value from its domain. Each constraint specifies allowed combinations of assignments of values to a subset of the variables. A *solution* to a CSP is an assignment to all the variables which satisfies all the constraints. Solutions can be found for CSPs through a process of backtrack search with an inference step at each node.

During the search for a solution of a CSP, constraint *propagation* algorithms are used. These propagators make inferences, recorded as domain reductions, based on the domains of the variables that are constrained. If at any point these inferences result in any variable having an empty domain then search backtracks and a new branch is considered. Through this process of systematic search many problems can be tackled, including Maximum Density Still Life.

10.4 Constraint Programming Models for Still Life

In constraint programming, *modelling* is the process of representing a problem as a CSP. It is straightforward to form a simple CP model based closely on the conditions for a still life. This model has a binary variable x_{ij} for the cell in row i and column j of the $N \times N$ board: $x_{ij} = 0$ if the cell is dead and $x_{ij} = 1$ if it is live.

The three types of constraints echo the three rules of the game:

1. If the sum of the eight neighbouring squares of cell (i, j) equals 3, then $x_{ij} = 1$.
2. If the sum of the eight neighbouring squares of cell (i, j) is less than 2, then $x_{ij} = 0$.
3. If the sum of the eight neighbouring squares of cell (i, j) is greater than 3, then $x_{ij} = 0$.

The above model was proposed by Bosch and Trick [3]. However, the model is inefficient as very little propagation occurs between the variables. This means that solving the problem essentially comes down to an exhaustive search. In order to counter these inefficiencies, Smith [14] re-modelled the problem.

Re-modelling is one of the methods most used by CP practitioners to increase the efficiency of solving a problem. Commonly used practices include introducing *auxiliary variables* which aid propagation, and introducing *implied constraints*. In addition, changing the order in which variables and values are assigned during search can increase search efficiency.

Auxiliary variables are variables which do not need to be in the model for it to be a full representation of a given problem. Their inclusion, along with the inclusion of appropriate constraints embracing these variables, may link search variables that had either no direct constraints connecting them in the basic model, or not enough of a connection for propagation to be forthcoming.

In the basic model for Still Life the constraints between a cell variable x_{ij} and its eight neighbouring variables are not helpful in guiding search. In order to aid this situation Smith [14] suggested looking at the *dual translation* of a CSP with non-binary constraints into a binary CSP. The constraints of the original problem become the variables of the dual CSP; the domain of a dual variable is the set of values that satisfy the original constraints.

In Still Life, the dual variables represent a 3×3 square of cells from the original board, called a *super-cell*. Each super-cell variable y_{ij} corresponds to an x_{ij} cell variable and its eight neighbours. The value of a super-cell variable can be represented as a 9-bit number: $y_{i,j} = x_{i,j} + 2x_{i,j+1} + 4x_{i,j-1} + 8x_{i+1,j} + 16x_{i+1,j+1} + 32x_{i+1,j-1} + 64x_{i-1,j} + 128x_{i-1,j+1} + 256x_{i-1,j-1}$. The domain of y_{ij} is the set of all the allowable values under the original constraints for that super-cell. This domain is different for each of the four corners, the four sides, and the middle of the $N \times N$ board. These variables are added as auxiliary variables to increase the propagation between the underlying x_{ij} cell variables, rather than using them to replace the cell variables. This is a far better model than the simple one previously expressed in this section. However, by using more efficient solving methods and models, as described in the next section, solutions for larger board sizes can be found.

Fig. 10.2 The optimal solution for the 69 × 69 board (courtesy of [5])

10.5 Solving Still Life

Although simple to state, the MDSL problem is computationally challenging for any but the smallest sizes of board. Especially difficult is to prove that the maximum number of live cells has been found. The raw search space for a maximum density stable pattern on a board of $N * N$ is $\mathcal{O}(2^{N^2})$ in size.

Various approaches have been employed. Elkies [7] relaxed the finite board restriction, and applied graph-theoretic techniques for the case of the infinite board. It is known that the maximal density converges to $\frac{1}{2}$ as N tends to infinity.

The first systematic approach for the full, finite-board problem was by Bosch [1, 2], who employed *Integer Programming* (IP) techniques to the problem. The optimal solutions up to size $N = 7$ were already known within Game of Life folklore (Fig. 10.1) [2, 13, 16]. Bosch found and proved the maximal solution for board sizes up to $N = 10$.

Bosch and Trick extended this approach to a hybrid of IP and CP [3]. Their most successful approach found and proved the maximal density for board sizes up to $N = 15$.

Smith [14] developed a pure CP approach in which the straightforward CP model was reformulated as a *dual* representation of 3×3 'super-cells', as described earlier. Finding all unique, i.e., non-symmetric, solution patterns was pursued further by Petrie et al. [11]. They extended Smith's modelling approach with the IP–CP hybrid of Bosch and Trick, to solve to uniqueness the problem up to board size $N = 10$.

Larrosa et al. [9] employed a space-intensive method in order to reduce the time complexity to $\mathcal{O}(N^2 2^{3N})$. Their most successful approach, which combined *bucket elimination* and search, found and proved the maximal density for board sizes up to $N = 20$.

Chu et al. [5] returned to a pure CP approach. They reformulated the problem in terms of *wastage*. Instead of counting how many live cells there are in a partial solution, they count how much space has been 'wasted' by not packing in live cells tightly enough. This perspective enabled strong upper bounds to be obtained based on a theoretical analysis.

By combining powerful lookahead techniques based on static relaxations of the problem with selected incomplete search, Chu et al. [4, 5] solved or nearly solved[1] board sizes up to $N = 50$. These results have been extended, obtaining and proving the maximum density of boards for all N to within a window of 1 live cell. These best-known results for small N are given in Table 10.1.

10.5.1 Density as Board Size Increases

Although the maximal number of live cells $A(N)$ follows no obvious pattern as the board size N increases, the *wastage count* $W(N)$, as defined in [5], has been proven to grow linearly asymptotically with N. The two are related by $A(N) = \frac{N^2}{2} + N - \frac{W(N)}{4}$. Hence the maximal density $D(N)$ is of order $\frac{1}{2} + \frac{1}{N} - \frac{1}{4N}$. This formula implies that density will asymptotically become $\frac{1}{2}$ as N tends to infinity, just as predicted by theory [7], as Fig. 10.3 shows.

The optimal solutions for MDSL for large N fall into 54 modulus classes [4]. All board sizes for $N \leq 54$ are covered in the results given in Table 10.1. For $N \geq 55$, the asymptotic formula for $A(N)$ is proven to be

$$\left\lfloor \frac{N^2}{2} + \frac{17}{27}N - 2 \right\rfloor \leq A(N) \leq \left\lceil \frac{N^2}{2} + \frac{17}{27}N - 2 \right\rceil \tag{10.1}$$

thus either defining the maximal number of live cells exactly, or bounding it to within 1 cell.

---

[1]The found upper bound is either attained (i.e., instance is solved), or is attained by one fewer live cell (i.e., instance solved up to ± 1).

Table 10.1 Best known results for MDSL (from [4, 5]). *Bold* denotes cases which are not completely solved at the time of writing

N	Lower	Upper	N	Lower	Upper	N	Lower	Upper
1	0	0	21	232	232	41	864	864
2	4	4	22	253	253	42	907	907
3	6	6	23	276	276	43	**949**	**950**
4	8	8	24	301	301	44	993	993
5	16	16	25	326	326	45	1,039	1,039
6	18	18	26	352	352	46	**1,084**	**1,085**
7	28	28	27	379	379	47	**1,131**	**1,132**
8	36	36	28	**406**	**407**	48	1,181	1,181
9	43	43	29	437	437	49	1,229	1,229
10	54	54	30	466	466	50	1,280	1,280
11	64	64	31	497	497	51	1,331	1,331
12	76	76	32	530	530	52	**1,382**	**1,383**
13	90	90	33	563	563	53	1,436	1,436
14	104	104	34	598	598	54	**1,489**	**1,490**
15	119	119	35	632	632	55	1,545	1,545
16	136	136	36	668	668	56	1,602	1,602
17	152	152	37	706	706	57	**1,658**	**1,659**
18	171	171	38	**743**	**744**	58	1,717	1,717
19	190	190	39	782	782	59	1,776	1,776
20	210	210	40	824	824	60	**1,835**	**1,836**

10.6 Conclusion

Solving instances of the Maximum Density Still Life problem has taxed researchers in combinatorial optimization for more than a decade. New insights combined with powerful CP techniques have now almost entirely solved the MDSL problem, proving the maximum density to within 1 cell for all N and allowing arbitrarily large optimal solutions to be constructed (Fig. 10.2).

The Still Life problem has stimulated developments in hybrid techniques, symmetry breaking, and constraint programming. The advances in known solutions have come more from improved approaches than from increased computational power. Once more, the complexity and richness underlying the simplicity of Conway's Game of Life is seen.

Acknowledgements   For helpful comments on earlier versions of parts of this chapter, we thank particularly Ian Gent, Mike Trick, Peter van Beek, and especially Barbara Smith. Peter J. Stuckey prepared the solution for the 69 × 69 board with help from Michael Wybrow. Parts of the work described were performed when some of the authors were at the University of Huddersfield and Imperial College London.

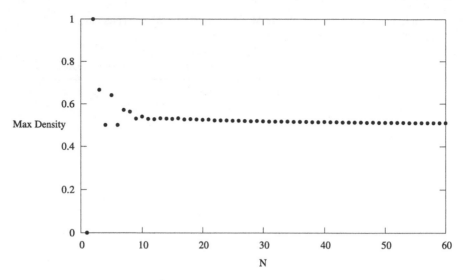

Fig. 10.3 Density converges to $\frac{1}{2}$ as the board size increases

References

1. Bosch, R.: Integer programming and Conway's Game of Life. SIAM Rev. **41**(3), 596–604 (1999)
2. Bosch, R.A.: Maximum density stable patterns in variants of Conway's Game of Life. Oper. Res. Lett. **27**, 7–11 (2000)
3. Bosch, R., Trick, M.: Constraint programming and hybrid formulations for three Life designs. Ann. Oper. Res. **130**, 41–56 (2004)
4. Chu, G., Stuckey, P., de la Banda, M.G.: A complete solution to the Maximum Density Still Life problem (in preparation)
5. Chu, G., Stuckey, P., de la Banda, M.G.: Using relaxations in Maximum Density Still Life. In: Proc. of Fifteenth Intl. Conf. on Principles and Practice of Constraint Programming (CP'09), pp. 258–273 (2009).
6. Cook, M.: Still life theory. http://www.paradise.caltech.edu/~cook/Warehouse/index.html. Accessed September 2005
7. Elkies, N.D.: The still-life density problem and its generalizations. In: Voronoi's Impact on Modern Science, Book I. Proc. of the Institute of Mathematics of the National Academy of Sciences of Ukraine, vol. 21, pp. 228–253 (1998)
8. Gardner, M.: The fantastic combinations of John Conway's new solitaire game. Sci. Am. **223**, 120–123 (1970).
9. Larrosa, J., Morancho, E., Niso, D.: On the practical use of variable elimination in constraint optimization problems: 'still-life' as a case study. J. Artif. Intell. Res. **23**, 421–440 (2005)
10. McIntosh, H.V.: Life's still lifes. http://delta.cs.cinvestav.mx/~mcintosh/newweb/marcostill.html (September 1988). Accessed September 2005
11. Petrie, K.E., Smith, B.M., Yorke-Smith, N.: Dynamic symmetry breaking in constraint programming and linear programming hybrids. In: Proc. of 2nd European Starting AI Researcher Symposium (STAIRS'04), pp. 96–106 (2004)
12. Rossi, F., P. van Beek, Walsh, T.: Handbook of Constraint Programming. Foundations of Artificial Intelligence, vol. 2. Elsevier, New York (2006)
13. Silver, S.: Dense stable patterns. http://www.argentum.freeserve.co.uk/dense.htm. Accessed September 2005

14. Smith, B.M.: A dual graph translation of a problem in 'Life'. In: Proc. of Eighth Intl. Conf. on Principles and Practice of Constraint Programming (CP'02), pp. 402–414 (2002)
15. Wallace, M.G.: Practical applications of constraint programming. Constraints 1(1/2), 139–168 (1996).
16. Weisstein, E.: Density. In: Eric Weisstein's Treasure Trove of Life C.A. http://www.ericweisstein.com/encyclopedias/life/Density.html. Accessed September 2005

Part III
Asynchronous, Continuous and Memory-Enriched Automata

Chapter 11
Larger than Life's Extremes: Rigorous Results for Simplified Rules and Speculation on the Phase Boundaries

Kellie Michele Evans

Larger than Life (LtL), is a four-parameter family of two-dimensional cellular automata that generalizes John Conway's *Game of Life* (Life) to large neighborhoods and general birth and survival thresholds. LtL was proposed by David Griffeath in the early 1990s to explore whether Life might be a clue to a critical phase point in the threshold-range scaling limit. The LtL family of rules includes Life as well as a rich set of two-dimensional rules, some of which exhibit dynamics vastly different from Life. In this chapter we present rigorous results and conjectures about the ergodic classifications of several sets of "simplified" LtL rules, each of which has a property that makes the rule easier to analyze. For example, these include symmetric rules such as the threshold voter automaton and the anti-voter automaton, monotone rules such as the threshold growth models, and others. We also provide qualitative results and speculation about LtL rules on various phase boundaries and summarize results and open questions about our favorite "Life-like" LtL rules.

11.1 Introduction and History

John Conway's *Game of Life* (Life) is the most famous example of a cellular automaton (CA), due in part to the fact that its update rule is very simple yet it generates extremely complicated dynamics [2, 15, 16]. For Life, as is the case with most CA rules, the initial state has an enormous impact on the resulting dynamics. When Life is started from a random initial state with an appropriate density of occupied cells, complex structures emerge. For example, *gliders* appear and their trajectories take them across the infinite lattice (see Fig. 11.1). If not stopped by some other Life pattern, they will walk on forever.

The glider is Life's most commonly occurring *spaceship*, which is a finite pattern that reappears (without additions or losses) after a number of generations and is displaced by a nonzero distance [26]. Spaceships are able to carry *information* across long spatial distances. This ability leads to interesting *reactions* when spaceships collide with other *coherent structures*. Some of these reactions are crucial ingredients in the proof that Life is *universal* [2]. New reactions and their outcomes

A. Adamatzky (ed.), *Game of Life Cellular Automata*,
DOI 10.1007/978-1-84996-217-9_11, © Springer-Verlag London Limited 2010

Fig. 11.1 The glider's
trajectory starting in the
northwest and moving along
the diagonal to the southeast

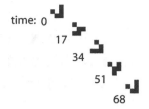

continue to be designed and tested by a group of "Life enthusiasts" who are fascinated by the many challenging open problems posed by Life. For example, various
missing details in Conway's proof that Life is universal have been constructed (*e.g.*
[3, 22, 24]). Some of these results are posted regularly to the website [27] and the
relatively new "Life wiki" has been created to catalogue and regularly update what
is known about Life [29].

The complex interactions among Life's coherent structures make it difficult to
classify the rule's *global dynamics*, meaning the limiting dynamics of the rule on an
infinite system starting from a random initial state. Various researchers have done
empirical studies that resulted in different answers [16]; nevertheless, Life seems
to be "between" rules whose global dynamics are *periodic*, meaning every site on
the lattice is eventually periodic with probability one or *aperiodic*, meaning the
sequence of 0's and 1's that occurs at each site of the lattice never cycles with probability one. Life's apparent location on a phase boundary led David Griffeath in the
early 1990s to wonder whether Life might be a clue to a critical phase point in the
threshold-range scaling limit. In order to explore this question, he proposed *Larger
than Life* (LtL), which is a four-parameter family of two-dimensional cellular automata that generalizes Life to large neighborhoods and general birth and survival
thresholds [19].

The LtL family of rules was also motivated by two mathematical prototypes for
continuous time models of population growth: the *Lotka–Volterra model*, which is
the simplest model of predator–prey interactions and the *logistic map*, which is a
simple population growth model. LtL is the simplest CA analogue of these famous
nonlinear models. That is, LtL is a spatial version of nonlinear population dynamics
in which both space and time are discrete.

The LtL family of rules includes Life as well as a rich set of two-dimensional
rules, some of which exhibit dynamics vastly different from Life [9]. For example,
the *threshold growth models* (TGM), the *threshold voter automaton* (TVA), and
the *anti-voter automaton* (AVA) all comprise subsets of LtL parameter space. The
update rules for the TGM depend on a fixed threshold and occupied sets can only
grow [16, 17]. For the TVA a site changes state iff its neighborhood contains *at least*
τ sites of the opposite state, where τ is a fixed threshold [6]. For the AVA, a site
changes state iff its neighborhood contains *at most* τ sites of the opposite state,
where τ is a fixed threshold. A family of "Life-like" rules comprises a much larger
subset of LtL parameter space than initially imagined by Griffeath [10, 11]. Like
Life, these rules appear to "self-organize" over time and coherent structures emerge.
Gravner and Griffeath describe the behavior of this type of CA, which includes Life,

as being, "so exotic that, at least for now, empirical study seems the only avenue to understanding" [18].

The most popular CAs to study are those which have some "nice" property, making them amenable to fairly standard mathematical analysis. For example, in the case of two-state CAs, such as those in the LtL family, there are rules which are symmetric in the states. That is, the conditions for changing from each state to the other are the same. Examples include the TVA and the AVA. Another fairly understandable class of two state rules are those which are monotone in live sites. This means that if A and B are sets of live sites and $A \subset B$ then after one time step, the set of live sites generated from having A as an initial state will be contained in the set of live sites generated if B is the initial state. The TGM is monotone nondecreasing in live sites. Another simplifying feature is to work with small neighbor sets such as the von Neumann neighborhood; in that case, each site in the grid sees its neighbors to the north, south, east, and west.

Most LtL rules have no such regularity properties. As a consequence, the mathematics behind many of the most intriguing LtL rules is rather impenetrable. It is therefore exceedingly difficult to obtain rigorous results for them. The question thus arises: Should we study only the restricted class of rules that have sufficient regularity, so that rigorous results are within reach? If we were to take this stance then for starters we would miss out on Life itself. Ever faster clock speeds on personal computers enable us to make ever greater empirical progress in understanding how such systems behave. At present, however, mathematically rigorous methods are not growing as quickly as the technological advances. As mathematicians, we are conflicted: do we stop seeking nonlinearity and rules that attempt to model real-life problems, for the sake of rigor, or do we seek complexity and consequently work harder to attain rigor? In this chapter, we attempt to do both. That is, we provide rigorous results and conjectures about the ergodic classifications of several sets of "simplified" LtL rules, each of which has a property that makes the rule easier to analyze. For example, these include symmetric rules such as TVA and AVA, monotone rules such as TGM, and others. We also provide qualitative results and speculation about LtL rules on various phase boundaries and summarize results and open questions about our favorite "exotic" LtL rules.

11.2 Larger than Life: Definition and Notation

Larger than Life (LtL) generalizes Life to large neighborhoods and general birth and survival thresholds as follows: Each site of the *two-dimensional lattice* $\mathbf{Z}^2$ is in one of two *states*, *live* (1) or *dead* (0). This is the *initial configuration* of the system. The *neighborhood* $\mathcal{N}$ of a site consists of the $(2\rho + 1) \times (2\rho + 1)$ sites in the box surrounding and including it. That is, the neighborhood of the origin is $\mathcal{N} = \{y \in \mathbf{Z}^2 : \|y\|_\infty \leq \rho\}$ (ρ a natural number), so that its translate $x + \mathcal{N}$ is the neighborhood of site $x \in \mathbf{Z}^2$. $\mathcal{N}$ is called the *generalized Moore* or *"range ρ" box* neighborhood. Each *time step*, all of the sites *update* (meaning change states or not) simultaneously according to the deterministic LtL *rule*, which in words is:

- **Birth:** A site that is dead at time t will become live at time $t + 1$ if and only if the number of live sites in its neighborhood at time t is in the closed interval $[\beta_1, \beta_2]$, $0 \leq \beta_1 \leq \beta_2$.
- **Survival:** A site that is live at time t will remain live at time $t + 1$ if and only if the number of live sites in its neighborhood (itself included) at time t is in the closed interval $[\delta_1, \delta_2]$, $1 \leq \delta_1 \leq \delta_2$.
- **Death:** A site that is dead at time t and does not become live at time $t + 1$ will remain dead at time $t + 1$. A site that is live at time t and does not remain live at time $t + 1$ will become dead at time $t + 1$.

Let us define the rule and notation precisely.

- Let $\mathscr{T}$ denote the CA rule. That is, $\mathscr{T} : \{0, 1\}^{\mathbf{Z}^2} \mapsto \{0, 1\}^{\mathbf{Z}^2}$.
- Let $\xi_t(x) \in \{0, 1\}$ denote the state of the site $x = (x_1, x_2) \in \mathbf{Z}^2$ at time t.
- Let ξ_t denote the system at time t. The collection of 1's in ξ_t comprises some set Λ, which is contained in $\mathbf{Z}^2$. As is customary in this area we will confound this configuration, consisting of all 1's on Λ, with the set Λ itself. Hence, if $\Lambda = \{x \in \mathbf{Z}^2 : \xi_t(x) = 1\}$, we write $\xi_t = \Lambda \subset \mathbf{Z}^2$. Suppose that the initial set of 1's, ξ_0, lies on the set Λ and that everywhere else (on Λ^c) there are 0's. Then we write $\xi_0 = \Lambda$. We use $\xi_t^\Lambda = B$ to mean that starting with $\xi_0 = \Lambda$ and updating t time steps yields a set of 1's that lies on the set B. In other words, $\xi_t^\Lambda = \mathscr{T}^t(\Lambda) = B$.
- The update rule for Larger than Life is given by:

$$\xi_{t+1}(x) = \begin{cases} 1 & \text{if } \xi_t(x) = 0 \text{ and } |(x + \mathcal{N}) \cap \xi_t| \in [\beta_1, \beta_2] \\ & \text{or} \\ & \text{if } \xi_t(x) = 1 \text{ and } |(x + \mathcal{N}) \cap \xi_t| \in [\delta_1, \delta_2]; \\ 0 & \text{otherwise.} \end{cases}$$

Thus, if $\Lambda \subset \mathbf{Z}^2$ is a set of $1's$ (on a background of $0's$), then the mapping of the LtL rule $\mathscr{T}$ is defined by $\mathscr{T}(\Lambda) = \{x \in \Lambda^c : \beta_1 \leq |(x + \mathcal{N}) \cap \Lambda| \leq \beta_2\} \cup \{x \in \Lambda : \delta_1 \leq |(x + \mathcal{N}) \cap \Lambda| \leq \delta_2\}$.

The LtL cellular automata form a four-parameter family of rules indexed by the endpoints of the intervals which determine each rule: β_1, β_2, δ_1, and δ_2. As such, LtL can be viewed as a four-dimensional hyperspace with points $(\beta_1, \beta_2, \delta_1, \delta_2)$ representing 2-dimensional cellular automaton rules.

In what follows, the cardinality, $|\cdot|$ of $\mathcal{N}$ is $|\mathcal{N}| = (2\rho + 1)^2$ since $\mathcal{N}$ is a box with side length $2\rho + 1$. When discussing a particular rule, we will write it as $(\rho, \beta_1, \beta_2, \delta_1, \delta_2)$. Thinking about LtL this way yields a five-parameter family of rules. However, we usually fix the range and cruise around the hyperspace that it determines thus dealing with a four-parameter family. In this framework, Life has parameters $(\rho, \beta_1, \beta_2, \delta_1, \delta_2) = (1, 3, 3, 3, 4)$.

11.3 Initial States

Since the update rule of a CA is deterministic, once the initial state is set its eventual dynamics are also set. Thus, if one is interested in capturing the dynamics that a particular CA would generate on an infinite system, it is essential to use an appropriate

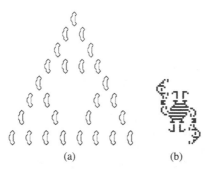

(a)   (b)

Fig. 11.2 (**a**) One-dimensional replicator for LtL rule $(5, 3, 3, 3, 3)$. Times $4k$, $k = 0, 1, 2, \ldots, 7$ of the space–time diagram are shown with time 0 depicted in the *top row*. The space–time diagram actually updates along a diagonal, but has been rotated here to illustrate its connection with Pascal's triangle mod 2. (**b**) Life's Max

initial state. We are interested in the possible dynamics of LtL rules starting from both random and deterministic initial states. Our choices for random initial states may be either translation invariant or non-translation invariant. The former means that if the configuration on the lattice is translated, or shifted, it has the same statistical properties that it had prior to translation. The lack of regularity in the latter usually makes analysis hopelessly complicated. Thus, since the dynamics generated by LtL rules are inscrutable enough to begin with, we typically restrict our attention to translation invariant initial states. We often restrict our attention even further to product measures, because they represent a fairly rich subset of all translation invariant initial states and, in many contexts, capture the ergodic theory of the more general translation invariant states. Since our rules have only two states, the product measures, μ_p, that we use consist of independent, identically distributed random variables that take on the value 1 with probability p ($0 < p < 1$) and 0 with probability $1 - p$. Such initial states are also convenient because they are relatively easy to compute with and well-suited to simulations.

We are also interested in the time evolutions of collections of finite deterministic initial configurations. In some cases, such interest is motivated by the exploration of a rule started from a random initial state. A good example of this is the development of our study of *replicators* [12]. These are configurations of live sites whose space–time diagrams are isomorphic to Pascal's Triangle Mod 2 (see Fig. 11.2a for an example). The first replicator we found emerged from a random initial configuration. However, understanding the replicator's space–time trajectory required performing experiments that started from finite deterministic initial configurations. Understanding the latter increased our knowledge of what the infinite system had the potential to do when started from a translation invariant initial state. In other cases, the motivation began with the deterministic initial state. For example, we studied the behavior of a set of rules started from finite lattice circles. One such rule, known as *Life Without Death* (LWoD), grew a complex snowflake-like pattern. The implication was that, started from a random initial configuration in the infinite system, the rule's limiting state would be spatially aperiodic. (Griffeath and Moore have since proved that LWoD is **P**-complete [21]. See Sect. 11.6.6 for further discussion and a range 2 LtL example with some of LWoD's properties.) Thus, although the behavior of CA dynamics which are started from deterministic and translation invariant initial configurations pose distinct problems, they are very much intertwined.

Due to the nonlinearity of most LtL rules, when conducting experiments on a finite lattice starting from product measure, the choice of an appropriate density, p, becomes an art form. This is because an initial state with too high (or low) a density of live sites might yield total death and/or the impossibility of birth due to overcrowding (respectively loneliness). However, an initial state with just the right density of live sites might yield complex dynamics. Empirical results indicate that there is an interval, $\mathscr{I} \subset [0, 1]$ such that if $p \in \mathscr{I}$ then the limiting state of ξ_t run on product measure with density p will be nontrivial (if such a nontrivial limiting state exists at all). Otherwise, the dynamics will yield a trivial limiting state even if a nontrivial one exists. We are interested in all *possible* dynamics and so seek the appropriate p (if it exists) that gives rise to these nontrivial limiting states.

We add the caveat that sometimes computer experimentation using random initial configurations indicates that a particular rule has only the trivial limiting state, when in fact there exists a finite configuration that gives rise to a nontrivial limiting state. In such a situation, the monkeys-at-the-typewriter principle implies that somewhere out there in an infinite random initial configuration (a la primordial soup), there will exist a configuration that gives rise to the nontrivial limiting state. Hence, for each rule we seek not only the appropriate p for a random initial state, but *any* initial configuration that yields a nontrivial limiting state.

There are also situations for which it is easy to find a p that gives rise to a nontrivial limiting state. However, the resulting dynamics may mask interesting local behavior. For example, there may exist configurations that survive on backgrounds of all 0's, but get completely destroyed by the dynamics that emerge from random initial conditions. In some situations these configurations are of little use when discussing the behavior of the infinite system, but they can be interesting in their own right. (See Sect. 11.6.3 for an example.) In other situations, they may be used to demonstrate the existence of a nontrivial limiting state with a much higher density of live sites than that which arises when the system is started from a random initial configuration. For example, Life has a finite initial pattern, named Max (Fig. 11.2b), that generates a density $1/2$ infinite still life (or *fixed* pattern; see Sect. 11.4.1 for a precise definition). The infinite still life is unstable on a background other than all 0's (see [8] for details).

Bosco's rule which is a quadratic scaling of Life to LtL's range 5 [13] and has LtL parameters $(5, 34, 45, 34, 58)$ also supports an infinite still life (or still life measure) which has density $1/2$. This still life consists of alternating stripes of live and dead sites, each with width 5 and is a scaled up version of Life's density $1/2$ infinite still life. It is easy to show that this infinite pattern is unstable. For example, if a radius 2 circle of live sites is inserted in the center of a 349×371 copy of the pattern with periodic boundary conditions and the rule updates, the pattern is quickly destroyed. By time 34 the center pattern has been destroyed and by time 300 there is no evidence that the pattern of stripes was ever there (see Fig. 11.3). By time 564

Fig. 11.3 From *left* to *right*: still life with density 1/2 supported by Bosco's rule after a radius 2 circle of live sites has been inserted in the figure's center and run for 34 time steps, 134 time steps, and 300 time steps, respectively (using periodic boundary conditions). The pattern of stripes is completely destroyed by time 300 and by time 809 (not shown), the eventual state of 4 oscillators and 6 still lifes is reached

(not shown), only 4 *oscillators*, 4 finite *still lifes* and 2 *bugs* remain (see Sect. 11.4.1 for definitions). Each bug winds up colliding with a still life at time 750 so that by time 809 (not shown), the eventual state of 4 oscillators and 6 still lifes is reached.

We claim that there exists a finite initial pattern like Life's Max that will generate Bosco's density 1/2 still life. The reader is encouraged to find it (or prove us wrong).

The dependence on initial conditions illustrated by the previous example is one illustration that finite and infinite nonlinear systems can behave very differently, especially when the dynamics are non-monotone. The instability of the alternating stripes of 1's and 0's should further illustrate the unlikely event that one would see that configuration emerge from a random initial state on such a finite system as one's computer screen.

11.4   LtL Definitions

In this section, we present two sets of definitions which will be used in Sects. 11.5 and 11.6. The first definitions are for LtL configurations that are essentially local. A *local* configuration is a set of sites that is periodic under some LtL rule but that may be unstable when put on backgrounds other than all 0's. In many cases, the existence of local configurations has nothing to do with the behavior of the infinite system. This is because there may be global dynamics that take over and destroy all local configurations. The second set of definitions addresses the global ergodic behavior of LtL rules when started from translation invariant initial configurations. That is, what is the tendency of the dynamics? For example, do they destroy all local configurations, or are local configurations the only sets of 1's that survive in the long run?

These definitions arose through the study of the LtL family of rules. However, they apply to any two-state CA rules. When the definitions overlap with already

(a)   (b)   (c)

Fig. 11.4 (**a**) Range 5 still life supported by LtL rule $(5, 42, 61, 28, 61)$. (**b**) Range 5 crystal supported by LtL rule $(5, 42, 61, 28, 61)$. (**c**) Range 2 periodic object with $n = 4$ supported by rule $(2, 6, 9, 6, 7)$. From *left* to *right* are times 0, 1, 2, 3

Fig. 11.5 Range 100 strobe with period $n = 33$ supported by rule $(100, 11{,}018, 16{,}360, 11{,}018, 18{,}197)$. From *left* to *right* are times 0, 1, 2, 3, and 4. Not all phases are depicted due to their large sizes

defined Life patterns, we attempt to use the Life terminology from the *Life lexicon* [26]. In a few cases, the names are slightly adapted in order to bring out characteristics specific to LtL. For the following, let $\mathscr{T}$ be a CA rule from the LtL family and ξ_t the system at time t.

11.4.1 Local Configurations

First we need the following notion that was introduced by Gravner and Griffeath in [16]. Say that a set $\Lambda \subset \mathbf{Z}^2$ is *completely symmetric* if it is invariant with respect to switching the sign of the ith coordinate and transposing the ith and jth coordinates $(i, j = 1, 2, \ldots, d)$. For $\Lambda \subset \mathbf{Z}^2$ this just means that Λ is symmetric with respect to the coordinate axes and 45 degree lines.

For each of the following definitions assume that $\Lambda \subset \mathbf{Z}^2$ is a set of 1's. For each definition we include an example along with the LtL rule that admits it.

- A *still life* is a configuration Λ which is fixed under $\mathscr{T}$. That is, $\mathscr{T}^t(\Lambda) = \Lambda$ for all $t \geq 0$. See Fig. 11.4a.
- A *crystal* is a completely symmetric finite still life. See Fig. 11.4b.
- A *periodic object* is a finite configuration Λ which is not a still life, but for which there exists a positive, finite integer n so that $\xi_t(x) = \xi_{t+n}(x)$ for all $t \geq 0$ and all $x \in \Lambda$. The smallest such n is called the *period* of Λ. See Fig. 11.4c. A *blinker* is a periodic object with period 2.
- A *strobe* is a completely symmetric periodic object. See Fig. 11.5.
- Suppose Λ is a finite configuration of 1's for which there exists a finite time, τ, and a vector, v, such that $\mathscr{T}^\tau(\Lambda) = \Lambda + v$. If v is some multiple of e_1, e_2, or $e_1 + e_2$ call Λ a *bug* (e_1 and e_2 are the unit vectors $[1, 0]$ and $[0, 1]$, respectively). Otherwise, call it a *disoriented bug*. Λ has period τ, mod translation, in

the direction of v. If $\tau = 1$, then the bug is *invariant* mod translation. Bugs are synonymous with Life's spaceships; they are so-named due to their distinct geometries, which will be discussed further in Sect. 11.6.9.

There are various bug varieties, including *disoriented bugs* and *jitter bugs*. These are defined in [10] along with examples. *Bug makers*, which are analogous to Life's *glider guns* are discussed in Sect. 11.6.8. Other local configurations include *replicators*; one example is depicted in Fig. 11.2a. For a definition, additional examples and theory see [12].

11.4.2 Ergodic Classifications for the Infinite System

In our extensive experimentation with LtL rules, we have discovered limiting states that can be described using fairly standard, quantitative definitions. We present those definitions here. We have also discovered more complex limiting states that are difficult to quantify. We give qualitative definitions for some of those in Sect. 11.6.

- *Global death* is almost sure convergence to the limiting state of all 0's. That is, $\lim_{t\to\infty} \xi_t(x) = 0$ for all $x \in \mathbf{Z}^2$ with probability one. Denote this state by $\underline{0}$.
- *Global survival* is almost sure convergence to the limiting state of all 1's. That is, $\lim_{t\to\infty} \xi_t(x) = 1$ for all $x \in \mathbf{Z}^2$ with probability one. Denote this state by $\underline{1}$.
- ξ_t *fixates* if for each x, $\xi_t(x)$ eventually has period 1 in t, with probability one. That is, $\lim_{t\to\infty} \xi_t(x)$ exists for all x, so each site changes state only finitely many times.
- ξ_t *is periodic* if for each x, $\xi_t(x)$ is eventually periodic in t with probability one. That is, for each x, there is a positive finite integer n and a large N such that $\xi_t(x) = \xi_{t+n}(x)$ for all $t > N$. We point out that any finite system is periodic. This holds since there are only a finite (perhaps very large) number of possible configurations the system can attain. Thus, eventually the system will comprise some configuration for a second time. Since the dynamics are deterministic, the system will cycle as it did after the first time it passed through that particular configuration. We are working with a two state system so a universe consisting of n sites admits 2^n possible configurations. This is the upper bound on the period of the system. Needless to say, we are mainly interested in cases when the infinite system is periodic.
- ξ_t *is uniformly periodic* if it is periodic and the eventual period of each site is the same. To distinguish this from a system that fixates, we assume additionally that the common eventual period is strictly greater than one. If the system is periodic and the eventual periods of the sites vary, we say that it is *nonuniformly periodic*. If a limiting state is periodic, either uniformly or nonuniformly, the state of each site settles into a periodic orbit eventually. Local periodicity additionally requires that correlations between distinct sites be uniformly controlled in space and time. In other words,
- ξ_t *is locally periodic* if it is periodic, and $\inf_{t, x \neq y, a, b} P(\xi_t(x) = a, \xi_t(y) = b) > 0$.

- ξ_t *is uniformly locally periodic* if it is locally periodic and the common eventual period of the sites is the same and strictly greater than one. If the system is locally periodic and the eventual period of each site varies, we say that it is *nonuniformly locally periodic*.
- ξ_t *generates aperiodic dynamics* if it is not periodic. Such rules, though deterministic, behave like traditional stochastic processes. Specifically, with probability one, for any site x in the system, the sequence of 0's and 1's that occurs at that site (*i.e.* $\{\xi_t(x)\}$, $t = 0, 1, 2, \ldots$) never cycles.
- ξ_t *is clustering* if for each x and y in a finite box, $\lim_{t \to \infty} P(\xi_t(x) = \xi_t(y)) = 1$. In words, there is a clustering of the trivial phases of all 0's or all 1's. That is, given any finite box B the probability that all $x \in B$ are in the same state goes to 1 as $t \to \infty$.

11.5 LtL Rules with Simplifying Features

In this section, we describe regions of LtL space where the rules have some sort of regularity property. In several cases, some of the LtL rules are well known. In those cases, we interpret the known rules in terms of LtL parameters and cite known results. We then perturb some of the known results to larger LtL subspaces, establish new results, and make conjectures about nearby LtL rules.

11.5.1 Symmetric LtL Rules

A simplifying feature for a two state CA rule is *symmetry* in the states. This means that the conditions for switching from each state to the other are the same. In the case of LtL rules, this means that the conditions for birth are the same as the conditions for death. Since LtL rules are defined by connected intervals for which birth and survival take place, it is easy to check that there are only two sets of symmetric LtL rules. These are the sets of rules for which both the birth and survival intervals are at one extreme or the other. That is, these are the range ρ LtL rules with parameters that satisfy:

1. for $\tau \geq 1$, $\beta_1 = \tau$, $\beta_2 = |\mathcal{N}| - 1$, $\delta_1 = |\mathcal{N}| - (\tau - 1)$, and $\delta_2 = |\mathcal{N}|$ or
2. for $\tau \geq 0$, $\beta_1 = 0$, $\beta_2 = \tau$, $\delta_1 = 1$, and $\delta_2 = |\mathcal{N}| - (\tau + 1)$.

By (1), a site switches state if there are *at least* τ sites of the other state in its neighborhood. On the other hand, by (2), a site changes state if it sees *at most* τ sites of the other state in its neighborhood. In that sense, these rules are complementary. (1) is called the *threshold voter automaton* (TVA) and (2) is the *anti-voter automaton* (AVA).

Fig. 11.6 When the LtL rule (5, 0, 30, 1, 80) updates on an initial radius 49 circle of dead sites on a background of 1's, sites in the connected regions of 1's and 0's flip flop every time step, except on the boundary, where some sites become fixed

Threshold Voter Automaton and Anti-voter Automaton

The TVA is typically defined as follows: The *threshold voter automaton* is a deterministic discrete time process in which, at each time t, the voter at x examines the opinions of her neighbors in $x + \mathcal{N}$, and changes her state iff at least τ neighbors have the opposite opinion [6]. As may be clear from the name, in the TVA switching states is thought of as switching political parties (assuming only *two* political parties; see [19] for a discussion of the multi-party case).

Durrett and Steif have proved a theorem and posed several conjectures about the TVA using the neighborhoods $\mathcal{N} = \{y : \|y\|_p \leq \rho\}$, $p \in [1, \infty]$, where $\|y\|_p = (|y_1| + |y_2| + \cdots + |y_d|)^{1/p}$, $p \in [1, \infty)$, $\|y\|_\infty = \sup_i |y_i|$ [6]. We are interested in the two-dimensional case ($d = 2$) with range ρ box neighborhoods ($p = \infty$). As such, Durrett and Steif's results apply to range ρ LtL rules with parameters given above in (1). (See [9] for a summary of Durrett and Steif's results interpreted in terms of these LtL parameters.)

On the other hand, the AVA interpreted in terms of range ρ LtL rules has parameters given in (2) above. The AVA can be interpreted as: a voter changes her state iff *at most* τ of her neighbors have the opposite opinion. For this rule, we make three conjectures. These are based on empirical data as well as comparison with Durrett and Steif's TVA results. The first conjecture is the AVA analogue to a TVA conjecture made by Durrett and Steif.

Conjecture 1 Given the range ρ LtL rule with parameters $\beta_1 = 0$, $\beta_2 = \tau$, $\delta_1 = 1$, and $\delta_2 = |\mathcal{N}| - (\tau + 1)$ ($\tau \geq 0$):

1. starting from any initial configuration, the system is uniformly locally periodic if $\tau \geq (|\mathcal{N}| - 1)/2$;
2. starting from product measure with density $p \in [3/10, 3/5]$ the system fixates if $\tau \leq (|\mathcal{N}| - 1)/4$.

In the previous conjecture, for $\tau \leq (|\mathcal{N}| - 1)/4$, if the system begins with product measure where the initial density is very low or very high then all sites quickly become the same state at the same time and then flip flop from live to dead and back again ad infinitum. For other initial states, the system can become locally periodic, with large connected regions flip flopping from live to dead and back again and some sites on the boundaries of these regions fixating. See Fig. 11.6 for an example.

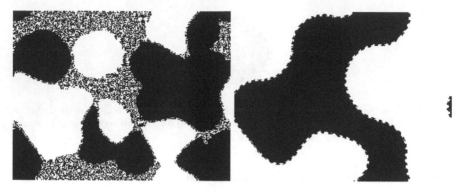

Fig. 11.7 From *left* to *right* are times 10 and 53 of rule $(5, 0, 50, 1, 70)$ started from product measure with density $1/2$ on a 256×200 lattice with periodic boundary conditions

Fig. 11.8 Pattern is fixed for LtL rule $(5, 0, 50, 1, 70)$

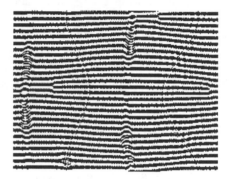

For the remaining values of τ in the previous conjecture, we claim that the system either fixates or is locally periodic, depending on the initial state. For example, starting from product measure with density $p = 1/2$ the LtL rule $(5, 0, 50, 1, 70)$ creates patches of 1's and 0's, both of which then oscillate from 0's to 1's and back again and these connected regions of one state grow. The boundaries between the solid patches of 1's and 0's remain jagged unless one of the colors fills the lattice. Some sites on the boundaries are fixed while all other sites flip flop from live to dead and back again ad infinitum. We leave it as a little exercise for the reader to determine the eventual state of Fig. 11.7. Does one color take over so that all sites eventually flip flop ad infinitum? Or do the two colors coexist, each becoming the other infinitely often (except for some boundary sites, which may become fixed)? Given the right initial configuration, this rule also has the ability to fixate (see Fig. 11.8).

The next two conjectures about the AVA are about sites that are eventually period two or that fixate in the same state; both require the range to go to ∞.

Conjecture 2 Let A_k be the event that for each x with $\|x\|_2 \leq k$, $\xi_t(x)$ eventually has period 2 in t. Suppose $\tau \geq (|\mathcal{N}| - 1)/2$. If we start from product measure with density $1/2$ then for all k, $P(A_k) \to 1$ as $\rho \to \infty$.

Fig. 11.9 LtL rule $(5, 49, 120, 71, 121)$ starting from product measure with density $p = 9/19$ on the *left* and density $p = 99/200$ on the *right*. In both cases, all sites are fixed except those on the boundary between live and dead sites, which are period 2. Observe that the majority of sites on the *left* are dead, while the majority on the *right* are live

Conjecture 3 Let C_k be the event that all sites x with $\|x\|_2 \leq k$, $\xi_t(x)$ fixate in the same state. Suppose $\tau = \theta(|\mathcal{N}| - 1)$ with $\theta \in [0, 1/4]$. If we start from product measure with density $1/2$ then for all k, $P(C_k) \to 1$ as $\rho \to \infty$.

LtL Rules near the TVA or AVA: Fixation and Local Periodicity

Now we extend Durrett and Steif's TVA results and the AVA conjectures in the previous subsection to near-by LtL rules. The first conjecture is about LtL rules near the TVA.

Conjecture 4 Suppose the range ρ LtL rule has parameters $\beta_2 = |\mathcal{N}| - 1$ and $\delta_2 = |\mathcal{N}|$. Then starting from any random initial configuration, the system:

1. is uniformly locally periodic if $\beta_1 \leq (|\mathcal{N}| - 1)/4$ and $\delta_1 \geq (3/4)(|\mathcal{N}| - 1) + 2$;
2. fixates if $\beta_1 \geq (|\mathcal{N}| - 1)/2 + 1$ and $\delta_1 \leq (|\mathcal{N}| + 1)/2$.

For the remaining values of β_1 and δ_1, experimental results suggest that the system either fixates or is locally periodic, depending on the initial configuration. For example, the LtL rule $(5, 49, 120, 71, 121)$, seems to converge to $\underline{1}$ starting from product measure with density $p \geq 1/2$. However, starting from product measure with density $p = 9/19$ on a 256×200 lattice, the system can fixate (the limiting state converges to $\underline{0}$) or become locally periodic. The image on the left of Fig. 11.9 is fixed everywhere but on the boundary between live and dead sites; those sites are period two. Similarly, starting from product measure with density $p = 99/200$ on a 256×200 lattice, the system can fixate (the limiting state converges to $\underline{1}$) or become locally periodic. The image on the right of Fig. 11.9 is fixed everywhere but on the boundary between live and dead sites; those sites are period two.

These rules are also interesting visually, for example see Fig. 11.10. For additional graphics (some with color), visit [7].

The next conjecture is about LtL rules near the AVA.

Conjecture 5 Suppose the range ρ LtL rule has parameters $\beta_1 = 0$ and $\delta_1 = 1$.

Fig. 11.10 Starting from a radius 49 circle of live sites, the LtL rule (5, 30, 120, 92, 121) becomes uniformly locally periodic with period 2

1. Starting from any random initial configuration, the system is uniformly locally periodic if $\beta_2 \geq (|\mathcal{N}| - 1)/2$ and $\delta_2 \leq (|\mathcal{N}| - 1)/2$;
2. starting from product measure with density $p \in [1/4, 3/4]$ the system fixates if $\beta_2 \leq (|\mathcal{N}| - 1)/4$ and $\delta_2 \geq (3/4)(|\mathcal{N}| - 1)$.

In the previous conjecture, for the remaining values of β_2 and δ_2 we claim that the system either fixates or is locally periodic, depending on the initial state.

The following theorem is an extension of Durrett and Steif's theorem (in [6]) to LtL rules near the TVA and a generalization to initial states with arbitrary density p ($0 < p < 1$).

Theorem 1 *Let B_k be the event that all sites x with $\|x\|_2 \leq k$ never change. Suppose the system starts from product measure with density p and that $\beta_2 = |\mathcal{N}| - 1$ and $\delta_2 = |\mathcal{N}|$. If $p \leq 1/2$, $\beta_1 > (1 - p/2)|\mathcal{N}|$ and $\delta_1 < (p/2)|\mathcal{N}|$ or if $p \geq 1/2$, $\beta_1 > ((1 + p)/2)|\mathcal{N}|$, and $\delta_1 < ((1 - p)/2)|\mathcal{N}|$ then for all k, $P(B_k) \to 1$ as $\rho \to \infty$.*

To do the proof, we need the following lemma from [6].

Lemma 1 (*Durrett and Steif* [6]) *Suppose $b < 1/2$. There are constants R_0 and ρ_0 so that if $r \geq R_0$ and $\rho \geq \rho_0$, then each site x in $B_2(0, r\rho)$ has $|(x + \mathcal{N}) \cap B_2(0, r\rho)| \geq b|\mathcal{N}|$, where $B_2(0, r\rho) = \{x: \|x\|_2 \leq r\rho\}$.*

Lemma 1 essentially says that a large ball is locally flat. This means that sites inside the ball and near the boundaries see almost as many sites inside as they would had they been inside a half-space. Let us prove Theorem 1.

Proof Consider a site $x \in B_2(0, r\rho)$ that is in state 0. Pick b and c so that $2(1 - \beta_1/|\mathcal{N}|) < 2c < b < 1/2$. We also need the following large deviations result from Chap. 1 of [5]:

Let S_n be the sum of n i.i.d. random variables that take the value 1 with probability p and 0 with probability $1 - p$. Then for $c < p$, $\lim_{n \to \infty}(1/n) \log P(S_n < cn) = -\gamma < 0$, $P(S_n < cn) \leq e^{-\gamma n}$ for all n. (In the spirit of Durrett, γ is an unimportant constant that changes from line to line.)

Since $b < 1/2$, the above gives that for $c < p/2$, $\lim_{n\to\infty}(1/n) \times \log P(S_{bn} < cn) = -\gamma < 0$. Combining this with Lemma 1 gives that if $r \geq R_0$, then the probability that all sites in $B_2(0, r\rho)$ have at least $c|\mathcal{N}|$ neighbors of each type approaches 1 as $\rho \to \infty$. Thus, x will have at most $|\mathcal{N}| - c|\mathcal{N}|$ 1's in its neighborhood. By the choice of c, we have $|\mathcal{N}| - c|\mathcal{N}| < \beta_1$. Thus, no 0 will become a 1 provided $c < p/2$. This implies that $2(1 - \beta_1/|\mathcal{N}|) < p$. Solving for β_1 gives $\beta_1 > (1 - p/2)|\mathcal{N}|$. Observe also that the choice of $2c < 1/2$ imposes the condition $(1 - \beta_1/|\mathcal{N}|) < 1/4$; i.e. $\beta_1 > (3/4)|\mathcal{N}|$. Both conditions will be met if $p \leq 1/2$ and $\beta_1 > 1 - p/2$ or if $p \geq 1/2$ and $\beta_1 > ((1 + p)/2)|\mathcal{N}|$.

Now consider a site $y \in B_2(0, r\rho)$ that is in state 1. Pick b and C so that $2\delta_1/|\mathcal{N}| < 2C < b < 1/2$. Arguing as we did above, we see that y will have at least $C|\mathcal{N}|$ 1's in its neighborhood and since $C|\mathcal{N}| > \delta_1$, no 1 will become a 0, provided $C < p/2$. This imposes the condition that $2\delta_1/|\mathcal{N}| < p$. Observe also that the choice of $2C < 1/2$ imposes the condition $\delta_1/|\mathcal{N}| < 1/4$; i.e. $\delta_1 < (1/4)|\mathcal{N}|$. Both conditions will be met if $p \leq 1/2$ and $\delta_1 < p|\mathcal{N}|/2$, or if $p \geq 1/2$, and $\delta_1 < ((1 - p)/2)|\mathcal{N}|$. $\square$

We point out that as the density of the initial state moves away from $1/2$, the restriction on the rules for which we have shown that the sites never change becomes more strict. Thus, starting from density $1/2$, as Durrett and Steif did in the TVA case, gives the result on the largest set of rules.

Let us briefly mention what the above says about finite ranges. Fix a finite range ρ. Let B be a box with side length n, where n is a large integer that depends on ρ. The previous theorem says that with high probability, say $1 - \epsilon$ ($\epsilon > 0$ and depends on n), none of the sites in B will ever change state. By the ergodic theorem (see Chap. 6 of [5]) with probability one at most proportion ϵ of the sites in the lattice ever change.

For LtL rules near the AVA, the analogue to the previous theorem is:

Theorem 2 *Let B_k be the event that all sites x with $\|x\|_2 \leq k$ never change. Suppose the system starts from product measure with density p and that $\beta_1 = 0$ and $\delta_1 = 1$. If $p \leq 1/2$, $\delta_2 > (1 - p/2)|\mathcal{N}|$ and $\beta_2 < (p/2)|\mathcal{N}|$ or if $p \geq 1/2$, $\delta_2 > ((1 + p)/2)|\mathcal{N}|$, and $\beta_2 < ((1 - p)/2)|\mathcal{N}|$ then for all k, $P(B_k) \to 1$ as $\rho \to \infty$.*

The proof is essentially the same as that for Theorem 1 except, in this case, the role of β_1 is played by δ_2 and that of δ_1 is played by β_2. For completeness, we illustrate these differences.

Proof Consider a site $x \in B_2(0, r\rho)$ that is in state 1. Pick b and c so that $2(1 - \delta_2/|\mathcal{N}|) < 2c < b < 1/2$. As we found in the proof of Theorem 1, the probability that all sites in $B_2(0, r\rho)$ have at least $c|\mathcal{N}|$ neighbors of each type approaches 1 as $\rho \to \infty$. Thus, x will have at most $|\mathcal{N}| - c|\mathcal{N}|$ 1's in its neighborhood. By the choice of c, we have $|\mathcal{N}| - c|\mathcal{N}| < \delta_2$. Thus, no 1 will become a 0 provided $c < p/2$, which means that $2(1 - \delta_2/|\mathcal{N}|) < p$. Solving for δ_2 gives $\delta_2 > (1 - p/2)|\mathcal{N}|$.

Observe also that the choice of $2c < 1/2$ imposes the condition $(1 - \delta_2/|\mathcal{N}|) < 1/4$; i.e. $\delta_2 > (3/4)|\mathcal{N}|$. Both conditions will be met if $p \leq 1/2$ and $\delta_2 > 1 - p/2$ or if $p \geq 1/2$ and $\delta_2 > ((1 + p)/2)|\mathcal{N}|$.

Now consider a site $y \in B_2(0, r\rho)$ that is in state 0. Pick b and C so that $2\beta_2/|\mathcal{N}| < 2C < b < 1/2$. Arguing as we did above, we see that y will have at least $C|\mathcal{N}|$ 1's in its neighborhood and since $C|\mathcal{N}| > \beta_2$, no 0 will become a 1, provided $C < p/2$. This imposes the condition that $2\beta_2/|\mathcal{N}| < p$. Observe also that the choice of $2C < 1/2$ imposes the condition $\beta_2/|\mathcal{N}| < 1/4$; i.e. $\delta_1 < (1/4)|\mathcal{N}|$. Both conditions will be met if $p \leq 1/2$ and $\beta_2 < p|\mathcal{N}|/2$, or if $p \geq 1/2$, and $\beta_2 < ((1 - p)/2)|\mathcal{N}|$.   □

The following theorem is another analogue of Theorem 1, for LtL rules near the AVA. In this case, however, all sites flip-flop every time step.

Theorem 3 *Let C_k be the event that for each site x with $\|x\|_2 \leq k$ the state of that site, $\xi_t(x)$ has period 2 in t. Suppose the system starts from product measure with density p and that $\beta_1 = 0$ and $\delta_1 = 1$. If $p \leq 1/2$, $\beta_2 > (1 - p/2)|\mathcal{N}|$ and $\delta_2 < (p/2)|\mathcal{N}|$ or if $p \geq 1/2$, $\beta_2 > ((1 + p)/2)|\mathcal{N}|$, and $\delta_2 < ((1 - p)/2)|\mathcal{N}|$ then for all k, $P(C_k) \to 1$ as $\rho \to \infty$.*

The proof for Theorem 3 is exactly like that of Theorem 1 with β_2 playing the role that β_1 played and δ_2 taking over for δ_1.

Majority Vote Rule

Note that the famous Majority Vote rule (where the voter changes her opinion iff she sees a majority of the opposite opinion in her neighborhood) is a TVA. It is generally believed that starting from product measure with density $1/2$, the probability that all sites in a large ball fixate in the same state goes to 1 as the range goes to infinity. This is one of the conjectures made by Durrett and Steif in [6]. (Conjecture 3 is its AVA analogue.) However, no one has been able to prove this.

Some progress has been made, however, on understanding the dynamics of the Majority Vote rule. Let us illustrate times 1, 2, 3, and 40 of the range five version of the rule starting from a random initial state with equal densities of the two opinions (black represents the Republicans and white the Democrats).

As we see in Fig. 11.11, after just one time step, there is massive self-organization. After another update, there is a dramatic smoothing of the edges between the colors. At time three, there is a two-color tessellation of the array, and within the resolution of a range 5 discretization, it suggests that the boundaries are continuously differentiable. At times four and after, there is convexification, and erosion of bounded regions, according to dynamics that approximate a continuous flow known as *motion by mean curvature*. (For a better approximation to the continuous flow one must use a larger range.) By depicting time 40, we illustrate how this process continues until the boundaries achieve uniformly small curvature, at which

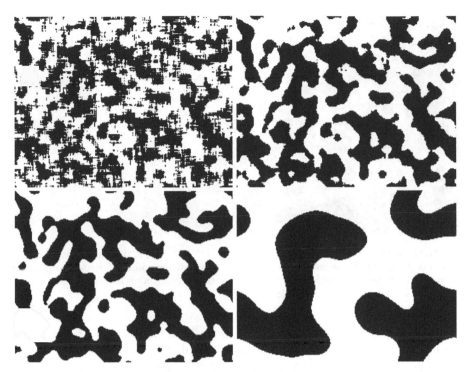

Fig. 11.11 Range 5 Majority Vote, which is the LtL rule with parameters $(5, 61, 120, 61, 121)$. Initial state: product measure with density $p = 0.5$. Times 1 and 2 are *left* to *right*, respectively, in the *top row* and times 3 and 40 are in the *bottom row*

point the system fixates. We chose to stop the evolution prior to fixation. Can the reader guess who wins in this case? Will it be the Republicans? The Democrats? Or will there be a stand-off between the two? We leave this as a little puzzle. In the style of Riemann, these CA approximations, suitably scaled, converge to Euclidean dynamics that cluster indefinitely. Results along these lines, as well as progress on the fixation of Majority Vote, are currently being formulated by Gravner and Griffeath (D. Griffeath, pers. comm.).

Before moving on, we make a few comments about the continuous time version of the TVA, known as the *threshold voter model*. It is defined in [6] as follows:

There is an independent rate one Poisson process $\{T_n^x, n \geq 1\}$ for each lattice point $x \in \mathbf{Z}^d$. At time T_n^x, the voter at x examines the points in her neighborhood $\{y: y - x \in \mathcal{N}\}$. If at least τ neighbors have the opposite opinion then the opinion at x changes, otherwise it stays the same.

The following conjecture about this model is made in [6]:

Conjecture 6 (Durrett and Steif [6]) If $\mathcal{N} = \{y: \|y\|_p \leq \rho\}$ and $\tau = \theta|\mathcal{N}|$ with $\theta \in (1/4, 1/2)$, then for large ρ the system clusters starting from product measure with density $1/2$.

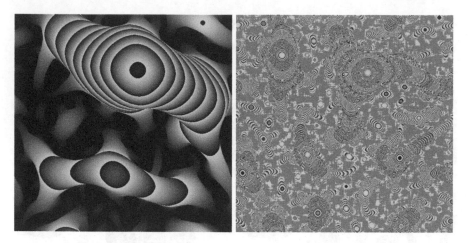

Fig. 11.12 Majority Vote rule with LtL parameters $(7, 113, 224, 113, 225)$. In each case, the initial state was uniform product measure on a 600×600 lattice with wrap around boundary conditions; the rule ran until the lattice filled with 1's. The difference in the figures is due to different random initial states and different grayscale palettes

By adding randomness to the updating of the TVA, there is hope for clustering. On the other hand, there do not seem to be any LtL rules that cluster. Indeed, for a rule to cluster, it is necessary that large sets of 0's grow, as do large sets of 1's. This requires a rule that is symmetric in 0's and 1's. Thus, if there were an LtL rule that clustered, it would be a TVA or an AVA. Of those rules, Majority Vote seems to be the most likely candidate for clustering. As we have discussed, Majority Vote is known (empirically) to fixate, so we are out of luck. However, we still include the discussion because of the interesting paradox it poses: If a two state rule clusters, then regardless of which state a large connected component of one state is in, it will grow, but if it encounters a larger set of the other state, it will get devoured. The resolution is that although a large nonconvex set has the capability to grow, there is always something bigger of the other state out there that will beat it in a competition.

The Majority Vote rule is also interesting visually when color is added to track the time at which each site in the lattice becomes a 1. A gray scale example is shown in Fig. 11.12. For more colorful examples, see [7].

11.5.2 Monotone LtL Rules

Another simplifying feature for a two state CA rule $\mathscr{T}$ to have is *monotonicity* or *attractiveness*. $\mathscr{T}$ is said to be *monotone nondecreasing* or *attractive* in live sites if it maps larger sets of 1's to larger sets of 1's. In other words, if A and B are sets of live sites and $A \subset B$ then $\mathscr{T}(A) \subseteq \mathscr{T}(B)$. On the other hand, $\mathscr{T}$ is said to be *monotone nonincreasing* in live sites if it maps larger sets of live sites to smaller sets of live sites. In other words, if C and D are sets of live sites and $C \subset D$ then

$\mathscr{T}(C) \supseteq \mathscr{T}(D)$. Monotonicity makes a rule easier to rigorously analyze; however, the majority of LtL rules are *not* monotone. In fact, each LtL rule may be defined in terms of four specific monotone rules as follows:

Let Λ be a set of live sites. Let $\mathscr{T}_1(\Lambda) = \{x: |(x + \mathscr{N}) \cap \Lambda| \geq \beta_1\}$. In other words, site x will be live next time iff there are at least β_1 live sites, including x, in its neighborhood this time. Thus, $\mathscr{T}_1$ is determined by β_1. Similarly, let $\mathscr{T}_2(\Lambda) = \{x: |(x + \mathscr{N}) \cap \Lambda| \leq \beta_2\}$, so that site x will be live next time iff there are at most β_2 live sites, including x, in its neighborhood this time. $\mathscr{T}_2$ is thus determined by β_2. The next two maps are just like the previous two, but determined instead by δ_1 and δ_2 respectively. Let $\mathscr{T}_3(\Lambda) = \{x: |(x + \mathscr{N}) \cap \Lambda| \geq \delta_1\}$ and $\mathscr{T}_4(\Lambda) = \{x: |(x + \mathscr{N}) \cap \Lambda| \leq \delta_2\}$.

The rules $\mathscr{T}_1$ and $\mathscr{T}_3$ are *monotone nondecreasing* in live sites, while the rules $\mathscr{T}_2$ and $\mathscr{T}_4$ are *monotone nonincreasing* in live sites. Let $\mathscr{T}$ be an LtL rule and Λ a set of live sites. Then $\mathscr{T}(\Lambda) = (\mathscr{T}_1(\Lambda) \cap \mathscr{T}_2(\Lambda) \cap \Lambda^c) \cup (\mathscr{T}_3(\Lambda) \cap \mathscr{T}_4(\Lambda) \cap \Lambda)$. Now $\mathscr{T}_2(\Lambda) \cap \Lambda^c$ is nonincreasing and $\mathscr{T}_3(\Lambda) \cap \Lambda$ is nondecreasing so every LtL rule is formed by four monotone rules.

A well known monotone nondecreasing LtL rule is the discrete threshold growth model (TGM). Let us discuss known results for the TGM that are relevant to our current discussion of LtL rules.

Discrete Threshold Growth Model

A CA from the TGM has range ρ LtL parameters $\beta_1 = \theta$, $\beta_2 = |\mathscr{N}| - 1$, $\delta_1 = 1$, and $\delta_2 = |\mathscr{N}|$ ($\theta \geq 1$).

Gravner and Griffeath have determined that each of the TGM rules generates one of three types of dynamics: *subcritical*, *critical*, or *supercritical*. Let us present these definitions, which are from [17].

Definition 1 (Gravner and Griffeath [17]) Let $A_\infty = \mathscr{T}^\infty(A_0) = \bigcup_{t=0}^\infty A_t$.

1. Say that the dynamics are *supercritical* if there exists a finite A_0 such that A_t eventually occupies every site in $\mathbf{Z}^2$; *i.e.* $A_\infty = \mathbf{Z}^2$.
2. Say that the dynamics are *critical* if $A_\infty \neq \mathbf{Z}^2$ for every finite A_0 but, for any A_0 with finite *complement*, $A_\infty = \mathbf{Z}^2$.
3. Say that the dynamics are *subcritical* if there exists a non-empty finite set, H (a *hole*), so that the dynamics cannot fill H, even when started from the initial set $A_0 = H^c$.

Gravner and Griffeath have shown (in [17]) that a range ρ LtL rule with parameters $\beta_1 = \theta$, $\beta_2 = |\mathscr{N}| - 1$, $\delta_1 = 1$, and $\delta_2 = |\mathscr{N}|$ ($\theta \geq 1$), has dynamics which are:

1. supercritical iff $\theta \leq 2\rho^2 + \rho$;
2. critical if $\theta \in [2\rho^2 + \rho + 1, 2\rho^2 + 2\rho]$;
3. subcritical iff $\theta > 2\rho^2 + 2\rho$.

In particular, this says that if $\theta \leq 2\rho^2 + \rho$, then the limiting state is $\underline{1}$. In addition, if $\theta \in [2\rho^2 + \rho + 1, 2\rho^2 + 2\rho]$, then started from product measure with any density $p > 0$, the limiting state is $\underline{1}$ a.s.

LtL Rules near TGM: Global Survival and Global Death

This section contains theorems which prove that certain LtL rules near the sets of monotone nondecreasing or nonincreasing rules (and, in some cases, in these sets of monotone rules) result in global death or global survival, starting from any random initial configuration. It is not a coincidence that these results apply to sets of rules that lie close to the boundaries of the parameter space (meaning that at least some of the parameters are at, or near, their extremes). It is near those extremes that we find the least nonlinear, and hence, most tractable rules.

The two theorems of this section specify LtL parameters that ensure global survival and global death, respectively. In order to prove them, we adapt to the LtL rules some of the formalism used in [17] to prove the results described above for the TGM.

Recall that, if $\Lambda \subset \mathbf{Z}^2$ is a set of 1's (on a background of 0's), we define

$$\mathscr{T}(\Lambda) = \left\{ x \in \Lambda^c : \beta_1 \leq \left| (x + \mathscr{N}) \cap \Lambda \right| \leq \beta_2 \right\}$$
$$\cup \left\{ x \in \Lambda : \delta_1 \leq \left| (x + \mathscr{N}) \cap \Lambda \right| \leq \delta_2 \right\}.$$

Let $\tilde{\Lambda} \subset \mathbf{R}^2$ be any subset of 1's (on a background of 0's). Define

$$\widetilde{\mathscr{T}}(\tilde{\Lambda}) = \left\{ x \in \tilde{\Lambda}^c : \beta_1 \leq \left| (x + \mathscr{N}) \cap \tilde{\Lambda} \right| \leq \beta_2 \right\}$$
$$\cup \left\{ x \in \tilde{\Lambda} : \delta_1 \leq \left| (x + \mathscr{N}) \cap \tilde{\Lambda} \right| \leq \delta_2 \right\},$$

where $\mathscr{N}$ is still the range ρ box neighborhood of the origin, and $| \cdot |$ continues to be cardinality.

Then $\mathscr{T}$ and $\widetilde{\mathscr{T}}$ are *conjugate*. That is, if $B \subset \mathbf{R}^2$ is any set of 1's (on a background of 0's), then $\widetilde{\mathscr{T}}(B) \cap \mathbf{Z}^2 = \mathscr{T}(B \cap \mathbf{Z}^2)$.

Now we define the *speeds* of half-spaces, filled either with 1's or 0's, assuming (in both cases) that their complements consist of any possible configurations of 1's. Suppose u is a two-dimensional unit vector in the sphere $S^1 \subset \mathbf{R}^2$. Denote the half-space $H_u^- = \{ x \in \mathbf{R}^2 : x \cdot u \leq 0 \}$ ($\cdot$ is the Euclidean dot product).

Definition 2 Let $\tilde{\xi}_0 \subset \mathbf{R}^2$ be the set of 1's at time 0.

1. Define the *speed*, $w^1(u)$ of H_u^- by $w^1(u) = \max\{\lambda \in \mathbf{R} : H_u^- + \lambda u \subseteq \widetilde{\mathscr{T}}(\tilde{\xi}_0)\}$ for every $\tilde{\xi}_0$ such that $H_u^- \subseteq \tilde{\xi}_0$. The maximum may not exist, since all sites in H_u^- may become 0's next time, or all sites in the system may become 1's. If all sites in H_u^- become 0's next time, define $w^1(u) \equiv -\infty$. If all sites in the system become 1's, define $w^1(u) \equiv \infty$.
2. Define the *speed*, $w^0(u)$ of H_u^- by $w^0(u) = \max\{\lambda \in \mathbf{R} : (H_u^- + \lambda u)^c \supseteq \widetilde{\mathscr{T}}(\tilde{\xi}_0)\}$ for every $\tilde{\xi}_0$ such that $(H_u^-)^c \supseteq \tilde{\xi}_0$. Again the maximum may not exist, since all

sites in H_u^- may become 1's next time, or all sites in the system may become 0's. If all sites in H_u^- become 1's next time, define $w^0(u) \equiv -\infty$. If all sites in the system become 0's, define $w^0(u) \equiv \infty$.

Lemma 2 *For any u, the following hold, provided $\rho \geq 1$.*

1. $1 \leq |\mathscr{N} \cap \{x: x \cdot u = 0\}| \leq 2\rho + 1$.
2. $\rho(2\rho + 1) \leq |\mathscr{N} \cap \{x: x \cdot u < 0\}| \leq 2\rho(\rho + 1)$.
3. $2\rho^2 + 2\rho + 1 \leq |\mathscr{N} \cap \{x: x \cdot u \leq 0\}| \leq 2\rho^2 + 3\rho + 1$.

Proof Let $E = \{|\mathscr{N} \cap l|: l$ is a line through the origin$\}$. For any u, since $\mathscr{N}$ is the range ρ box neighborhood, $1 \leq E \leq 2\rho + 1$. This proves (1). Combining the symmetry of $\mathscr{N}$ with (1) gives

$$(1/2)\big((2\rho + 1)^2 - (2\rho + 1)\big) \leq |\mathscr{N} \cap \{x: x \cdot u < 0\}| \leq (1/2)\big((2\rho + 1)^2 - 1\big),$$

which is (2). Again combining the symmetry of $\mathscr{N}$ and (1), we get

$$1 + (1/2)\big((2\rho + 1)^2 - 1\big) \leq |\mathscr{N} \cap \{x: x \cdot u \leq 0\}| \quad \text{and}$$
$$|\mathscr{N} \cap \{x: x \cdot u \leq 0\}| \leq 2\rho + 1 + (1/2)\big((2\rho + 1)^2 - (2\rho + 1)\big),$$

which is (3). $\square$

Proposition 1 *Assume $\rho \geq 1$.*

1. *If $\delta_1 \leq 2\rho^2 + 2\rho + 1$ and $\delta_2 = (2\rho + 1)^2$, then $w^1(u) \geq 0$ for every u.*
2. *If $\beta_1 > 2\rho(\rho + 1)$ then $w^0(u) \geq 0$ for every u.*

Proof First we prove (1). If every site in H_u^- remains a 1 at the next time step, then $w^1(u) \geq 0$ for every u. (This holds since we are assuming that $H_u^- \subset \tilde{\xi}_0$.) Let $y \in \mathbf{R}^2$ be in a small neighborhood of the origin. Then for any u, $|(y + \mathscr{N}) \cap \{x: x \cdot u \leq 0\}| = |\mathscr{N} \cap \{x: x \cdot u \leq 0\}| \geq 2\rho^2 + 2\rho + 1$. The inequality holds by Lemma 2, part 3. Thus, if $\delta_1 \leq 2\rho^2 + 2\rho + 1$ and $\delta_2 = (2\rho + 1)^2$, then every site in H_u^- will remain a 1 next time.

Now we prove (2). If every site in H_u^- remains a 0 at the next time step, then $w^0(u) \geq 0$ for every u. (This holds since we are assuming that $(H_u^-)^c \supset \tilde{\xi}_0$.) Let $y \in \mathbf{R}^2$ be in a small neighborhood of the origin. Then for any u, $|(y + \mathscr{N}) \cap \{x: x \cdot u > 0\}| = |\mathscr{N} \cap \{x: x \cdot u > 0\}| \leq 2\rho(\rho + 1)$. The inequality holds by combining the symmetry of $\mathscr{N}$ with Lemma 2, part 2. Thus, if $\beta_1 > 2\rho(\rho + 1)$ then every site in H_u^- will remain a 0 next time. $\square$

We say that LtL dynamics are *space-filling* if there exists a finite A_0 such that A_t eventually occupies every site in $\mathbf{Z}^2$; *i.e.* $A_\infty = \mathbf{Z}^2$. In particular, this implies that A_∞ exists. We say that the dynamics are *space-emptying* if there exists a co-finite set C_0, such that C_t eventually occupies no site in $\mathbf{Z}^2$; *i.e.* $C_\infty = \emptyset$. The first definition represents the case where sets of 1's do supercritical growth, and the second where sets of 0's do supercritical growth (see [16] and [17]). These are relevant to the LtL family because a large proportion of the rules result in global survival or global death.

Proposition 2

1. *LtL dynamics are space-filling iff $w^1(u) > 0$ for every u.*
2. *LtL dynamics are space-emptying iff $w^0(u) > 0$ for every u.*

Proof First we prove (1). $w^1(u) > 0$ implies that there exists a bounded subset $A \subset \mathbf{R}^2$ so that $\widetilde{\mathcal{F}}^t(A) \uparrow \mathbf{R}^2$ (see Lemma 1 on p. 853 of [16]). The converse is obvious since an edge speed of 0 constrains the spread of 1's to a corresponding half-space.

Now we prove (2). $w^0(u) > 0$ implies that there exists a co-finite subset $B \subset \mathbf{R}^2$ so that $\widetilde{\mathcal{F}}^t(B) \downarrow \emptyset$ (see Lemma 1 on p. 853 of [16]). The converse is obvious since an edge speed of 0 constrains the spread of 0's to a corresponding half-space. $\square$

The above shows that if one wants to prove that a large ball of 1's (or 0's) does supercritical growth, it suffices to show that any half-space grows (or shrinks) linearly.

Theorem 4 *If $\rho \geq 1$, $\beta_1 \leq \rho(2\rho + 1)$, $\beta_2 = 4\rho(\rho + 1)$, $\delta_1 \leq \rho(2\rho + 1) + 1$, and $\delta_2 = (2\rho + 1)^2$, then starting from any random initial state, the limiting state is $\underline{1}$.*

Proof By Proposition 2, it suffices to show that $w^1(u) > 0$ for every u. Since $\delta_1 \leq \rho(2\rho + 1) + 1 \leq 2\rho^2 + 2\rho + 1$, and $\delta_2 = (2\rho + 1)^2$, Proposition 1 gives $w^1(u) \geq 0$ for every u. Thus, we need only show that $w^1(u) \neq 0$ for every u. Since we are working with $w^1(u)$, we assume $H_u^- \subset \widetilde{\xi}_0$. Let $y \in \mathbf{R}^2$ be in a small neighborhood of the origin. Then, for any u, $|(y + \mathcal{N}) \cap \{x: x \cdot u < 0\}| = |\mathcal{N} \cap \{x: x \cdot u < 0\}| \geq \rho(2\rho + 1)$. *I.e.* there are at least $\rho(2\rho + 1)$ occupied sites in y's neighborhood. Thus, if $y \in (H_u^-)^c$ was a 0, it will become a 1 if $\beta_1 \leq \rho(2\rho + 1)$. If $y \in (H_u^-)^c$ was a 1, it will remain a 1 if $\delta_1 \leq \rho(2\rho + 1) + 1$. $\square$

Theorem 5 *If $\rho \geq 1$, $2\rho^2 + 3\rho + 1 \leq \beta_1 \leq \beta_2 \leq 4\rho(\rho + 1)$ and $2\rho^2 + 3\rho + 1 < \delta_1 \leq \delta_2 \leq (2\rho + 1)^2$, then starting from any random initial state, the limiting state is $\underline{0}$.*

Proof By Proposition 2, it suffices to show that $w^0(u) > 0$ for every u. Since $\beta_1 > \rho(2\rho + 1)$, Proposition 1 gives $w^0(u) \geq 0$ for every u. Thus, we need only show that $w^0(u) \neq 0$ for every u. Since we are working with $w^0(u)$, we assume $(H_u^-)^c \supset \widetilde{\xi}_0$. Let $y \in \mathbf{R}^2$ be in a small neighborhood of the origin. Then, for any u, $|(y + \mathcal{N}) \cap \{x: x \cdot u \geq 0\}| = |\mathcal{N} \cap \{x: x \cdot u \geq 0\}| \leq 2\rho^2 + 3\rho + 1$. *I.e.* there are at most $2\rho^2 + 3\rho + 1$ occupied sites in y's neighborhood. Thus, if $y \in (H_u^-)^c$ was a 1, it will become a 0 if $\delta_1 > 2\rho^2 + 3\rho + 1$. If $y \in (H_u^-)^c$ was a 0, it will remain a 0 if $\beta_1 \geq 2\rho^2 + 3\rho + 1$ (knowing a y is a 0 allows us to subtract one site from the count). $\square$

Corollary 1 *If $\rho \geq 1$, $2\rho^2 + 3\rho + 1 \leq \beta_1 \leq \beta_2 \leq 4\rho(\rho + 1)$ and $\delta_1 = \delta_2 = 0$, then starting from any random initial state the limiting state is $\underline{0}$.*

Proof This follows from Theorem 5 since in this case $1 \to 0$ automatically so survival is even more difficult. $\qquad\qquad\qquad\qquad\qquad\qquad\qquad\qquad\qquad\qquad\qquad\qquad\qquad$ □

Conjecture 7 If $\rho \geq 1$, $\beta_1 \in [2\rho^2 + \rho + 1, 2\rho^2 + 2\rho]$, $\beta_2 = 4\rho(\rho + 1)$, $\delta_1 \in [2, 2\rho^2 + 2\rho + 1]$, and $\delta_2 = (2\rho + 1)^2$, then for $p \in (p(\delta_1), 1]$ starting from product measure with density p the limiting state is $\underline{1}$ a.s. ($p(\delta_1) \geq 0$ depends on δ_1; if, as in threshold growth $\delta_1 = 1$, then $p(\delta_1) = 0$).

Conjecture 7 is like a proposition from [17], which was restricted to the TGM; however, in our case survival is not guaranteed. Thus we cannot use the bootstrap methods (see [1]) that are crucial to Gravner and Griffeath's proof. They used bootstrap methods to show that a large set of 1's can grow, provided it gets some "help" from the 1's in the initial product measure surrounding it. The argument thus relies on the survival of sparse sets of 1's from the random initial state. However, by choosing the $p(\delta_1)$ of our conjecture carefully, we may still be able to adapt some of Gravner and Griffeath's ideas. The following conjecture is similar to the previous one. However, it is about sets of 0's (rather than 1's) bootstrapping their way to global death (rather than global survival).

Conjecture 8 If $\rho \geq 1$, $2\rho^2 + 2\rho + 1 \leq \beta_1 \leq \beta_2 \leq 4\rho(\rho + 1)$ and $2\rho^2 + 2\rho + 2 \leq \delta_1 \leq \delta_2 \leq (2\rho + 1)^2$ then for $p \in [0, p(\beta_1))$ starting from product measure with density p, the limiting state is $\underline{0}$ a.s. ($p(\beta_1) \leq 1$ depends on β_1).

What proportion of all LtL rules have we shown result in global survival? What about global death? Recall that for a fixed range ρ, there are $(2k^2 + 3k + 1)^2$ possible LtL rules, where $k = 2\rho(\rho + 1)$. We have shown that for all rules such that $\beta_1 \leq \rho(2\rho + 1)$, $\beta_2 = 4\rho(\rho + 1)$, $\delta_1 \leq \rho(2\rho + 1) + 1$, and $\delta_2 = (2\rho + 1)^2$, the limiting state is $\underline{1}$. There are $(\rho(2\rho + 1) + 1)(\rho(2\rho + 1) + 2)$ such rules. Thus, such rules represent $(\rho(2\rho + 1) + 1)^2/(2k^2 + 3k + 1)^2$ ($k = (1/2)(2\rho + 1)^2$) of the total. Observe that $\lim_{\rho \to \infty}(\rho(2\rho + 1) + 1)^2/(2k^2 + 3k + 1)^2 = 0$.

In the case of global death, we have made more progress. We have shown that for all rules such that $2\rho^2 + 3\rho + 1 \leq \beta_1 \leq \beta_2 \leq 4\rho(\rho + 1)$ and $2\rho^2 + 3\rho + 1 \leq \delta_1 \leq \delta_2 \leq (2\rho + 1)^2$, the limiting state is $\underline{0}$. There are $(2l^2 + 3l + 1)^2$ such rules, where $l = \rho^2 + (1/2)\rho - 1/2$. Thus, such rules represent $(2l^2 + 3l + 1)^2/(2k^2 + 3k + 1)^2$ ($l = \rho^2 + (1/2)\rho - (1/2)$, $k = 2\rho(\rho + 1)$) of the total. In this case, $\lim_{\rho \to \infty}((2l^2 + 3l + 1)^2/(2k^2 + 3k + 1)^2) = 1/16$. We also showed that an additional $2\rho^2 + \rho$ rules have limiting state $\underline{0}$. However, since this is quadratic in ρ, it does not change the above proportion, which is our interest.

The limiting state $\underline{1}$ will be attained only if $\delta_2 = (2\rho + 1)^2$. Thus, the set of LtL rules with this limiting state is contained in a three-dimensional cross section of LtL parameter space. Hence, it makes sense that the proportion of such rules goes to 0 as the range gets large. On the other hand, $\underline{0}$ does not have such a requirement, and we have shown that the set of rules with this limiting state live in a four-dimensional subspace. The moral? The rules are Larger than Life, but global survival is difficult. Perhaps a corollary to the above is: Death is Larger than Life.

Fig. 11.13 Ergodic classification of range ρ LtL rules with parameters $\beta_2 = |\mathcal{N}| - 1$, $\delta_2 = |\mathcal{N}|$. "$*$" indicates that the limiting behavior has been proved

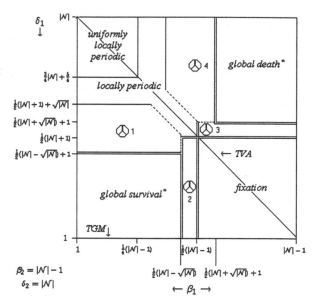

How have the results obtained in this and Sect. 11.5.1 helped us better understand the geometry of LtL space in terms of the ergodic classifications of the rules? To answer this question let us depict the LtL cross section of rules with $\beta_2 = |\mathcal{N}| - 1$ and $\delta_2 = |\mathcal{N}|$ in terms of the ergodic classification at each point. (The cross section includes the set of LtL rules that are monotone nondecreasing in live sites.) The diagram in Fig. 11.13 indicates the TVA set of rules with a diagonal line that cuts through the picture. The TGM rules are also indicated with a line at the bottom of the picture.

We indicate with a "$*$" the regions for which we have proved global death and global survival. There is also a "$*$" in the upper left corner of the diagram because we are going to prove in the next section (see Theorem 6) that the rule with parameters $\beta_1 = 1$, $\beta_2 = |\mathcal{N}| - 1$ and $\delta_1 = \delta_2 = |\mathcal{N}|$ is uniformly locally periodic. We indicate also the regions we believe to have locally periodic or fixed limiting states.

For the regions closer to the middle of the diagram, the ergodic classification is more complicated. In those cases, we use a warning icon to indicate the uncertainties. We believe a large portion of the rules in region 1 result in global survival. Gravner and Griffeath have shown that the limiting state for the rules in the intersection of region 2 and the TGM line result in global survival when started from product measure of any positive density. In Conjecture 7 we state that this is also the case for the remaining rules in region 2, but only if the initial density is large enough. If the density is not large enough, we believe the limiting state is fixation. It may turn out that, in any case, the limiting state is fixation (see the discussion after Conjecture 7). In Conjecture 8 we state that the rules in region 3 result in global death provided the initial product measure has a small enough density. Otherwise we believe those rules fixate or are locally periodic. We believe that some of the rules in region 4 result in global death while others are locally periodic. The dashed lines indicate the regions for which the boundaries are especially murky.

11.5.3 Birth or Death Only LtL Rules

Another simplifying feature of two state CA rules is for sites in one of the states to automatically switch to the other state at the next time step. There are two families of such LtL rules: those for which 0's automatically become 1's and those for which 1's automatically become 0's. Range ρ LtL rules of the first type have parameters $\beta_1 = 0$, $\beta_2 = 4\rho(\rho + 1)$, δ_1, δ_2. Range ρ rules for which 1's automatically become 0's have LtL parameters β_1, β_2, $\delta_1 = \delta_2 = 0$. Such rules are known as the two state *Greenberg–Hastings Model* (GHM).

Two-State Greenberg–Hastings Model

In this section we will need the definition of a *stable periodic object* (SPO), which is a finite configuration of 0's and 1's, comprising a set $A \subset \mathbf{Z}^2$, that cycles indefinitely and cannot be disturbed by external influences. More specifically for an LtL rule, a finite configuration on a set $A \subset \mathbf{Z}^2$ is an SPO iff for all x in A with $\xi_t(x) = 0$, $\beta_1 \leq |(x + \mathcal{N}) \cap A \cap \xi_t| \leq \beta_2$, and for all x in A with $\xi_t(x) = 1$, either $|(x + \mathcal{N}) \cap A \cap \xi_t| < \delta_1$ or $|(x + \mathcal{N}) \cap A \cap \xi_t| > \delta_2$ for all times t. A *minimal SPO* contains *no* proper subpattern that is also an SPO.

The following theorem is the *two-state* version of a lemma in [14].

Theorem 6 *If* $\rho \geq 1$, $\beta_1 = 1$, $\beta_2 = 4\rho(\rho + 1)$, *and either* $\delta_1 = \delta_2 = (2\rho + 1)^2$ *(this is the* cyclic cellular automaton *(CCA)), or* $\delta_1 = \delta_2 = 0$ *(GHM), then starting from product measure with density* p $(0 < p < 1)$ *the rule is locally periodic of period* 2 *with probability one.*

Proof We need to show that

1. each site is period 2 eventually in t and
2. $\inf_{t, x \neq y, a, b} P(\xi_t(x) = a, \ \xi_t(y) = b) > 0$.

To show (1), let C be the set of sites that have period 2 eventually in t. Observe that the configuration consisting of a 1 within range ρ of a 0 is an SPO and thus in C; hence, $C \neq \emptyset$. Suppose $x \notin C$, $y \in C$, and $\|x - y\| \leq \rho$. Then $\exists$ a first time τ when $\xi_\tau(x) \neq \xi_\tau(y)$. But then, after time τ, the configuration consisting of x and y is an SPO. Hence, $x \in C$, which is a contradiction. Thus, every site belongs to C as claimed.

Now we show (2). Given any time t, distinct sites x and y, and states a and b, the event in question will occur as long as there exists a pair of disjoint minimal SPOs through x and y in ξ_0. The probability of this is at least $(1/pq)^4 > 0$, where $q = 1 - p$. $\qquad\square$

The proof of Theorem 6 showed that the existence of a minimal SPO was enough to enslave the other sites in the system into periodic cycles of length two. Are there other LtL rules which are uniformly locally periodic and admit minimal SPOs? If so, can we use the SPOs, as we did in the proof of Theorem 6, to prove local periodicity? Attempting to answer these questions led to the discovery of SPOs for various LtL rules; these are given in Proposition 3. It also led to the following two conjectures about uniform and nonuniform local periodicity. However, proving that all sites become enslaved to minimal SPOs becomes very tricky when β_1 is greater than one. We are convinced, nonetheless, that the conjectures are true.

Proposition 3 *The configuration consisting of exactly k 1's inside a box with side length $\rho + 1$ ($k \le \rho^2 + 2\rho$) is an SPO with period 2 for the LtL rule with parameters $\beta_1 \le \min(k, (\rho + 1)^2 - k)$, $\beta_2 = 4\rho(\rho + 1)$ and*

1. $1 \le \delta_1 \le \delta_2 \le \min(k, (\rho + 1)^2 - k) - 1$ *or*
2. $\max((2\rho + 1)^2 - k, 3\rho^2 + 2\rho + k) + 1 \le \delta_1 \le \delta_2 \le (2\rho + 1)^2$.

Note that since there are only two colors, symmetry implies that the configuration consisting of exactly k 0's inside the box described in Proposition 3 is the alternate phase of the proposition's SPO.

Proof Let B be a box with side length $(\rho + 1)^2$ and containing exactly k 1's. At time 0, each $x \in B$ has $k \le |(x + \mathcal{N}) \cap \xi_0| \le 3\rho^2 + 2\rho + k$. If $\xi_0(x) = 1$ then, since in (1) $k > \delta_2$ and in (2) $\delta_1 > 3\rho^2 + 2\rho + k$, $\xi_1(x) = 0$. If, on the other hand, $\xi_0(x) = 0$ then since $\beta_1 \le k$, $\xi_1(x) = 1$. Thus, there will be exactly $(\rho + 1)^2 - k$ sites in state 1 in the box of side length $\rho + 1$ at time 1. Thus, at time 1, each $x \in B$ has $(\rho + 1)^2 - k \le |(x + \mathcal{N}) \cap \xi_0| \le (2\rho + 1)^2 - k$. If $\xi_1(x) = 1$ then since in (1) $(\rho + 1)^2 - k > \delta_2$ and in (2) $\delta_1 > (2\rho + 1)^2 - k$, $\xi_2(x) = 0$. If, on the other hand, $\xi_1(x) = 0$ then since $\beta_1 \le (\rho + 1)^2 - k$, $\xi_1(x) = 1$ as desired. □

Conjecture 9 If $\rho \ge 1$, $k \le \rho^2 + 2\rho$, $\beta_1 \le \min(k, (\rho + 1)^2 - k)$, $\beta_2 = 4\rho(\rho + 1)$, and

1. $1 \le \delta_1 \le \delta_2 \le \min(k, (\rho + 1)^2 - k) - 1$ or
2. $\max((2\rho + 1)^2 - k, 3\rho^2 + 2\rho + k) + 1 \le \delta_1 \le \delta_2 \le (2\rho + 1)^2$,

then starting from product measure with density p ($0 < p < 1$) the rule is locally periodic of period 2, with probability one (in other words, it is uniformly locally periodic).

Observe that part (2) of our conjecture contains some of the TVA rules and those near the TVA (that is, when $\delta_2 = (2\rho + 1)^2$), and agrees with the conjectures about those rules, given in Sect. 11.5.1.

Conjecture 10 If $\rho \ge 1$, $\beta_1 \le 2\rho^2 + 2\rho$, $\beta_2 = 4\rho(\rho + 1)$, and

1. $1 \le \delta_1 \le \delta_2 \le \beta_1 - 1$ or
2. $(2\rho + 1)^2 - \beta_1 + 1 \le \delta_1 \le \delta_2 \le (2\rho + 1)^2$,

then starting from product measure with density p ($0 < p < 1$) the rule is locally periodic, with probability one (note that many sites may actually have period one).

Again part (2) of our conjecture contains some of the TVA rules (that is, when $\delta_2 = (2\rho + 1)^2$), and those near the TVA agree with the earlier conjectures made about them.

11.6 Snapshots from the Boundaries

In this section we present snapshots of various LtL rules whose ergodic classifications were described quantitatively in Sect. 11.4.2. Each of these rules warrant *qualitative* distinctions, however, either because of the dynamics they exhibit on the way to their limiting states, or because their limiting states provide interesting geometries not captured in the quantitative definitions. We seek techniques, perhaps along the lines of those used in [4], which will enable us to make these qualitative notions more precise. We also present more speculative results, including qualitative descriptions of rules near those that are Life-like and references to results and open questions related to Conway's original interest in Life. In addition, we discuss a couple of our favorite threshold-range scaling problems.

11.6.1 Bootstrapping

The first snapshot comes from the rule $(2, 11, 18, 6, 18)$ which we conjecture has an aperiodic limiting state. Empirical results suggest that the rule lies on a phase boundary between aperiodic and locally periodic dynamics. If the density of the initial state is large enough, regions of aperiodic dynamics are able to spread by getting help from nearby sets of 1's, or "bootstrapping." Figure 11.14 illustrates this. On the other hand, if the initial density is small, disjoint regions of activity are not able to spread. Rather they are confined to convex sets and eventually become locally periodic. Figure 11.15 illustrates this.

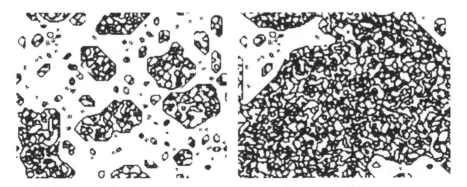

Fig. 11.14 From *left* to *right* are times 100 and 500 of the rule $(2, 11, 18, 6, 18)$; the initial state was product measure with density $p = 5/21$ on a 256×200 lattice with periodic boundary conditions

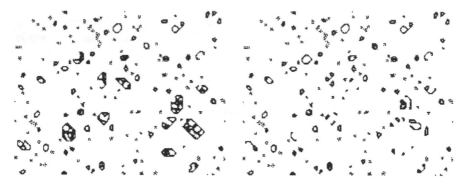

Fig. 11.15 From *left* to *right* are times 100 and 3,000 of the rule $(2, 11, 18, 6, 18)$; the initial state was product measure with density $p = 11/51$ on a 256×200 lattice with periodic boundary conditions

Fig. 11.16 From *left* to *right* are times 2, 5, and 11 of the rule $(13, 10, 100, 100, 320)$; the initial state was product measure with density $p = 1/2$ on a 512×400 lattice with periodic boundary conditions

11.6.2 Self-organized Batik

The next snapshot comes from the range 13 rule $(13, 10, 100, 100, 320)$. We conjecture that the limiting state of this rule started from product measure with density $p = 1/2$ is nonuniformly locally periodic. Additionally empirical results suggest that the eventual period of every site is either 1, 2, or 4. The intrigue is the nucleation as well as the geometry that occurs when the rule is started from a random initial state.

As can be seen in Fig. 11.16 starting from product measure with $p = 1/2$ the system quickly relaxes to all 0's except for five seeds that form the centers of nucleating target patterns. The target patterns fill in almost all of the space with concentric waves of alternating 0's and 1's that are fixed. In the gaps between the waves residual pockets of activity continue to fluctuate, some for a very long time, before locking into periodic cycles (see Fig. 11.17).

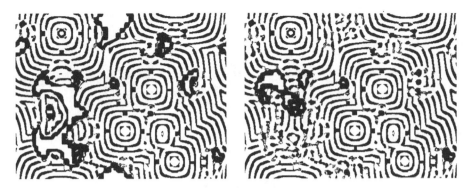

Fig. 11.17 From *left* to *right* are times 25 and 75 of the rule $(13, 10, 100, 100, 320)$; the initial state was product measure with density $p = 1/2$ on a 512×400 lattice with periodic boundary conditions

11.6.3 Global Tilings and Waterbed Dynamics

This snapshot is on a boundary between periodic and aperiodic dynamics. Before describing it, we need two definitions.

- Say that a rule's limiting state is a *global tiling* if it consists of a fixed or locally periodic pattern that occupies almost the entire lattice. Many LtL rules have global tilings, the majority of which consist of approximately horizontal and vertical regions of alternating stripes of 0's and 1's. One such example appears in Sect. 11.6.2, Fig. 11.17. Another example will be presented in Sect. 11.6.4, Fig. 11.21.

- Say that a rule has *waterbed dynamics* if a global tiling almost covers the lattice but regions between the tiling either remain unstable or fixate causing other regions that had been fixed to become unstable. The dynamics remind one of putting together almost all of the pieces of a puzzle and finding that the last few pieces necessary for completion force one to remove some of the pieces already in place. This process continues ad infinitum so that the puzzle can never be completed. The dynamics are so named because the aperiodic regions remind one of the bubbles that can be seen moving around inside the mattress of a waterbed. The limiting state seems neither to fixate nor to become entirely aperiodic. However, it is unclear whether these dynamics are a metastable state and might eventually settle down. It is possible that the dynamics of the infinite system settle down, just not in our lifetimes. Another possibility is that there is a topological deformity in the fixed regions. That is, perhaps the geometry required for fixation is impossible for the rule to attain so that the puzzle analogy made above is actually what is taking place.

Let us illustrate waterbed dynamics with the example $(2, 7, 7, 7, 11)$. (See Fig. 11.18.) Starting from product measure with density $p = 1/5$, at time 25 patterned regions have started growing and are larger by time 125. Between these patterned regions, "Life-like" coherent local structures such as still lifes, blinkers, and

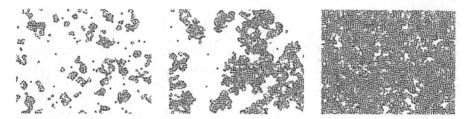

Fig. 11.18 From *left* to *right* are times 25, 125, and 600 of rule (2, 7, 7, 7, 11) started from product measure with density 1/5 on a 256 × 200 lattice with periodic boundary conditions

Fig. 11.19 From *left* to *right* one phase each of a: blinker, period 2 orthogonal bug, and period 4 diagonal bug, supported by LtL rule (2, 7, 7, 7, 11). These coherent structures emerged from an initial state consisting of product measure with density 1/5

bugs survive (see Fig. 11.19). However, the patterned regions dominate and destroy the local structures. By time 600 the fixed patterned regions have nearly filled in the lattice; however, isolated pockets between the fixed regions remain unstable. This continues to be the case at time 600,000.

11.6.4 Slow Convergence

We present the next set of snapshots to illustrate an important theme that occurs throughout LtL phase space: the closer a rule is to a phase boundary the longer it takes to reach its limiting state. The example we use is the range 2 rule (2, 9, 16, 9, 20) which empirical results suggest lies on a boundary between aperiodic and locally periodic dynamics. Its limiting state is locally periodic but it takes significantly longer to attain that state than nearby rules. For example, if we use product measure with density $p = 1/5$ as the initial state on a 256 × 200 lattice with periodic boundary conditions, then it takes this rule 2,814 time steps to reach its locally periodic limiting state; this is illustrated in Figs. 11.20 and 11.21. On the other hand, using the same initial state the rule (2, 9, 15, 9, 20) takes fewer than 150 time steps to reach its locally periodic limiting state and the rule (2, 9, 16, 9, 21) takes fewer than 125 time steps.

Let us illustrate LtL rule (2, 9, 16, 9, 20)'s march toward periodicity. As can be seen in Fig. 11.20, by time 5 all live sites are in finite disjoint configurations. These configurations grow so that by time 25 the density of live sites has increased. The stripes that appear are regions where the live sites are fixed but not necessarily for all times. This is because the regions where the dynamics are aperiodic continue to grow. By time 300 the regions of stripes are dispersed throughout the lattice. Between them the regions consist of aperiodic dynamics. In Fig. 11.21 we see that at time 600 some of the fixed regions from time 300 have grown while others have been taken over by aperiodic regions. The majority of the fixed regions from time

Fig. 11.20 From *left* to *right* are times 5, 25, and 300 of the rule $(2, 9, 16, 9, 20)$; initial state was product measure with density $p = 1/5$ on a system size of 256×200, and wrap around boundary conditions

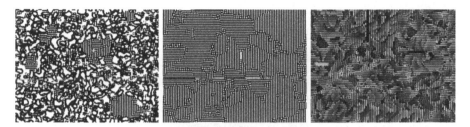

Fig. 11.21 The *first two* images, from *left* to *right* are times 600 and 2,814 of the rule $(2, 9, 16, 9, 20)$; the initial state was product measure with density $p = 1/5$ on a system size of 256×200, and wrap around boundary conditions. The *rightmost* image also started with density $1/5$ product measure, but the time at which each site became live is tracked using distinct shades of gray. The final limiting state was reached at time 3,852 and is depicted

600 survive, however, and eventually the aperiodic dynamics succumb to the fixed pattern, which is apparent at time 2,814.

The geometry of the limiting state of rule $(2, 9, 16, 9, 20)$ is noteworthy as well. As can be seen in Fig. 11.21, at time 2,814 the system's pattern consists of large regions of vertical and horizontal stripes of live and dead sites locked together like pieces of a puzzle. We have shown in [8] that an infinite pattern composed of either the vertical or horizontal stripes is an infinite still life for the rule. To illustrate how the eventual pattern of the limiting state evolved, we ran the rule on another initial state of product measure with density $p = 1/5$ and kept track of the time at which each site became live. We did this by assigning distinct shades of gray to the sites depending on when they turned on. Using this initial product measure, it took 3,852 time steps for the limiting state to be reached (this is even longer than before, which is not uncommon). That limiting state is depicted in the far right of Fig. 11.21. For additional (colorful) images, go to [7].

11.6.5 Slow Growth

This snapshot is included because of its interesting growth pattern starting from nonrandom initial conditions. For example, Fig. 11.22 shows the LtL rule (3, 14, 19,

Fig. 11.22 LtL rule
(3, 14, 19, 21, 49) starting
from a radius 50 circle of live
sites on a 300 × 300 lattice,
pictured at time 1,050.
Difference from *left* to *right*
is the palette, which uses
shades of gray to denote time
at which live sites turn on

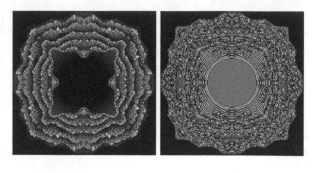

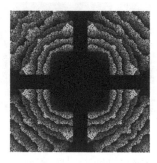

Fig. 11.23 LtL rule (3, 14, 19, 21, 49) starting from a radius 57 circle of live sites on a 300 × 300 lattice, pictured at time 1,000. At time 675 two rectangles of live sites, each with dimensions 5 × 300 were inserted into the picture's center and then the rule updated for another 325 time steps to generate the depicted figure, which reminds one of the contours which represent solutions to a differential equation

21, 49) starting from a radius 50 circle of live sites, after it has run 1,050 time steps. Shades of gray are used to indicate the time at which live sites turn on. Initially, the growth is rapid with the outermost edge of live sites growing at a rate of 1 each time step. (Another way to put this is $c/3$, where c is the *speed of light* or "the greatest speed at which any effect can propagate" [25]; here $c = 3$, since this is a range 3 rule.) However, the growth rate quickly slows down and concave regions fill in. The boundary becomes more convex and then concave regions again are formed and this growth pattern continues. By time 1,050 the configuration of live sites is contained in a 300 × 300 box. Thus, the average growth rate is less than 2/51.

Figure 11.23 shows the same LtL rule (3, 14, 19, 21, 49) starting from a radius 57 circle of live sites on a 300 × 300 lattice, pictured at time 1,000. At time 675 two rectangles of live sites, each with dimensions 5 × 300 were inserted into the picture's center and then the rule updated for another 325 time steps to generate the depicted image, which reminds one of the contour solutions to a differential equation. It would be interesting to study these boundary "curves" as time evolves.

Fig. 11.24 LtL rule
(2, 9, 9, 11, 25) started from
$p = 13/23$ product measure
after 150 time steps (*left*) and
600 time steps (*right*).
Diagonal, horizontal, and
vertical ladders are shown
along with patches of fixed
and periodic regions

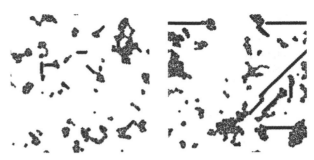

11.6.6 Ladders

Before discussing the next snapshot, which is depicted in Fig. 11.24, we need the
following definition from Griffeath and Moore [21]: "We will call a *ladder* a one-
dimensional periodic structure, which grows in one direction in a straight horizontal
or vertical line."

Figure 11.24 was generated by the rule (2, 9, 9, 11, 25) from a random initial
state with density 13/23. The "ladders" it generates are viable provided they do not
collide with any other configurations of live sites. The range one rule known as *Life
Without Death* (LWoD) [21], which inspired the ladder definition given above admits
analogous range one ladders, but only those that grow in the horizontal (or vertical)
direction. As seen in Fig. 11.24 this range two rule has that variety of ladders and
also those that grow in diagonal directions.

Starting from product measure with a suitably chosen density (close to $p =
13/23$) of live sites, the rule's dynamics seem to become nonuniformly locally
periodic rather quickly and ladders appear to be the only sets of 1's that grow
past a finite box. On the other hand the LWoD rule is more robust, generating
intricate snowflake-like patterns which might turn out to grow without bound.
(Dean Hickerson has shown that many LWoD patterns grow quadratically (D. Hickerson, pers. comm.); however, the "speed limit" question remains open (see
http://11011110.livejournal.com/174137.html, by D. Eppstein).)

We point out that, on a finite lattice with periodic boundary conditions, an initial product measure with $p = 13/23$ allows ladders to emerge from randomness.
However, if the density is too high (*e.g.* $p = 16/23$) then the initial live sites crowd
out ladders (*i.e.* no seeds have space to grow into ladders) and if p is too low (*e.g.*
$p = 8/23$) then with high probability everything dies (*i.e.* there aren't enough live
sites to form ladder seeds; of course, with an infinite random initial condition, regardless of how small p is, there will be a positive density of ladder seeds). This
is another illustration that, when conducting experiments on finite lattices, selecting
an appropriate p is an art form.

As mentioned in the introduction to this chapter, Griffeath and Moore have
proved that Ladder CAs can express arbitrary Boolean circuits, which shows the
problem of predicting such a CA for a finite amount of time is **P**-complete, the
question of whether a finite configuration grows to infinity is **P**-hard, and the long-
term behavior of initial conditions with a periodic background is undecidable [21].

Fig. 11.25 The seed is depicted on *top* and the orthogonal ladder it generates is shown *below* at time 150

Fig. 11.26 The seed is depicted on *top* and the diagonal ladder it generates is shown *below* at time 150

In their paper, they ask "Are there other ladder CA rules? [Besides LWoD.]" We claim that the range 2 LtL rule discussed above is a ladder CA. To show this, let us first recall Griffeath and Moore's definitions from [21]:

"Define a *ladder CA* as one in which:

1. There are finite seeds which give birth to horizontal or vertical ladders and nothing else.
2. Ladders can be turned: there are finite structures with which they can collide, producing a new ladder 90 degrees from the old one and nothing else.
3. One ladder can block another: a horizontal ladder (say) can collide with a vertical one which is already there in such a way that it stops without generating anything else."

Griffeath and Moore also "assume that the CA rule is rotationally symmetric, so that ladders and turns work in all four lattice directions. If these collisions are sensitive to the ladders' spatial or temporal phase, then we also require the following:

• Ladders can be shifted or delayed so that their phase can be adjusted as desired."

Griffeath and Moore state, "Two more behaviors follow from these requirements:

• Ladders can be delayed by arbitrary times, by executing a sequence of turns in a square spiral.
• Ladders can end."

Let us show that the LtL rule $(2, 9, 9, 11, 25)$ is a ladder CA. Since this range 2 CA also supports diagonal ladders, we will use *orthogonal ladder* to describe both horizontal and vertical ladders. Figure 11.25 shows that there is a finite seed which gives birth to an orthogonal ladder and nothing else. The seed is depicted alone to the north and below it is the ladder it forms after 150 time steps. Figure 11.26 shows that there is a finite seed which gives birth to a diagonal ladder and nothing else. The seed is depicted to the north and the diagonal ladder it generates after 150 time steps is depicted below. Thus, our CA satisfies Griffeath and Moore's axiom (1).

Now let us show that this range 2 CA satisfies Griffeath and Moore's axiom (2). The image on the left of Fig. 11.27 shows an orthogonal ladder and finite structures to its left and right. The image on the right shows the setup after 150 time steps. The ladder has been turned into a diagonal ladder. Figure 11.28 depicts a diagonal ladder and a finite structure below it (on the left). The image on the right of Fig. 11.28

Fig. 11.27 A horizontal ladder and the finite object that turns it into a diagonal ladder is depicted on the *left*. The same figure after the rule updates 150 time steps appears on the *right*

Fig. 11.28 A diagonal ladder and the finite object that turns it into a ladder is depicted on the *left*. The same figure after the rule updates 150 time steps appears on the *right*

Fig. 11.29 A horizontal ladder and a vertical ladder are depicted on the *left*. The result after the rule is run for 150 time steps is on the *right* and shows that the horizontal ladder collides with the vertical one and is stopped without generating anything else

Fig. 11.30 An unimpeded horizontal ladder is depicted on *top* of each image. The latter meets a finite structure which turns it into a diagonal ladder (*left*) and then another finite object turns it back into a horizontal ladder. The final result is that the original horizontal ladder has been shifted and delayed

shows that, after 150 time steps, the diagonal ladder has been turned into a vertical one. Combining Figs. 11.27 and 11.28 shows that an orthogonal ladder can be turned; that is, there are finite structures with which it can collide, producing a new ladder 90 degrees from the old one and nothing else. Similar constructions can be created to show that diagonal ladders can also be turned 90 degrees. Rather than include the construction here, the reader is encouraged to do it herself.

Finally, Fig. 11.29 shows that one ladder can block another; that is, the horizontal ladder depicted on the left is aimed at the vertical ladder to its right. After 150 time steps, the horizontal ladder collides with the vertical ladder which was already there in such a way that it stopped without generating anything else. Thus Griffeath and Moore's axiom (2) is satisfied. Similarly, a diagonal ladder can be blocked by another diagonal ladder and the reader is encouraged to show it.

As is the case with LWoD, the collisions supported by this range 2 CA are sensitive to the ladders' spatial or temporal phase, thus more work is required. First let us show that a ladder can be shifted and delayed. The image on the left of Fig. 11.30

Fig. 11.31 On the *left*: a ladder and a copy of itself with a strategically placed oscillator and on the *right* the same configuration after the rule has updated 150 times. Interaction with the oscillator pushed the original ladder ahead of where it otherwise would have been

shows an unimpeded horizontal ladder on top and the same ladder after it has been turned into a diagonal ladder (see Fig. 11.27 for details on the turn). The image on the right shows the same unimpeded ladder gaining ground as the diagonal ladder is turned back into a horizontal one (see Fig. 11.28 for details on the turn). The final result is that the original horizontal ladder has been shifted and delayed. Finally, Fig. 11.31 shows that a strategically placed oscillator can "push" an orthogonal ladder ahead of where it otherwise would have been.

This range 2 ladder CA has properties not exhibited by LWoD, namely diagonal ladders. On the other hand, empirical studies suggest that it is not capable of growing "chaotic lava" nor does it appear to support "parasitic shoots" that move along ladders, both of which are crucial aspects of LWoD's dynamics that make it difficult to predict. Various questions arise:

• Do diagonal ladders give the range 2 rule discussed here any properties that LWoD does not have?
• Is this ladder CA somehow easier to predict than LWoD?

The reader is encouraged to answer these questions.

11.6.7 Bugs with Trails

This snapshot is of the "Life-like" LtL rule $(5, 9, 9, 5, 5)$ which is distinct from other Life-like rules discussed in the previous section and in [10] and [11] because its bugs, which emerge from random initial configurations, generate *trails* of live sites. On finite lattices, empirical results suggest that the dynamics eventually fixate; however, the bugs' trails provide clues as to how the eventual pattern arose. (Note that the Life terminology for a "bug with a trail" is a *puffer*. More formally, according to the *Life lexicon*, "a puffer is an object that moves like a spaceship, except that it leaves debris behind" [25].)

Figure 11.32 shows that starting with density $p = 1/10$ product measure, by time 22 there are bugs heading west, north, and east, each leaving a trail, which appears as a sequence of parallel line segments. By time 58 the bug that had been heading east has met its demise in the form of a collision with a still life. The northbound bug has wrapped around the lattice to collide with a trail and vanish. The westbound bug has wrapped around the lattice and can be seen at time 58 heading west; the other northbound bug seen at time 58 emerged from the time 22 debris and is heading north. Can the reader guess what will happen next? (Hint: Fixation is reached by time 87.)

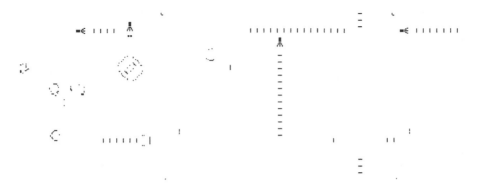

Fig. 11.32 From *left* to *right* are times 22 and 58 of the rule $(5, 9, 9, 5, 5)$; initial state was product measure with density $p = 1/10$ on a system size of 256×200, and periodic boundary conditions

Bugs with trails remind us of ladders; that is, the moving part in "front" leaves behind a fixed pattern. The questions thus arise:

- do these bugs with trails satisfy Griffeath and Moore's ladder axioms?
- how does this rule's computational power compare to that of a ladder CA?

11.6.8 Bug Logic, Bosco, and Open Questions

As is well known, Conway's initial interest in Life had to do with finding a CA with a simple update rule capable of simulating a *universal machine*; that is, a machine that can be programmed to perform any desired calculation. An early 50 dollar question was whether a finite pattern could generate an infinite population of live sites. Bill Gosper won the 50 dollars with the first "glider gun" and this provided Conway with a crucial building block in his proof that Life is universal. (A *gun* is any stationary pattern that emits spaceships forever [26].) In his proof, Conway showed how to define Life patterns that can imitate computers. That is, he showed that Life's patterns can be configured spatially to create *glider guns* as well as AND, OR, and NOT *logical gates*. He also described an *auxiliary storage device*, capable of holding arbitrarily large numbers [2].

In 1994 we showed several Life-like LtL rules to Conway and he challenged us to prove that some of them (which he called "interval rules") are universal. The most promising LtL rule (to date) is Bosco's rule, which is named for its period 166 oscillator and was discussed briefly in Sect. 11.3. Bosco's trajectory is depicted in Fig. 11.33: starting at time 0, moving northeast along the diagonal to the phase depicted at time 25 and then turning around at time 50 to move southwest along the diagonal. Observe that Bosco's phases at times 83 and 108 are rotated translations of those that are depicted at times 0 and 25, respectively. At time 133, a rotated translation of Bosco's time 50 turn would appear, but depicted instead is time 137, which is Bosco along with its spark (see Fig. 11.34). At time 166 Bosco is back to the starting position.

Fig. 11.33 Times 0, 25, 50, 83, 108, and 137 of Bosco's trajectory

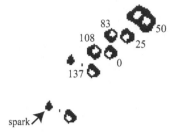

Fig. 11.34 Bosco depicted with its spark. The *large arrow* is pointing to Bosco's spark, which appears northwest of Bosco. The two adjacent cells closer to Bosco are also part of the spark

Fig. 11.35 Period 166 bug maker (or "gun") supported by Bosco's rule. At time 0 (not depicted), two copies of the period 166 oscillator named Bosco are strategically placed (in the west) so that every 166 time steps one additional bug emerges and heads east. By time 376 (depicted), three bugs have emerged. Eventually an infinite stream of bugs is generated

We have shown that Bosco's rule supports patterns such as those Conway used to prove that Life is universal. Specifically, we built a *sliding block memory*, similar to the auxiliary storage device Conway described and claimed could be built [13]. Our construction was based on Life's sliding block memory designed and built by Dean Hickerson in 1990 [22]. Because a *bug maker*, which is synonymous with a *glider gun* (but for those less violently inclined) is one of the crucial ingredients, we include a picture of a period 166 bug maker supported by Bosco's rule in Fig. 11.35.

Paul Rendell implemented a Turing Machine in Conway's Game of Life in 2000 [24]. The Minsky Register Machine (MRM) is a model of computation equivalent in power to the Turing Machine [23]. According to Paul Chapman, a Universal MRM (URM) can be built which can emulate any other MRM by encoding its instructions and register contents as initial values of the URM's registers. In 2002 Chapman constructed a Life pattern, based on Hickerson's Sliding Block Memory, which implements the actions of a URM [3]. Chapman claims his construction is a true model of universal computation in Life and that, unlike the finite tape of Rendell's Turing Machine, the values in the URM's registers are unbounded.

The question remains: Is Bosco's rule universal? What about other large range Life-like LtL rules? If Life is universal, then these Life-like rules most likely are as well. What are the necessary and sufficient ingredients required to prove it? A goal is to create a set of axioms, much like the one Griffeath and Moore have for Ladder CAs, such that, if a specific CA satisfies the axioms, then one can say that the rule is **P**-complete, universal, and other things. We expect that rules which support bugs,

Fig. 11.36 From *left* to *right*: Bug supported by LtL rule (5, 34, 45, 34, 58), which has been enlarged. This translation invariant bug was scaled up to ranges 25, 50, 100, and 200. Those bugs appear from *left* to *right* and are supported by the respective rules: (25, 706, 958, 706, 1,216), (50, 2,773, 3,755, 2,773, 4,738), (100, 10,983, 14,849, 10,983, 18,764), and (200, 43,712, 59,102, 43,712, 74,685). The range 25, 50, and 100 bugs are also translation invariant

but whose dynamics relax quite quickly (*i.e.* rules not as robust as Life) are probably not as powerful as Life-like rules such as Bosco's rule. On the other hand, there are also Life-like LtL rules that seem to support no finite still lifes; however, they support puffers and/or rakes. What can be said about the computational powers of such rules?

We add that Bosco's spark, which is depicted in Fig. 11.34 is useful because it may interact with other live sites without affecting Bosco. (More generally, a *spark* is a pattern that dies. The term is typically used to describe a collection of cells periodically thrown off by an oscillator or bug [26].) Bosco's spark is located quite a distance away, which is key to the numerous reactions needed for constructing bug makers and other reactions required in the construction of the sliding block memory. A similar periodic object with a spark was used by Gosper to build Life's first glider gun. The question thus arises: is a periodic object which has a spark one of the sufficient ingredients for proving a "Life-like" CA from the LtL family is universal?

11.6.9 LtL's Bugs: Threshold-Range Scaling

We end this section with our favorite LtL question about the threshold-range scaling limit of LtL's bugs. That is, given a bug from a fixed range, we have shown that there is a sequence of bugs, found by rescaling the given bug to increasing ranges, that appears to converge to a limit [11]. For example, see Fig. 11.36. What is the geometry of this limiting Euclidean shape? Observe that the bugs appear to be mathematically "similar" in the sense of scaling proportionally. That is, they seem to have the same shape, but different sizes. There are numerous other bug sequences of "similar" bugs; for example, there is a sequence for the bug depicted in Fig. 11.37, which we include because of its noteworthy, nearly symmetric shape. See [11] for additional examples and details.

Fig. 11.37 Bug supported by LtL rule (100, 10,983, 18,479, 10,983, 18,564), which is noteworthy due to its nearly symmetric shape. Nevertheless, it does move north. We leave it to the reader
to determine its period and speed

Fig. 11.38 From *left* to *right*: period 2 bug supported by LtL rule with "ball" neighborhood and
parameters (100, 8,826, 12,220, 8,826, 15,740) and period 3 bug supported by LtL rule with "diamond" neighborhood and parameters (100, 5,612, 7,585, 5,612, 9,776)

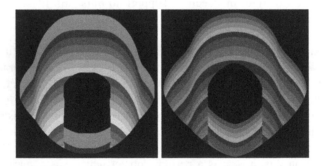

Fig. 11.39 From *left* to *right* are range 200 bugs supported by LtL rules (200, 43,712, 59,102,
43,712, 74,685) and (200, 44,444, 70,666, 44,444, 77,898), respectively. In both bugs' pictures,
the time at which each site became live is depicted using distinct shades of gray. The color illustrates the bugs' aerodynamic geometries

The "bugs with stomachs" depicted in Figs. 11.36 and 11.37 are also supported
by large range LtL rules with "ball" neighborhoods, which are discrete approximations of circles, and "diamond" neighborhoods, which are generalizations of
von Neumann neighborhoods. For example, see the range 100 bugs depicted in
Fig. 11.38.

The current record holders for largest range bugs are those from range 200 depicted in Fig. 11.39 (the one on the left is also depicted in black on the right
of Fig. 11.36). Both were found using the threshold-range scaling algorithm described in [11]. That is, both rules were run on the initial configuration depicted in
Fig. 11.40, which is a circular region of 1's that contains an off-center "hole" of
0's. The rules then "sculpted" the initial state into the range 200 bugs depicted in

Fig. 11.40 Initial
configuration for bugs
depicted in Fig. 11.39

Fig. 11.41 "Alien" bug
supported by LtL rule
(100, 9, 350, 12, 360,
10, 350, 15, 740)

Fig. 11.39. In both pictures, the time at which each site became live is depicted using distinct shades of gray. The color illustrates the bugs' aerodynamic geometries.

Finally, the record holder for most complex bug geometry is the range 100 bug depicted in Fig. 11.41. It is especially intriguing due to its apparent "wings." Its period remains an open question. Observe that the rule that supports it has $\beta_1 \neq \delta_1$. Perhaps there is a sequence of bugs each of which is mathematically similar to this one and that converges to a Euclidean bug with wings (and a stomach).

11.7 Software

The experimental nature of this research makes it impossible without appropriate software. Thus far we have used *WinCA* and *Mirek's Celebration* (or MCell). WinCA is an interactive modeling environment for Windows created by Bob Fisch and David Griffeath in the early 1990s and is freeware available from Griffeath's website [20]. It is on the older side; for example, using Vista, it must run in compatibility mode for one of the earlier versions of Windows (XP with service pack 2 is lightning fast). However, it *is* aging like Mick Jagger: not as young as it used to be, but still rockin'. In fact, as clock speeds on personal computers increase, so too does WinCA's ability to run large range LtL rules on large lattice sizes. MCell is a CA modeling environment, created by Mirek Wojtowicz, which allows one to create detailed initial states on the fly and is freeware also available for download at Griffeath's website [20]. The largest LtL range currently handled by MCell is 10.

Creating intricate initial configurations with MCell or WinCA still requires an external paint program. Newer "paint" programs emphasize digital photography, making interaction with WinCA more complicated than it was in the last millennium. This research would be enhanced by new software that runs LtL rules, combines the best qualities of WinCA and MCell, and interacts well with current photo-editing software. Perhaps one of the fantastic Life programs, such as Golly, could be modified accordingly [28].

11.8 Conclusion and Long-Term Goals

In this chapter, we presented rigorous and speculative results about the LtL family of CA rules. There are vast regions of LtL space that have yet to be explored and numerous long-term goals. For example, we aim to understand the geometry of LtL space in terms of the ergodic classifications of its rules as well as local coherent structures the rules support (*e.g.* does the rule support bugs and other structures that can be used to construct bug makers and other ingredients needed to show the rule has various computational powers?). For a fixed range ρ, we would like to create an interactive four-dimensional Rubik's-like cube, where each rule is represented by a small cube. Each such small cube is colored according to the rule's known properties (*e.g.* green might be "Life-like"). The interactive cube would allow the user to look at various one-, two- and three-dimensional cross sections. If the user "zooms out" diagrams such as the one in Fig. 11.13 would be shown. If the user "zooms in" far enough, the user would be able to run the rule in real time (and the user would be able to set the initial condition, conduct experiments, and more). The larger goal is to determine which LtL aspects scale coherently, specifically in the limit as the range approaches ∞. In this limit, the "lattice" becomes $\mathbf{R}^2$ so the big question is: what are the dynamics and coherent structures supported by the limiting "Euclidean" family of rules that live in $\mathbf{R}^2$?

Acknowledgements Thanks to Andrew Adamatzky for his interest in this work and to Carl, Duncan, Jadyn, Joan, and Mike for assistance finding time for (and/or tolerating) Larger than Life.

References

1. Aizenman, M., Lebowitz, J.: Metastability effects in bootstrap percolation. J. Phys. A, Math. Gen. **21**, 3801–3813 (1988)
2. Berlekamp, E., Conway, J., Guy, R.: What is Life? In: Winning Ways for Your Mathematical Plays, vol. 2, Chap. 25. Academic Press, New York (1982)
3. Chapman, P.: Life universal computer. http://www.igblan.free-online.co.uk/igblan/ca/ (2002). Cited 6 December 2009
4. Crutchfield, J., Hanson, J.: Turbulent pattern bases for cellular automata. Physica D **69**, 279–301 (1993)
5. Durrett, R.: Probability: Theory and Examples. Wadsworth and Brooks Cole, Pacific Grove (1991)

6. Durrett, R., Steif, J.: Fixation results for threshold voter systems. Ann. Probab. **21**(1), 232–247 (1993)
7. Evans, K.: Larger than Life's extremes: a gallery of images. http://www.csun.edu/~kme52025/extremes.html
8. Evans, K.: Larger than Life's invariant measures. In: Electronic Notes in Theoretical Computer Science, vol. 252, pp. 55–75. Elsevier, Amsterdam (2009)
9. Evans, K.: Larger than Life: it's so nonlinear. Ph.D. Dissertation, University of Wisconsin–Madison (1996). http://www.csun.edu/~kme52026/thesis.html
10. Evans, K.: Larger than Life: digital creatures in a family of two-dimensional cellular automata. In: Cori, R., Mazoyer, J., Morvan, M., Mosseri, R. (eds.) Discrete Models: Combinatorics, Computation, and Geometry, DM-CCG 2001. Discrete Mathematics and Theoretical Computer Science Proceedings AA, pp. 177–192. DMTCS, LORIA, Nancy (2001)
11. Evans, K.: Threshold-range scaling of Life's coherent structures. Physica D **183**, 45–67 (2003)
12. Evans, K.: Replicators and Larger than Life examples. In: Griffeath, D., Moore, C. (eds.) New Constructions in Cellular Automata, pp. 119–159. Oxford University Press, London (2003)
13. Evans, K.: Is Bosco's rule universal? In: Margenstern, M. (ed.) Machines, Computations, and Universality. Lecture Notes in Computer Science, vol. 3354, pp. 188–199. Springer, Berlin (2005). WWW companion page to paper: http://www.csun.edu/~kme52026/bosco/bosco.html
14. Fisch, R., Gravner, J., Griffeath, D.: Cyclic cellular automata in two dimensions. In: Alexander, K.S., Watkins, J.C. (eds.) Spacial Stochastic Processes; A Festschrift in Honor of the Seventieth Birthday of T.E. Harris, pp. 171–185. Birkhauser, Boston (1991)
15. Gardner, M.: Mathematical Games — The fantastic combinations of John Conway's new solitaire game, Life. Sci. Am. **223**, 120–123 (1970)
16. Gravner, J., Griffeath, D.: Threshold growth dynamics. Trans. Am. Math. Soc. **340**(2), 837–870 (1993)
17. Gravner, J., Griffeath, D.: First passage times for discrete threshold growth dynamics. Ann. Probab. **24**(4), 1752–1778 (1996)
18. Gravner, J., Griffeath, D.: Cellular automaton growth on $\mathbf{Z}^2$: theorems, examples, and problems. Adv. Appl. Math. **21**, 241–304 (1998). http://psoup.math.wisc.edu/extras/r1shapes/r1shapes.html. Cited 10 December 2009
19. Griffeath, D.: Self-organization of random cellular automata: four snapshots. In: Grimmett, G. (ed.) Probability and Phase Transitions, pp. 49–67. Kluwer Academic, Dordrecht/Norwell (1994)
20. Griffeath, D.: Primordial Soup Kitchen. http://math.wisc.edu/~griffeat/kitchen.html. Cited 10 December 2009
21. Griffeath, D., Moore, C.: Life Without Death is P-complete. Complex Syst. **10**, 437–447 (1998). http://psoup.math.wisc.edu/java/lwodpc/lwodpc.html. Cited 1 December 2009
22. Hickerson, D.: Description of sliding block memory. http://www.radicaleye.com/lifepage/patterns/sbm/sbm.html (1990). Cited 20 November 2009
23. Minsky, M.: Computation: Finite and Infinite Machines, Sect. 15.1. Prentice Hall, New York (1967)
24. Rendell, P.: Turing machine implemented in Conway's Game of Life. http://rendell-attic.org/gol/tm.htm (2000). Cited 20 November 2009
25. Silver, S.: Steven Silver's Life Page. http://www.argentum.freeserve.co.uk/life.htm. Cited 1 December 2009
26. Silver, S.: Life Lexicon. Compiled by Silver, S.A. http://www.argentum.freeserve.co.uk/lex.htm. Cited 1 December 2009
27. Summers, J.: Game of Life status page. http://home.mieweb.com/jason/life/status.html. Cited 1 December 2009
28. Trevorrow, A., Rokicki, T. (with code contributions by Hutton, T., Greene, D., Summers, J.): Golly: open source, cross-platform application for exploring Conway's Game of Life and other cellular automata. http://golly.sourceforge.net/. Cited 9 December 2009
29. LifeWiki, the wiki for Conway's Game of Life. http://www.conwaylife.com/wiki/index.php?title=Main_Page. Cited 1 December 2009

Chapter 12
RealLife

Marcus Pivato

Let $D \in \mathbb{N}$, let λ be the D-dimensional Lebesgue measure on $\mathbb{R}^D$, and let $\mathbf{L}^\infty :=$ $\mathbf{L}^\infty(\mathbb{R}^D, \lambda)$. Let $\mathscr{A} := \{0, 1\}$ and let $\mathscr{A}^{\mathbb{R}^D} \subset \mathbf{L}^\infty$ be the set of all Borel-measurable functions $\mathbf{a} : \mathbb{R}^D \longrightarrow \mathscr{A}$, which we refer to as *configurations*. If $v \in \mathbb{R}^D$, then we define the shift map $\sigma^v : \mathscr{A}^{\mathbb{R}^D} \longrightarrow \mathscr{A}^{\mathbb{R}^D}$ by $\sigma^v(\mathbf{a}) := \mathbf{a}'$, where $\mathbf{a}'(x) := \mathbf{a}(x + v)$ for all $x \in \mathbb{R}^D$. A *Euclidean automaton* (EA) is a function $\Phi : \mathscr{A}^{\mathbb{R}^D} \longrightarrow \mathscr{A}^{\mathbb{R}^D}$ such that:

- $\Phi \circ \sigma^v = \sigma^v \circ \Phi$ for all $v \in \mathbb{R}^D$ (i.e. Φ commutes with all shifts); and
- there is some compact neighbourhood $\mathbf{K} \subset \mathbb{R}^D$ of zero such that, if $\mathbf{a}, \mathbf{a}' \in \mathscr{A}^{\mathbb{R}^D}$, and $\mathbf{a}|_\mathbf{K} = \mathbf{a}'|_\mathbf{K}$, then $\Phi(\mathbf{a})(0) = \Phi(\mathbf{a}')(0)$ (i.e. Φ is determined by local information).

Thus, a Euclidean automaton is like a cellular automaton, but the underlying 'space' is the D-dimensional continuum $\mathbb{R}^D$ rather than the discrete integer lattice $\mathbb{Z}^D$. One way to visualize a Euclidean automaton is as follows. Imagine a sequence of cellular automata $\{\Phi_n\}_{n=1}^\infty$ acting on $\mathscr{A}^{\mathbb{Z}^D}$ with larger and larger neighbourhoods. Suppose $\{\Phi_n\}_{n=1}^\infty$ all have 'qualitatively similar' behaviour when the 'lattice spacing' of $\mathbb{Z}^D$ is rescaled so that neighbourhoods of $\{\Phi_n\}_{n=1}^\infty$ are all the same size and shape (e.g. they all look like the unit ball). As $n \to \infty$, the 'lattice spacing' goes to zero, and the sequence $\{\Phi_n\}_{n=1}^\infty$ might converge to a 'continuum limit' — a transformation Φ on $\mathscr{A}^{\mathbb{R}^D}$. In this case, Φ is a Euclidean automaton.

In particular, suppose $\{\Phi_n\}_{n=1}^\infty$ is a sequence of **Larger than Life** CA with larger and larger radius, but all of which have the same birth/survival thresholds (see Chap. 11 in this volume). Then the 'continuum limit' of the sequence $\{\Phi_n\}_{n=1}^\infty$ is a Euclidean automaton called RealLife, which we now describe.

Let $\mathscr{K} := \{\kappa \in \mathbf{L}^\infty(\mathbb{R}^D; [0, \infty)); \kappa \text{ has compact support, and } \int_{\mathbb{R}^D} \kappa = 1\}$. For any $\kappa \in \mathscr{K}$ and $\mathbf{a} \in \mathscr{A}^{\mathbb{R}^D}$, we define the *convolution* of $\mathbf{a}$ by κ as follows:

$$\kappa * \mathbf{a}(x) := \int_{\mathbb{R}^D} \kappa(y) \cdot \mathbf{a}(x - y) \, d\lambda[y].$$

For example, let $\mathbf{K} \subset \mathbb{R}^D$ be a compact neighbourhood of zero (e.g. a ball or a cube), and let $\kappa := \lambda[\mathbf{K}]^{-1} \mathbf{1}_\mathbf{K}$ be the indicator function of $\mathbf{K}$, normalized to have

A. Adamatzky (ed.), *Game of Life Cellular Automata*,
DOI 10.1007/978-1-84996-217-9_12, © Springer-Verlag London Limited 2010

integral 1. Then $\kappa \in \mathscr{K}$, and $\kappa * \mathbf{a}(x) = \lambda[\mathbf{K}]^{-1} \int_{\mathbf{K}} \mathbf{a}(x - k) \, d\lambda[k]$ is the average value of $\mathbf{a}$ near x. If $0 < s_0 \leq b_0 < b_1 \leq s_1 \leq 1$, then we define the corresponding *RealLife* Euclidean automaton $\maltese = {}_\kappa \maltese_{b_0,s_0}^{b_1,s_1} : \mathscr{A}^{\mathbb{R}^D} \longrightarrow \mathscr{A}^{\mathbb{R}^D}$ as follows:

$$\forall \mathbf{a} \in \mathscr{A}^{\mathbb{R}^D}, \quad \maltese(\mathbf{a})(x) := \begin{cases} 1 & \text{if } \mathbf{a}(x) = 1 \text{ and } s_0 \leq \kappa * \mathbf{a}(x) \leq s_1; \\ 1 & \text{if } \mathbf{a}(x) = 0 \text{ and } b_0 \leq \kappa * \mathbf{a}(x) \leq b_1; \quad (12.1) \\ 0 & \text{otherwise.} \end{cases}$$

Let $\Theta := \{(s_0, b_0, b_1, s_1); \ 0 < s_0 \leq b_0 < b_1 \leq s_1 \leq 1\}$ be the set of *threshold four-tuples*. The Euclidean automaton $\maltese$ depends upon the choice of kernel $\kappa \in \mathscr{K}$ and the four-tuple $(s_0, b_0, b_1, s_1) \in \Theta$ (we will normally suppress this dependency in our notation).

In her empirical explorations of Larger than Life CA, Evans observed the emergence of complex, compactly supported fixed points (called *still lifes*), periodic solutions (called *oscillators*) and propagating structures called (called *bugs*), analogous to the *gliders* and *space-ships* in the Game of Life. Intriguingly, the still lifes, oscillators, and bugs found in large-radius Larger than Life CA appear to be rescaled, 'high resolution' versions of those found in small-radius Larger than Life CA. Evans conjectured that these still lifes (resp. oscillators, bugs) converged to 'continuum limits', which were still lifes (resp. oscillators, bugs) for some Euclidean automaton.

In [5], the author defined the RealLife family of EA, established some basic continuity properties for them, and then proved Evans' conjecture (using a suitable definition of 'continuum limit'). The proofs of these results involve some rather technical real analysis. However, once these basic results are established, RealLife EA are actually *easier* to analyse than Larger than Life CA, because we can apply the entire toolbox of modern continuum mathematics. For example, life forms in RealLife often have smooth boundaries, so they can be analyzed using differential calculus; this technique was used in [5] to prove existence theorems for several classes of still lifes in RealLife, and to show that these still lifes are stable under small perturbations.

In Sect. 12.1, we summarise the main results of [5]. In Sect. 12.2, we offer a heuristic explanation for the morphology of the typical 'bug' in RealLife and Larger than Life.

12.1 What Happens in RealLife

Let $0 < s_0 \leq b_0 < b_1 \leq s_1 \leq 1$ be the parameters introduced above. We refer to $[b_0, b_1]$ as the 'birth interval' and $[s_0, s_1]$ as the 'survival interval'. If we define $\mathfrak{b} := \mathbf{1}_{[b_0, b_1]}$ and $\mathfrak{s} := \mathbf{1}_{[s_0, s_1]}$, then, for any $\mathbf{a} \in \mathscr{A}^{\mathbb{R}^D}$ and $x \in \mathbb{R}^D$, we can rewrite Eq. 12.1 as

$$\maltese(\mathbf{a})(x) := \mathbf{a}(x) \cdot \mathfrak{s}(\kappa * \mathbf{a}(x)) + (1 - \mathbf{a}(x)) \cdot \mathfrak{b}(\kappa * \mathbf{a}(x)). \quad (12.2)$$

Recall that cellular automata are continuous in the Cantor space metric on $\mathscr{A}^{\mathbb{Z}^D}$. Likewise, it is easy to show that Euclidean automata are continuous in the 'compact-open' metric on $\mathscr{A}^{\mathbb{R}^D}$ (the obvious analogy of the Cantor metric: $d(\mathbf{a}, \mathbf{a}')$ is 'small' if $\mathbf{a}$ and $\mathbf{a}'$ agree in a neighbourhood around the origin). However, this metric is not shift-invariant, and is somewhat counterintuitive: $d(\mathbf{a}, \mathbf{a}')$ can be large even if $\mathbf{a}$ and $\mathbf{a}'$ differ on a set of tiny measure. We will find it more suitable to work in the L^1 metric.

Let $\mathbf{L}^1 := \mathbf{L}^1(\mathbb{R}^D, \lambda) = \{f : \mathbb{R}^D \longrightarrow \mathbb{R}; \int |f| < \infty\}$; thus, $^1\mathscr{A}^{\mathbb{R}^D} := \mathscr{A}^{\mathbb{R}^D} \cap \mathbf{L}^1$ is the set of configurations whose support has finite measure. Note that $\maltese(^1\mathscr{A}^{\mathbb{R}^D}) \subseteq {}^1\mathscr{A}^{\mathbb{R}^D}$ (because $0 < b_0$). Extend $\maltese$ to a function $\maltese : \mathbf{L}^1 \longrightarrow \mathbf{L}^1$ by applying Eq. 12.2 in the obvious way. For any $\mathbf{a} \in {}^1\mathscr{A}^{\mathbb{R}^D}$, let $M(\mathbf{a}) := \lambda[\alpha^{-1}\{s_0, b_0, s_1, b_1\}]$, where $\alpha := \kappa * \mathbf{a}$. We define

$$^0\mathscr{A}^{\mathbb{R}^D} := \{\mathbf{a} \in {}^1\mathscr{A}^{\mathbb{R}^D}; \ M(\mathbf{a}) = 0\}.$$

(Note that the definition of $^0\mathscr{A}^{\mathbb{R}^D}$ depends on (s_0, b_0, s_1, b_1) and κ.) The operator $\maltese$ is L^1-continuous on $^0\mathscr{A}^{\mathbb{R}^D}$, which is generally a 'large' subset of $^1\mathscr{A}^{\mathbb{R}^D}$. To explain this, we must define 'thin' sets. If $\mathbf{T} \subset \mathbb{R}^D$, and $\gamma > 0$, let $\mathbb{B}(\mathbf{T}, \gamma) := \{x \in \mathbb{R}^D; d(x, \mathbf{T}) < \gamma\}$. Say $\mathbf{T}$ is *thin* if $\lim_{\gamma \to 0} \lambda[\mathbb{B}(\mathbf{T}, \gamma)] = 0$. For example, any compact, piecewise smooth $(D - 1)$-submanifold of $\mathbb{R}^D$ is thin. The kernel κ is *almost continuous* if there is a thin set $\mathbf{T} \subset \mathbb{R}^D$ such that, for any $\gamma > 0$, κ is uniformly continuous on $\mathbb{B}(\mathbf{T}, \gamma)^{\complement}$. For example:

- If κ is continuous, then κ is almost continuous (κ has compact support, so continuity implies uniform continuity).
- If $\mathbf{K}$ is an open set and $\partial \mathbf{K}$ is thin (e.g. $\partial \mathbf{K}$ is a piecewise smooth manifold), then $\kappa := \lambda\lfloor\mathbf{K}\rfloor^{-1}\mathbf{1}_{\mathbf{K}}$ is almost continuous.

Theorem 1 *Let $(s_0, b_0, b_1, s_1) \in \Theta$, and let $\kappa \in \mathscr{K}$.*

(a) $\maltese$ *is L^1-continuous on $^0\mathscr{A}^{\mathbb{R}^D}$.*

(b) *If κ is almost-continuous, then $^0\mathscr{A}^{\mathbb{R}^D}$ is a shift-invariant, dense $G\delta$ subset of $^1\mathscr{A}^{\mathbb{R}^D}$.*

Proof [5, Theorems 1.1 and 1.2]. $\square$

(Note: $^0\mathscr{A}^{\mathbb{R}^D}$ is generally a strict subset of $^1\mathscr{A}^{\mathbb{R}^D}$, and $\maltese$ is *not* L^1-continuous on all of $^1\mathscr{A}^{\mathbb{R}^D}$; see [5, §1] for details.)

Let Φ be a Euclidean automaton and let $\mathbf{a} \in \mathscr{A}^{\mathbb{R}^D}$. We say $\mathbf{a}$ is a *still life* for Φ if $\Phi(\mathbf{a}) = \mathbf{a}$. If $p \in \mathbb{N}$, then $\mathbf{a}$ is a *p-oscillator* if $\Phi^p(\mathbf{a}) = \mathbf{a}$ (a still life is thus a 1-oscillator). If $p \in \mathbb{N}$ and $v \in \mathbb{R}^D$, then $\mathbf{a}$ is a *p-periodic bug* with *velocity* v if $\Phi^p(\mathbf{a}) = \sigma^{pv}(\mathbf{a})$. Still lifes, oscillators, and bugs are all referred to as *life forms*.

Recall that ✝ is determined by the parameters (s_0, b_0, b_1, s_1) and κ. A small perturbation of in these parameters should yield a small change in ✝, and a small change in its life forms. In particular, if $\{\kappa\}_{n=1}^{\infty}$ is a sequence of kernels converging to κ, and $\{(s_0^n, b_0^n, b_1^n, s_1^n)\}_{n=1}^{\infty}$ is a sequence of four-tuples converging to the four-tuple (s_0, b_0, b_1, s_1), then the corresponding sequence $\{✝_n\}_{n=1}^{\infty}$ of RealLife EA should converge to the RealLife EA ✝ determined by (s_0, b_0, b_1, s_1) and κ, and the life forms of $\{✝_n\}_{n=1}^{\infty}$ should 'evolve' toward life forms for ✝.

Formally, let $\{\Phi_n\}_{n=1}^{\infty}$ be a sequence of Euclidean automata, and let $\mathfrak{A} \subseteq \mathscr{A}^{\mathbb{R}^D}$ be a shift-invariant subset. The sequence $\{\Phi_n\}_{n=1}^{\infty}$ *evolves to* Φ *on* $\mathfrak{A}$ if, for any $\{\mathbf{a}_n\}_{n=1}^{\infty} \subset \mathscr{A}^{\mathbb{R}^D}$ such that $\mathbf{L}^1\text{-}\lim_{n \to \infty} \mathbf{a}_n = \mathbf{a} \in \mathfrak{A}$, the following holds:

(a) If $\Phi_n(\mathbf{a}_n) = \mathbf{a}_n$ for all $n \in \mathbb{N}$, then $\Phi(\mathbf{a}) = \mathbf{a}$.

(b) Let $P \in \mathbb{N}$, and suppose $\Phi^p(\mathbf{a}) \in \mathfrak{A}$ for all $p \in [0 \ldots P)$.

    [i] If $\Phi_n^P(\mathbf{a}_n) = \mathbf{a}_n$ for all $n \in \mathbb{N}$, then $\Phi^P(\mathbf{a}) = \mathbf{a}$.

    [ii] If $\{v_n\}_{n=1}^{\infty} \subset \mathbb{R}^D$ and $\lim_{n \to \infty} v_n = v \in \mathbb{R}^D$, and $\Phi_n^P(\mathbf{a}_n) = \sigma^{Pv_n}(\mathbf{a}_n)$ for all $n \in \mathbb{N}$, then $\Phi^P(\mathbf{a}) = \sigma^{Pv}(\mathbf{a})$.

Theorem 2 *Fix* $(s_0, b_0, b_1, s_1) \in \Theta$ *and* $\kappa \in \mathscr{K}$. *Let* ✝ *be the* RealLife *EA defined by* (s_0, b_0, b_1, s_1) *and* κ.

- *Let* $\{\kappa_n\}_{n=1}^{\infty} \subset \mathscr{K}$. *For all* $n \in \mathbb{N}$, *let* ✝$_n$ *be the* RealLife *EA defined by* (s_0, b_0, b_1, s_1) *and* κ_n. *Suppose* $\mathbf{L}^1\text{-}\lim_{n \to \infty} \kappa_n = \kappa$. *Then*

(a) $\mathbf{L}^1\text{-}\lim_{n \to \infty} ✝_n(\mathbf{a}) = ✝(\mathbf{a})$, *for all* $\mathbf{a} \in {}^0\mathscr{A}^{\mathbb{R}^D}$.

(b) *If* $\sup_{n \in \mathbb{N}} \|\kappa_n\|_{\infty} < \infty$, *then* $\{✝_n\}_{n=1}^{\infty}$ *evolves to* ✝ *on* ${}^0\mathscr{A}^{\mathbb{R}^D}$.

- *Let* $\{(s_0^n, b_0^n, b_1^n, s_1^n)\}_{n=1}^{\infty} \subset \Theta$. *For each* $n \in \mathbb{N}$, *let* ✝$_n$ *be the* RealLife *EA defined by* $(s_0^n, b_0^n, b_1^n, s_1^n)$ *and* κ. *Suppose* $\lim_{n \to \infty}(s_0^n, b_0^n, b_1^n, s_1^n) = (s_0, b_0, b_1, s_1)$. *Then*

(c) $\mathbf{L}^1\text{-}\lim_{n \to \infty} ✝_n(\mathbf{a}) = ✝(\mathbf{a})$, *for all* $\mathbf{a} \in {}^0\mathscr{A}^{\mathbb{R}^D}$.

(d) $\{✝_n\}_{n=1}^{\infty}$ *evolves to* ✝ *on* ${}^0\mathscr{A}^{\mathbb{R}^D}$.

Proof [5, Theorems 1.3 and 1.4].   □

Example 3 Fix $(s_0, b_0, b_1, s_1) \in \Theta$. For any $n \in \mathbb{N} \sqcup \{\infty\}$, let $\mathbf{K}_n \subset \mathbb{R}^D$ be a compact set, let $\kappa_n := \lambda[\mathbf{K}_n]^{-1} \mathbf{1}_{\mathbf{K}_n}$, and let ✝$_n$ be the RealLife EA defined by (s_0, b_0, b_1, s_1) and κ_n. Suppose $\lim_{n \to \infty} \lambda[\mathbf{K}_n \triangle \mathbf{K}_\infty] = 0$, as shown in Fig. 12.1. Then $\mathbf{L}^1\text{-}\lim_{n \to \infty} ✝_n(\mathbf{a}) = ✝_\infty(\mathbf{a})$, for all $\mathbf{a} \in {}^0\mathscr{A}^{\mathbb{R}^D}$, and $\{✝_n\}_{n=1}^{\infty}$ evolves to ✝$_\infty$ on ${}^0\mathscr{A}^{\mathbb{R}^D}$ [5, Corollary 1.5].

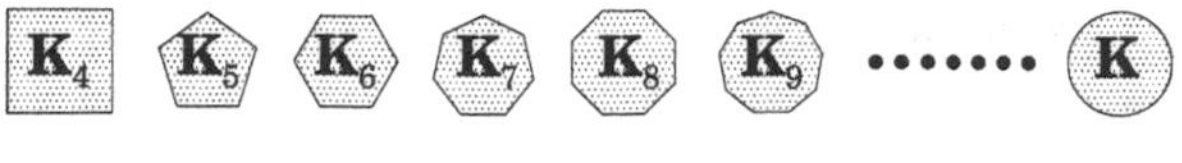

Fig. 12.1 Example 3

Fig. 12.2 The cells of the sigma-algebra $\mathscr{B}_\varepsilon$

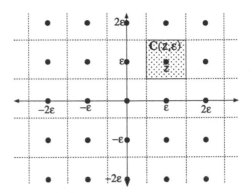

12.1.1 RealLife vs. Larger than Life

In what sense is a RealLife EA the 'continuum limit' of a sequence of Larger than Life CA? Fix $\varepsilon > 0$, and let $\varepsilon \mathbb{Z}^D := \{\varepsilon z; \ z \in \mathbb{Z}^D\}$. For any $z \in \mathbb{Z}^D$, let $\mathbf{C}(z, \varepsilon) := \varepsilon z + [\frac{-\varepsilon}{2}, \frac{\varepsilon}{2})^D$ be the half-open cube of sidelength ε, centred at εz. Let $\mathscr{B}_\varepsilon$ be the sigma algebra generated by $\{\mathbf{C}(z, \varepsilon); \ z \in \mathbb{Z}^D\}$ (see Fig. 12.2). Let $\mathbf{L}_\varepsilon^1 := \mathbf{L}^1(\mathbb{R}^D, \mathscr{B}_\varepsilon, \lambda)$ (that is, $\mathbf{L}_\varepsilon^1$ is the set of all $f \in \mathbf{L}^1$ such that f is constant on $\mathbf{C}(z, \varepsilon)$ for each $z \in \mathbb{Z}^D$). Let $\ell^1 := \{f : \mathbb{Z}^D \longrightarrow \mathbb{R}; \ \sum_{z \in \mathbb{Z}^D} |f(z)| < \infty\}$. Then there is a natural isomorphism $\mathbf{L}_\varepsilon^1 \cong \ell^1$. If $^\varepsilon \mathscr{A}^{\mathbb{R}^D} := {}^1 \mathscr{A}^{\mathbb{R}^D} \cap \mathbf{L}_\varepsilon^1$, then $^\varepsilon \mathscr{A}^{\mathbb{R}^D} \cong {}^1 \mathscr{A}^{\mathbb{Z}^D}$, where $^1 \mathscr{A}^{\mathbb{Z}^D} := \mathscr{A}^{\mathbb{Z}^D} \cap \ell^1$ is the set of $\mathbb{Z}^D$-indexed configurations with finite support. Thus, any cellular automaton on $\mathscr{A}^{\mathbb{Z}^D}$ which preserves $^1 \mathscr{A}^{\mathbb{Z}^D}$ can be represented as a transformation of $^\varepsilon \mathscr{A}^{\mathbb{R}^D}$ in a natural way. In particular, given a RealLife automaton $\maltese$ with neighbourhood radius 1, and given $\varepsilon > 0$, we shall construct a 'discrete approximation' $\maltese_\varepsilon$ of $\maltese$ in the form of a Euclidean automaton of radius 1, which is isomorphic to a Larger than Life cellular automaton with neighbourhood radius of order $\mathcal{O}(1/\varepsilon)$.

Let $\mathscr{M}(\varepsilon \mathbb{Z}^D)$ be the space of signed measures on $\varepsilon \mathbb{Z}^D$. For any $z \in \mathbb{Z}^D$, let $\delta_{\varepsilon z}$ be the point mass at εz. For any $\kappa \in \mathscr{K}$, we define $\bar{\kappa}_\varepsilon \in \mathscr{M}(\varepsilon \mathbb{Z}^D)$ by

$$\bar{\kappa}_\varepsilon := \sum_{z \in \mathbb{Z}^D} k_z \delta_{\varepsilon z}, \quad \text{where } k_z := \int_{\mathbf{C}(z, \varepsilon)} \kappa(c) \, d\lambda[c], \quad \text{for any } z \in \mathbb{Z}^D. \quad (12.3)$$

Now define $\bar{\maltese}_\varepsilon : \mathbf{L}_\varepsilon^1 \longrightarrow \mathbf{L}_\varepsilon^1$ by

$$\bar{\maltese}_\varepsilon(\mathbf{a}) := \mathbf{a} \cdot \mathfrak{s} \circ (\bar{\kappa}_\varepsilon * \mathbf{a}) + (1 - \mathbf{a}) \cdot \mathfrak{b} \circ (\bar{\kappa}_\varepsilon * \mathbf{a}), \quad \text{for all } \mathbf{a} \in \mathbf{L}_\varepsilon^1. \quad (12.4)$$

Note that $\bar{\maltese}_\varepsilon(^\varepsilon \mathscr{A}^{\mathbb{R}^D}) \subseteq {}^\varepsilon \mathscr{A}^{\mathbb{R}^D}$ so $\bar{\maltese}_\varepsilon$ restricts to a transformation on $^\varepsilon \mathscr{A}^{\mathbb{R}^D}$.

For all $\mathbf{a} \in \mathbf{L}^1$, let $\bar{\mathbf{a}}_\varepsilon \in \mathbf{L}_\varepsilon^1$ be *conditional expectation* of $\mathbf{a}$ given $\mathscr{B}_\varepsilon$. That is, for any $x \in \mathbb{R}^D$:

$$\bar{\mathbf{a}}_\varepsilon(x) := \frac{1}{\varepsilon^D} \int_{\mathbf{C}(x, \varepsilon)} \mathbf{a}(c) \, d\lambda[c],$$

where $\mathbf{C}(x, \varepsilon)$ is the unique ε-cube in $\mathscr{B}_\varepsilon$ which contains x. We extend $\bar{\mathbf{t}}_\varepsilon$ to a transformation $\mathbf{t}_\varepsilon : \mathbf{L}^1 \longrightarrow \mathbf{L}^1_\varepsilon$ by defining $\mathbf{t}_\varepsilon(\mathbf{a}) := \bar{\mathbf{t}}_\varepsilon(\bar{\mathbf{a}}_\varepsilon)$ for all $\mathbf{a} \in \mathbf{L}^1$. Note that $\mathbf{t}_\varepsilon(\mathbf{a}) = \bar{\mathbf{t}}_\varepsilon(\mathbf{a})$ for all $\mathbf{a} \in \mathbf{L}^1_\varepsilon$ (because $\bar{\mathbf{a}}_\varepsilon = \mathbf{a}$ for any $\mathbf{a} \in \mathbf{L}^1_\varepsilon$). We will suppress the distinction between $\mathbf{t}_\varepsilon$ and $\bar{\mathbf{t}}_\varepsilon$, and write both as "$\mathbf{t}_\varepsilon$".

Let $\tilde{\kappa}_\varepsilon := [k_\mathbf{z}]_{\mathbf{z} \in \mathbb{Z}^D} \in \ell^1 \subset \ell^\infty(\mathbb{Z}^D)$. Define the operator $\tilde{\mathbf{t}}_\varepsilon : \ell^1 \longrightarrow \ell^1$ by

$$\tilde{\mathbf{t}}_\varepsilon(\mathbf{a}) := \mathbf{a} \cdot \mathbf{s} \circ (\tilde{\kappa}_\varepsilon * \mathbf{a}) + (1 - \mathbf{a}) \cdot \mathbf{b} \circ (\tilde{\kappa}_\varepsilon * \mathbf{a}), \quad \text{for all } \mathbf{a} \in \ell^1.$$

Theorem 4 *Fix* $(s_0, b_0, s_1, b_1) \in \Theta$ *and* $\kappa \in \mathscr{K}$. *Let* $\mathbf{t}$ *be the resulting* RealLife *EA.*

(a) *For any* $\varepsilon > 0$, $\tilde{\mathbf{t}}_\varepsilon(\mathscr{A}^{\mathbb{Z}^D}) \subseteq \mathscr{A}^{\mathbb{Z}^D}$, *and* $\tilde{\mathbf{t}}_\varepsilon : \mathscr{A}^{\mathbb{Z}^D} \longrightarrow \mathscr{A}^{\mathbb{Z}^D}$ *is a* Larger than Life *CA, with birth interval* $[b_0, b_1]$ *and survival interval* $[s_0, s_1]$. *(For example: suppose* $\kappa = \lambda[\mathbf{K}]^{-1} \mathbf{1}_\mathbf{K}$, *where* $\mathbf{K} = [-1, 1]^D \subset \mathbb{R}^D$. *If* $\varepsilon = 1/n$ *and* $\mathbb{K}_n := [-n \ldots n]^D \subset \mathbb{Z}^D$, *then* $\tilde{\mathbf{t}}_\varepsilon$ *is a* Larger than Life *CA with neighbourhood* $\mathbb{K}_n$.)

(b) *The natural isomorphism from* $\mathbf{L}^1_\varepsilon$ *to* ℓ^1 *is a dynamical isomorphism from* $(\mathbf{L}^1_\varepsilon, \mathbf{t}_\varepsilon)$ *to* $(\ell^1, \tilde{\mathbf{t}}_\varepsilon)$, *which restricts to a dynamical isomorphism from* $(^\varepsilon\mathscr{A}^{\mathbb{R}^D}, \mathbf{t}_\varepsilon)$ *to* $(^1\mathscr{A}^{\mathbb{Z}^D}, \tilde{\mathbf{t}}_\varepsilon)$.

(c) *If* $\mathbf{a} \in {}^0\mathscr{A}^{\mathbb{R}^D}$, *then* $\mathbf{L}^1\text{-}\lim_{\varepsilon \to 0} \mathbf{t}_\varepsilon(\mathbf{a}) = \mathbf{t}(\mathbf{a})$.

(d) *If* $\lim_{n \to \infty} \varepsilon_n = 0$, *then* $\{\mathbf{t}_{\varepsilon_n}\}_{n=1}^\infty$ *evolves to* $\mathbf{t}$ *on* $^0\mathscr{A}^{\mathbb{R}^D}$.

Proof [5, Theorem 2.1 and Proposition 2.4]. □

Remarks (a) We cannot simulate RealLife on computer; we can only simulate large-radius Larger than Life CA. Theorem 4(c) says this will yield 'good approximation' of RealLife.

(b) Evans [3] empirically found that Larger than Life CA of increasingly large radii have life forms that are virtually identical after rescaling. This, together with Theorem 4(d), suggests (but doesn't prove) that RealLife EA have life forms which are morphologically similar to those seen by Evans in Larger than Life.

12.1.2 Still Lifes

Empirically, Larger than Life CA exhibit many still lifes whose shapes resembles disks or annuli (distorted by lattice anisotropy). We can prove the existence of analogous still lifes for RealLife. For simplicity we state this result for 2-dimensional RealLife. (There is a similar result for higher dimensions.)

Proposition 5 *Let* $D = 2$. *Let* $\| \bullet \|_*$ *be norm on* $\mathbb{R}^2$. *For all* $r > 0$, *let* $\bigodot(r) := \{x \in \mathbb{R}^D; \|x\|_* \leq r\}$. *Let* $\mathbf{K} := \bigodot(1)$, *and* $\kappa := \lambda[\mathbf{K}]^{-1} \cdot \mathbf{1}_\mathbf{K}$. *Let* $s_0 \leq \frac{1}{4}$, *and let* $r < \min\{\sqrt{b_0}, \frac{1}{2}\}$. *If* $\mathbf{A} \subseteq \bigodot(r)$, *and* $s_0 \cdot \lambda[\mathbf{K}] \leq \lambda[\mathbf{A}]$ *then* $\mathbf{a} := \mathbf{1}_\mathbf{A}$ *is a still life.* (For example: if $\bigodot(\sqrt{s_0}) \subseteq \mathbf{A} \subseteq \bigodot(R)$, then $\mathbf{a}$ is a still life; see Fig. 12.3(a)).

Proof [5, Proposition 3.4]. □

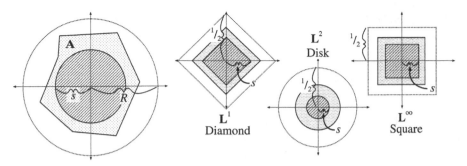

Fig. 12.3 Ball-shaped still lifes. (A) Proposition 5. (B) Example 6

Example 6 Let $s_0 \leq \frac{1}{4} < b_0$, and $R < \frac{1}{2}$. Figure 12.3(b) portrays three still lives:

ℓ^1 norm: For any $r > 0$, let $\mathbf{D}(r) := \{x = (x_1, x_2) \in \mathbb{R}^D; |x_1| + |x_2| \leq r\}$ (diamond). Let $\kappa := \frac{1}{2}\mathbf{1}_{\mathbf{D}(1)}$. Then $\mathbf{1}_{\mathbf{D}(r)}$ is a still life, for all $r \in [\sqrt{s_0}, \frac{1}{2})$.

ℓ^2 norm: For any $r > 0$, let $\mathbf{B}(r) := \{x \in \mathbb{R}^2; |x| \leq r\}$ (disk). Let $\kappa := \frac{1}{\pi}\mathbf{1}_{\mathbf{B}(1)}$. Then $\mathbf{1}_{\mathbf{B}(r)}$ is a still life for any $r \in [\sqrt{s_0}, \frac{1}{2})$.

ℓ^∞ norm: For any $r > 0$, let $\mathbf{C}(r) := [-r, r]^2$ (square). Let $\kappa := \frac{1}{4}\mathbf{1}_{\mathbf{C}(1)}$. Then $\mathbf{1}_{\mathbf{C}(r)}$ is a still life for any $r \in [\sqrt{s_0}, \frac{1}{2})$.

We say that the kernel κ is *rotationally symmetric* if there exists a function $K : [0, \infty) \longrightarrow [0, \infty)$ such that $\kappa(x) = K(|x|)$ for all $x \in \mathbb{R}^D$. If $R > 0$, let $\mathbf{B}(R) := \{x \in \mathbb{R}^D; |x| \leq R\}$ and $\mathbf{b}_R :- \mathbf{1}_{\mathbf{B}(R)} \in {}^1\mathscr{A}^{\mathbb{R}^D}$. If $r \in [0, R]$, let $\mathbf{A}(r, R) := \{x \in \mathbb{R}^D; r \leq |x| \leq R\}$ be the *bubble* with inner radius r and outer radius R (e.g. if $D = 2$, then $\mathbf{A}(r, R)$ is an annulus). Let $\mathbf{a}_{r,R} := \mathbf{1}_{\mathbf{A}(r,R)} \in {}^1\mathscr{A}^{\mathbb{R}^D}$.

Proposition 7 *Suppose ♟ has rotationally symmetric kernel κ.*

(a) *There are differentiable increasing functions $\underline{S}_1 : [0, \infty) \longrightarrow [0, \infty)$ and $\beta : [0, \infty) \longrightarrow [0, 1]$ such that, for any $R > 0$, if $s_0 \leq \beta(R) < b_0$ and $\underline{S}_1(R) \leq s_1$, then $\mathbf{b}_R$ is a still life.*

(b) *Let $\Delta := \{(r, R); 0 < r < R\}$. There are differentiable functions $\beta, \underline{B}_0, \overline{B}_1, \underline{S}_1 : \Delta \longrightarrow [0, 1]$ such that, for any $(r, R) \in \Delta$, if $s_0 \leq \beta(r, R) < b_0$, $\underline{S}_1(r, R) \leq s_1$, and either $\underline{B}_0(r, R) < b_0$ or $b_1 < \overline{B}_1(r, R)$ then $\mathbf{a}_{r,R}$ is a still life.*

Proof [5, Proposition 3.6(b,c)].   □

One can also prove the existence of still lifes shaped like thin, infinitely extended, gently undulating 'ribbons' (if $D = 2$) or 'curtains' (for $D \geq 3$) [5, Propositions 3.5 and 3.6(a)].

12.1.3 Robustness of Still Lifes in the Hausdorff Metric

If $\mathbf{X}, \mathbf{Y} \subset \mathbb{R}^D$ are closed sets, then the *Hausdorff distance* from $\mathbf{X}$ to $\mathbf{Y}$ is defined

$$d_H(\mathbf{X}, \mathbf{Y}) := \frac{1}{2} \sup_{x \in \mathbf{X}} \inf_{y \in \mathbf{Y}} d(x, y) + \frac{1}{2} \sup_{y \in \mathbf{Y}} \inf_{x \in \mathbf{X}} d(y, x).$$

We define the metric d_* on $^1\mathscr{A}^{\mathbb{R}^D}$ as follows: for any $\mathbf{a}, \mathbf{b} \in {}^1\mathscr{A}^{\mathbb{R}^D}$, if $\mathbf{a} = 1_\mathbf{A}$ and $\mathbf{b} = 1_\mathbf{B}$ (for some $\mathbf{A}, \mathbf{B} \subset \mathbb{R}^D$), then $d_*(\mathbf{a}, \mathbf{b}) := \|\mathbf{a} - \mathbf{b}\|_1 + d_H(\partial \mathbf{A}, \partial \mathbf{B})$.

The operator $\maltese$ is *not* d_*-continuous [5, Remark (a), §4]. However, under certain conditions, a still life in $^1\mathscr{A}^{\mathbb{R}^D}$ will be surrounded by a d_*-neighbourhood of other still lifes. If $\mathbf{a} \in {}^1\mathscr{A}^{\mathbb{R}^D}$, we say that $\mathbf{a}$ is κ-*smooth* if the function $\kappa * \mathbf{a}$ is $\mathscr{C}^2$ on an open dense subset of $\mathbb{R}^D$. For example, if $\kappa \in \mathscr{C}^2$, then *all* elements of $\mathscr{A}^{\mathbb{R}^D}$ are κ-smooth [because, if $\mathbf{a} \in \mathscr{A}^{\mathbb{R}^D}$, then for any $c, d \in [1 \dots D]$, $\partial_c \partial_d (\kappa * \mathbf{a})(x) = (\partial_c \partial_d \kappa) * \mathbf{a}(x)$ is defined and continuous at all $x \in \mathbb{R}^D$]. Even when κ is not smooth (e.g. $\kappa = \lambda[\mathbf{K}]^{-1} 1_\mathbf{K}$ for $\mathbf{K} \subset \mathbb{R}^D$), the property of κ-smoothness is still fairly common. For example, call a subset $\mathbf{A} \subset \mathbb{R}^D$ *smoothly open* if $\mathbf{A}$ is open and $\partial \mathbf{A}$ is a piecewise smooth manifold. Thus, an open ball is smoothly open, because its boundary is a sphere, and an open cube is smoothly open, because its boundary is a union of finitely many flat faces. We have:

Lemma 1 *Let* $\mathbf{K}$ *be smoothly open and let* $\kappa := \lambda[\mathbf{K}]^{-1} 1_\mathbf{K}$. *If* $\mathbf{A} \subset \mathbb{R}^D$ *is also smoothly open, then* $\mathbf{a} := 1_\mathbf{A}$ *is* κ-*smooth*.

Proof [5, Lemma 4.1].   □

If $\mathbf{a}$ is κ-smooth, and $\mathfrak{s} \circ (\kappa * \mathbf{a}) = 1_\mathbf{S}$ and $\mathfrak{b} \circ (\kappa * \mathbf{a}) = 1_\mathbf{B}$ for some $\mathbf{S}, \mathbf{B} \subset \mathbb{R}^D$, then let

$$M_\infty(\mathbf{a}) := \sup_{x \in \partial \mathbf{S} \cup \partial \mathbf{B}} \frac{1}{|\nabla(\kappa * \mathbf{a})(x)|}.$$

Theorem 8 *Suppose* $\mathbf{a} \in {}^1\mathscr{A}^{\mathbb{R}^D}$ *is* κ-*smooth still life, with* $M_\infty(\mathbf{a}) < \infty$. *If* $\mathrm{cl}(\mathbf{B}) \subset \mathrm{int}(\mathbf{A})$ *and* $\mathrm{cl}(\mathbf{A}) \subset \mathrm{int}(\mathbf{S})$ *(Fig. 12.4(a)), then there is some* $\varepsilon > 0$ *such that, for any* $\mathbf{a}' \in {}^1\mathscr{A}^{\mathbb{R}^D}$, *if* $d_*(\mathbf{a}, \mathbf{a}') < \varepsilon$ *then* $\mathbf{a}'$ *is also a still life*.

Proof [5, Theorem 4.2].   □

12.2 The Anatomy of Bugs

The many 'bugs' discovered by Evans in Larger than Life CA are intriguing for several reasons:

- Bugs appear in Larger than Life CA for a wide range of values of the parameters s_0, b_0, b_1, s_1 and kernel κ.

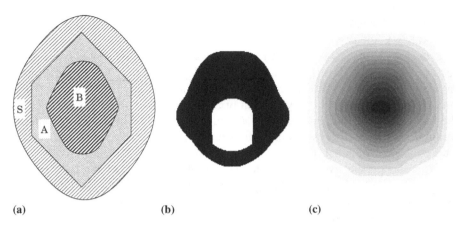

(a)   **(b)**   **(c)**

Fig. 12.4 (a) Theorem 8. (b) A morphologically typical bug in Larger than Life (moving upwards). In this case, we have threshold parameters $s_0 = b_0 = \frac{706}{2,601} < b_1 = \frac{958}{2,601} < s_1 = \frac{1,216}{2,601}$, and neighbourhood $\mathbb{K} = [-75 \ldots 75]^2 \subset \mathbb{Z}^2$, but other parameter values produce similar bugs. (c) Greyscale contour map of $\alpha := \kappa * \mathbf{a}$, where $\mathbf{a} \in {}^1\mathscr{A}^{\mathbb{R}^D}$ is a bug like the one in figure (b), and κ is the convolution kernel for a RealLife EA. (Here, black $= 1$, white $= 0$)

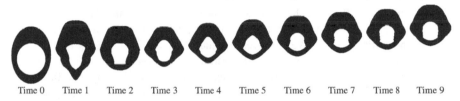

Time 0   Time 1   Time 2   Time 3   Time 4   Time 5   Time 6   Time 7   Time 8   Time 9

Fig. 12.5 Birth of a bug: an off-centre elliptical annulus self-organizes into a bug after 9 iterations of a Larger than Life CA

- For most of these parameter values, the bugs appear to have a similar morphology: a bilaterally symmetric, distorted annulus, with an off-centre hole, as in Fig. 12.4(b).
- Bug-containing configurations appear to be 'attractors': a random initial condition (or one containing an off-centre elliptical annulus) will often self-organize into one with a bug after a few iterations; see Fig. 12.5.

Theorem 4(d) implies that a sequence of period-p bugs for larger and larger radius Larger than Life CA will 'evolve' towards a period-p bug for RealLife. However, this isn't the only way that RealLife bugs can originate. For any fixed $N \in \mathbb{N}$, an $N \times N$ propagating structure in Larger than Life must be eventually periodic, because the set $\mathscr{A}^{[0 \ldots N]^2}$ is finite. However, propagating structures in RealLife need not be periodic, or even quasiperiodic, because the set $\mathscr{A}^{[0,1]^D}$ is uncountably infinite. Thus, a velocity-v propagating structure in RealLife is not necessarily a periodic orbit of the transformation $\maltese \circ \sigma^{-v}$; instead, it is best seen as an *attractor* of $\maltese \circ \sigma^{-v}$, confined to some small neighbourhood of an archetypal configuration similar to the one portrayed in Fig. 12.4(b).

Formally, for any $v \in \mathbb{R}^D$, we define an *aperiodic bug* with *velocity* v to be an open subset $\mathscr{G} \subseteq {}^1 \mathscr{A}^{\mathbb{R}^D}$ such that $\maltese \circ \sigma^{-v}(\mathscr{G}) \subseteq \mathscr{G}$. If $\Phi := \maltese \circ \sigma^{-v}$, then we have $\mathscr{G} \subseteq \Phi^{-1}(\mathscr{G}) \subseteq \Phi^{-2}(\mathscr{G}) \subseteq \Phi^{-3}(\mathscr{G}) \subseteq \cdots$. The *basin of attraction* of $\mathscr{G}$ is the open set $\mathscr{B} := \bigcup_{t=0}^{\infty} \Phi^{-t}(\mathscr{G})$. (For example, in Fig. 12.5, the 'time 0' configuration is in $\mathscr{B}$, while the 'time 9' configuration is probably in $\mathscr{G}$.) Let $R > 0$; the bug $\mathscr{G}$ has *radius* R if the support of every element of $\mathscr{G}$ (as a function on $\mathbb{R}^D$) is confined to the ball of radius R (around 0). If $\mathbf{a} \in \mathscr{G}$, and $\mathscr{G}$ is contained in an ε-ball around $\mathbf{a}$ in the d_* metric, and ε is small enough, then every configuration in $\mathscr{G}$ will appear 'morphologically similar' to $\mathbf{a}$.

From this perspective, it is easy to see why there exist aperiodic bugs with a morphology similar to the one in Fig. 12.4(b). Let $\mathbf{a} \in {}^1 \mathscr{A}^{\mathbb{R}^D}$ be the bug shown in Fig. 12.4(b), and let $\alpha := \kappa * \mathbf{a}$. A greyscale contour plot of α is shown in Fig. 12.4(c). The maximal values for α occur in the center of $\mathbf{a}$, around the top edge of the 'hole'. Indeed, in this region, α exceeds s_1 (the upper survival threshold), so that cells along the top edge of the hole in $\mathbf{a}$ 'die'. However, the hole in $\mathbf{a}$ also causes the value of α to drop sharply as we move downwards, so that $\alpha(x)$ is actually less than s_0 (the minimum survival threshold) for x along the bottom edge of $\mathbf{a}$; hence these cells also die off. Meanwhile, the values of α are within the 'birth interval' $[b_0, b_1]$ for some cells just above the top edge of $\mathbf{a}$, and also just above the bottom edge of the hole; hence these cells are brought to life. The net result of all this is that $\maltese(\mathbf{a})$ will be a structure morphologically similar to $\mathbf{a}$, but translated upwards by some vector v. We don't require that $\maltese(\mathbf{a}) = \sigma^v(\mathbf{a})$ — only that $\maltese(\mathbf{a})$ is 'close' to $\sigma^v(\mathbf{a})$. See Fig. 12.6 for an illustration of this argument.

At this point, the logic of Theorem 4 can be reversed. For any $\varepsilon > 0$, let $\maltese_\varepsilon$ and $\widetilde{\maltese}_\varepsilon$ be as in that theorem. Let $\mathscr{G}$ be an aperiodic bug for $\maltese$, let $\mathbf{a}^0 \in \mathscr{G}$, and let $\overline{\mathbf{a}}_\varepsilon^0 \in \mathbf{L}_\varepsilon^1$ be its conditional expectation. For all $t \in \mathbb{N}$, let $\mathbf{a}^t := \sigma^{-v} \circ \maltese^t(\mathbf{a})$ and let $\overline{\mathbf{a}}_\varepsilon^t := \sigma^{-v} \circ \maltese_\varepsilon^t(\overline{\mathbf{a}}_\varepsilon^0)$. Now, $\{\mathbf{a}_0, \mathbf{a}_1, \mathbf{a}_2, \ldots\} \subset \mathscr{G}$, so if $\mathscr{G}$ is contained in a δ-ball around $\mathbf{a}^0$, then $\mathbf{a}^t$ will be δ-close to $\mathbf{a}^0$ for all $t \in \mathbb{N}$. This suggests that $\overline{\mathbf{a}}_\varepsilon^t$ will be 'close' to $\overline{\mathbf{a}}_\varepsilon^0$ for all $t \in \mathbb{N}$ — in particular, the support of $\overline{\mathbf{a}}_\varepsilon^t$ should be confined in some compact neighbourhood $\mathbf{U} \subset \mathbb{R}^D$ around the origin. But the set of elements of $\mathbf{L}_\varepsilon^1$ with support confined to $\mathbf{U}$ is finite (because $\mathbf{L}_\varepsilon^1$ is isomorphic to ${}^1 \mathscr{A}^{\mathbb{Z}^D}$). Thus, the sequence $\{\overline{\mathbf{a}}_0^\varepsilon, \overline{\mathbf{a}}_1^\varepsilon, \overline{\mathbf{a}}_2^\varepsilon, \ldots\}$ must be eventually periodic, which means it is a bug for the Larger than Life CA $\widetilde{\maltese}_\varepsilon$. This is not a rigorous proof, but it suggests that the study of RealLife can shed light on phenomena in Larger than Life.

12.3 Open Questions

Simulations of RealLife show objects with smooth boundaries, which smoothly evolve over time. Can this motion be described by a suitable system of differential equations? Could these equations be solved to construct oscillators and/or bugs?

In the Game of Life, it is possible to build 'glider guns', 'glider reflectors' and 'glider-based' logic, yielding machines capable of universal computation and even

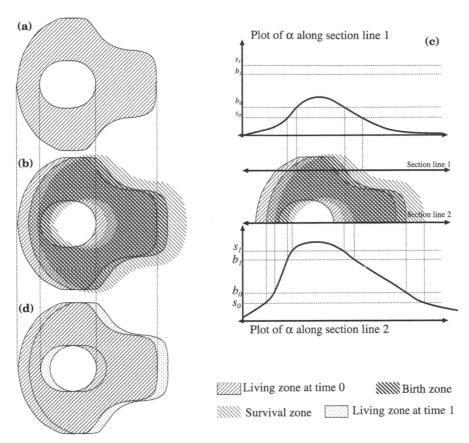

Fig. 12.6 The propagation of a bug. (**a**) The bug at time 0. (**b**) The 'survival zone' $\alpha^{-1}[s_0, s_1]$ and the 'birth zone' $\alpha^{-1}[b_0, b_1]$, superimposed on figure (**a**) (here, $\alpha := \kappa * \mathbf{a}$). (**c**) Plots of the function α along two horizontal 'sections' through figure (**b**), showing how α passes through the threshold values $s_0 \leq b_0 \leq b_1 \leq s_1$. (**d**) Use the 'survival zone' in figure (**b**) to trim some cells of the left-hand edges of the bug in figure (**a**), and use the 'birth zone' in figure (**b**) to add some cells to the right edge; we get the bug at time 1

self-replication; see [1, 2]. Evans [4] has shown that at least one **Larger than Life** CA (*Bosco's Rule*) exhibits similar universal computation (but not self-replication). Does any **RealLife** EA contain universal computers or self-replicators?

References

1. Berlekamp, E.R., Conway, J.H., Guy, R.K.: Winning Ways for Your Mathematical Plays, vol. 4, 2nd edn., Chap. 25: 'What Is Life?'. AK Peters, Wellesley (2004)
2. Durand, B., Róka, Z.: The game of life: universality revisited. In: Cellular Automata (Saissac, 1996). Math. Appl., vol. 460, pp. 51–74. Kluwer Academic, Dordrecht (1999)
3. Evans, K.M.: Larger than Life: threshold-range scaling of Life's coherent structures. Phys. D **183**(12), 45–67 (2003)

4. Evans, K.M.: Is Bosco's rule universal? In: Machines, Computations, and Universality. Lecture Notes in Comput. Sci., vol. 3354, pp. 188–199. Springer, Berlin (2005)
5. Pivato, M.: RealLife: the continuum limit of Larger than Life cellular automata. Theor. Comput. Sci. **372**(1), 46–68 (2007)

Chapter 13
Variations on the Game of Life

Ferdinand Peper, Susumu Adachi, and Jia Lee

The Game of Life is defined in the framework of Cellular Automata with discrete states that are updated synchronously. Though this in itself has proven to be fertile ground for research, it leaves open questions regarding the robustness of the model with respect to variations in updating methods, cell state representations, neighborhood definitions, etc. These questions may become important when the ideal conditions under which the Game of Life is supposed to operate cannot be satisfied, like in physical realizations. This chapter describes three models in which Game of Life-like behavior is obtained, even though some basic tenets are violated.

The first model [12] uses asynchronous updating as the timing scheme. Rather than all cells being updated simultaneously in each step, they are updated with a certain probability. Most cellular automaton models break down under such a condition, and the Game of Life is no exception to that [4]. An interesting question, however, is how undesired behavior can be repaired. The method followed in this chapter applies to cellular automata in general, and it boils down to simulating a synchronous cellular automaton on its asynchronous counterpart. The method comes at the cost of increased time and an increased number of cell states. In its basic form, as proposed in 1974 by Nakamura [14] (see also [5, 15, 16, 22, 23]), the method requires as much as $3n^2$ cells in the asynchronous model, compared to n states for the synchronous model. The improved version of the method, adopted in this chapter, only requires $n^2 + 2n$ states. Applied to the Game of Life, this improved method results in an asynchronous cellular automaton with eight states, which is four states less than with Nakamura's original algorithm.

The second model [1] adopts a different type of state representation. Rather than using discrete states, it uses states that are continuously valued between zero and one. The model employs a temperature parameter to reflect the degree to which its states are continuous as opposed to discrete; this offers an opportunity to study at which temperatures the original Game of Life is fully retained, and how it degrades on a continuous scale.

The third model [2] is not the real Game of Life in its strictest sense, but, when simulated, it looks remarkable similar, giving the viewer the typical excitement of gliders, blinkers, and stable configurations, be it that the experience lasts

A. Adamatzky (ed.), *Game of Life Cellular Automata*,
DOI 10.1007/978-1-84996-217-9_13, © Springer-Verlag London Limited 2010

an increased number of generations before stable or semi-stable configurations are reached. Additionally, configurations in this model tend to develop over larger areas, which is mainly due to the neighborhood being extended to distance two, even though it remains of Moore type. There is a wide variety of enlarged neighborhood transition rules that give fascinating behaviors, like the *Larger Than Life* models of Evans [9]. The third model belongs to this class, but its behavior is more in line with the original Game of Life, and, just as important, it is extremely simple. The similarities of this new model with its conceptual roots do not stop here: the same methods used to prove that the Game of Life is computationally universal are also applicable to this model, though some care needs to be taken to deal with spatial and phase shifts of signals.

13.1 Asynchronously Timed Game of Life

In synchronous cellular automata, like the Game of Life, all cells are updated simultaneously at discrete time steps. Events in nature, however, rarely take place at exactly the same time, and for this reason alternative updating methods have been considered. When updating times for individual cells are random with each having a fixed average update rate, we obtain Poisson processes; the resulting update scheme is usually called Poisson. Another type of updating scheme selects a random set of cells to be updated at each time step. Mixtures of deterministic and randomized updating schemes are also possible. All these methods, which relax the conditions of synchronous updating, are called *asynchronous updating*. Asynchronous versus synchronous schemes have been widely studied in cellular automata [3, 4, 10, 11, 21], and in physical systems, e.g., coupled map lattices [13], asymmetric exclusion processes [19], and spin systems [6].

One of the first attempts to apply asynchronous updating to the Game of Life is the SGL(p) model [4], in which each cell has a certain probability p ($0 < p \leq 1$) to be updated at each time step. Obviously, SGL(1) is equal to the original Game of Life. If $p \to 0$, the updating scheme in SGL(p) becomes Poisson. It is well known that, when running on finitely-sized arrays, the Game of Life will eventually evolve from a random initial configuration into a steady state with a low-density of living cells, consisting of some simple periodic or stationary patterns, like the glider, the blinker, and the block. When p is large enough, the evolution of SGL(p) from a random initial configuration will settle down to a steady state with few living cells remaining. When p reaches 0 in the limit, however, the SGL(p) will evolve into alternating stripes of living and dead cells [4]. Another apparent difference between the Game of Life and SGL($p < 1$) concerns computational ability. Whereas the Game of Life is capable of universal computation, and hence, can emulate the behaviour of any other system [24], the SGL($p < 1$) is thought to be incapable of carrying out reliable computation even if p is arbitrarily close to 1. In general, due to the unexpected transitions of cells, it tends to be difficult to control the behavior of asynchronous Cellular Automata to the extent that computational functionality arises.

The above suggest that asynchrony of updating and reliability are at odds with each other, but this is not necessarily true. Another parameter in the model is its complexity in terms of the number of cell states, and by increasing this number, the 100 percent reliability of the original Game of Life can be regained, as we will show, no matter whether the update of cells is synchronous or asynchronous. Our approach [12] is based on the transformation of the Game of Life to an 8-state cellular automaton with Moore neighborhood. The method employed for this transformation is an improvement over Nakamura's method [14] (see also [5, 15, 16, 22, 23]), in the sense that less cell states are required (8 instead of 12 for the Game of Life).

When the (conventional) Game of Life has a state set $V = \{0, 1\}$ and a transition function $\delta : V^9 \to V$, then δ satisfies the condition that for the neighboring cells' states $q_1, \ldots, q_8 \in V$:

$$\text{if } 2 \leq \sum_{i=1}^{8} q_i \leq 3 \text{ then } \delta(1, q_1, \ldots, q_8) = 1,$$
$$\text{otherwise } \delta(1, q_1, \ldots, q_8) = 0;$$
$$\text{if } \sum_{i=1}^{8} q_i = 3 \text{ then } \delta(0, q_1, \ldots, q_8) = 1,$$
$$\text{otherwise } \delta(0, q_1, \ldots, q_8) = 0.$$

The synchronous Game of Life is transformed into an asynchronous model with eight states through the method detailed in [12]. The resulting Asynchronous Game of Life retains the original model's neighborhood, but its state set is extended to $V_a = V \cup V \times V \cup V'$, whereby the subsets are:

Basic states $V = \{0, 1\}$
Intermediate states $V \times V = \{(0, 0), (1, 0), (0, 1), (1, 1)\}$
Succession states $V' = \{0', 1'\}$

The basic states correspond to the original Game of Life's cell states. Assume $s : V \cup V \times V \to V$ is a mapping such that

$$\forall p, q \in V: s(p) = p \wedge s\big((p, q)\big) = p.$$

The transition function δ_a of the Asynchronous Game of Life is as follows.

(1) If $q_1, \ldots, q_8 \in V \cup V \times V$, then
$$\text{if } 2 \leq \sum_{i=1}^{8} s(q_i) \leq 3 \text{ then } \delta_a(1, q_1, \ldots, q_8) = (1, 1),$$
$$\text{else } \delta_a(1, q_1, \ldots, q_8) = (1, 0);$$
$$\text{if } \sum_{i=1}^{8} s(q_i) = 3 \text{ then } \delta_a(0, q_1, \ldots, q_8) = (0, 1),$$
$$\text{else if } \exists j (1 \leq j \leq 8), \ q_j \neq 0 \text{ then } \delta_a(0, q_1, \ldots, q_8) = (0, 0).$$

Informally, when a cell is in a basic state and no neighboring cell is in the succession state, the following happens. A cell in state 1 changes its state to $(1, 1)$ if two or three of the neighboring cells are in state 1, in state $(1, 0)$, or in state $(1, 1)$; otherwise the cell changes its state to $(1, 0)$. A cell in state 0 changes its state to $(0, 1)$ if three of the neighboring cells are in state 1, in state $(1, 0)$, or in state $(1, 1)$; otherwise it changes its state to $(0, 0)$, except when the eight neighboring cells are all in state 0, because then there is no need for the cell to change its state, as it is quiescent and surrounded by quiescent cells.

(2) If $q_1, \ldots, q_8 \in V \times V \cup V'$, then
$$\delta_a((0,0), q_1, \ldots, q_8) = 0'$$
$$\delta_a((1,0), q_1, \ldots, q_8) = 0'$$
$$\delta_a((0,1), q_1, \ldots, q_8) = 1'$$
$$\delta_a((1,1), q_1, \ldots, q_8) = 1'.$$

Informally, a cell in intermediate state $(0,0)$ or $(1,0)$ changes its state to $0'$ if no neighboring cell is in the basic state. A cell in intermediate state $(0,1)$ or $(1,1)$ changes its state to $1'$ if no neighboring cell is in the basic state.

(3) If $q_1, \ldots, q_8 \in V \cup V'$, then
$$\delta_a(0', q_1, \ldots, q_8) = 0$$
$$\delta_a(1', q_1, \ldots, q_8) = 1.$$

Informally, a cell in succession state $0'$ (resp. $1'$) changes its state to 0 (resp. 1) if no neighboring cell is in the intermediate state.

(4) In all other cases, $\delta_a(q_c, q_1, \ldots, q_8) = q_c$, where $q_c, q_1, \ldots, q_8 \in V_a$.
In other words, in the cases other than defined above, a cell will undergo a trivial transition, i.e., a transition that keeps its state unchanged.

In practical implementations, part (4) of the transition function is usually left out, and instead the rule is adopted that a cell is only updated if the domain of the transition function matches the states of the cell and its neighbors. If there is no such match, then the cell's state is left unchanged by default.

The change of the cell state from basic to intermediate mimics the cell state transition in the Game of Life. A cell in the intermediate state stores both the information about its new state after a state transition as well as the information about the previous state before the state transition, such as to allow each of its neighboring cells to read either the old or the new state depending on whether the neighbor lags behind the cell or not. Cells in intermediate states remain in their states until all of their neighboring cells enter intermediate or succession states. Succession states are the states assumed by cells before they are updated to basic states.

The above Asynchronous Game of Life operates under any arbitrary updating scheme. For our simulations we employ the stochastic updating scheme in [4], i.e., at each time step each cell undergoes a transition with probability p ($0 < p \leq 1$), and we denote the resulting model by AGL(p). Unlike the SGL(p) in [4], the behaviour of the Game of Life can always be recovered from that of the AGL(p) even for very small p. For example, since any arbitrary logic function can be constructed in the Game of Life by means of the collisions of gliders [8], it can also be done in AGL(p). Figure 13.1 shows a simulation of two gliders colliding in AGL(0.5). Due to the asynchrony of the model, there are many different ways in which the two gliders can evolve, but the end result is always the same: a dead cell space.

In the example in Fig. 13.2, the simulation starts from random initial configuration, which evolves into a configuration consisting of some simple periodic and stationary patterns (Fig. 13.2(b)). Self-organizing behaviour similar to this is found in AGL(p) with respect to various p values (Fig. 13.2(c)–(f)). It is easy to verify that the configurations in Fig. 13.2(c)–(f) composed of cells in state 1, state $1'$, state $(1,1)$, or state $(0,1)$ resemble the configuration in Fig. 13.2(b).

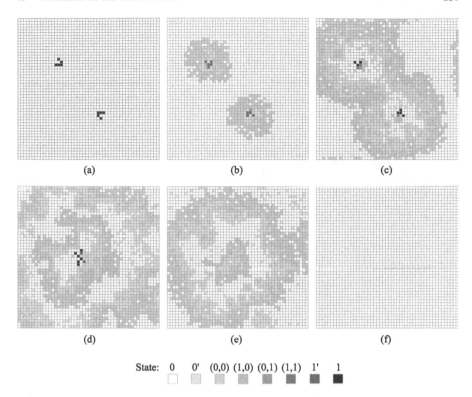

State: 0   0'   (0,0) (1,0) (0,1) (1,1)   1'   1

Fig. 13.1 The collision of two gliders in AGL(0.5) on a 50×50 array with periodic boundary conditions [12]. (**a**) The initial configuration consists only of cells in state 0 or 1. (**b**) Two gliders move toward each other, while waves of cells in state $(0, 0)$ or $0'$ (displayed as *light gray cells*) emerge from them. These waves, called *synchronization waves*, indicate which cells are in pace with each other. (**c**) The two waves fuse when they meet, which indicates that the two gliders are more or less synchronized. (**d**) After the two gliders collide with each other, (**e**) they vanish as they also would in the original Game of Life, and (**f**) eventually all cells in the cellular space become quiescent

Although the behaviour of the Game of Life can be recovered from that of the AGL(p), the evolution of AGL(p) is more complicated than that of the Game of Life. The increased complexity of the AGL(p)'s behaviour results not only from the use of transition rules and cell states that are more complex than in the Game of Life, but also from the stochastic nature of the state transitions that are either asynchronous if $p < 1$, or synchronous if $p = 1$. Usually, the less the value of p is, the more time steps the AGL(p) will take to reach a configuration resembling a configuration of the Game of Life evolved from the same initial configuration. Though all configurations in Fig. 13.2(c)–(f) contain patterns resembling the one in Fig. 13.2(b), strictly speaking they are more dynamic due to the presence of synchronization waves, which may spread disorderly to the very end, because of the asynchronous update of the cells.

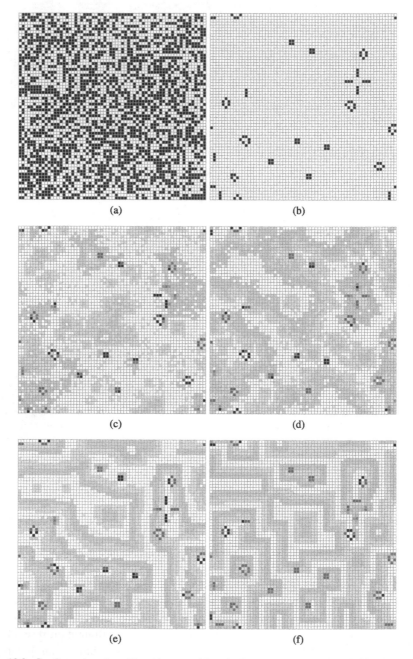

Fig. 13.2 Configurations in a 64×64 array with periodic boundary conditions [12]. (**a**) Random initial configuration. (**b**) Configuration evolved from the initial configuration in the conventional Game of Life. (**c**) Configuration evolved from the initial configuration in AGL(0.3) at time step $t = 80,000$, and similar configurations in (**d**) AGL(0.7), (**e**) AGL(0.99), and (**f**) AGL(1.0)

The resemblance of the asynchronous Game of Life to its classical counterpart is underlined by spectral analysis in [12], which reveals $1/f$ noise in the power spectrum at lower frequencies at both models [17]. Higher frequencies tend to show a Lorentzian noise spectrum in the asynchronous model, where this is less the case in the synchronous model. This may be caused by synchronization waves, which have relatively high frequencies.

13.2 Game of Life with Continuous States

The cellular automaton described in this section employs a transition function based on a continuously valued expression that includes a positive temperature parameter T. In the limit $T \to 0$ this transition function coincides with that of the Game of Life. States in this model are continuously valued, and cell state transitions, though deterministic, become increasingly fuzzy as the temperature parameter rises [1].

The use of continuous cell states enables the formulation of the transition function as a simple nonlinear expression. Assuming that the state of cell i at time t is $S_i(t)$, that $N(i)$ is the neighborhood of cell i, and that the cells undergo transitions at discrete values of time t (synchronous update), we define the transition function through the following equations:

$$S_i(t+1) = F\big(E_i(t)\big) \tag{13.1}$$
$$F(z) = 1/\big[1 + \exp(-2z/T)\big] \tag{13.2}$$
$$E_i(t) = E_0 - \big[x_i(t) - \chi_0\big]^2 \tag{13.3}$$
$$x_i(t) = S_i(t) + 2n_i(t) \tag{13.4}$$
$$n_i(t) = \sum_{j \in N(i)} S_j(t) \tag{13.5}$$

where T is the *temperature*, $E_i(t)$ is the *local energy* of cell i, and E_0 and χ_0 are the *energy shift* parameter and the *state shift* parameter, respectively. The value $n_i(t)$ is the sum of the states of the eight cells that are neighbors of cell i in a distance-1 Moore neighborhood. The values of $x_i(t)$ at which cell i becomes alive or stays alive are:

- 5 when cell i is alive and two of its neighbors are alive,
- 6 when cell i is dead and three of its neighbors are alive,
- 7 when cell i is alive and three of its neighbors are alive.

The values of E_0 and χ_0 will be chosen such that $E_i(t)$ becomes positive when $x_i(t)$ assumes the values 5, 6, or 7. $E_i(t)$ being positive means that $F(E_i(t))$ tends to 1 when T is sufficiently close to 0, according to (13.2). To derive appropriate values for E_0 and χ_0, we note that the following condition on $x = x_i(t)$ ensures a positive value of $E_i(t)$:

$$\chi_0 - \sqrt{E_0} < x < \chi_0 + \sqrt{E_0}$$

and this should correspond to values of x being 5, 6, or 7. On the other hand a negative value of $E_i(t)$ is ensured if

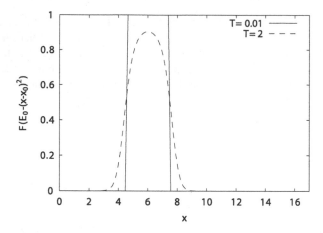

Fig. 13.3 Transition function as a function of x [1]. Temperatures are set to $T = 0.01$ and 2, and the other parameters are set to $E_0 = 2.25$, and $\chi_0 = 6$

$$x < \chi_0 - \sqrt{E_0} \quad \text{or} \quad x > \chi_0 + \sqrt{E_0}$$

and this should correspond to values of x being 4 or less or 8 or more. The cut-off threshold between these two conditions is determined by the equations:

$$4 < \chi_0 - \sqrt{E_0} < 5 \quad \text{and} \quad 7 < \chi_0 + \sqrt{E_0} < 8.$$

Choosing values at the centers of these ranges gives us:

$$E_0 = 2.25 \quad \text{and} \quad \chi_0 = 6.$$

These values of E_0 and χ_0 are assumed throughout the remainder of this section, though other values are also possible (see [1] for a discussion). The cutoff threshold becomes sharper as $T \to 0$. Figure 13.3 shows the values of x for the temperatures $T = 0.01$ and $T = 2$. For $T = 0.01$ the cutoff between $F \approx 0$ and $F \approx 1$ is abrupt, whereas for $T = 2$ it is quite smooth.

By simulations we investigate the influence of variations in the temperature T on the model's behavior. We first observe the time evolution at different temperatures of a *glider*. Figure 13.4 shows this pattern as it progresses for the two different temperatures $T = 0.7$ and $T = 0.8$. When the temperature is $T = 0.7$ the glider progresses without losing its shape, but when the temperature is $T = 0.8$ it changes into different patterns due to the accumulation of errors.

Next, we consider a *glider gun*. Figure 13.5 shows its evolution over time for the temperatures $T = 0.61$ and $T = 0.62$. When the temperature is $T = 0.61$ gliders are generated from the glider gun as usual, but when the temperature is a little bit higher, $T = 0.62$, the glider gun decays by the time $t = 30$ because of the accumulation of errors.

Cell states are continuous numbers in the range $[0, 1]$, and their deviations from the "ideal" value 0 or 1 (whichever is nearer) can be regarded as errors caused by thermal fluctuations. The reason for patterns to decay in the high temperature region is that these errors accumulate over time. On the other hand, the non-decay

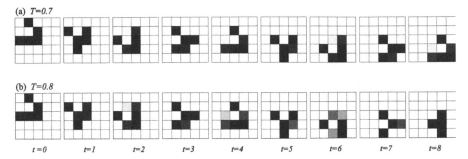

Fig. 13.4 Evolution of a glider from time $t = 0$ to $t = 8$ [1]. The cell's states are displayed by *gray-scales* — state 0 indicated by *white*, state 1 by *black*, and a state in between by *gray-scales*. (**a**) $T = 0.7$: periodical pattern of the glider. Most cells of the glider appear *black*, because their states are very close to 1. (**b**) $T = 0.8$: errors accumulate, causing the pattern to diverge after $t = 4$

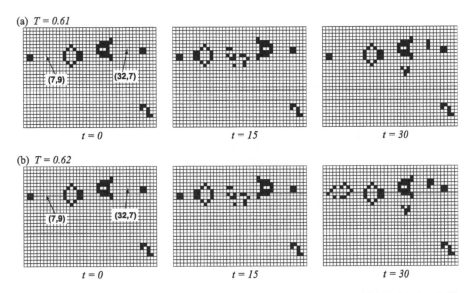

Fig. 13.5 Evolution over time of a glider gun in a cellular array of 40×30 [1]. (**a**) $T = 0.61$: gliders are generated periodically. (**b**) $T = 0.62$: errors accumulate at cells $(7, 9)$ and $(32, 7)$, which are indicated at $t = 0$, and the glider gun decays by $t = 30$

of patterns in the low temperature region suggests that these errors remain below a threshold at which they can be corrected due to the bistability of the transition function. A better idea about the decay of patterns in the glider gun configuration can be obtained by observing the behavior of a few of its individual cells over time. Figures 13.6(a) and (b) show the evolutions of the states of cells $(7, 9)$ and $(32, 7)$, respectively, both at the temperatures $T = 0.61$ and $T = 0.62$. When $T = 0.61$, the same cell states appear periodically after $t = 15$, i.e., $S_i(t + 30) = S_i(t)$, but for $T = 0.62$ the trajectories of the cell states are more irregular.

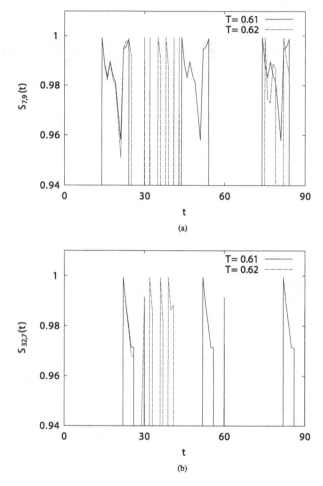

Fig. 13.6 Evolutions over time [1] at temperatures $T = 0.61$ and $T = 0.62$ of (**a**) the state of cell $(7, 9)$ and of (**b**) the state of cell $(32, 7)$

The temperatures at which patterns decay tend to be different for different patterns. Table 13.1 shows the decay temperature T_0 of several patterns. The beehive, the paper clip, the loaf, the boat, the eater, the barge, the pond, the snake, the tub, the block and the ship are stable patterns, and the clock, the beacon and the blinker are patterns with a cycle of 2 generations. Each pattern has a different decay temperature, which is as low as around 0.54 for the beehive and as high as approximately 0.99 for the ship, so the beehive pattern is more susceptible to thermal fluctuations than the other patterns. By setting the temperature sufficiently low — lower than, say, the decay temperature of the beehive — we obtain a model with dynamics similar to that of the Game of Life.

Table 13.1 Decay temperatures T_0 of several patterns, as observed in simulations employing 64-bit floating point representations [1]

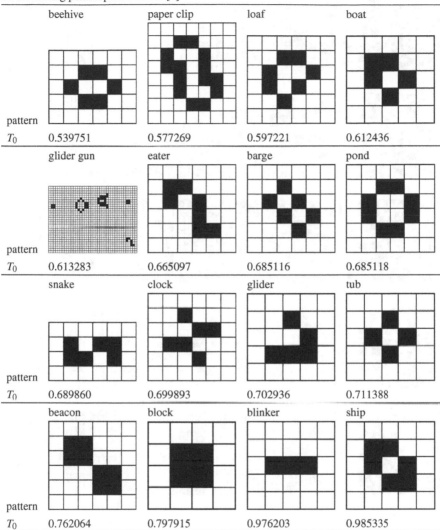

	beehive	paper clip	loaf	boat
pattern				
T_0	0.539751	0.577269	0.597221	0.612436
	glider gun	eater	barge	pond
pattern				
T_0	0.613283	0.665097	0.685116	0.685118
	snake	clock	glider	tub
pattern				
T_0	0.689860	0.699893	0.702936	0.711388
	beacon	block	blinker	ship
pattern				
T_0	0.762064	0.797915	0.976203	0.985335

To investigate the decay temperature for patterns in general, we consider the evolution over time of the random pattern in Fig. 13.7 for various temperatures. The evolutions at temperatures $T = 0.53$ and $T \to 0$ are very similar, suggesting that in the region $0 < T \le 0.53$ the model behaves like the Game of Life. The higher temperature regions see an increase in the number of gray cells, but still there is convergence to static or periodic patterns. At very high temperatures — typically $T \ge 2.0$ — the model fails to converge to such patterns. This means that the relaxation time is infinitely large at high temperatures, leading to dynamical behavior resembling chaos.

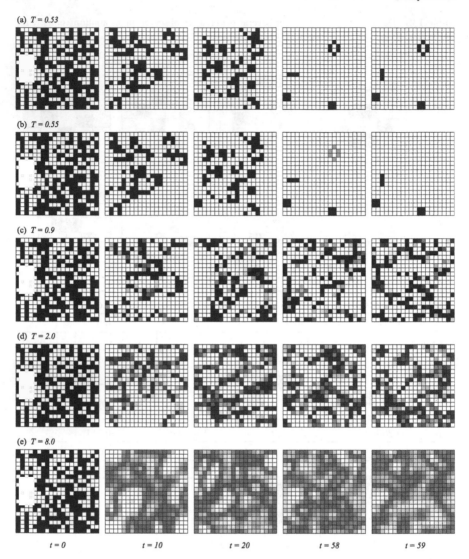

Fig. 13.7 Evolution over time of a 20×20 random pattern [1] for the temperatures (**a**) $T = 0.53$, (**b**) $T = 0.55$, (**c**) $T = 0.9$, (**d**) $T = 2.0$, and (**e**) $T = 8.0$. The other parameters are set to $E_0 = 2.25$ and $x_0 = 6$. The evolution for temperatures near zero, which corresponds to the Game of Life, is similar to the evolution for $T = 0.53$. The initial random pattern happens to develop a beehive, which eventually decays when $T \geq 0.55$. The decay temperatures for other random patterns tend to be higher

The above simulations suggest that the Game of Life is reproduced by the continuous model if temperatures are sufficiently low. Analysis of the power spectrum, conducted in [1], confirm this. Temperatures below $T \approx 0.7$ tend to give $1/f$ noise — typical of the Game of Life [17] — but this changes to a Lorentzian spectrum for temperatures higher than $T \approx 0.9$, which is indicative of chaotic behavior.

The continuous model of the Game of Life described in this section is remarkably simple, in the sense that the transition function can be expressed as a continuous nonlinear function, which allows the model to be studied over a continuum of temperatures. Another approach to the modeling of the Game of Life by a continuous nonlinear model is by Cellular Neural Networks in [7, 18]. This approach appears less straightforward, and it leaves less room for variations in the model's parameters.

13.3 Game of Life with an Enlarged Neighborhood

Within the class of distance-1 Moore neighborhood totalistic cellular automata there is limited choice in Game of Life-like models. A greater variety of models can be obtained by increasing the size of the neighborhood. This includes the fascinating *Larger than Life* (*LtL*) class of Cellular Automata [9], which have a rich dynamics, but also quite complex transition rules and large neighborhoods.

This section shows a Cellular Automaton that, though strictly speaking a member of the LtL class, has a transition rule with remarkable simplicity, while exhibiting dynamics similar to the Game of Life. It is this similarity, as well as the resemblance to patterns in kaleidoscopes, that has earned this model its name *Kaleidoscope of Life* [2]. Cells in this model can be in one of two states, like in the Game of Life, i.e., 0 (dead) or 1 (alive). Each cell has a neighborhood containing the cells at orthogonal or diagonal distances one or two from the cell. This neighborhood covers a total of 24 cells. We assume that each cell in the cell space is identified by a unique integer, and that $\sigma_i(t)$ is the state of cell i at time t and $N(i)$ is the neighborhood of cell i. The transition rule of the model is defined in terms of the states of the cells in the neighborhood of cell i:

$$\sigma_i(t+1) = \begin{cases} 1 & \text{if } \sum_{j \in N(i)} \sigma_j(t) = 4, \\ 0 & \text{otherwise.} \end{cases} \tag{13.6}$$

In other words, a cell becomes or stays alive if the number of living cells in its neighborhood is 4; otherwise the cell dies or remains dead. All cells in the cell space undergo transitions simultaneously, so the model is synchronously timed. The transition rule is outer-totalistic, since it is the sum of states of cells in the neighborhood of a cell that determines the cell's next state. Unlike in the Game of Life, a cell's state itself is irrelevant to its next state. Because of this reason the rule is called *inner-independent*.

The above simple transition rule is able to generate a rich variety of patterns. When we start, for example, from an initial configuration that is randomly initialized with a probability of 0.2 for a cell being alive, we obtain dynamic behaviors like those in Fig. 13.8. The pattern at time $t = 220$ exhibits two circles, which, as we will see later, are oscillating with a period of 8. This pattern will also occur in the second simulation, which starts from two 2×2 blocks of living cells (Fig. 13.9). The pattern of two blocks diffuses toward the right and the left symmetrically, and then at time $t = 20$ four blocks appear, so the blocks have effectively been duplicated in 20 generations.

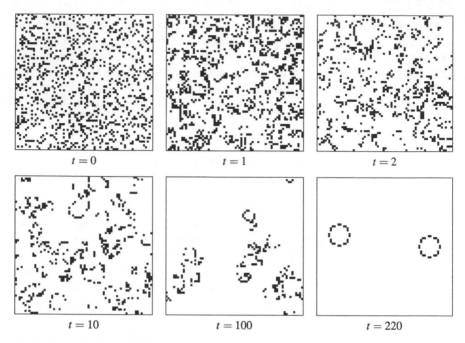

Fig. 13.8 Evolution over 220 time steps of a random initial pattern in the 72×72 Cellular Automaton based on rule (13.6) with periodic boundary conditions [2]. Cells are initialized randomly with each cell having a probability of 0.2 that it is alive

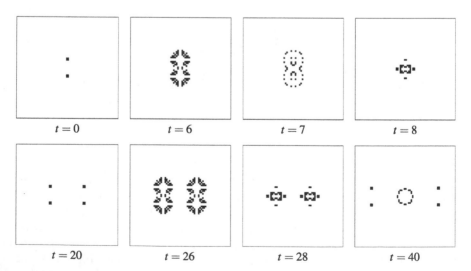

Fig. 13.9 Evolution over time of two 2×2 blocks at a distance of 12 cells from each other [2]. The size of the cell space is 72×72. The pattern diffuses toward the right and the left, resulting in 4 blocks after 20 steps

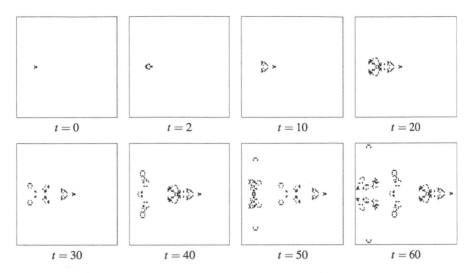

Fig. 13.10 A gunship pattern propagating to the right over time in a space of 120×120 cells [2]. The shape of the gunships' head is the same at times $t = 20$, $t = 40$ and $t = 60$, so it has a period of 20 steps

The last simulation shows a so-called *gunship* as it evolves over time (Fig. 13.10). The gunship launches other patterns, somewhat resembling exhaust gas, as it moves ahead.

The patterns encountered in the above examples strongly resemble those in the Game of Life: we have stable patterns, oscillating patterns, and patterns moving across the cell space (gliders). The most frequently encountered patterns in the Kaleidoscope of Life are:

- Stable patterns: *C-block*, *X-block*, *L-block*, *K-block*, *Eyes*, *Mustache*, and *Stand*, ordered in a decreasing frequency of appearance (see Fig. 13.11).
- Oscillators: *Mouth*, *Blinker*, *Pulsar*, *Rotary*, *Jaws* all with a period of 2 steps, *Ninjastar* with a period of 4 steps, and *Hanabi* (meaning "fireworks" in Japanese) with a period of 8 steps (see Fig. 13.11). Frequently appearing patterns are Mouth, Blinker, Pulsar, and Hanabi.
- Gliders: four types of gliders of which we highlight the *A-glider* (Fig. 13.12), and the *C-glider* (Fig. 13.13). The other two gliders are described in [2]. The C-glider has two versions, which are mirror images of each other. The left version is called C_L-*glider*, and the right version C_R-*glider*. The A-glider has a period of 2 and the C-glider has a period of 12. Both gliders travel 3 cells per 2 time steps.

The Kaleidoscope of Life is computationally universal [2], and the proof of this is based — like in the Game of Life — on the design of logic circuits on the cell space. Two configurations play a prominent role in this context: C_L-gliders are used as signals in circuits and Hanabi configurations are used to conduct operations on these signals. The effect of a Hanabi on a C_L-glider depends on its position and phase relative to the glider. One useful role of the Hanabi is that of an Eliminator,

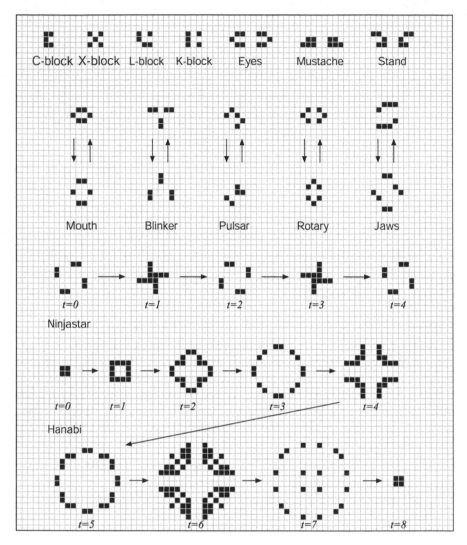

Fig. 13.11 Seven stable patterns (*top row*), five oscillators with a period of 2 (*second row*), a Ninjastar with a period of 4 (*third row*), and a Hanabi with a period of 8 (*bottom row*) [2]

which eliminates a C_L-glider colliding into it (provided this collision is from the proper angle and in the proper phase). Another use of the Hanabi is to turn C_L-gliders to the left, or — by using three left turns in series — to turn them to the right. The Hanabi is also used for the conversion between glider types. Particular useful in this context is the conversion from an A-glider to a C_L-glider, for reasons that we explain below.

Another type of interaction takes place between C_L-gliders. Depending on the relative positions and phases two C_L-gliders have towards each other, they interact in different ways, like annihilating each other (Fig. 13.14).

Fig. 13.12 An *A*-glider has a period of 2 and travels 3 cells per 2 steps [2]

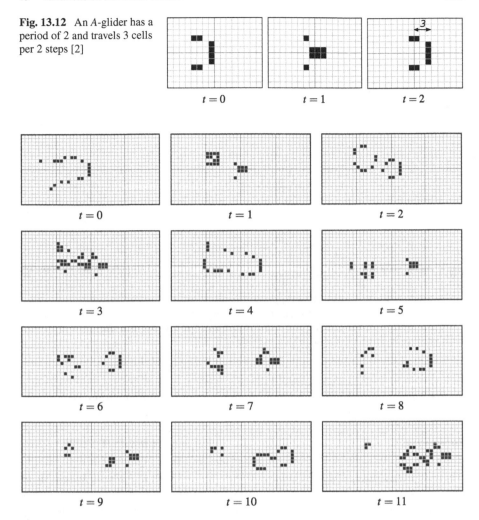

Fig. 13.13 A C_L-glider has a period of 12 and travels 3 cells per 2 steps [2]. The C_L-glider is the left version of its (*right*) mirror image, the C_R-glider

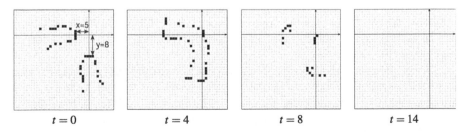

Fig. 13.14 Two gliders annihilating each other upon colliding [2]. At $t = 0$, the glider moving to the east is 5 cells away from the collision point, and the glider moving to the north 8 cells. Both gliders are in the same phase, and they completely annihilate each other in 14 steps

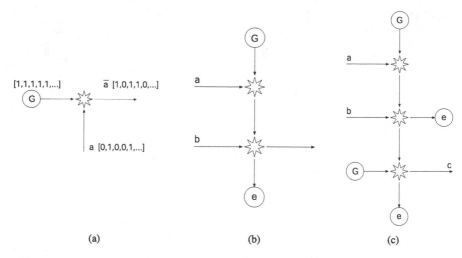

Fig. 13.15 Schematic diagrams of (**a**) the NOT gate, (**b**) the AND gate and (**c**) the OR gate [2]. The symbols G and e denote a glider gun and an eliminator, respectively. The *star* at the intersections of two streams of gliders is where gliders are annihilated when they collide. A *1-symbol* in a sequence in the NOT-gate stands for the presence of a C_L-glider, and a *0-symbol* for its absence. There are 552 time steps between two subsequent elements in a sequence. A sequence input to the NOT-gate from below is inverted and output to the right side

A noteworthy interaction occurs when two C_L-gliders collide in such a way that one A-glider and two C_L-gliders are created as a result. When the A-glider is converted by a Hanabi into a C_L-glider, we then have in effect an operation that produces three C_L-gliders upon receiving two C_L-gliders as input. This effect is exploited in [2] to design a Fan-Out, which produces two C_L-gliders for every single C_L-glider input to it, and which has an internal C_L-glider running around in it. In theory a Fan-Out can be used as the basis for a configuration that emits a steady stream of gliders by feeding back one of the two outputs to the input terminal, but unfortunately this does not work due to phase differences between the Fan-Out's input and output [2]. For this reason, as many as four Fan-Outs are used in [2] that feed their outputs into each other's inputs; in combination with three Eliminators to kill off three streams of excess gliders, this results in a kind of *glider gun* that produces one C_L-glider every 552 time steps. Like in the Game of Life, a glider gun acts as a clock signal, and it is thus an important ingredient in the construction of logic gates (Fig. 13.15).

Though the approach outlined above allows us to construct individual logic gates, it is necessary to place the gates on the cell space such that their inputs and outputs are compatible. The period of the C_L-glider (12 time steps) is different from that of the Hanabi (8 time steps) and, to make things worse, an operation on a signal may cause various shifts in position on the cell space and in the signal's phase. A left turn for example shifts the phase of a C_L-glider by three time steps. Guaranteeing that C_L-gliders can be redirected to the exact locations where they are expected while being in the proper phase requires additional circuit elements. A rough method to adjust positions and phases of signals is to redirect them along four subsequent left turns, but this allows only shifts in horizontal or orthogonal positions of multiples of 18 cells and shifts in phase by multiples of 12 time steps. Obviously, for more refined adjustments, a different method is required. To this end, a so-called *Shifter* is constructed in [2], which can adjust paths of C_L-gliders by one cell at a time, as well as adjust their phases by one time step at a time. The construction of the Shifter exploits the large number of ways — with many variations in shifts in space and phase — in which two C_L-gliders can annihilate each other (see details in [2]). The Shifter is the cog in the wheel that enables us to prove the computational universality of the Kaleidoscope of Life.

The Kaleidoscope of Life is unusual in the sense that its transition rule is inner-independent. A cell's state being independent of its previous state gives the model characteristics resembling those of classical spin systems [2].

The similarities of the Kaleidoscope of Life to the Game of Life do not stop at the level of patterns, but also extend to statistical properties. The Kaleidoscope of Life's spectrum shows roughly $1/f$ noise at lower frequencies, like in the Game of Life [17], though the exponent of f is somewhat lower than in the Game of Life, probably due to larger variety of oscillators in the Kaleidoscope of Life [2].

13.4 Conclusions and Discussion

The variations of the Game of Life outlined in this chapter are in the use of asynchronous updating instead of synchronous, the use of continuously valued states instead of discrete, and the use of extended neighborhoods.

The asynchronously timed Game of Life model employs an extended state set of cells for time bookkeeping, which holds a cell in approximate lock-step with its neighbors. This results in a semi-loose correspondence between the asynchronous model's behavior and the Game of Life's: a similar cell pattern occurs over time in both models in a similar order, but various areas of the cell space may be out of lockstep by a degree proportionally to the distance between those areas. Though ultimately both models do not differ fundamentally in behavior, there is the interesting occurrence of synchronization waves that are used in the asynchronous model to keep cells in approximate lock-step. These waves are unrelated to the dynamics of the Game of Life and merely reflect the disparities in update speed of different areas due to the random nature of update.

The Game of Life model with continuously valued states behaves similar to its discrete counterpart, provided the temperature is below a certain threshold. Depending on the configuration, the threshold below which behavior remains unchanged will differ, but there appears to be a safe range of temperatures at which the continuous dynamics closely resembles the original Game of Life's. Temperatures in this range make the sigmoid function F in Sect. 13.2 sufficiently steep to guarantee a clear separation between (almost-)0 states and (almost-)1 states, thus effectively making the update function resemble a discrete one. The free temperature parameter allows the study of the Game of Life in a wide range of regimes, from models that

are virtually discrete to more fuzzy models. This may lead to a deeper understanding
of the Game of Life's characteristics, like its robustness to noise.

The model with an enlarged neighborhood differs substantially from the Game of
Life, but the patterns emerging from its dynamics have similar characteristics: there
are completely static patterns, as well as oscillators, and gliders. There appears to
be no "naturally" occurring glider gun, though it is possible to design an artificial
one. The larger neighborhood — as compared to the Game of Life — yields an
increased number of combinations of living cells that rewrite to a living cell in a state
transition, and consequently there is quite a variety of patterns in the Kaleidoscope
of Life. This variety is of great help in the design of logic gates, as well as in laying
them out on the cell space, such that output terminals of gates are correctly aligned
with input terminals. As a result of the flexibility in laying out patterns, it is possible
to build arbitrary logic circuits — the hallmark of computational universality.

What other variation on the Game of Life could one conceive of? The same
method used to transform the Game of Life into an asynchronous model, as outlined in Sect. 13.1, is also applicable to the Kaleidoscope of Life, so a combination
of asynchronous timing and an enlarged neighborhood is within the realm of possibilities.

It is also possible to apply the method in Sect. 13.2 on the Kaleidoscope of Life
to obtain its continuous equivalent. For this to work it is necessary to define the
continuous state of cell i at time t by $S_i(t)$, and to define the function $n_i(t)$ in (13.5)
as follows:

$$n_i(t) = \sum_{j \in N(i)} S_j(t)$$

whereby $N(i)$ is the neighborhood of a cell in the Kaleidoscope of Life. Furthermore, replacing the function in (13.4) by $x_i(t) = n_i(t)$ and setting the values of the
parameters to $E_0 = 0.25$ and $\chi_0 = 4$ according to a similar method as in Sect. 13.2,
and adopting the other equations (13.1), (13.2), and (13.3) unchanged, we obtain a
version of the Kaleidoscope of Life with continuously valued states.

Combining asynchronous updating with continuously valued states may be less
easy to achieve, since the transformation into continuous states works best when the
original model has binary states, rather than the eight or more states an asynchronous
model would have.

Another possible variation on the Game of Life may be the choice of the underlying grid, for example a triangular or hexagonal grid instead of a square grid. Some
search on the internet will reveal some previous work done in this field, though there
appears to be a lack of published results. Work has also been done on the Game Of
Life on hyperbolic tilings [20]. More research on these variations may uncover yet
other intriguing phenomena in the Game of Life.

References

1. Adachi, S., Peper, F., Lee, J.: The Game of Life at finite temperature. Physica D **198**, 182–196
   (2004)

2. Adachi, S., Lee, J., Peper, F., Umeo, H.: Kaleidoscope of Life: a 24-neighborhood outer-totalistic cellular automaton. Physica D **237**, 800–817 (2008)
3. Bersini, H., Dectours, V.: Asynchrony induces stability in cellular automata based models. In: Brooks, R.A., Maes, P. (eds.) Artificial Life IV, pp. 382–387. MIT, Cambridge (1994)
4. Blok, H.J., Bergersen, B.: Synchronous versus asynchronous updating in the "game of life". Phys. Rev. E **59**, 3876–3879 (1999)
5. Capcarrère, M.: Cellular automata and other cellular systems: design & evolution. Ph.D. Thesis, Swiss Federal Institute of Technology, Lausanne (2002)
6. Choi, M.Y., Huberman, B.A., Digital dynamics and the simulation of magnetic systems. Phys. Rev. B **28**, 2547–2554 (1983)
7. Chua, L.O., Roska, T., Venetianer, P.L.: The CNN is universal as the Turing machine. IEEE Trans. Circuits Syst.-I **40**(4), 289–291 (1993)
8. Berlekamp, E.R., Conway, J.H., Guy, R.K.: Winning Ways for Your Mathematical Plays, vol. 2. Academic Press, San Diego (1982)
9. Evans, K.M.: Larger than Life: threshold-range scaling of Life's coherent structures. Physica D **183**, 45–67 (2003)
10. Ingerson, T.E., Buvel, R.L.: Structures in asynchronous cellular automata. Physica D **10** 59–68 (1984)
11. Le Caër, G.: Comparison between simultaneous and sequential updating in $2^{n+1} - 1$ cellular automata. Physica A **157**, 669–687 (1989)
12. Lee, J., Adachi, S., Peper, F., Morita, K.: Asynchronous Game of Life. Physica D **194**, 369–384 (2004)
13. Lumer, E.D., Nicolis, G., Synchronous versus asynchronous dynamics in spatially distributed systems. Physica D **71** 440–452 (1994)
14. Nakamura, K.: Asynchronous cellular automata and their computational ability. Syst. Comput. Controls **5**(5), 58–66 (1974)
15. Nakamura, K.: Synchronous to asynchronous transformation of polyautomata. J. Comput. Syst. Sci. **23**, 22–37 (1981)
16. Nehaniv, C.L.: Evolution in asynchronous cellular automata. In: Standish, R.H., Bedau, M.A., Abbass, H.A. (eds.) Artificial Life VIII, pp. 65–73. MIT, Cambridge (2003)
17. Ninagawa, S., Yoneda, M., Hirose, S.: $1/f$ fluctuation in the "Game of Life". Physica D **118**, 49–52 (1998)
18. Pazienza, G.E., Gomez-Ramirez, E., Vilasi's-Cardona, X.: Polynomial cellular neural networks for implementing the game of life. In: Marques de Sá, J., Alexandre, L.A., Duch, W., Mandic, D.P. (eds.) Proc. ICANN 2007. LNCS, vol. 4668, pp. 914–923. Springer, Berlin (2007)
19. Rajewsky, N., Santen, L., Schadschneider, A., Schreckenberg, M.: The asymmetric exclusion process: comparison of update procedures. J. Stat. Phys. **92**, 151–194 (1998)
20. Reiter, C.A.: The Game of Life on a hyperbolic domain. Comput. Graph. **21**(5), 673–683 (1997)
21. Schönfisch, B., de Roos, A.: Synchronous and asynchronous updating in cellular automata. Biosystems **51**, 123–143 (1999)
22. Toffoli, T.: Integration of the phase-difference relations in asynchronous sequential networks. In: Ausiello, G., Böhm, C. (eds.) Automata, Languages, and Programming. LNCS, vol. 62, pp. 457–463. Springer, Berlin (1978)
23. Toffoli, T., Margolus, N.: Cellular Automata Machines. MIT, Cambridge (1987)
24. S. Wolfram: Cellular Automata and Complexity. Addison–Wesley, Reading (1994)

Chapter 14
Does *Life* Resist Asynchrony?

Nazim Fatès

14.1 Introduction

Undoubtedly, Conway's Game of Life — or simply Life — is one of the most amazing inventions in the field of cellular automata. Forty years after its discovery, the model still fascinates researchers as if it were an inexhaustible source of puzzles. One of the most intriguing questions is to determine what makes this rule so particular among the quasi-infinite set of rules one can search. According to W. Poundstone, *"Conway wanted to create a game that would be as unpredictable as possible, yet with the simplest possible rules"* [18]. Still, what kind of intuition struck Conway when he designed the model? Was it mere chance or did he use a method for finding a rule that is simple, yet leads to a surprising behaviour? Is the Life rule somewhat related to real life, and if so, can we discover models of equal richness by observing natural phenomena?

Our long-term objective is to analyse how the properties of the Life rule are related to the properties of natural systems. In other words, we ask whether Conway's Life resists structural perturbations, where *structural* means that the perturbations modify the interactions of the components interaction instead of modifying their state. Our aim is to examine whether the Life rule verifies an important property of living organisms: their robustness to asynchrony, *i.e.*, their ability to keep a constant behaviour when the updating of their components is modified. We thereby study the behaviour of an asynchronous version of Life on a regular and an irregular topology.

To date, these issues haven't received much attention. One reason to explain this lack of interest is that Conway's model is generally considered as a simple metaphor without any serious connection to real life. It is thought of as an example that illustrates how basic local rules can produce complex and unpredictable behaviours. However, even if the model is metaphor, we want to know if asynchrony allows us to discover novel interesting properties. The next section introduces some bibliographical landmarks to see how this issue has been tackled so far.

A. Adamatzky (ed.), *Game of Life Cellular Automata*,
DOI 10.1007/978-1-84996-217-9_14, © Springer-Verlag London Limited 2010

14.2 A Brief History of the Problem

The idea to test how Life resists noise dates from as early as 1978 with a
paper by Schulman and Seiden [20]. They examined how the introduction of a
stochastic element in the local evolution rule would perturb the long term evolution of the system. They replaced the deterministic transitions $p_{k,s}$ — which are
equal to 0 or 1 if a cell in state k with s lives neighbours dies or leaves — by
$p_{k,s}(T)$ which are the probabilities to be in state 1 at the next iteration. They take
$p_{k,s}(T) = (p_{k,s} + d.T)/(1+T)$, where d is the current density and T is the stochastic parameter, named "temperature" by analogy with the physical parameter. Interestingly, this parameter is chosen such as not to influence the evolution towards increasing or decreasing the density. The main drawback with this approach is that the
definition of the probabilities of transition $p_{k,s}$ is not local: the density should first
be computed using the state of the whole grid before the transitions are determined.
Nevertheless, the authors analyse the perturbed model with first order and improved
mean-field analysis. Their findings can be summarised by saying that: (a) The mean-
field analysis fails to predict the evolution of the classical deterministic Life but
succeeds with high-temperature models. (b) More surprisingly, the analysis shows
that a phase transition exists which separates two regimes: depending on the value
of T, the system is either attracted to the null density or to an attractor of density
$d^* = 0.37$. However, a deeper analysis of this phase transition remains yet to be
achieved.

Adachi et al. also proposed a modified version of Life. Their model takes into
account a "temperature" as a stochastic parameter and uses continuous values for
the cells' state [1]. The authors show that most patterns are stable if the temperature
is sufficiently low.

Tackling the question of how Life is affected by asynchronous updating,
Bersini and Detours [4] proposed a qualitative study of the model under fully asynchronous updating, i.e., when the updating is sequential and cells are chosen randomly and uniformly without any memory (see [12] for a classification of rules
with this updating scheme). The authors observed that Life's behaviour was qualitatively altered by the change of updating. They deduced that asynchrony had a stabilising effect since they could observe the emergence of "labyrinth" patterns that
would eventually allow the system to reach a fixed point. However, no systematic
study of the phenomenon was conducted; in particular, the authors did not examine
how this stabilising effect scales with the size of the grid.

The first quantitative study of the change of behaviour was conducted by Blok
and Bergersen [5]. They examined Life when each cell has a probability α to be
updated at each time step, the so-called α-asynchronous updating. They identified
a continuous change of behaviour of the model depending on α. More precisely,
they found that Life displays a second-order phase transition which belongs to the
directed percolation universality class (see below). In this chapter, we will carry out
similar experiments but with a different protocol.

Finally, let us mention the stochastic variant of Life which was introduced by
Monetti and Albano: they modified the local rule by adding probabilities p_s and

p_b for survival and birth, respectively [17]. This change produces a first-order irreversible phase transition between an "extinct" state where no cell is alive and a "live" state where some structures persist. An improved mean-field analysis was developed to support the experimental observations [16].

We now present formally our model, where, as a first step, the perturbations apply only to the updating procedure (and *not* to the transition rule). Our starting point is thus similar to the study by Blok and Bergersen, but with the scope of analysis extended to a greater range of conditions.

14.3 Asynchronous `Life`

Let us consider the square grid $\mathscr{L} = \{1, \ldots, L\} \times \{1, \ldots, L\}$ with periodic boundary conditions (the space is a torus). The state of a cell $c \in \mathscr{L}$ at time t is denoted by $\sigma_c^t \in \{0, 1\}$.

Each cell $c \in \mathscr{L}$ is associated with its *neighbourhood* $\mathscr{N}(c) \subset \mathscr{L}$. We use the 8-nearest neighbours topology: $\mathscr{N}(c) = \{c' \in \mathscr{L}, \ |c'_x - c_x| = 1 \text{ or } |c'_y - c_y| = 1\}$.

The *local rule* is the function which gives the next state of a cell according to the state of its neighbours. In the case of `Life` and its variants, it is defined as an *outer-totalistic* function, *i.e.*, the application of the local rule on a cell c is a function of only two variables: σ_c^t and $s = \sum_{c' \in \mathscr{N}(c)} \sigma_{c'}^t$. A general expression of the local transition rule of `Life` is stated with two threshold rules:

- *birth rule*: a cell in state in state 0 becomes a 1 iff $s \in [B_l, B_h]$;
- *survival rule*: a cell in state 0 remains a 1 iff $s \in [S_l, S_h]$;

where the thresholds B_l, B_h, S_l, S_h are non-zero integers. The original `Life` is defined with the thresholds $B_l = 3$, $B_h = 3$, $S_l = 2$ and $S_h = 3$.

Classically, the model is defined with a *synchronous* updating, *i.e.*, all cells are updated at each time step. In this chapter, we also consider *asynchronous* updating, *i.e.*, only a fraction of cells are updated at each time step. There are various ways of considering asynchronous updating in cellular automata; we consider two simple schemes:

(a) *α-synchronous updating* [11]: at each time step, each cell is updated with probability α and left unchanged with probability $1 - \alpha$;
(b) *fully asynchronous updating* [12]: at each time step, one cell is chosen uniformly at random and updated, the other cells are left unchanged.

The probability α is called the *synchrony rate*. When $\alpha = 1$ we have classical synchronous updating; the system is deterministic. For $\alpha < 1$, the system becomes stochastic. In order to compare different simulations, we use the *rescaled time* τ defined as $\tau = t \cdot \alpha$ for α-asynchronous updating and $\tau = t/L^2$ for fully-asynchronous updating. Intuitively, for $\alpha \to 0$, both α-synchronous updating and fully-asynchronous updating become equivalent under time rescaling. Indeed, the probability that two neighbouring cells are simultaneously updated becomes negligible, which allows the two processes to be mapped onto each other by an appropriate transformation. However, note that this is valid only for a *bounded* rescaled simulation time; if we consider infinite-time evolutions of the system, the equivalence may not hold any more.

Formally, we model asynchrony by specifying the set of cells that are updated at each time t. This is done by using a function $\Delta : \mathbb{N} \to \mathscr{P}(\mathscr{L})$, called the *updating scheme*. The *global* transition function $F_\Delta : \{0, 1\}^{\mathscr{L}} \to \{0, 1\}^{\mathscr{L}}$ gives the sequence of configurations $(\sigma_c^t)_{t \in \mathbb{N}}$ with the iterative scheme: $\sigma_c^{t+1} = F_\Delta(\sigma_c^t)$, where:

$$\forall t \in \mathbb{N}, \ \forall c \in \mathscr{L}, \quad \sigma_c^{t+1} = \begin{cases} f(\sigma_c^t, s) & \text{if } c \in \Delta(t) \\ \sigma_c^t & \text{otherwise} \end{cases}$$

and where

$$s = \sum_{c' \in \mathscr{N}(c)} \sigma_{c'}^t,$$

and $f(q, s)$ is the *local* transition function defined by:

$$f(0, s) = \begin{cases} 1 & \text{if } s = 3 \\ 0 & \text{otherwise} \end{cases} \quad \text{and} \quad f(1, s) = \begin{cases} 1 & \text{if } s \in [2, 3] \\ 0 & \text{otherwise} \end{cases}.$$

Using this formalism, we reformulate our objective as an analysis of the behaviour of the global transition rule F_Δ, which depends on the updating scheme Δ. The word "behaviour" has yet to be defined more precisely; in this chapter we use two functions for quantifying changes in behaviour:

- the density $d(x)$ is the number of 1s in x over the size of x;
- the activity $\chi(x)$ is the number of unstable cells in x over the size of x.

These two parameters are only rough projections of a configuration $x \in \{0, 1\}^{\mathscr{L}}$ on the [0, 1] interval and clearly they capture only a small part of what one generally calls a "behaviour". However, we will see that these functions give sufficient information to observe interesting phenomena, especially when qualitative changes of the behaviour of the system need to be detected.

14.4 Assessing Life's Robustness to Asynchrony

We now turn our attention to the evaluation of the robustness of the Life rule. We proceed in three steps: (1) an informal observation of the changes induced by asynchrony, (2) a quantification of these changes, (3) an analysis of phase transitions with the tools from statistical physics.

14.4.1 First Experiment

Figure 14.1 shows Life's steady state typical configurations for three different updating schemes. It illustrates the influence of the updating scheme on the behaviour of the system: we see that for asynchronous updating, a new type of pattern has

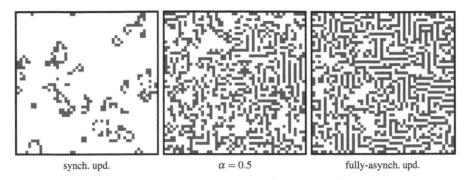

synch. upd. $\alpha = 0.5$ fully-asynch. upd.

Fig. 14.1 Snapshots of configurations obtained with three updating schemes: synchronous (*left*), α-asynchronous (*middle*), fully asynchronous (*right*). Configurations are taken at time $\tau = 100$ (*top*) and $\tau = 200$ (*bottom*)

emerged. We call these patterns "*labyrinths*"; their presence on the grid depends on the synchrony rate α: for $\alpha > 0.9$, no labyrinth pattern appears, for $\alpha < 0.9$, the labyrinth patterns will be increasingly dense with a lower synchrony rate. How can we explain this change?

We repeated the previous experiment by decreasing α by steps of 5% and we observed that the labyrinth shapes progressively appears when $\alpha < 0.9$. For $\alpha < 0.8$, the system rapidly reaches a steady state where the density becomes stable. As α decreases, blank spaces in the steady state disappear in favour of the labyrinth shapes; for $\alpha < 0.5$, these shapes almost cover the whole grid (see Fig. 14.1).

14.4.2 A Quantification of the Changes

How does this steady state vary as a function of the synchrony rate? A possible method for quantifying the variation is to measure the average activity of the steady-state. Figure 14.2 shows the steady state activity χ_∞ as a function of α. This parameter expresses the average activity for an infinite size system and an infinite time evolution. Approximations of χ_∞ are obtained by running a system for a stabilisation time of 1,000 steps and then averaging the activity of the system for another 1,000 steps. It confirms that a qualitative change of behaviour occurs for $\alpha \sim 0.9$.

As noted by Blok and Bergersen, there also exists a "jump" near $\alpha = 1$. This discontinuity results from a modification of behaviour which appear when a small amount of noise is introduced: some patterns with a short cycle stability (*e.g.*, the blinkers) are destroyed by the asynchronous updating. Noteworthy is the analogy with similar phenomena observed in 1D Elementary Cellular Automata (ECA), see [11]. Although for `Life`, the "jump" is relatively small (of the order of 0.01), it can be much larger in the case of other cellular automata.

If it is easy to explain the change of behaviour near $\alpha = 1$, the modification which occurs near $\alpha = 0.9$ is rather puzzling at first sight. What kind of transformation does the system undergo at this critical value?

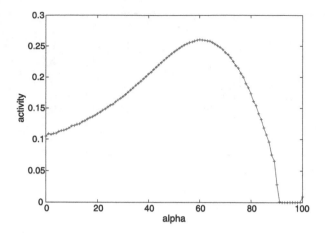

Fig. 14.2 Steady state activity χ_∞ vs. synchrony rate α. Grid size is 100×100. Averages obtained with 12 samples

14.4.3 A Second-Order Phase Transition

The first quantitative study of the abrupt change of behaviour in the asynchronous Game of Life is due to Blok and Bergersen [6]. The authors showed that the phenomenon was a second-order phase transition: the macroscopic functions that describe the global behaviour as a function of α are continuous but their derivative are discontinuous at the critical point $\alpha_c \sim 0.9$. The critical threshold α_c separates two distinct macroscopic behaviours called *phases*. The first phase ($\alpha > \alpha_c$) is the *frozen* phase, in which the systems evolves with low-density patterns and quickly stabilises to a fixed point. The second phase ($\alpha < \alpha_c$) is the *labyrinth* phase, it is characterised by a steady state with higher density and the absence of a stabilisation at a fixed point (at least for large grid sizes, *e.g.*, for $L > 20$).

Note that this description is ideal in the sense that it only applies to infinite systems. As simulation imposes to use finite systems, how then do we know that we are in the presence of a phase transition? This problem is complex and there exists a great range of methods to answer it. In general, the system is observed near the critical threshold and we measure how a macroscopic description, the *order parameter*, varies as a function of the lattice size L. Ideally, one should observe that the effect of L become less and less important, *i.e.*, the order parameter tends to converge to an asymptotic value for $L \to \infty$.

Concomitantly, we should also observe that the order parameter obeys power laws as it gets closer to the threshold. The origin of these power laws comes from the divergence of the system's spatial length scale at the critical point: the system becomes self-similar and its structure follows a fractal pattern. A major discovery in statistical physics is that the power laws that describe the system are not arbitrary: it was observed that very different systems have power laws with the same *critical exponents*. The set of systems which exhibit the same critical exponents is called a *universality class*. The identification of a universality class is thus a trusted method to show that a brutal change of behaviour is a phase transition (and thus not a mere continuous change in the system's behaviour).

By measuring the value of the steady state activity χ_∞ as a function of the synchrony rate α, Blok and Bergersen demonstrated that asynchronous `Life`'s phase transition belonged to the universality class of *directed percolation* (see [14] for a detailed description). Their protocol consisted in measuring a single critical exponent, *i.e.*, they showed that near criticality, the system obeys:

$$\chi_\infty(\alpha) \sim (\alpha - \alpha_c)^\beta$$

with β being predicted by directed percolation theory. Although this allowed them to locate the critical threshold at $\alpha_c = 0.9083$, it is well-known in the literature that using this method alone does not give precise results [14]. Indeed, the closer we get to the critical threshold, the longer we need to wait for the system to stabilise to a steady state. This phenomenon, known as the *critical slowing down*, is difficult to counterbalance. It often introduces systematic biases in the measures, which can not be "seen" by merely computing the uncertainties in the statistical data. To avoid the biases of this method, we follow another protocol which was used by Grassberger for linear stochastic cellular automata [13].

The protocol consists of an interactive process: (a) we fix α and start with a random configuration, (b) we monitor the evolution of the order parameter for a long simulation time until we observe a sub-critical or super-critical behaviour; (c) we repeat the experiment with a value closer to the critical point until we are not able to observe the difference between sub- and super-critical behaviour. As we expect the evolution of the order parameter to be a power law near criticality, that is: $\chi(t) \sim t^{-\delta}$, $\chi(t)$ should thus appear as a straight line on a log–log plot. The critical threshold is found by observing the changes of convexity in the plots displayed in log–log scale: a plot with a positive or negative curvature corresponds to super-critical or sub-critical behaviour, respectively.

We tested both the density and the activity as an order parameter and observed that the activity gave much better results. Figure 14.3 displays different evolutions of the steady state activity $\chi(t)$ for different synchrony rates. We find that for $\alpha = 0.912$, the curve bends downwards, which indicates an sub-critical regime ("frozen phase"); while for $\alpha = 0.910$, the curve bends upward, which indicates a super-critical regime ("labyrinth phase"). We thus locate the phase transition at the point $\alpha_c = 0.911 \pm 10^{-3}$.

Note that this value, although close to the value given by Blok and Bergersen is *not* in agreement with their uncertainty range. As said before, we believe that the small discrepancy comes from the biases that exist in the method they have used. Other experiments are now needed in order to settle this value with greater precision. This could be done for example by measuring the so-called dynamical exponents, which are obtained by starting from an initial condition close to the absorbing phase (here the fixed points). In any case, we observed a good agreement between the measured slope and the expected value $\delta = 0.451$ which corresponds to the directed percolation critical exponent.

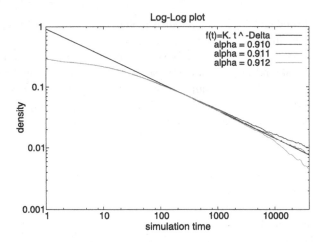

Fig. 14.3 Evolution of the activity χ vs. time; grid size is 800×800, averages are taken with 50 samples. The *straight line* has slope $-\delta = -0.451$

14.5 Initial Conditions and Asymptotic Behaviour

So far we used only uniform random initial conditions, *i.e.*, configurations where each cell had a probability $d_{ini} = 1/2$ to be in state 1. For the great majority of dynamical systems, the choice of the initial condition generally has an effect in the outcome of a trajectory. What about the asynchronous Life? For $\alpha < \alpha_c$, do we always observe the labyrinth phase regardless of the initial condition? To answer this question, we first observe the system behaviour from an experimental point of view. Part of this behaviour is then explained using a mean-field analysis and another part is explained with a close-up on how the labyrinth phase develops from specific locations in the lattice.

14.5.1 Experimental Approach

Bagnoli et al. were among the first authors to examine in detail the importance of the initial condition in the evolution of the Game of Life [3]. For classical synchronous updating, they detected two sharp transitions in the asymptotic density d_∞ as a function of the initial density d_{ini}. Figure 14.4 shows $d_\infty = f(d_{ini})$ for the synchronous and asynchronous case ($\alpha = 0.5$). The asymptotic density is approximated by taking a 100×100 grid and measuring the average density during 1,000 steps after a stabilisation time of 5,000 steps.

The curve obtained for the synchronous case $\alpha = 1$ displays two sharp transitions. The first one is for a small d_{ini}: there exists a threshold which separates the initial conditions which lead to the "extinct" state ($d = 0$) from the initial conditions which lead to the "frozen" state ($d > 0$). Above this threshold, we observe a saturation phenomenon which indicates that the initial density might not have a great importance in the evolution of the system. A symmetrical situation is observed for $d_{ini} > 0.8$ where the system is again lead to the "extinct" state. Now, if we have a

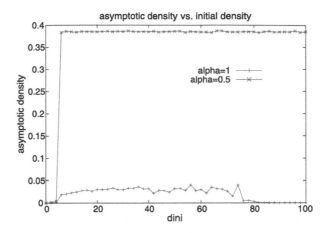

Fig. 14.4 Asymptotic density vs. initial density; grid size is 100×100

look at the asynchronous updating for $\alpha = 0.5$, we notice that: (a) the sharp transition for low d_{ini} is still present, which shows that the labyrinth phase is not always attained, even for $\alpha < \alpha_{\text{c}}$, (b) the sharp transition for high d_{ini} has disappeared. How can we explain these two phenomena?

14.5.2 Mean-Field Analysis

Since the evolution of many cellular automata is difficult or impossible to predict analytically, one might need some approximations to make such a mathematical analysis possible. One of the simplest approximations is the mean-field method, which consists in neglecting all spatial correlations. The mean-field analysis is equivalent to considering an infinite system were the cells would be redistributed randomly at each time step.

Let us assume that the density is d at time t and let us calculate the density d' at time $t + 1$. Since the updating is asynchronous only a fraction α of the cells is updated. In this subset of updated cells, a cell in state 0 transforms into a 1 (probability p_{01}) if and only if it has 3 neighbours (among 8) in state 1. Similarly, a cell that in state 1 turns into a 0 (probability p_{10}) if and only if it has *not* 2 or 3 neighbours. Consequently, we obtain the following set of equations:

$$d' = d + \alpha\big[(1 - d).p_{01} - d.p_{10}\big].$$

We have

$$p_{01} = C_8^3 d^3 (1 - d)^5$$

and

$$p_{10} = 1 - \big[C_8^3 d^3 (1 - d)^5 + C_8^2 d^2 (1 - d)^6\big].$$

This results in:

$$d' = d + \alpha d\big[84 d^2 (1 - d) + 56 d^3 (1 - d)^5 - 1\big].$$

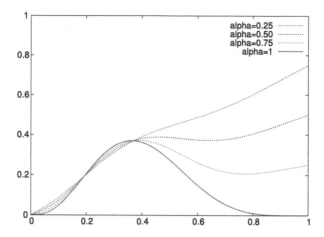

Fig. 14.5 Mean-field analysis: d' vs. d for different values of α

We can remark that the solutions of this equation ($d' = d$) are the same in synchronous and asynchronous case. This implies that the mean-field analysis does not give us any information on how the system is affected by the asynchronous updating! Intuitively, we can observe that if the system is at equilibrium, no matter how many cells are updated since the density of updated cells will not vary. From the mean-field viewpoint, the synchrony rate α acts only as a "dilution factor" and this why we need to study experimentally the effect of α.

Figure 14.5 displays d' as a function of d for different values of α. For $\alpha = 1$, we obtain the well-known mean-field analysis with two solutions: $d^- = 0.19$ is a unstable point while $d^* = 0.37$ is a stable one. In the limit $\alpha \to 0$, we simply obtain the identity function; intermediate values result in a "hybrid" function between these two extremal cases. Obviously, the decrease in the synchrony rate mainly affects the evolution of initial configurations with high densities. This observation explains the difference in behaviour between low and high values of α: for a small α, only a small fraction of the cells are updated, the density progressively decreases until it reaches the first stable point. By contrast, when α is large, the density goes in one step below the stability limit d^* and the system is attracted to the frozen state $d = 0$.

Now that we have explained what happens for high densities, let us turn our attention to the other side of the curve. Is there a phase transition for small values of d_{ini}? Recall that a necessary condition for having phase transitions is to observe a threshold whose position is independent of the grid size.

14.5.3 A Close-up on Small Initial Densities

Figure 14.6 displays the scaling behaviour of $d_\infty = f(d_{\text{ini}})$ for different grid sizes and for $\alpha = 0.5$. The estimation of d_∞ is obtained by averaging 200 samples for which we let the system evolve during 1,000 time steps as a stabilisation time and then take the average value of the density for another 1,000 steps. The curves ob-

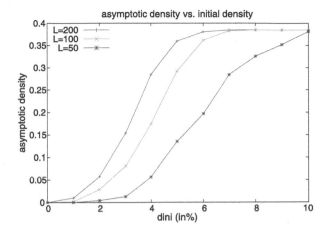

Fig. 14.6 $d_\infty = f(d_{ini})$, close up for small initial densities with $\alpha = 0.5$. Averages obtained with 200 samples for the grid sizes $L = 50, 100, 200$

tained for the fully-asynchronous updating are similar, although the density saturates at $d_{max} = 0.43$ instead of $d_{max} = 0.38$.

The observations confirm the existence of a strong increase of the asymptotic density d_∞ as a function of d_{ini}. However, there is here no clear "signature" of the presence of a phase transition as we see no infinite slope appearing for a given threshold value of d_{ini}. Instead, as the grid size increases, the slope of the curve also increases but the threshold simultaneously seems to be "translated" to the left. Moreover, starting from low-d_{ini} initial conditions, it is possible to observe that the labyrinth phase develops from very specific parts of the lattice that we call *germs*, by analogy with crystallography where the germs refers to the elements which can initiate crystal growth. In order to support our claim that there is no phase transition here, let us analyse how the labyrinth phase develops on the grid.

14.5.4 When Germs Colonise the Grid

In a previous work [10], we showed that it was possible to predict the shape of the curve $d_\infty = f(d_{ini})$ by studying how the germs developed on an empty grid (see Fig. 14.7). We considered all 3×3 germs and examined what was the probability of seeing each germ expand and fill the whole grid, *i.e.*, be the source of the labyrinth phase. The probability of seeing the labyrinth phase was then obtained by grouping the germs according to their number of 1s, and by relating this number to their probability of presence on the grid as a function of the initial density d_{ini}.

We now propose a slightly different method, which is less demanding in terms of simulation time. Instead of studying the $2^{3*3} = 512$ potential germs one by one, we will simply set the initial density d_{ini} and then initialise a 3×3 block of cells with the central cell set in state 1 and the 8 neighbouring cells initialised randomly with initial density d_{ini}. All the other cells of the grid are left in state 0. Note that the resulting configuration does *not* have density d_{ini}.

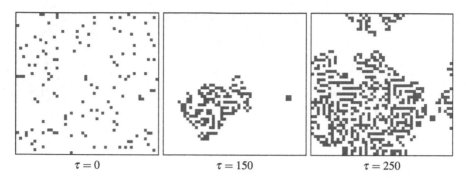

$\tau = 0$ $\qquad\qquad\qquad\qquad$ $\tau = 150$ $\qquad\qquad\qquad\qquad$ $\tau = 250$

Fig. 14.7 Colonisation of the grid by a germ. The grid size is 50×50, the synchrony rate is $\alpha = 0.5$, random initial configuration was obtained with initial density $d_{\text{ini}} = 0.10$

We then let the system run until it either stabilises at a fixed point or it reaches the labyrinth phase. To decide if the labyrinth phase is reached, we use the following criterion: the activity χ is higher than the threshold value 0.1. We verified experimentally that this criterion was satisfied when the labyrinth phase appeared on the grid. The number of labyrinth phases attained over the number of samples is $P_{\text{germ}}(d_{\text{ini}})$. From this ratio, we derive the global probability P_{LP} that the labyrinth phase emerges as a function of the initial density d_{ini}. To calculate P_{LP}, we assume the different 3×3 patterns are uncorrelated. The labyrinth phase then emerges if at least one of the germs is the source of the labyrinth phase. As there are $d_{\text{ini}} \cdot L^2$ such germs on average, an approximation of the probability of emergence of the labyrinth phase P_{LP} is:

$$P_{\text{LP}}(d_{\text{ini}}) = 1 - \left[1 - P_{\text{germ}}(d_{\text{ini}})\right]^{d_{\text{ini}} L^2}.$$

Figure 14.8 displays $P_{\text{LP}} = f(d_{\text{ini}})$ for the fully asynchronous updating scheme and for different lattice sizes. We see that the shape of the curves obtained are similar to the curves of Fig. 14.6. The curves can be divided into three parts: a slow increase for small d_{ini}, followed by a sharp increase for a small range of d_{ini} and then a saturation for higher values of d_{ini}. As the steady state density reached for $\alpha = 0.5$ is close to 0.38, we observe a qualitative agreement between the plots of Fig. 14.6 and those of Fig. 14.8.

This agreement pleads in favour of the non-existence of a phase transition: indeed, if the germ hypothesis is correct, as soon as we have $d_{\text{ini}} > 0$, the probability that the labyrinth phase appears from a germ which "invades" the whole lattice is non-zero, and this probability increases (non-linearly) as a function of the lattice size. These observations can be extended to any $\alpha < \alpha_c$: the probability to observe the labyrinth phase tends to 1 as the grid size tends to infinity. In short, for $\alpha < \alpha_c$, there exists no phase transition with regard to the variation of d_{ini}. We leave the question open for $\alpha > \alpha_c$, in particular for the synchronous case $\alpha = 1$ (the question was raised by Bagnoli et al. [3]).

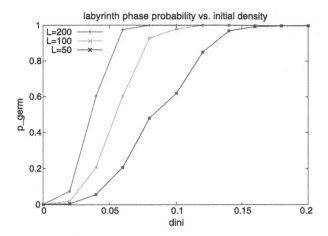

Fig. 14.8 Estimated probability P_{LP} to observe the labyrinth phase vs. initial density d_{ini}; we use the "germ" approximation for $\alpha = 0.5$ and $L = 50, 100, 200$

14.6 Extensions of the Asynchronous Game of Life

Our examination so far was limited to a `Life` rule with a asynchronous updating but with a regular topology and with its classical definition. In this section, we examine how the behaviour of the asynchronous `Life` model is affected either by a change of topology or by a change in the thresholds that define the local transition rule.

14.6.1 How Important Is a Regular Topology?

In a similar way that perfect synchrony is not a realistic assumption, it is also possible to question if a perfectly connected grid where each cell has exactly the same number of neighbours is "realistic" (see [8] for a short survey related to this question). The question whether `Life`'s phase transition is still present when the links between a cell and its neighbours are randomly destroyed was examined in an experimental study [10]. The authors observed that this perturbation had a surprising effect: the location of the critical threshold α_c was lowered as the rate of missing links ε^- was increased. Simultaneously, the phase transition became less and less visible and for $\varepsilon \sim 10\%$, the phase transition was no longer observable [10].

Following the same protocol as described in Sect. 14.4.3, we performed simulations to test how missing links perturb the phase transition. We verified that directed percolation was conserved, at least for $\varepsilon^- < 6\%$. For this range of perturbation, the critical threshold α_c decreased linearly with ε^-. For a greater amount of perturbations ($\varepsilon^- > 6\%$), we observed a "blurring effect", *i.e.*, the phase transition becomes more and more difficult to observe and the critical threshold is hardly located. Similar phenomena were recently observed on a stochastic version of the Greenberg–Hasting cellular automaton [9]. It is an open question to explain how this loss of precision results from the lattice perturbations.

We also observed that the phase transition was perturbed when links were randomly *added* in a local way, *i.e.*, when some neighbours are counted twice. Similarly, Huang et al. observed that rewiring the links randomly in the lattice (a link has a given probability to be rewired to a cell chosen uniformly in the lattice) produces the same type of phase transition as the asynchronous updating [15]. It is an interesting problem to determine how non-local topological perturbations and asynchronous updating combine.

14.6.2 Are Life-Like Rules Affected by Asynchrony?

It is now time to go back to our initial question: what is there so special about the Life rule? We saw that it displayed a great sensitivity to its updating scheme and that the evolution of the synchronous and asynchronous systems strongly depended on the initial density. Can this be considered as a "signature" of complex behaviour? If this is the case, then we would have a novel method to search the infinite space of CA rules for Life-like behaviour. We leave the question open to stimulate further research, but before coming to the conclusion of this chapter, let us highlight a few facts.

The Kaleidoscope of Life rule, which was identified by Adachi et al. as being a Turing-universal rule [2] also has a second-order phase transition which is produced by asynchronous updating. The critical threshold appears at $\alpha_c \sim 0.89$ and separates an "extinct" phase, where patterns tend to disappear, and an "active-sparse" phase with more or less regularly dispersed living sites (see Fig. 14.9). For this rule, the presence of a phase transition induced by asynchrony raises a challenging question: is it a mere coincidence or is there some possibility of establishing an "equivalence" of behaviour between the two models? (Also note the proximity of the two thresholds.)

To have a larger panel of rules, we briefly explored the larger space of Life-like rules, where the thresholds are varied. Let us denote by $\mathbb{L}abcd$ the CA rule defined with birth thresholds $B_l = a$ and $B_h = b$ and survival thresholds $S_l = c$ and $S_l = d$. We found out numerous rules which display asynchrony-induced phase transitions. We indicate for instance the rules: $\mathbb{L}1756$, $\mathbb{L}3643$, $\mathbb{L}3312$, $\mathbb{L}3636$, $\mathbb{L}3666$, etc. although there seems to be some "regions" in the rule space where phase transitions are more densely found, we have no clue yet whether there exists a common property shared by these rules. Figure 14.10 shows different snapshots which display the qualitative changes of behaviour when varying the synchrony rate α. We observe that various type of transitions are seen. For $\mathbb{L}1756$, we even observe *two* phase transitions: from a "sparse" phase to a "dense-labyrinth" phase and then to an "order-striped" phase. Interested readers may consult the paper by Regnault et al. [19] for similar observations and the study by de la Torre and Mártin [7] for other Extended-Life behaviour in the synchronous case.

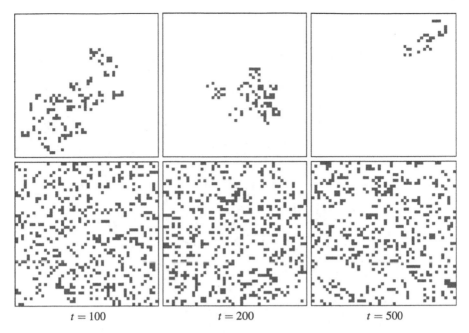

$t = 100$ $\qquad\qquad$ $t = 200$ $\qquad\qquad$ $t = 500$

Fig. 14.9 Three snapshots of the evolution of `Kaleidoscope Life`: (*top*) $\alpha = 0.90$ critical synchrony rate; (*bottom*) $\alpha = 0.50$ super-critical rate

14.7 Discussion and Openings

We studied an asynchronous version of Conway's Game of Life where the cells were updated with partial or full asynchrony. We showed that the variation of the synchrony rate α and initial density d_{ini} leads to the evolution of the system into qualitatively different steady states, the phases, which we identified as the extinct, the frozen and the labyrinth phases. There exists a critical synchrony rate α_c, which separates the frozen phase ($\alpha > \alpha_c$) from the labyrinth phase ($\alpha < \alpha_c$). This change of behaviour was identified as belonging to the directed percolation universality class, a sign which indicates that the behaviour near the critical point is difficult to predict with pure analytical methods. Using numerical simulations, the critical threshold α_c was estimated as $\alpha_c = 0.911$.

We also analysed the dependence of `Life` on the initial condition and showed that the initial density d_{ini} significantly influences the evolution of the system. Here also, for a given lattice size, there exists an abrupt change that depends on the value of d_{ini}: for $\alpha < \alpha_c$, the emergence of the labyrinth phase is a random process that depends on a complex mechanism, which is only partially understood. We proposed to analyse this mechanism by focusing on "germs", *i.e.*, small "islands" of cells in state 1 in a "sea" of cells in state 0. The analysis allowed us to understand qualitatively how the labyrinth phase emerges from the germs, but the quantitative predictions have still to be improved. At any rate, these results suggest — by contrast with asynchrony — that the abrupt change of behaviour for d_{ini} is *not* a phase tran-

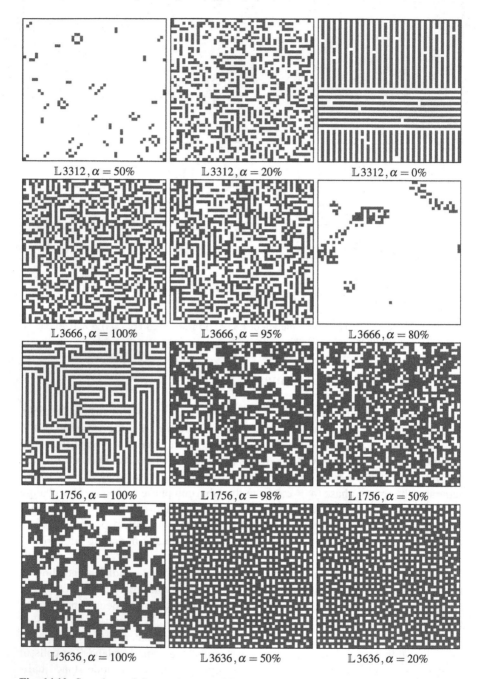

Fig. 14.10 Snapshots of the steady-state of Extended Life rules (radius 1 rules with different thresholds)

sition, *i.e.*, there exists no critical initial density that would separate the labyrinth phase from the frozen phase. Finally, we examined how these observations could be generalised to systems close to the classical `Life`. We found that perturbing the topology affected the phase transition by shifting the value of the critical threshold while making it more difficult to observe this threshold (the "blurring effect"). We provided examples of rules similar to `Life` where asynchrony also induced phase transitions.

In short, examining the robustness of `Life` to desynchronisation led us to discover many non-linearities in its behaviour depending on the degree of synchrony and on the initial condition. The mystery of `Life` is still out there and it is a challenging question to determine if similar rules can be found in the infinite space of CA rules.

References

1. Adachi, S., Peper, F., Lee, J.: The Game of Life at finite temperature. Physica D **198**, 182–196 (2004)
2. Adachi, S., Lee, J., Peper, F., Umeo, H.: Kaleidoscope of life: A 24-neighbourhood outer-totalistic cellular automaton. Physica D **237**(6), 800–817 (2008)
3. Bagnoli, F., Rechtman, R., Ruffo, S.: Some facts of life. Physica A **171**, 249–264 (1991)
4. Bersini, H., Detours, V.: Asynchrony induces stability in cellular automata based models. In: Brooks, R.A., Maes, P. (eds.) 4th International Workshop on the Synthesis and Simulation of Living Systems, Artificial Life IV, pp. 382–387. MIT, Cambridge (1994)
5. Blok, H.J., Bergersen, B.: Effect of boundary conditions on scaling in the "game of life". Phys. Rev. E **55**, 6249–6252 (1997)
6. Blok, H.J., Bergersen, B.: Synchronous versus asynchronous updating in the "game of life". Phys. Rev. E **59**(4), 3876–3879 (1999)
7. de la Torre, A.C., Mártin, H.O.: A survey of cellular automata like the "game of life". Physica A: Stat. Theor. Phys. **240**(34), 560–570 (1997)
8. Fatès, N.: Critical phenomena in cellular automata: perturbing the update, the transitions, the topology. Acta Phys. Pol. B **3**(2), 315–325 (2010)
9. Fatès, N., Berry, H.: Robustness of the critical behaviour in a discrete stochastic reaction–diffusion medium. In: Peper, F., et al. (eds.) Proceedings of IWNC 2009. PICT, vol. 2, pp. 141–148. Springer, Berlin (2010)
10. Fatès, N., Morvan, M.: Perturbing the topology of the game of life increases its robustness to asynchrony. In: Sloot, P.M.A., Chopard, B., Hoekstra, A.G. (eds.) Proceedings of the 6th International Conference on Cellular Automata for Research and Industry. LNCS, vol. 3305, pp. 111–120. Springer, Berlin (2004)
11. Fatès, N., Morvan, M.: An experimental study of robustness to asynchronism for elementary cellular automata. Complex Syst. **16**, 1–27 (2005)
12. Fatès, N., Morvan, M., Schabanel, N., Thierry, E.: Fully asynchronous behavior of double-quiescent elementary cellular automata. Theor. Comput. Sci. **362**, 1–16 (2006)
13. Grassberger, P.: Synchronization of coupled systems with spatiotemporal chaos. Phys. Rev. E **59**(3), R2520 (March 1999)
14. Hinrichsen, H.: Nonequilibrium critical phenomena and phase transitions into absorbing states. Adv. Phys. **49**, 815–958 (2000)
15. Huang, S.-Y., Zou, X.-W., Tan, Z.-J., Jin, Z.-Z.: Network-induced nonequilibrium phase transition in the "game of life". Phys. Rev. E **67**, 026107 (2003)
16. Monetti, R.A.: First-order irreversible phase transitions in a nonequilibrium system: mean-field analysis and simulation results. Phys. Rev. E **65**, 016103 (2001)

17. Monetti, R.A., Albano, E.V.: Critical edge between frozen extinction and chaotic life. Phys. Rev. E **52**(6), 5825 (1995)
18. Poundstone, W.: The Recursive Universe. William Morrow and Company, New York (1985). ISBN 0-688-03975-8
19. Regnault, D., Schabanel, N., Thierry, É.: On the analysis of simple "2d" stochastic cellular automata. In: Proceedings of LATA. LNCS, vol. 5196, pp. 452–463. Springer, Belin (2008)
20. Schulman, L.S., Seiden, P.E.: Statistical mechanics of a dynamical system based on Conway's Game of Life. J. Stat. Phys. **19**, 293 (1978)

Chapter 15
LIFE with Short-Term Memory

Ramón Alonso-Sanz

This chapter considers an extension to the standard framework of cellular automata in which, cells are endowed with memory of their previous state values. The effect of short-term memory, i.e., memory of only the latest states, in the (formally unaltered) Life rule is described in this work.

15.1 Cellular Automata with Memory

Conventional cellular automata (CA) are ahistoric (memoryless): i.e., the new state of a cell depends on the neighborhood configuration solely at the preceding time step. Thus, if $\sigma_i^{(T)}$ is taken to denote the value of cell i at time step T, the site values evolve by iteration of the mapping:

$$\sigma_i^{(T+1)} = \phi(\{\sigma_j^{(T)}\} \in \mathcal{N}_i),$$

where ϕ is an arbitrary function which specifies the cellular automaton *rule* operating on the cells in the neighborhood ($\mathcal{N}$) of the cell i.

The standard framework of CA by can be extended by implementing memory capabilities in cells:

$$\sigma_i^{(T+1)} = \phi(\{s_j^{(T)}\} \in \mathcal{N}_i),$$

with $s_j^{(T)}$ being a state function of the series of states of the cell j up to time-step T:

$$s_j^{(T)} = s(\sigma_j^{(1)}, \ldots, \sigma_j^{(T-1)}, \sigma_j^{(T)}).$$

Thus in CA with memory, while the mapping ϕ remain unaltered, historic memory of all past iterations is retained by featuring each cell by a summary of its past states. So to say, cells *canalize* memory to the map ϕ.

A simple scenario will be taken for illustration purposes in Table 15.1: that of a two-dimensional automaton with two possible state values: $\sigma \in \{0, 1\}$, with the transition rule operating on nearest neighbors, in which a cell becomes (or remains)

A. Adamatzky (ed.), *Game of Life Cellular Automata*,
DOI 10.1007/978-1-84996-217-9_15, © Springer-Verlag London Limited 2010

Table 15.1 The *speed of light* cellular automaton. Ahistoric formulation in the *upper row* of patterns, and two scenarios with majority memory *below*

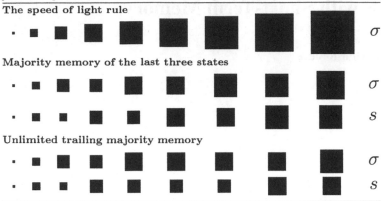

alive if any cell in its neighborhood is alive, but becomes (or remains) dead otherwise. The perturbation in Table 15.1 spreads as fast as possible, i.e., at the *speed of light*.

One simple memory implementation is that of keeping track of the most frequent of the last three states: $s_i^{(T)} = mode(\sigma_i^{(T-2)}, \sigma_i^{(T-1)}, \sigma_i^{(T)})$, with initial conditions $s_i^{(1)} = \sigma_i^{(1)}, s_i^{(2)} = \sigma_i^{(2)}$. Such majority memory exerts a characteristic inertial effect that tends to restrain the CA dynamics, as shown in Table 15.1 regarding the spread from a single seed. Increasing the length of memory increases the inertial effect, as is shown in Table 15.1 with unlimited trailing memory $s_i^{(T)} = mode(\sigma_i^{(1)}, \ldots, \sigma_i^{(T)})$, with $s_i^{(T)} = \sigma_i^{(T)}$ in case of a tie: $card\{1\} = card\{0\}$.

The effect of unlimited trailing memory on LIFE has been studied in [10]. This study focus on the effect of short-term memory, mainly memory limited to the last three states, often referred as $\tau=3$-memory. With this kind of memory endowed in cells, the patterns in Table 15.1 remain stable for two time-steps, thus the speed is halved.

The memory mechanism considered here differs from that of other CA with memory reported in the literature. In particular with that referred to as *higher order* (in-time) CA, such as for example the second order reversible formulation: $\sigma_i^{(T+1)} = \phi(\{\sigma_j^{(T)}\} \in \mathcal{N}_i) \ominus \sigma_i^{(T-1)}$, which remains reversible in the form: $\sigma_i^{(T+1)} = \phi(\{s_j^{(T)}\} \in \mathcal{N}_i) \ominus \sigma_i^{(T-1)}$ [1]. Other possible memory implementations, not considered here, are, (i) to refer a cell of the neighborhood *delayed* a number of time steps in the past. So, for example, in the context of elementary CA the central cell may be referenced not at T−1: $\sigma_i^{(T+1)} = \phi(\sigma_{i-1}^{(T)}, \sigma_i^{(T-1)}, \sigma_{i+1}^{(T)})$, (ii) the transition rule may be *extended* to *explicitly* consider the influence at time T−1, e.g., $\sigma_i^{(T+1)} = \psi(\sigma_i^{(T-1)}, \sigma_{i-1}^{(T)}, \sigma_i^{(T)}, \sigma_{i+1}^{(T)})$.

To the best of our knowledge the study of the effect of memory on CA has not been paid the attention that it deserves. Thus, as two symptomatic examples, Wuensche and Lesser [17, p. 15], just mention the possibility of *historical time reference*, which is excluded from their general study as "it would result in a qualitatively different behavior", whereas S. Wolfram [16] in the context of higher order linear CA, refers to "somewhat involved analysis not performed here".

We have studied the effect of memory embedded in cells in several CA [2, 3] and Boolean networks [6, 7] scenarios. The effect of short-term memory in particular, has been addressed in one-dimensional cellular automata in [8, 9]. We have recently studied the effect of memory in a CA-like formulation of the spatialized prisoner's dilemma [4, 5].

15.2 Life with Short-Term Majority Memory

Following one of the primary features of Cellular Automata, Conway's Game of Life (LIFE)[1] is ahistoric (memoryless): no memory of previous iterations, but the last one is taken into account. In this chapter we construct a variation of *LIFE* featuring cells by their most frequent states, not necessarily the last one.

Stable forms under LIFE remain stable with majority memory (see in Gardner [11, 12] the commonest stable forms). Also those forms that reproduce in T=3 the initial form will remain unaltered. So do the period two (flip-flops) forms collected in Table 15.2. This is so because, starting from any of this kind of structures, the most frequent and actual states are coincident, i.e., $\sigma_i^{(T)} = s_i^{(T)}$, $\forall i$, T.

A single *mutant* appearing in an stable rich configuration (*agar*) can lead to its destruction in the ahistoric model, whereas its effect tends to be restricted to its proximity with memory. Thus, the stable configurations in Table 15.3 suffer an inexorable destruction without memory due to a sole error in its central part (upper), whereas with $\tau=3$ memory (lower), the advance of the damage is notably restrained. More examples on the effect of long-term memory on agars may be found in [10].

Figure 15.1 shows the dynamics of the glider (under conventional LIFE), with $\tau=3$ majority memory. The glider evolves, from T=5 at half the velocity as in the conventional LIFE rule, repeating twice every live structure. The R-pentomino generated as most-frequent state configuration following T=3 in Fig. 15.1 is, as in conventional LIFE, a *Methuselah* when seeded as initial configuration, generating five $\tau=3$ gliders. The actual configuration at T=4 in Fig. 15.1 has also a notable activity in LIFE with $\tau=3$ majority memory, when seeded at T=1, but does not produce gliders.

Table 15.2 Period-two oscillators in LIFE

---

[1]A cell that was dead will come alive iff it is adjacent to exactly three live neighbours. A live cell will dead unless it is adjacent to either two or three live neighbors.

Table 15.3 The effect of a virus in the tub (up) and the block (down) LIFE agars. Ahistoric (*upper*) and $\tau=3$ majority memory (*lower*)

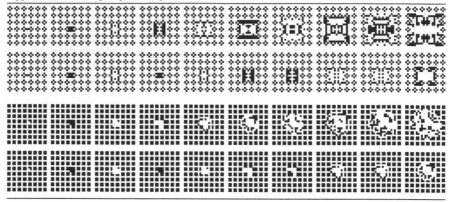

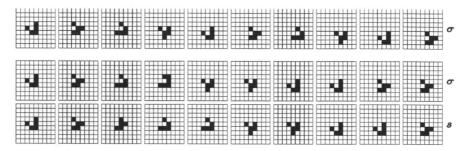

Fig. 15.1 The glider up to T=10. Ahistoric LIFE (up) and with $\tau=3$ majority memory (down) dynamics

In most cases, the gliders interact in LIFE with $\tau=3$ majority memory as do in conventional LIFE, though at half velocity, repeating every pattern. Figures 15.2 and 15.3 show examples of two gliders travelling in the same direction and in orthogonal directions respectively. In Fig. 15.4, four gliders collide. Conventional LIFE reaches extinction, a blinker (second scenario of Fig. 15.2), and a still-life *loaf* (lower scenario of Fig. 15.2) as in Figs. 15.2–15.4, but without the repetition of patterns from T=5.

No mobile configurations, other than the glider, have been found in LIFE with $\tau=3$ majority memory. The so-called *spaceships* in particular, that travel in conventional LIFE, do not move but soon stabilize, with majority memory.

Figure 15.5 shows a well-known glider-gun in conventional LIFE at T=70, and that configuration in LIFE with $\tau=3$ majority memory. In the latter case, only one glider is generated up to T=70. Later on, a second glider, gliding in parallel with generated by T=70, emerges, but no more gliders are generated. Thus, the glider-gun with $\tau=3$ majority memory is not stable and only shots two *bullets*.

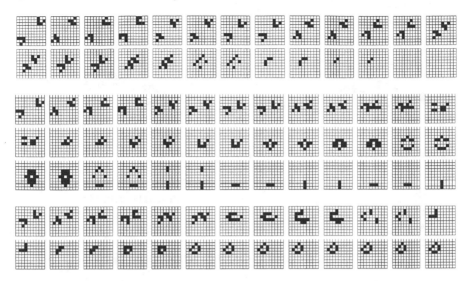

Fig. 15.2 Collisions of gliders travelling in the same direction

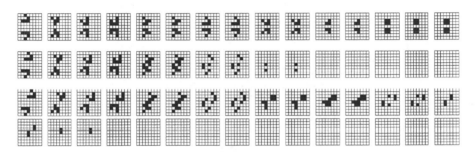

Fig. 15.3 Glider collisions travelling in orthogonal directions

Fig. 15.4 A collision of four gliders

A simple measure of the activity of a given cellular automaton is the measure of its changing rate:

$$c = \frac{1}{N \times N} \sum_{i=1}^{N} \sum_{j=1}^{N} \sigma_{i,j}^{(T-1)} \oplus \sigma_{i,j}^{(T)}.$$

When starting at random, the conventional Life rule shows three phases of evolution. The first phase is a relatively short transient phase — at most ten or tens of generations — in which excessively high or low initial densities adjust themselves; the second phase may last for thousands of generations in which nothing seems to

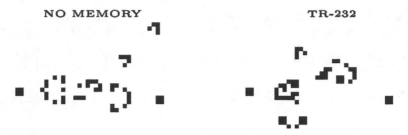

Fig. 15.5 The conventional glider-gun (*left*) and that with $\tau=3$ majority memory (*right*) at T=70.

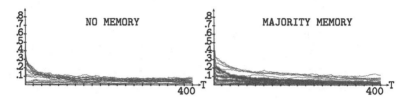

Fig. 15.6 (Color online) Density (*red*) and changing rate (*blue*) up to T=400

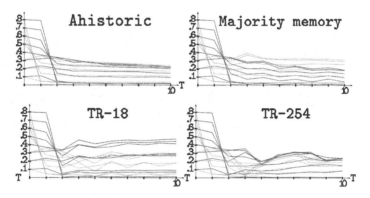

Fig. 15.7 Density and changing rate up to T=10

be definite; followed by the third and final phase in which isolated groups of cells go through predictable cycles of evolution.

Figure 15.6 shows the evolution of the density (red) and changing rate (blue) in conventional and $\tau=3$ majority memory LIFE, when starting from random configurations in a 100×100 lattice. In both cases both parameters tend to decline, but in the scenario with memory in a lower extent. Thus, the patterns starting from a random configuration tend to be more populated in LIFE with $\tau=3$ majority memory than in conventional LIFE. Besides, clusters of changing cells remain active when implementing memory in a way that is not seen in conventional LIFE. Figure 15.7 zooms the first ten time-steps, showing the initial plummeting in the simulations starting from a high density of live cells.

Partial memory can be implemented in LIFE either as inner memory:

Outer memory

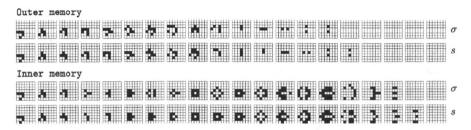

Inner memory

Fig. 15.8 The glider with partial memory

$$\sigma_i^{(T+1)} = \phi\left(s_i^{(T)}, \sum_{j \in \mathcal{N}_i} \sigma_j^{(T)}\right),$$

or as outer memory:

$$\sigma_i^{(T+1)} = \phi\left(\sigma_i^{(T)}, \sum_{j \in \mathcal{N}_i} s_j^{(T)}\right).$$

Figure 15.8 shows how the glider is extinguished soon in both scenarios. No further analysis will be made in this chapter on partial memory.

15.3 Life with Elementary Rules as Memory

Elementary rules are one-dimensional ($d = 1$) CA with two possible values ($k = 2$) at each site ($\sigma \in \{0, 1\}$), with rules operating on nearest neighbors ($r = 1$). Following Wolfram's notation, these rules are characterized by a sequence of binary values (β) associated with each of the eight possible triplets ($\sigma_{i-1}^{(T)}, \sigma_i^{(T)}, \sigma_{i+1}^{(T)}$):

$$111 \ 110 \ 101 \ 100 \ 011 \ 010 \ 001 \ 000$$
$$\beta_1 \ \ \beta_2 \ \ \beta_3 \ \ \beta_4 \ \ \beta_5 \ \ \beta_6 \ \ \beta_7 \ \ \beta_8 \ .$$

The *rule number* of elementary CA, $\mathcal{R} = \sum_{i=1}^{8} \beta_i 2^{8-i}$, ranges from 0 to 255. *Legal* rules are *reflection symmetric* (so that 100 and 001 as well as 110 and 011 yield identical values), and *quiescent* ($\beta_8 = 0$). These restrictions leave 32 possible *legal* rules of the form: $\beta_1\beta_2\beta_3\beta_4\beta_2\beta_6\beta_40$.

Elementary CA rules (f) can in turn act as memory rules:

$$s_i^{(T)} = f\left(\sigma_i^{(T-2)}, \sigma_i^{(T)}, \sigma_i^{(T-1)}\right).$$

Figure 15.9 shows the evolving patterns from a 3×3 square, under conventional LIFE and in LIFE with elementary legal rules as memory rules, referred as TR-rule number. In ahistoric LIFE, the 3×3 square becomes at T=6 a set of four blinkers that generates a simple period-two oscillator, named *traffic light*. The actual and trait series show nil patterns at several time-steps in Fig. 15.9, which does not imply extinction after any of them. This kind of cataleptic episodes are unfeasible in the

R. Alonso-Sanz

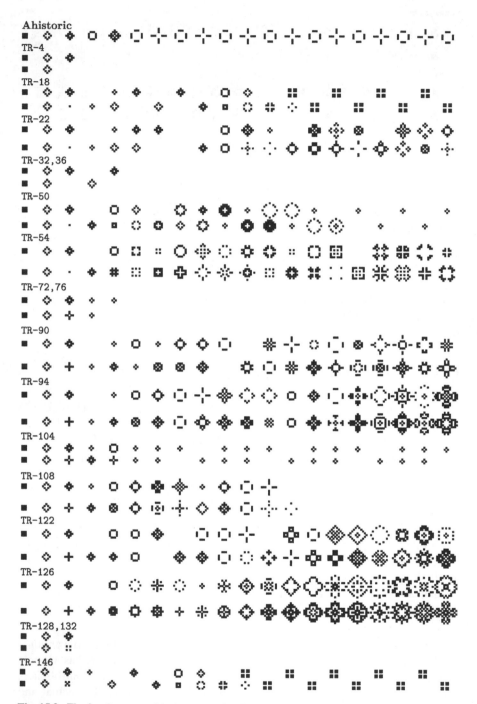

Fig. 15.9 The 3 × 3 square with elementary legal memory rules in LIFE

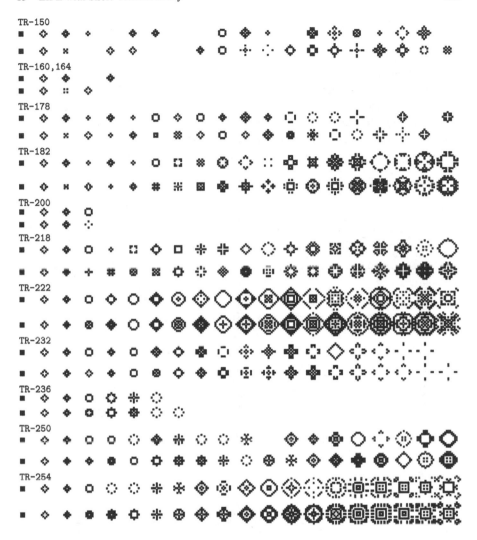

Fig. 15.9 (continued)

ahistoric context, but they are not rare with elementary CA rules acting as memory. They are more probable in early time-steps, though they also occur, at lesser extent, later in time.

Figure 15.10 shows the evolution of the density of live cells (red) and of the changing rate (blue) in LIFE with cells endowed with legal elementary rules in a 100 × 100 lattice. Rule 18 seems to induce a fairly unexpected changing rate when acting as memory rule. The seemly simple rule 18(00010010) (only dead cells with exactly one living neighbor to become alive, living cells die) was one of the first rules carefully analyzed [14], following its *intriguing* properties [15]. It seems that rule 18 also supports some kind of *intriguing* dynamics when acting as memory rule on LIFE.

The main features of the dynamics in Fig. 15.9 are indicative of the dynamics starting at random as in Fig. 15.10. Thus, for example, rules 4, 32, 36, 72, 76, 128, 132, 160, 164 soon extinguish starting at random, as do in Fig. 15.9. Rule 104 acting as memory induces a strong decay in the activity of LIFE, that by T=20 is reduced to small active clusters in some simulations. The not shown rule 200, much behaves as rule 104 when acting as memory rule. Rule 108 is an exception as it does not induces extinction, as may be expected from Fig. 15.9. On the contrary, rule 108 activates

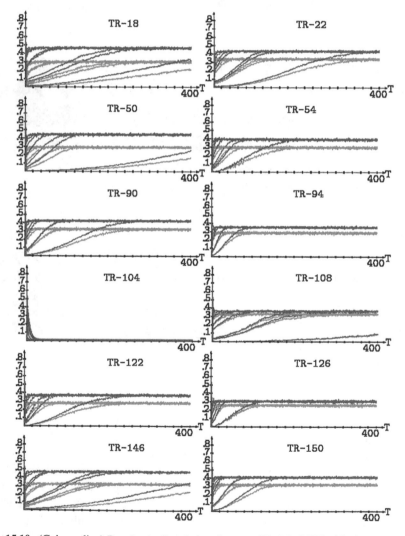

Fig. 15.10 (Color online) Density (*red*) and changing rate (*blue*) in LIFE with elementary legal memory rules

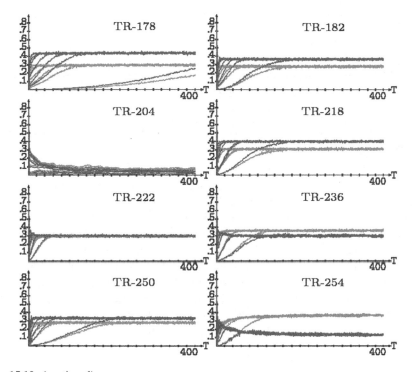

Fig. 15.10 (continued)

LIFE in most simulations, with a case of very slow, though monotone, increase in both density and change rates, that by T=400 reaches only a 10%. As a norm, the rules that induce activity in Fig. 15.9 show also fairly high density and change rates in Fig. 15.10. This is so, for example, with the linear rules 90 and 150 and with rule 22(00010110), considered as the one-dimensional LIFE analog, which induces a notable activity in LIFE. Rule 22, together with rule 18, has been used to illustrate the spontaneous generation of complex structure, in contrast to superficial evidence that would suggest that this simple rule would lead to fairly simple behavior [13, 18]. Rule 22 seems to corroborate this appreciation supporting, as rule 18 does, a high activity in LIFE when acting as memory rule.

In most cases in Fig. 15.10, the changing rate curve is above that of the density once reached their almost-steady levels. Rule 222 stabilizes both parameters into a fairly coincident rate close to 0.3. Oddly, rule 236 and, notably, rule 254 show the density curve above the changing rate.

The majority of the asymmetric quiescent rules that support activity when acting as memory rules in LIFE evolve to reach plateaus for both changing rate density, with the former above density, as in the most frequent evolution in Fig. 15.10. Just to cite an example, the density and changing rate of the important rule 110 evolves much like, say, rule 54 in Fig. 15.10. The *traffic* rule 184 does no support a permanent activity, and both parameters decay much as with majority memory in Fig. 15.6, albeit with changing rate above density. No-quiescent memory rules tend to anni-

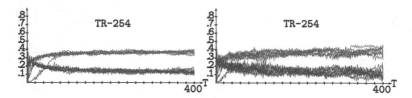

Fig. 15.11 (Color online) Density (*red, upper curves*) and changing rate (*blue, lower curves*) in LIFE with rule 254 as memory rule. The lattice sizes are 50×50 in *left panel* and 25×25 in the *right* one

hilate the LIFE dynamics: after the initial decay most trait states are the live one at T=3, which means almost extinction by overcrowding at T=4.

The main characteristics of the effect of elementary rules acting as memory, i.e., their *signature*, remain in lattices of smaller size than in that of 100×100 size chosen here. The main effect of decreasing the lattice size is the increase in the oscillation of both density and changing rates, which is fairly small in Fig. 15.10 once reached the (almost) steady levels. Figure 15.11 shows how this applies for rule 254 acting as memory in 50×50 and 25×25 lattices: the most important characteristic of rule 254 acting as memory in LIFE found in the simulations in the bigger lattice, i.e., changing rate above density, remains in the smaller ones.

15.4 Life with Minimal Memory

The lowest degree of memory conceivable is that of featuring cells by binary functions of their last two states: $s_i^{(T)} = f(\sigma_i^{(T)}, \sigma_i^{(T-1)})$. These mappings are characterized by a sequence four binary values:

$$\begin{matrix} 11 \ 10 \ 01 \ 00 \\ \beta_1 \ \beta_2 \ \beta_3 \ \beta_4 \end{matrix} \equiv \sum_{i=1}^{4} \beta_s 2^{4-i} = \mathscr{R}.$$

The rule number varies in the [0, 15] interval, with $\mathscr{R} = 12$ being the identity rule $s_i^{(T)} = \sigma_i^{(T)}$, and $\mathscr{R} = 6$ being the parity rule: $s_i^{(T)} = \sigma_i^{(T)} \oplus \sigma_i^{(T-1)}$. The majority rule is $\mathscr{R} = 8$ in this context.

Every two-bit rule may be related of an elementary three-bit one, in which the former operates regardless the value of the third bit. Thus, for example, the two-bit parity rule 6(0110) relates to the three-bit rule 102:

$$\begin{matrix} 111 \ 110 \ 101 \ 100 \ 011 \ 010 \ 001 \ 000 \\ 0 \quad 1 \quad 1 \quad 0 \quad 0 \quad 1 \quad 1 \quad 0 \end{matrix},$$

or to rule 60(00111100) with $\sigma_i^{(T-1)}$ playing the role of $\sigma_{i-1}^{(T)}$ instead of that of $\sigma_{i+1}^{(T)}$. In the same vein, the two-bit rule 14(1110) relates to the three-bit rule 238(11101110), or to rule 252(11111100) in the alternative case. These four three-bit rules, i.e., 60, 102, 238, and 252, are no legal rules.

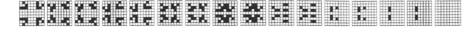

Fig. 15.12 A collision of four gliders in LIFE with $\tau=2$ rule 10 memory

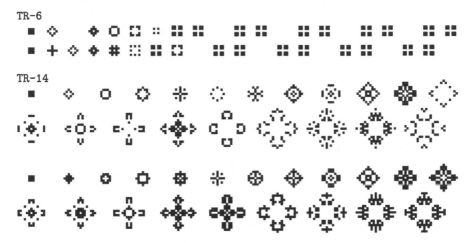

Fig. 15.13 The 3×3 square in LIFE with two $\tau=2$ memory rules. The actual and trait sates are shown in the figure

But the patterns generated by related rules in the two and three time-step memory scenarios are not coincident, because the dynamics differs initially at T=3. In the three-state memory scenario, memory becomes activated for the first time following T=3, thus the pattern at this time-step is the same as in the ahistoric model, whereas in the two-state memory scenario, memory is activated already following T=2, thus the pattern at T=3 is (likely) not the same as in the ahistoric model.

Rule 10(1010) acting as memory replicates twice every pattern, because

$$s_i^{(T)} = \left(\sigma_i^{(T)} \cap \sigma_i^{(T-1)}\right) \cup \left(\overline{\sigma}_i^{(T)} \cap \sigma_i^{(T-1)}\right) = \sigma_i^{(T-1)} \cap \left(\overline{\sigma}_i^{(T)} \cup \sigma_i^{(T)}\right) = \sigma_i^{(T-1)}.$$

An example is given in Fig. 15.12, a figure showing the collision of four gliders starting as in Fig. 15.4. The glider-gun in Fig. 15.5 remains stable with rule 10 acting as memory, generating gliders as in the conventional ahistoric LIFE formulation, but at half the velocity. The two-bit rule 10 relates to the three-bit rules 170(10101010) and 240(11110000), i.e., the shift rules: $\sigma_i^{(T+1)} = \sigma_{i+1}^{(T)}, \sigma_i^{(T+1)} = \sigma_{i-1}^{(T)}$, when acting as spatial rules.

Figure 15.13 shows the evolution of the initial 3×3 square in LIFE with two different minimal memory rules. The parity rule (TR-6) soon generates a period-three oscillator one of whose components is the nil-pattern, whereas TR-14, i.e., cell considered alive if was not dead in the two last time-steps, soon generates a sophisticated dynamics.

Accordingly to the expectation from Fig. 15.13, TR-14 leads to a fairly high density and change rates, close to 0.3, when starting from a 3×3 square in a 100×100 lattice, as shown in the TR-14 panel of Fig. 15.14. This plateau is seemly coincident

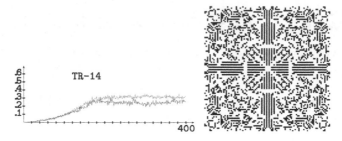

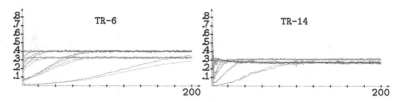

Fig. 15.14 Evolution of the 3 × 3 square in LIFE with the τ=2 rule 14 memory. Density and changing rates up to T=400 (*left*), and pattern at T=400 in a 100 × 100 lattice

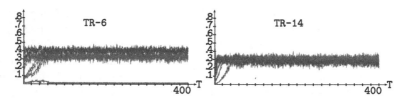

Fig. 15.15 Density and changing rate up to T=200 in LIFE with two τ=2 memory rules in a 100 × 100 lattice

Fig. 15.16 Density and changing rate in LIFE with the two τ=2 memory rules of Fig. 15.15 in a 25 × 25 lattice

with that reached starting at random, as shown in Fig. 15.15. In fact both parameters slightly oscillate close to 0.3, with the density a bit greater than the changing rate. Density above changing rate relates the effect of the two-bit rule 14 to that of its qualitatively equivalent three-bit rule 254.

On the contrary, the effect of TR-6 on LIFE when starting at random is not pointed in Fig. 15.13. Thus, Fig. 15.15 shows a plateau for the density again circa 0.3, and a higher level of the changing rate, stated at 0.4.

As already pointed in the three-time step memory scenario, decreasing the lattice size increases the oscillation of both density and changing rates, but the signature of the effect of the rule memory remains. Figure 15.16 gives an example with the same rules as in Fig. 15.15. Incidentally, a simulation collapses in the TR-6 panel of Fig. 15.16.

15.5 Conclusion

A variation of Conway's Game of Life (LIFE) cellular automaton is presented in this chapter. The rules of LIFE remain unaltered but, unlike in conventional LIFE, they are applied to cells featured with trait states that are binary functions of their most recent state values.

This kind of memory embedded in cells, notably alters the dynamics of the LIFE rule compared to its conventional ahistoric formulation. Particular attention is paid to the study of the effect of the majority of the last three-state rule acting as memory. It only preserves a mobile configuration: the (conventional) glider, which travels with majority memory at half the velocity than in conventional LIFE.

The effect of legal elementary rules, as well as that of keeping memory of only the two latest states, is also scrutinized. Some of these rules generate a notable permanent activity in LIFE with memory (far of the tendency to stabilization in conventional LIFE), that deserves further study.

Acknowledgements   This work was supported under the MICINN Spanish grant no MTM2009-14621-C02-02.

References

1. Alonso-Sanz, R.: Reversible cellular automata with memory. Physica D **175**, 1–30 (2003)
2. Alonso-Sanz, R.: A structurally dynamic cellular automaton with memory. Chaos Solitons Fractals **32**(4), 1285–1295 (2007)
3. Alonso-Sanz, R.: Cellular automata with memory. In: Meyers, R. (ed.) Encyclopedia of Complexity and Systems Science, pp. 823–848. Springer, New York (2009), and the references therein
4. Alonso-Sanz, R.: Spatial order prevails over memory in boosting cooperation in the iterated prisoner's dilemma. Chaos **19**(2), 023102 (2009)
5. Alonso-Sanz, R.: Memory versus spatial disorder in the support of cooperation. Biosystems **97**, 90–102 (2009)
6. Alonso-Sanz, R., Bull, L.: Boolean networks with memory. Int. J. Bifurc. Chaos **18**(12), 3789–3797 (2008)
7. Alonso-Sanz, R., Cardenas, J.P.: One-dimensional $K = 4$ Boolean networks with memory. Physica D **273**(23), 3099–3108 (2008)
8. Alonso-Sanz, R., Martin, M.: One-dimensional cellular automata with memory in cells of the most recent value. Complex Syst. **15**, 203–236 (2005)
9. Alonso-Sanz, R., Martin, M.: One-dimensional cellular automata with elementary memory rules in cells: the case of linear rules. J. Cell. Autom. **1**(1), 70–86 (2006)
10. Alonso-Sanz, R., Martin, M.C., Martn, M.: Historic Life. Int. J. Bifurc. Chaos **116**(6), 1665–1682 (2001)
11. Gardner, M.: The fantastic combinations of John Conway's new solitaire game 'life'. Sci. Am. **223**(4), 120–123 (1971)
12. Gardner, M.: On cellular automata, self-reproduction, the Garden of Eden and the game "life". Sci. Am. **224**(2), 112–117 (1971)
13. Grassberger, P.: Chaos and diffusion in deterministic cellular automata. Physica D **10**, 52–58 (1984)
14. Grassberger, P.: Long-range effects in an elementary cellular automaton. J. Stat. Phys. **45**(5–6), 27–40 (1986)

15. Ito, H.: Intriguing properties of global structure in some classes of finite cellular automata. Physica D **31**, 318–338 (1988)
16. Wolfram, S.: Algebraic properties of cellular automata. Physica D **10**, 1–35 (1984)
17. Wuensche, A., Lesser, M.: The Global Dynamics of Cellular Automata. Addison–Wesley, Reading (1992)
18. Zabolyzky, J.G.: Critical properties of rule 22 elementary cellular automata. J. Stat. Phys. **50**(5–6), 1255–1262 (1988)

Chapter 16
Localization Dynamics in a Binary Two-Dimensional Cellular Automaton: The Diffusion Rule

Genaro J. Martínez, Andrew Adamatzky,
and Harold V. McIntosh

We study a two-dimensional cellular automaton (CA), called *Diffusion Rule*, which exhibits diffusion like dynamics of propagating patterns. In computational experiments we discover a wide range of mobile and stationary localizations (gliders, oscillators, glider guns, puffer trains), analyze spatio-temporal dynamics of collisions between gliders, and discuss possible applications in unconventional computing.

16.1 Introduction

In our previous studies on minimal cellular automaton (CA) models of reaction–diffusion chemical system we constructed a binary-cell-state eight-cell neighborhood 2D CA model of a quasi-chemical system with one substrate and one reagent [6]. In that model chemical reactions were represented by semi-totalistic transition rules. Every cell switches from state 0 to state 1 depending on whether sum of neighbors in state 1 belongs to some specified interval or not. A cell remains in state 1 if sum of neighbors in state 1 belongs to another specified interval.

From 1,296 cell-state transition rules, we selected a set of rules with complex behavior [6]. Amongst the complex rules, namely in G-class of morphological classification [6], we located so-called *Diffusion Rule*. CA governed by this rule often exhibits slowly non-uniformly growing patterns, resembling diffusive patterns in chemical systems with non-trivial coefficients of diffusion, or reaction-dependent diffusion coefficients, so the name of the rule.

The rule simulates sub-excitable [34] medium-like mode of perturbation propagation — cell in state 0 takes state 1 if there are exactly two neighbors in state 1, otherwise the cell remains in state 0, and, conditional inhibition — cell in state 1 remains in state 1 if there are exactly seven neighbors in state 1, otherwise the cell switches to state 0.

In present chapter we are trying to answer the following questions. Is there a reaction–diffusion binary-state CA that express complex dynamic? Can we demonstrate that CA exhibit non-stationary growth of reaction–diffusion patterns? Do stationary or mobile generators of localizations, glider guns, exist in binary-state

A. Adamatzky (ed.), *Game of Life Cellular Automata*,
DOI 10.1007/978-1-84996-217-9_16, © Springer-Verlag London Limited 2010

291

292   G.J. Martínez et al.

reaction–diffusion CA? Can the reaction–diffusion CA simulate an effective procedure and therefore be universal?

CA with space–time dynamics similar to that in spatially extended chemical
systems are studied from early days of CA theory and applications [33], however
most rules discovered so far lack minimality (some of the rules employ dozens of
cell-states). Methods of selecting the rules also widely vary depending on theoretical frameworks, e.g. probabilistic spaces [20, 36] and genetic algorithms [15, 31].
Therefore, we envisage a strong need for a systematic analysis of propagating patterns like those observed in the Diffusion Rule. The propagating patterns are of
upmost importance in modern computer science because such patterns play a vital role in developing novel and emerging computing paradigms and architectures,
particularly collision-based computing [2, 5, 23].

We must mention that various authors have already obtained pioneering results
in the studied rule. Magnier et al. discovered three primary gliders [24]. David Eppstein found four gliders known, already incorporated in our framework, and four
new gliders which were novel for us (Fig. 16.4(q), (t) and (u)).[1] The glider traveling along diagonals of the lattice (Fig. 16.4(v)) was firstly recorded by Amling in
2002 (see Eppstein's web site). Finally, a glider gun and three puffer trains were
discovered by Wótowicz.[2]

Diffusion Rule CA is just one of many complex CA[3] exhibiting mobile localizations. Other famous examples include semi-totalistic rules as the Game of Life [11,
18], Brian's-brain and Critters rules [32], High Life [10], Life 1133 [22], Life Without Death [19], and Life variant $B35/S236$.[4] Other variant is with Larger than Life
[17] and the Beehive and Spiral rules hexagonal CA [4, 7, 37], Life like CA with
extended neighbourhoods [1]. More recent candidates were proposed by George
Maydwell with Hexagonal Life and Hexagonal Long Life rules.[5] Amongst 3D binary state CA supporting gliders Life 4555 and Life 5766 by Carter Bays [8, 9] are
most widely known. In 1D there are Rule 110 [14, 27, 30, 35] and Rule 54 [12, 21,
26] CA, which support an impressive range of gliders.

Our chapter is structured as follows. In Sect. 16.2 we introduce basic concepts of
CA model under investigation. Section 16.3 introduces results of statistical analysis
of Diffusion Rule using mean field theory. In Sect. 16.4 we present basic structures
discovered in the Diffusion Rule CA. Section 16.5 compiles a catalogs of non-trivial
interactions between gliders, which could be used to designing basic elements of
collision-based computers. In Sect. 16.6 we highlight our achievements in analysis
of the Diffusion Rule and prospects for future studies.

---

[1] See Eppstein's findings at http://fano.ics.uci.edu/ca/rules/b2s7/.

[2] http://www.mirwoj.opus.chelm.pl/ca/rules/life_2.gif.

[3] http://uncomp.uwe.ac.uk/genaro/otherRules.html/.

[4] http://www.ics.uci.edu/~eppstein/ca/b35s236/.

[5] Several interesting candidates in hexagonal representation are proposed by Andrew Wuensche using *DDLAB* at http://www.ddlab.com/ and by Maydwell using *SARCASim* at http://www.
collidoscope.com/ca/.

16.2 Basic Notations

We study family of 2D binary-state cellular automaton (CA) defined by tuple $\langle \mathbb{Z}^2, \Sigma, u, f \rangle$, where $\mathbb{Z}$ is the set of integers, every cell $x \in \mathbb{Z}^2$ has eight neighbors, orthogonal and diagonal (i.e. classical Moore's neighborhood but without central cell) $u(x) = \{y \in \mathbb{Z}: y \neq x$ and $|x - y| \leq 1\}$, $\Sigma = \{0, 1\}$ is the set of *states*, and f is a local transition function defined as follows:

$$x^{t+1} = f(u(x^t)) = \begin{cases} 1, & \text{if } (x^t = 0 \text{ and } \sigma_x^t \in [\theta_1, \theta_2]) \text{ or } (x^t = 1 \text{ and } \sigma_x^t \in [\delta_1, \delta_2]) \\ 0, & \text{otherwise} \end{cases}$$

(16.1)

where $\sigma_x^t = |\{y \in u(x): y^t = 1\}|$, and $\theta_1, \theta_2, \delta_1, \delta_2$ are some fixed parameters such that $0 \leq \theta_1 \leq \theta_2 \leq 8$ and $0 \leq \delta_1 \leq \delta_2 \leq 8$.

We can write the rule as $R(\delta_1 \delta_2 \theta_1 \theta_2)$ or like $B\theta_1 \dots \theta_2 / S\delta_1 \dots \delta_2$ more traditional code. Also, the rule can be interpreted as a simple discrete model of a quasi-chemical system with substrate '0' and reagent '1', where $[\theta_1, \theta_2]$ is an interval of reaction, or association between substrate and reagent, and $[\delta_1, \delta_2]$ is an interval of dissociation. The family of rules includes Conway's Game Life, when intervals $[\delta_1, \delta_2]$ and $[\theta_1, \theta_2]$ are interpreted as intervals of survival and birth, respectively.

In our previous paper [6] we morphologically classified all 1,296 rules, and studied how changes in parameters $R(\delta_1 \delta_2 \theta_1 \theta_2)$ of cell-state transition rule influence space–time dynamics. For example, we discovered [6] a small subset of rules *Life* $2c22$,[6] $2 \leq c \leq 8$, which could be interpreted as quasi-chemical precipitating systems. For parameter set $[\theta_1, \theta_2] = [22]$, the system is transformed into 2^+-medium, CA model of excitable system in sub-excitable mode [2, 6].

We have found a cluster of semi-totalistic rules supporting structures of the Diffusion Rule. They are $B2/S2\dots 8$ called *Life* $dc22$[7] where d and c take values between 2 and 8, and $d \leq c$. Therefore, we found that the rule $B2/S7$ or $R(7722)$ exhibits richest dynamics of localized patterns amongst all the rules studied by us. Rules of the local transition are simple:

1. A cell in state 0 will take state 1 if it has exactly two neighbors in state 1, otherwise cell remains in state 0.
2. A cell in state 1 remains in state 1 if it has exactly seven neighbors in state 1, otherwise cell takes state 0.

16.3 Mean Field Approximation

Mean field theory is a proved technique for discovering statistical properties of CA without analyzing evolution spaces of individual rules [13, 20, 28]. The method

---

[6]http://uncomp.uwe.ac.uk/genaro/Diffusion_Rule/life_2c22.html.

[7]http://uncomp.uwe.ac.uk/genaro/Life_dc22.html.

assumes that elements of the set of states Σ are independent, uncorrelated between each other in the rule's evolution space. Therefore we can study probabilities of states in neighborhood in terms of probability of a next state (the state to which the neighborhood evolves), thus probability of a neighborhood is the product of the probabilities of each cell in the neighborhood. Using this approach we can construct mean field polynomial for a semi-totalistic evolution rule [25] as follow:

$$p_{t+1} = \sum_{v=\delta_1}^{\delta_2} \binom{n-1}{v} p_t^{v+1} q_t^{n-v-1} + \sum_{v=\theta_1}^{\theta_2} \binom{n-1}{v} p_t^{v} q_t^{n-v} \qquad (16.2)$$

where n represents the number of cells in neighborhood, v indicates how often state 1 occurs in Moore's neighborhood, $n - v$ shows how often state 0 occurs in the neighborhood, p_t is a probability of cell being in state 1, q_t (its complement $1 - p$) is a probability of cell being in state 0.

On the basis of outcomes of computational experiments we can suggest intervals of extreme densities of initial random configurations which leads to the emergence of localizations in the Diffusion Rule. In the lower limit best densities d are $0.004 < d < 0.015$ and in the upper limit they are $0.992 < d < 0.997$ for the first 15–20 steps of evolution (Fig. 16.1). CA starting its evolution in random configuration with lower density of 1-states exhibits stationary or mobile self-localizations (like gliders or oscillators) at the beginning of the evolution, however in many cases collisions between mobile localizations leads to catastrophes, when 1-state patterns spread all over the lattice. Random initial configurations with higher (upper limit) density of 1-states produce either vanishing reactions between localizations or symmetrical growing patterns emerged as unions of two or more gliders.

Thus, the mean field polynomial for the Diffusion Rule is following:

$$p_{t+1} = 8p_t^8 q_t + 28 p_t^2 q_t^7 \qquad (16.3)$$

The fixed point is 0.236 that represents configurations with large density of 1-states emerging from any random initial condition (we should note that the fixed point for Conway's Game of Life is 0.37); this represents global density of 1-states necessary for evolution dynamics to stabilize. Also, we can see an unstable fixed point 0.05 (Fig. 16.2), that implies the existence of regions with unpredictable behavior or complex dynamic [28].

Thus, $p = 0$ is a super-stable point, although it is quite close to the unstable point. The super-stable is important, as it means a quiescent substrate exists. Since mean field theory is just an initial approximation, it ought to be worthwhile to gather up some Monte Carlo approximations for a few generations just to see if the first estimate maintains itself, more or less.

We must also mention that the probability to find 'interesting' behavior is very low, about 0.05. Perhaps, this may be the reason why the Diffusion Rule was not studied before — when observing evolution from random configurations one more likely (with probability 0.3) to encounter a catastrophe (e.g. when placing three cells anywhere in Moore's neighborhood, just not in one line) then stable mobile localization.

Fig. 16.1 Three random
initial densities for the
Diffusion Rule: (**a**) 0.004,
(**b**) 0.013 and (**c**) 0.995
respectively, evolving late of
18 generations on lattices of
200×200

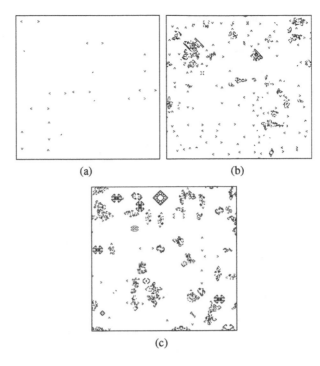

(a)   (b)

(c)

Fig. 16.2 Diagram of mean
field polynomial for the
Diffusion Rule

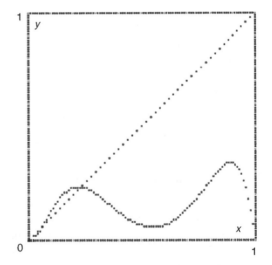

16.4 The Diffusion Rule Universe

In present section we study a range of basic structures, stationary and mobile localizations, generators of localizations and polymer-like structures formed of the
mobile localizations.

Table 16.1 Properties of gliders in the Diffusion Rule CA

Glider	Volume	Translation	Period	Speed	Weight	Move
g1	8	1	1	$c/1$	4	Orthogonal
g2	12	1	1	$c/1$	4	Orthogonal
g3	12	1	1	$c/1$	4	Orthogonal
g4	12	1	1	$c/1$	4	Orthogonal
g5	30	4	4	$c/1$	7	Orthogonal
g6	30	4	4	$c/1$	7	Orthogonal
g7	45	4	4	$c/1$	14	Orthogonal
g8	45	4	4	$c/1$	14	Orthogonal
g9	56	4	4	$c/1$	14	Orthogonal
g10	56	4	4	$c/1$	14	Orthogonal
g11	70	4	4	$c/1$	24	Orthogonal
g12	72	4	4	$c/1$	14	Orthogonal
g13	75	4	4	$c/1$	18	Orthogonal
g14	84	4	4	$c/1$	24	Orthogonal
g15	96	4	4	$c/1$	18	Orthogonal
g16	96	4	4	$c/1$	22	Orthogonal
g17	96	4	4	$c/1$	26	Orthogonal
g18	96	4	4	$c/1$	30	Orthogonal
g19	112	4	4	$c/1$	26	Orthogonal
g20	126	4	4	$c/1$	26	Orthogonal
g21	144	4	4	$c/1$	26	Orthogonal
g22	144	4	4	$c/1$	30	Orthogonal
g23	210	4	4	$c/1$	38	Orthogonal
g24	338	2	4	$c/2$	52	Orthogonal
g25	405	2	4	$c/2$	79	Orthogonal
g26	576	2	8	$c/4$	75	Diagonal

16.4.1 Mobile Self-localizations

In computational experiments with the Diffusion Rule CA we discovered 26 mobile self-localizations — gliders or particles — traveling orthogonally or diagonally in the lattice. Properties of the gliders, including volume, speed, direction of motion are listed in Table 16.1.

Configurations of minimal gliders and compound gliders are shown in Figs. 16.3 and 16.4, respectively. From Table 16.1, we can see that 96% of gliders move orthogonally, and that g23, g24, g25 and g26 are the largest gliders in the family of mobile localizations. First column in the table gives names of gliders. Second column in Table 16.1 represents glider's volume calculated as number of cells occupied by the glider. Third and fourth columns are translation and period. Fifth column

Fig. 16.3 Configurations of minimal, or primary, gliders in the Diffusion Rule: (**a**) g1 glider, (**b**) g2 glider, (**c**) g3 glider, and (**d**) g4 glider

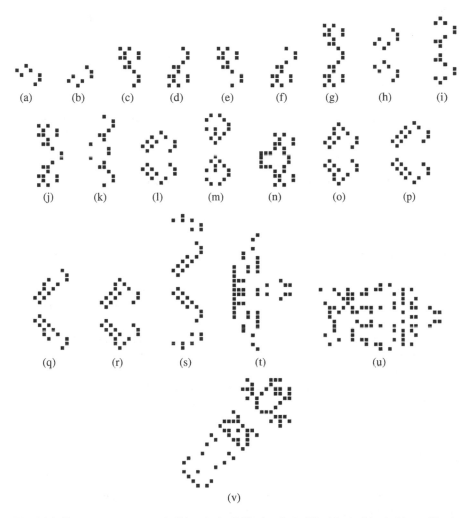

Fig. 16.4 Twenty two compound gliders in the Diffusion Rule CA: (**a**) g5, (**b**) g6, (**c**) g7, (**d**) g8, (**e**) g9, (**f**) g10, (**g**) g11, (**h**) g12, (**i**) g13, (**j**) g14, (**k**) g15, (**l**) g16, (**m**) g17, (**n**) g18, (**o**) g19, (**p**) g20, (**q**) g21, (**r**) g22, (**s**) g23, (**t**) g24, (**u**) g25, and (**v**) g26 gliders, respectively

shows the *speed* = *c*/*period*, where *c* is the maximum speed. Sixth column is the weight that represents the number of cells with state 1 within glider's volume. The last column indicates whether or not glider moves along orthogonally or diagonally.

Fig. 16.5 Five oscillators in the Diffusion Rule CA: (**a**) o1 and (**b**) o2 are flip-flops, (**c**) o3, (**d**) o4, and (**e**) o5 are blinkers, respectively

Table 16.2 Properties of oscillators in the Diffusion Rule CA

Oscillator	Volume	Period	Weight
o1	4	2	2
o2	9	2	3
o3	10	4	4
o4	16	4	4
o5	30	4	8

There are two types of gliders — *primary* and *compound* [27, 36]: a primary glider can not be decomposed into smaller mobile localizations, a compound glider is made of at least two primary gliders.

16.4.2 Oscillators

The Diffusion Rule CA exhibits five types of stationary localizations known as oscillators. The two most common flip-flops and three blinker patterns are shown in Fig. 16.5.

Flip-flop configurations are o1 and o2 oscillators both structures flipping at 45°. Blinkers of period four are o3, o4 and o5 oscillators. Table 16.2 shows basic characteristics of each oscillator.

Some of the oscillators and their assembles act as eaters, i.e. stationary patterns that annihilate gliders colliding to them (see examples of collisions in Sect. 16.5.3).

16.4.3 Avalanches

Avalanches are novel structures, have not described before in related studies. They are assembles of adjacent gliders that cause explosive growth of rhomboid shaped patterns with deterministic edges and quasi-chaotic interior. Avalanches can be constructed from various compositions of adjacent gliders.

Figure 16.6 shows an avalanche produced in composition of two g1 gliders, adjacent at 90°; this avalanche pattern grows diagonally inside the third quadrant. The minimal volume of an avalanche is 4×4 with eight cells in state 1.

One can use even number of gliders to construct symmetrical avalanches; two examples are shown in Fig. 16.7.

Fig. 16.6 Two g1 gliders
produce an avalanche pattern:
(**a**) initial condition,
(**b**) configuration after 15
steps of evolution and
(**c**) configuration after 200
steps of evolution

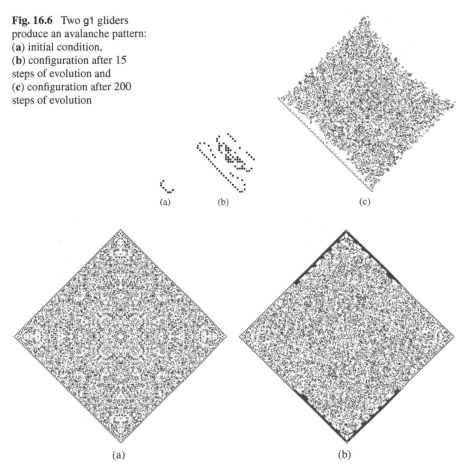

(a)   (b)   (c)

(a)   (b)

Fig. 16.7 Construction of the symmetrical avalanches: (**a**) symmetrical growth initiated by two
g4 gliders, configuration at 237th step of evolution, (**b**) avalanche pattern with non-trivial internal
symmetries produced by assembly of two g2 glider, and two g3 gliders, configuration at 237th step
of evolution

16.4.4 Puffer Trains

A puffer train is a mobile localization which generate (leaves traces) of stationary
localizations along its motion path. There are 16 known types of stable puffer trains
(which produce oscillators) in the Diffusion Rule CA. Basic properties of puffer
trains are shown in Table 16.3.

A particular case of puffer train is shown in Fig. 16.8. This puffer bears fragments
of g4 glider. All configurations of the puffer are displayed in Fig. 16.9, and there we
can see that configurations shown in Fig. 16.9(c), (d), (e) and (f) can be interpreted
as spaceships.

Moreover, the Diffusion Rule CA exhibits dozens of non-stable puffer trains. In
Fig. 16.10 we see five non-stable puffer trains, which produce asymmetrically grow-

Table 16.3 Characteristics of puffer trains in the Diffusion Rule CA

Puffer train	Produce	Volume	Translation	Period	Speed	Weight	Move
p1	o3	30	4	4	$c/1$	7	Orthogonal
p2	o1	35	4	4	$c/1$	8	Orthogonal
p3	o1	42	4	4	$c/1$	9	Orthogonal
p4	o1	42	4	4	$c/1$	9	Orthogonal
p5	o1	42	4	4	$c/1$	10	Orthogonal
p6	o1	54	4	4	$c/1$	13	Orthogonal
p7	o1	56	4	4	$c/1$	9	Orthogonal
p8	$2 \times$ o1	63	4	4	$c/1$	14	Orthogonal
p9	o1	63	4	4	$c/1$	15	Orthogonal
p10	$2n \times$ o1	65	4	4	$c/1$	14	Orthogonal
p11	$2 \times$ o1	80	4	4	$c/1$	17	Orthogonal
p12	o1	84	4	4	$c/1$	11	Orthogonal
p13	$2 \times$ o1	90	4	4	$c/1$	11	Orthogonal
p14	$2 \times$ o1	105	4	4	$c/1$	15	Orthogonal
p15	$2n \times$ o1	120	4	4	$c/1$	28	Orthogonal
p16	o1 $\vee \epsilon$	242	4	4	$c/1$	22	Orthogonal

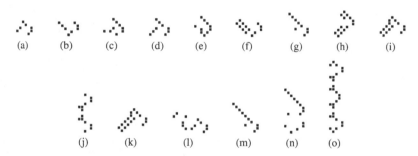

Fig. 16.8 Fifteen puffer trains observed in evolution of the Diffusion Rule CA: (**a**) p1, (**b**) p2, (**c**) p3, (**d**) p4, (**e**) p5, (**f**) p6, (**g**) p7, (**h**) p8, (**i**) p9, (**j**) p10, (**k**) p11, (**l**) p12, (**m**) p13, (**n**) p14, and (**o**) p15 puffer trains, respectively

ing, or quasi-chaotic, patterns. All discovered non-stable puffer trains have speed $1/c$ and period four.

16.4.5 Mobile Glider Guns

Glider guns are localized patterns that periodically lose their stability and give birth to traveling mobile localizations, gliders. In computational experiments we discovered twelve types of mobile glider guns in the Diffusion Rule CA.

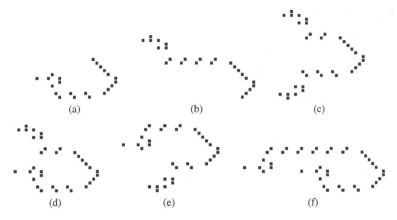

(a)   (b)   (c)

(d)   (e)   (f)

Fig. 16.9 Special p16 puffer train similar to spaceships

Table 16.4 Characteristics of glider guns in the Diffusion Rule CA. *Second column* of the table shows what type of glider each glider gun produces

Gun	Produce	Volume	Translation	Period	Speed	Weight	Move
gun1	g1	30	4	4	$c/1$	7	Orthogonal
gun2	g4	50	4	4	$c/1$	9	Orthogonal
gun3	g4	50	4	4	$c/1$	9	Orthogonal
gun4	g4	60	4	4	$c/1$	9	Orthogonal
gun5	g4	60	4	4	$c/1$	9	Orthogonal
gun6	g4	72	4	4	$c/1$	15	Orthogonal
gun7	g2 ∧ g3	72	4	4	$c/1$	16	Orthogonal
gun8	2 × g4	80	4	4	$c/1$	14	Orthogonal
gun9	2 × g4	90	4	4	$c/1$	14	Orthogonal
gun10	g1 ∧ g4	143	4	4	$c/1$	15	Orthogonal
gun11	2 × g1	154	4	4	$c/1$	24	Orthogonal
gun12	(2 × g4) ∨ g4	176	4	4	$c/1$	19	Orthogonal

Basic parameters of the twelve glider guns are shown in Table 16.4 and gun's configurations in Fig. 16.11. The most remarkable feature is that all primary gliders can be produced by glider guns. Some glider guns can generate two types of gliders at once, thus gun7 (Fig. 16.11(f)) generates g2 and g3 gliders at the same time, however both gliders travel coupled in pairs.

The generator gun12 can produce one or two g4 gliders, see examples in Fig. 16.12. This gun is also extendable, the extension is determined by the number of o1 oscillators (which should be more then two). The positions of oscillators determine number of glider streams generated by gun12.

Fig. 16.10 Some non-stable puffer trains

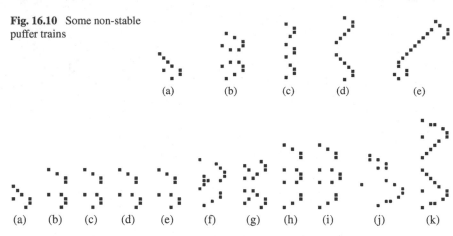

(a)   (b)   (c)   (d)   (e)

(a)   (b)   (c)   (d)   (e)   (f)   (g)   (h)   (i)   (j)   (k)

Fig. 16.11 Configurations of glider guns in the Diffusion Rule CA

(a)   (b)   (c)

Fig. 16.12 Expendable glider gun in the Diffusion Rule CA

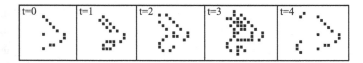

Fig. 16.13 Configurations of puffer-gun in the Diffusion Rule CA

So far we did not find glider guns which produce diagonally-moving streams of gliders, neither guns generating compound gliders or stationary guns.

16.4.6 Glider Gun and Puffer Train

There is at least one special mobile structure that combines in itself properties of both the glider gun and the puffer train. Figure 16.13 shows glider gun producing g1 glider and o1 oscillator each 4th step of CA evolution. This puffer-gun moves orthogonally, has a volume of 70 cells and weights 12 units.

16.4.7 Avalanche Gun

Avalanche gun is another remarkable example of mobile generators (Fig. 16.14). The mobile gun produces an avalanche every 4th step of CA evolution.

Fig. 16.14 Configuration of an avalanche gun in the Diffusion Rule CA: (**a**) production of two g1 gliders to 90°, (**b**) development of a quasi-chaotic reaction that destroys next avalanche conserving the avalanche gun

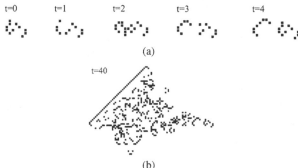

(a)

(b)

However, life-time of the gun producing each avalanche is short: when avalanche produced by the gun it grows and then destroys the next avalanche produced.

16.5 Collisions Between Localized Patterns

The Diffusion Rule CA combines high-degree unpredictability with enormously rich dynamics of collisions between mobile and stationary objects. This section studies the outcomes of the collision reactions.

16.5.1 Forming Diffusing Patterns by Collisions

Gliders colliding in the Diffusion Rule CA can produce an explosively growing diffusion-like pattern, the diffusive patterns do usually have non-stationary boundaries and they exhibit quasi-chaotic internal dynamics.

Two most distinct examples are shown in Fig. 16.15. In the first example (Fig. 16.15(a)), g4 glider collides with g3 and g2 gliders, diffusion pattern produced is 'lead' by three gliders and puffer train. In the second example (Fig. 16.15(b)), two g4 gliders collides with two g1 gliders. The reaction produces multiple gliders, puffer trains, oscillators, and even vertical glider guns during their collision dynamics.

16.5.2 Reactions Between Propagating Patterns

Soliton-Like Reaction

In certain initial conditions gliders collide similarly to solitons, namely they restore their structure and velocity vector after collisions. In Fig. 16.16 we see snapshots of the head-on collision dynamics between two g4 gliders (begin at even distance

Fig. 16.15 Examples of
diffusion-like patterns
produced in collision between
gliders in the Diffusion Rule
CA: (**a**) pattern produced at
260th step of evolution after
collision between g4, g2
gliders and g3 glider,
(**b**) pattern produced at 260th
of evolution after collision
between two g4 gliders, and
two g1 gliders

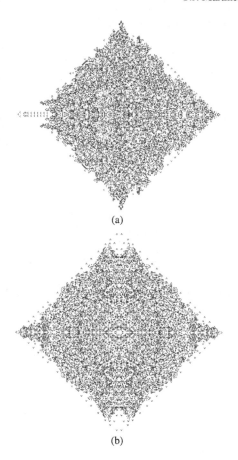

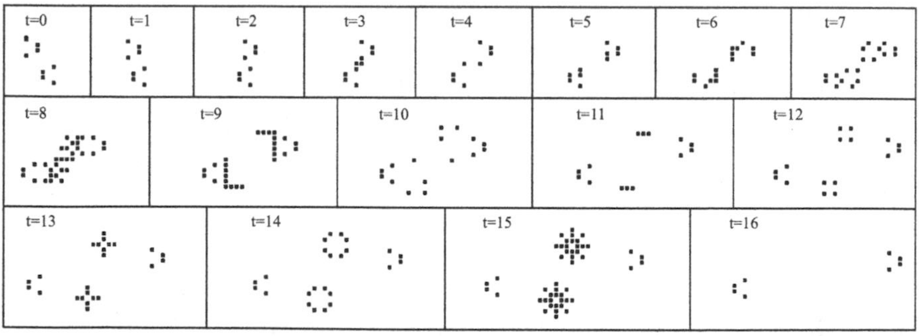

(a)

(b)

t=0	t=1	t=2	t=3	t=4	t=5	t=6	t=7
t=8		t=9		t=10		t=11	t=12
t=13		t=14		t=15		t=16	

Fig. 16.16 Soliton-like reaction between two g4 gliders

before collision). When the gliders collide they temporarily lose their stability, produce varieties of transient structures. After few steps of evolution the gliders are restored and transient structures are annihilated.

Fig. 16.17 Examples of eater
configurations in the
Diffusion Rule CA

(a)   (b)

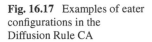

Fig. 16.18 Eater destroys gliders emitted by glider gun

Fig. 16.19 Delay reaction in
the Diffusion Rule CA:
(**a**) initial position of gliders
before collision, (**b**) final
result of delay operation, the
glider traveling west delayed
and translated southward

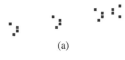

(a)

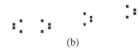

(b)

Eater Reaction

Eaters are stationary localizations which destroy gliders colliding into them. The
most simple eater is built with o1 oscillator. This eater destroys g1 gliders but is
shifted four cells along the glider's initial direction of motion. Both g1 and g2
gliders are destroyed when they collide with the eater built of o3 oscillator. In
Fig. 16.17(a) you can see an eater made of two o3 oscillators destroying g1 and
g4 gliders. A 'universal' eater, which destroys all types of primary gliders is illustrated in Fig. 16.17(b).

Using eaters one can control glider streams emitted by glider guns, thus in
Fig. 16.18 we can see how the eater eliminates g1 gliders produced by gun1. There
is at least one mobile eater of gliders, this is g24 glider that eats g1 gliders.

Delay Reaction

The basic delay reaction can be implemented when glider g3 collides with glider
g4 and is transformed to glider g4 in the result of collision. The original glider g4
is delayed for two time steps and it is translated southward one cell per collision
(Fig. 16.19).

Multiplication and Reduction Reactions

A typical reaction of glider multiplication is shown in Fig. 16.20(a): two g4 gliders
are involved in head-on collision, with odd distance between glider heads before
collision, four new g4 gliders are produced in result of the collision. Reduction is

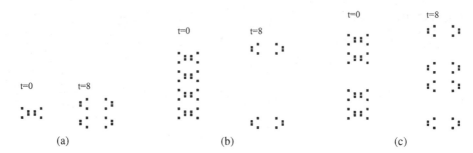

Fig. 16.20 Multiplication of gliders: (**a**) multiplication $1 \times 1 \rightarrow 4$ reaction of gliders in binary collision, (**b**) reduction $4 \times 4 \rightarrow 4$ reaction in glider collision, (**c**) conservative collision reaction

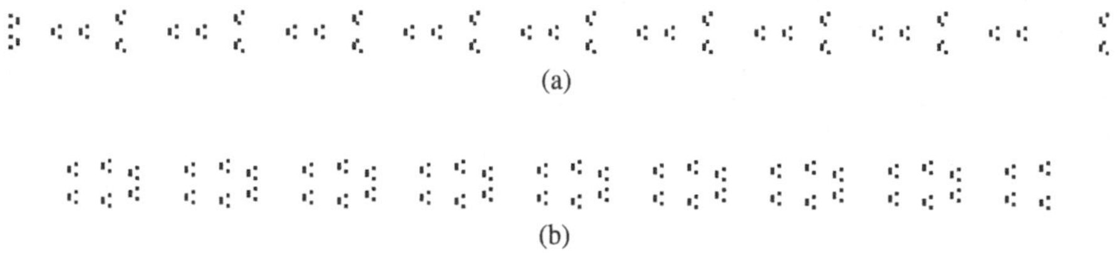

Fig. 16.21 Construction of packages of g4 gliders by multiplication reaction: (**a**) initial configuration, (**b**) final configuration, recorded after 186th steps of evolution

implemented as multiplication of gliders, where gliders in the multiplied columns are in proximity of each other. As shown in Fig. 16.20(b), when we collide two rows of g4 gliders, four gliders in each row (eight gliders in total are involved in the collision), then just four new gliders are produced. Adjusting distance between gliders in colliding columns of gliders we can achieve almost any (but odd) result of multiplication (Fig. 16.20(c)).

Using multiplication reactions we can also construct arbitrary packages of gliders. For example, to construct a stream of packages, six gliders per package, we collide stream of four-glider g4 packages traveling west with a pair of g4 gliders traveling east (Fig. 16.21(a)). Sequentially, all four-glider packages are transformed to six-glider packages (Fig. 16.21(b)), the operation is halted by pair g3 gliders.

Reflection Reaction

We discovered nine types of reflection-type collisions between stationary and mobile self-localizations. Let us discuss some examples shown in Fig. 16.22. When two g1 gliders collide with each other (head-on collision, even distance, with slight shift between gliders along south–north axis) two g4 gliders are generated; these g4 gliders move in the direction perpendicular to original trajectories of colliding gliders (Fig. 16.22(a)). Glider colliding with o1 oscillator is reflected at the angle 90°, as shown for collision of g2, g4 and g3 gliders with o1 oscillator (Fig. 16.22(b), (c), (d)).

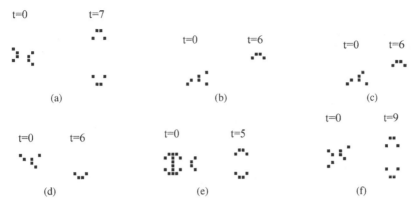

Fig. 16.22 Collisions leading to reflections: (**a**) two g1 gliders collide with each other, (**b**) g4 glider collides with o1 oscillator, (**c**) g2 glider collides with o1 oscillator, (**d**) g3 glider collides with o1 oscillator, (**e**) g1 glider collides with o5 oscillator, (**f**) g3 glider collides with o3 oscillator

Gliders can also be derived in addition of reflections, when colliding with stationary localizations. Thus, when a g1 glider collides with a o5 oscillator, two g1 gliders (going in opposite directions to each other) are generated and follow trajectories perpendicular to original trajectory of collided g1 gliders (Fig. 16.22(e)).[8] Similarly, when g3 glider collides with o3 oscillator, two g4 gliders are produced (Fig. 16.22(f)).

Annihilation Reaction

A significant amount of collisions between localizations in the Diffusion Rule CA leads to annihilation of colliding patterns. Few examples of initial configurations of colliding objects (leading to annihilation) are shown in Fig. 16.23, for binary collisions between gliders (Fig. 16.23(a)–(f)) and multiple collisions between gliders and oscillators (Fig. 16.23(g)–(k)).

16.5.3 Computation Capacities in the Diffusion Rule

Basic operations necessary to implement a functionally complete set of logical gates can be derived from collision dynamics presented in Fig. 16.22. Following paradigms of collision-based computing [3] we encode logical TRUE by presence of a glider or an oscillator, while absence of mobile or stationary objects corresponds to logical FALSITY.

---

[8]This reaction can synchronize multiple collisions as you can see in our example FANOUT.rle (Fig. 16.24) available from http://uncomp.uwe.ac.uk/genaro/Diffusion_Rule/diffusionLife.html.

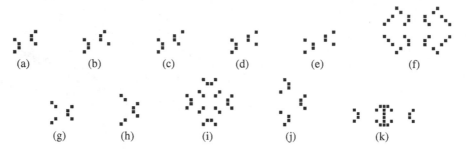

Fig. 16.23 Initial positions of colliding gliders and oscillators leading to annihilation reaction: (**a**)–(**f**) binary collisions, (**g**)–(**k**) multiple collisions

Namely, Fig. 16.22(b), (c), (d) demonstrate that stationary localizations, oscillators, can play a role of mirrors thus deflecting gliders from their original trajectory. The mirrors can be used to route signals. Signals can be deleted, erased by placing eaters along trajectories of gliders, representing the signals. Signals can be also delayed in collisions with other gliders or oscillators.

Collisions used to construct FANOUT gate are shown in Fig. 16.22(e), (f), a glider collides to stationary localization, and two new gliders are produced in result of the collision.

Dynamics displayed in Fig. 16.22(a) shows a typical collision gate, where inputs *x* and *y* are represented by trajectories of gliders traveling East and West, respectively. While trajectories of new gliders, traveling South and North, encode value of *x* AND *y*. Constant TRUE is made up of ceaseless stream of gliders, generated by glider guns.

FANOUT Gate Implementation

A FANOUT gate can be implemented in Diffusion Rule CA by means of multiple collisions (Fig. 16.24). Essentially the reaction g1 → o5 yield 2g1 gliders (Fig. 16.22(e)), and this reaction is synchronized symmetrically producing and deleting g1 gliders orthogonally. Finally also all information generated become deleted because four o3 oscillators are fixed for delete the last four g1 gliders.

Constructing a Memory Device

A memory in the Diffusion Rule CA can be constructed using basic interactions between g1 gliders and o1 oscillator, as illustrated in Fig. 16.25. A bit of information is represented in the memory unit by shaded 2 × 2 cells square in Fig. 16.25. The bit can be read by sending a g1 glider to the memory unit (top configurations in Fig. 16.25, glider travels East). When g1 glider collides with oscillator o1 oscillator, both glider and oscillator are annihilated (and one new o1 oscillator is formed at

Fig. 16.24 FANOUT gate implementation in Diffusion Rule also with halt condition determined by four o5 blinkers

some distance from the memory unit). Then the bit is erased as a result of the reading operation. To write down the bit one can send another g1 glider (in Fig. 16.25 this 'writing' glider travels West) toward now empty memory unit and associated o1 oscillator. Both g1 glider and associated oscillator are destroyed, however o1 oscillator is restored in memory unit (shaded region in Fig. 16.25), i.e., we write a bit again.

Asynchronous XNOR and XOR Gate

Exploiting some features of the interaction between g1 glider and o1 oscillator (Fig. 16.25) we can implement an *asynchronous* device which calculates XNOR and XOR operation at once (Fig. 16.26). Such a gate is designed by a scheme similar

Fig. 16.25 Construction of a memory device in the Diffusion Rule CA. Snapshots of the configurations. Time increases from right to left and from top to bottom. The domain where the bit of information (o1 oscillator) is written to is represented by a *shaded zone*

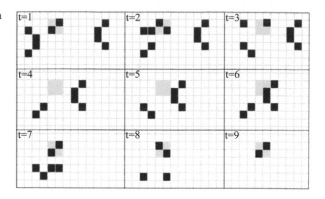

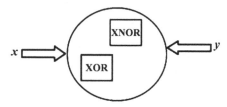

Fig. 16.26 Scheme of XNOR and XOR gate

to that outlined in [4]. Oscillator in position shaded by gray in Fig. 16.25 represents logical value TRUE and absence of the oscillator — value FALSE of logical operation XNOR, exclusive NOR operation. An auxiliary oscillator (generated when oscillator in shade position is annihilated) represents the result of XOR operation, exclusive OR operation.

Assuming that signals on inputs x and y can be generated at any time (no synchronization) except not at the same time, we obtain the following dynamics of the device (Fig. 16.26):

x	y	XNOR	XOR
0	0	o1	0
0	g1	0	o1
g1	0	0	o1
g1	g1	o1	0

where o1 and g1 stay for oscillator and glider, 0 means absence of the objects.

16.6 Discussion

The findings presented in the chapter are based on computational experiments with the Diffusion Rule CA and, particularly, exhaustive search of mobile and stationary localizations emerged in space–time dynamics of the automaton.

Amongst known 2D CA supporting localizations, the Diffusion Rule CA is the minimal model because cell-state transitions does not depend on intervals of 'cell sensitivity' but on singletons, i.e. transition $0 \rightarrow 1$ occurs if there is exactly two

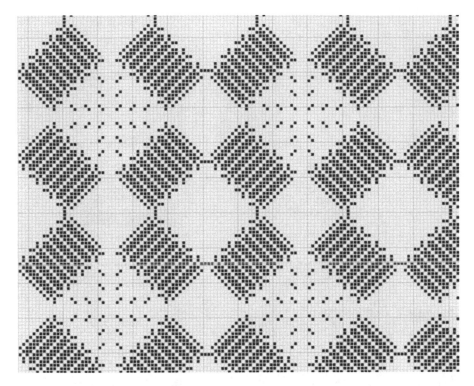

Fig. 16.27 Simulating a luminescence pattern with the Diffusion Rule

neighbors in state 1, and transition $1 \rightarrow 1$ if there exactly seven neighbors in state 1. Moreover, we are not aware of any other CA which exhibits so large variety of mobile localizations (gliders) and high diversity of outcomes of collisions between mobile and/or stationary localizations. Despite of trying to undertake exhaustive study of localization dynamics we nevertheless missed several important points that could form objectives of future studies.

A simple but interesting physical simulation was made setting a reaction like luminescence. Using packages of both diagonal lines of 50 cells everyone. The luminescence phenomenon was obtained over the evolution in densities of cells in small intervals as illustrated in Fig. 16.27. The final state is dominated by blinkers or oscillators.[9]

It is not a trivial problem to find large stable patterns constructed with big complex structures. Cell colonies damaged by a virus is one of such configurations (Fig. 16.28).

Diffusion Rule can simulate the evolution of a cell like the elemental cellular automata (ECA) Rule 18 or Rule 90. For example, we take a diagonal line with

---

[9]This simulations are developed by the group of researchers iGEM-Mexico at the MIT; http://www.fenomec.unam.mx/pablo/igem/.

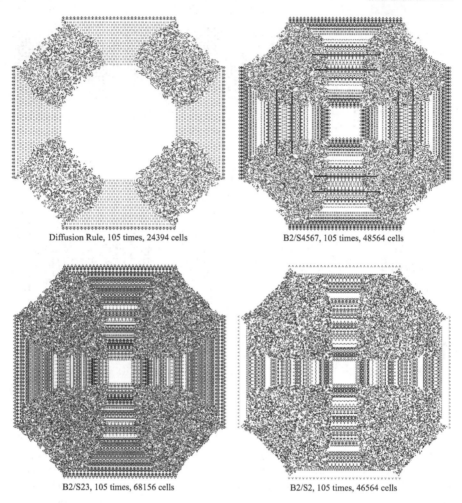

Diffusion Rule, 105 times, 24394 cells · B2/S4567, 105 times, 48564 cells · B2/S23, 105 times, 68156 cells · B2/S2, 105 times, 46564 cells

Fig. 16.28 Virus propagation in the Diffusion Rule and three mutations in the same initial condition. Evolution rules and data are given *below* snapshots

503 cells in the initial condition. During the evolution original line is multiplied in lines less and less small producing oscillators. Finally, the evolution space is dominated by oscillators that represent exactly a cell alive in 1D case. However, there is generated a second evolution of the same type as illustrated in Fig. 16.29. In this example, it took 512 steps of evolution to reach the configuration of 7,596 living cells.

The existence of stable configurations seems difficult to find in a rule which is generally chaotically producing super-nova explosions. Nevertheless, we have an example where four glider guns are synchronized to annihilate the gliders. In this case, two **g1** gliders and two **g4** gliders come into quadruple collision shown in Fig. 16.30.

Fig. 16.29 Simulating the evolution of a cell of the ECA Rule 18 or Rule 90 with the Diffusion Rule

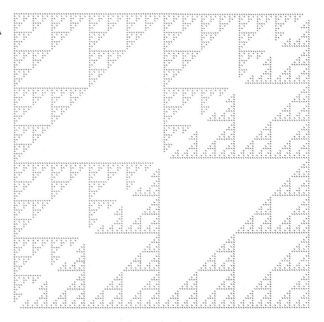

Fig. 16.30 Simultaneous annihilation across of four particles produced by four glider guns. The evolution shows a global configuration in 36 steps with 118 live cells

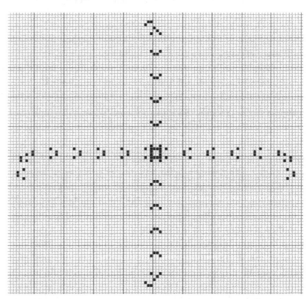

We envisage that problems to be solved in future include implementation of quasi-chemical reactions between gliders, studies of grammars derived from glider interactions and implementation of a full effective decision procedures based on glider collisions. It will be also worth to demonstrate intrinsic universality and self-reproduction. Another project would be to use de Bruijn diagrams [29] to check if there are any still undiscovered gliders with velocity one or still life configurations, and besides to use algorithms specialized in automatic search for complicated or big gliders [16], oscillators, glider guns or more configurations. Also we are planning to make an exploration of the cluster of semi-totalistic rules originated by the Diffusion Rule.[10] Last but not least open problem is to decide if all types of gliders can be constructed in collision with other gliders, a closure property with respect to set of gliders.

Acknowledgements   We thank Cris Moore, David Hillman, Paul Callahan and Markus Redeker for their valuable comments. We also grateful to Dave Greene for fruitful discussions about complex constructions in the Diffusion Rule. First author thanks EPSRC (grant EP/D066174/1) for their support. All simulations are done with Golly: http://golly.sourceforge.net/.

References

1. Adachia, S., Lee, J., Peper, F., Umeo, H.: Kaleidoscope of life: a 24-neighbourhood outer-totalistic cellular automaton. Physica D **237**, 800–817 (2004)
2. Adamatzky, A.: Computing in Nonlinear Media and Automata Collectives. Institute of Physics Publishing, Bristol and Philadelphia (2001)
3. Adamatzky, A. (ed.): Collision-Based Computing. Springer, Berlin (2002)
4. Adamatzky, A., Wuensche, A.: Computing in spiral rule reaction–diffusion cellular automaton. Complex Syst. **16**(4), 277–297 (2006)
5. Adamatzky, A., Costello, B.D.L., Asai, T.: Reaction–Diffusion Computers. Elsevier, Amsterdam (2005)
6. Adamatzky, A., Martínez, G.J., Seck-Tuoh-Mora, J.C.: Phenomenology of reaction–diffusion binary-state cellular automata. Int. J. Bifurc. Chaos **16** (10), 1–21 (2006)
7. Adamatzky, A., Wuensche, A., Costello, B.D.L.: Glider-based computing in reaction diffusion hexagonal cellular automata. Chaos Solitons Fractals **27**(2), 287–295 (2006)
8. Bays, C.: Candidates for the Game of Life in three dimensions. Complex Syst. **1**, 373–400 (1987)
9. Bays, C.: New game of three-dimensional life. Complex Syst. **5**, 15–18 (1991)
10. Bell, D.I.: High Life — An interesting variant of life. http://www.tip.net.au/~dbell/ (1994)
11. Berlekamp, E.R., Conway, J.H., Guy, R.K.: Winning Ways for Your Mathematical Plays, vol. 2, Chap. 25. Academic Press, San Diego (1982)
12. Boccara, N., Nasser, J., Roger, M.: Particle like structures and their interactions in spatio-temporal patterns generated by one-dimensional deterministic cellular automaton rules. Phys. Rev. A **44**(2), 866–875 (1991)
13. Chaté, H., Manneville, P.: Evidence of collective behavior in cellular automata. Europhys. Lett. **14**, 409–413 (1991)
14. Cook, M.: Universality in elementary cellular automata. Complex Syst. **15**(1), 1–40 (2004)
15. Das, R., Mitchell, M., Crutchfield, J.P.: A genetic algorithm discovers particle-based computation in cellular automata. In: Davidor, Y., Schwefel, H.-P., Männer, R. (eds.) Parallel Problem Solving from Nature-III. Lecture Notes in Computer Science, vol. 866, pp. 344–353. Springer, Berlin (1994)
16. Eppstein, D.: Searching for spaceships. In: Nowakowski, R.J. (ed.) More Games of No Chance. MSRI Publications, vol. 42, pp. 433–452. The Mathematical Sciences Research Institute, Berkeley (2002)

---

[10]http://uncomp.uwe.ac.uk/genaro/Life_dc22.html.

17. Evans, K.M.: Replicators and larger-than-life examples. In: Griffeath, D., Moore, C. (eds.) New Constructions in Cellular Automata. Santa Fe Institute Studies on the Sciences of Complexity. Oxford University Press, London (2003)
18. Gardner, M.: Mathematical Games — The fantastic combinations of John H. Conway's new solitaire game Life. Sci. Am. **223**, 120–123 (1970)
19. Griffeath, D., Moore, C.: Life Without Death is P-complete. Complex Syst. **10**, 437–447 (1996)
20. Gutowitz, H.A., Victor, J.D.: Local structure theory in more that one dimension. Complex Syst. **1**, 57–68 (1987)
21. Hanson, J.E., Crutchfield, J.P.: Computational mechanics of cellular automata: an example. Phys. D **103**(1–4), 169–189 (1997)
22. Heudin, J.-C.: A new candidate rule for the game of two-dimensional life. Complex Syst. **10**, 367–381 (1996)
23. Jakubowski, M.H., Steiglitz, K., Squier, R.: Computing with Solitons: A Review and Prospectus. Multiple-Valued Log. Special Issue on Collision-Based Computing **6** (5–6) (2001)
24. Magnier, M., Lattaud, C., Heudin, J.-K., Complexity classes in the two-dimensional life cellular automata subspace. Complex Syst. **11**(6), 419–436 (1997)
25. Martínez, G.J.: Teoría del Campo Promedio en Autómatas Celulares Similares a The Game of Life. Tesis de Maestría, CINVESTAV-IPN, México (2000)
26. Martínez, G.J., Adamatzky, A., McIntosh, H.V.: Phenomenology of glider collisions in cellular automaton Rule 54 and associated logical gates. Chaos Solitons Fractals **28**, 100–111 (2006)
27. Martínez, G.J., McIntosh, H.V., Seck-Tuoh-Mora, J.C.: Gliders in Rule 110. Int. J. Unconventional Computing **2**(1), 1–49 (2006)
28. McIntosh, H.V.: Wolfram's Class IV and a Good Life. Physica D **45**, 105–121 (1990)
29. McIntosh, H.V.: Phoenix. http://delta.cs.cinvestav.mx/~mcintosh/oldweb/pautomata.html (1994)
30. McIntosh, H.V.: Rule 110 as it relates to the presence of gliders. http://delta.cs.cinvestav.mx/~mcintosh/oldweb/pautomata.html (1999).
31. Mitchell, M.: Life and evolution in computers. Hist. Philos. Life Sci. **23**, 361–383 (2001)
32. Tommaso, T., Norman, M.: Cellular Automata Machines. MIT, Cambridge (1987)
33. von Neumann, J.: Theory of Self-reproducing Automata. Edited and completed by Burks, A.W. University of Illinois, Urbana and London (1966)
34. Sediña-Nadal, I., Mihaliuk, E., Wang, J., Pérez-Muñuzuri, V., Showalter, K.: Wave propagation in subexcitable media with periodically modulated excitability. Phys. Rev. Lett. **86**, 1646–1649 (2001)
35. Wolfram, S.: A New Kind of Science. Wolfram Media, Champaign (2002)
36. Wuensche, A.: Classifying cellular automata automatically. Complexity **4**(3), 47–66 (1999)
37. Wuensche, A.: Self-reproduction by glider collisions: the beehive rule. In: Alife9 Proceedings, pp. 286–291. MIT, Cambridge (2004)

Part IV
Non-orthogonal Lattices

Chapter 17
The Game of Life in Non-square Environments

Carter Bays

17.1 One Dimensional Rules

One dimensional cellular automata (CA) differ from CA in higher dimensions in that the restrictive "grid" (essentially a single line of cells) limits the number of rules that can be applied. Hence, many 1D CA involve neighborhoods that extend beyond the immediate two "touching" neighbors of a cell whose next generation status we wish to evaluate. Or more than the two states ("alive", "dead") may be utilized. Thus there are no bona fide one dimensional Game of Life (GoL) rules (see Chap. 1); nevertheless we shall briefly explore the simplest rules along with the concept of one-dimensional "gliders".

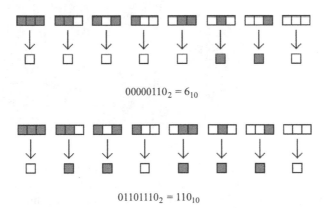

$$00000110_2 = 6_{10}$$

$$01101110_2 = 110_{10}$$

Fig. 17.1 The one dimensional rules "6" and "110" are depicted by the diagram shown. There are 8 possible states involving a center cell and its two immediate neighbors. The next generation state for the center cell depends upon the current configuration; each possible current state is given. The rule is specified by the binary number depicted by the next generation state of the center cell. This notation is standard for the simplest 1D CA and was introduced by Wolfram (see [8]), who also converts the binary representation to its decimal equivalent. There are 256 possible rules, but most are not as interesting as rule 110. Rule 6 is one of many that generate nothing but gliders

A. Adamatzky (ed.), *Game of Life Cellular Automata*,
DOI 10.1007/978-1-84996-217-9_17, © Springer-Verlag London Limited 2010

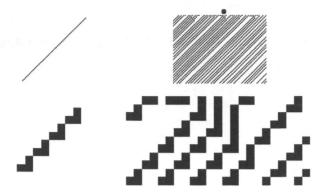

Fig. 17.2 Rule 6 (along with many others) creates only gliders. At the *upper left*, we have several generations starting with a single live cell (*top*). Each successive generation moves down one level on the page. At the *lower left* is an enlargement of the first few generations. By following the diagram for rule 6 in Fig. 17.1, the reader can see exactly how this configuration evolves. At the *top right*, we start with a random configuration; at the *lower right* we have enlarged the small area directly under the *large dot*. Very quickly, all initial random configurations lead solely to gliders heading West

Fig. 17.3 Evolution of rule 110 for the first 500 generations, given a random starting configuration. With 1D CA, we can depict a great many generations on a 2D display screen

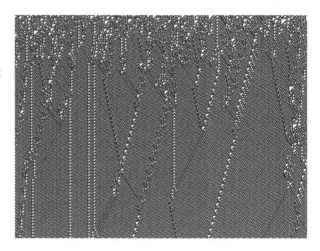

For our discussion we shall only look at rules involving just the two adjacent neighbors and two states. Unlike 2D (and higher) dimensions, we usually consider the relative position of the neighbors when giving a rule. Since three cells (center, left, right) are involved in determining the state for the next generation of the central cell, we have $2^3 = 8$ possible initial states, with each state leading to a particular outcome. And since each initial state causes a particular outcome (i.e. the cell in the middle "lives" or "dies" next generation) we thus we have 2^8 possible rules. The behavior of these 256 rules has been extensively studied by Wolfram [8] who also introduced a very convenient shorthand that completely describes each rule (Fig. 17.1).

Fig. 17.4 Rule 110 at generations 2,000–2,500. The structures that move vertically are stationary oscillators; slanted structures can be considered "gliders". Unlike higher dimensions, where gliders move in an unobstructed grid with no other live cells in the immediate vicinity, many 1D gliders reside in an environment of oscillating cells (the background pattern). The *black square* outlines an area depicted in the next figure

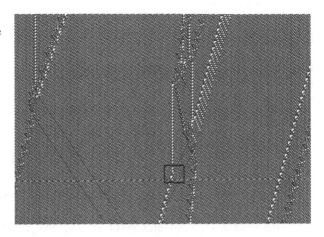

Fig. 17.5 An area from Fig. 17.4 is enlarged. One can carefully trace the evolution from one generation to the next. The background pattern repeats every seven generations

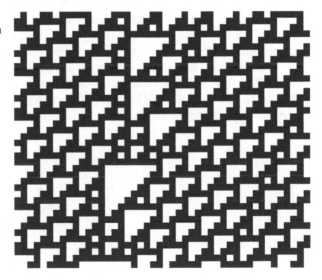

Gliders in 1D CA are very common (Figs. 17.2 and 17.3) although many gliders for "stable" rules exist against a uniform patterned background (Figs. 17.3–17.5) instead of a grid of non-living cells.

As we add to the complexity of defining 1D CA we greatly increase the number of possible rules. For example, just by having 3 states instead of two, we note that now, instead of 2^3 possible initial states, there are 3^3 (Fig. 17.6). This leads to 27 possible initial states, and we now can create 3^{27} unique rules — more than seven trillion! However Wolfram observed that even with more complex 1D rules, the fundamental behavior for all rules is typified by the simplest rules [8].

Fig. 17.6 There are 27 possible configurations when we have three states instead of two. Each configuration would yield some specific outcome; thus there would be three possible outcomes for each state, and hence 3^{27} possible rules

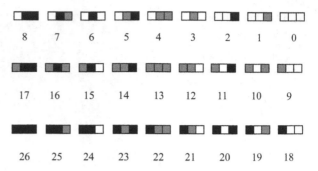

17.2 The Game of Life in Two Dimensional Non-square Grids

Although most 2D CA research involves a square grid, the triangular tessellation has been investigated somewhat. Here we have 12 touching neighbors; as with the square grid, they are all treated equally (Fig. 17.7). The increased number of neighbors allows for the possibility of more GoL rules and their gliders.

Figure 17.8 shows many of these gliders and their various GoL rules. The GoL rule 2,7,8/3 supports two rather unusual gliders (Figs. 17.9 and 17.12) and to date is the only known GoL rule other than Conway's original 2,3/3 Game of Life that exhibits glider guns. Figure 17.10 shows starting configurations for two of these guns and Fig. 17.11 exhibits evolution of the two guns after 800 generations.

Due to the extremely unusual behavior of the period 80 2,7,8/3 glider (Fig. 17.12), it is highly likely that other guns exist.

Fig. 17.7 Each cell in the triangular grid has 12 touching neighbors. The subject central cells can have two orientations, "*E*" and "*O*"

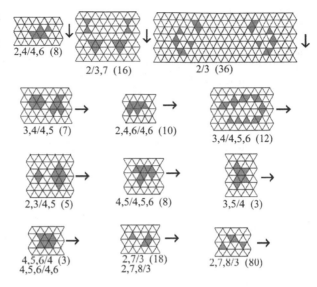

Fig. 17.8 Most of the known GoL rules and their gliders are illustrated. The period for each is given in *parentheses*

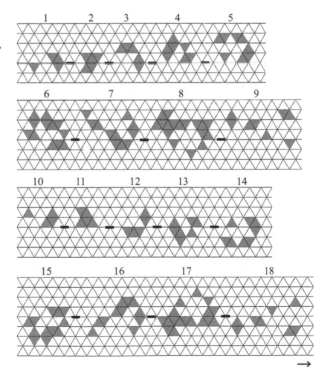

Fig. 17.9 The "small" 2,7,8/3 glider is shown. This glider also exists for the GoL rule 2,7/3

Fig. 17.10 The GoL rule 2,7,8/3 is of special interest in that it is the only known GoL rule besides Conway's rule that supports "glider guns" — configurations that spew out an endless stream of gliders. In fact, there are probably several such configurations under this rule. Here we illustrate two guns; the *top* one generates period 18 ("small") gliders and the *bottom* one creates period 80 ("large") gliders. Unlike Gosper's 2,3/3 gun, these guns translate across the grid in the direction indicated. In keeping with the fanciful jargon for names, translating glider guns are also called "rakes"

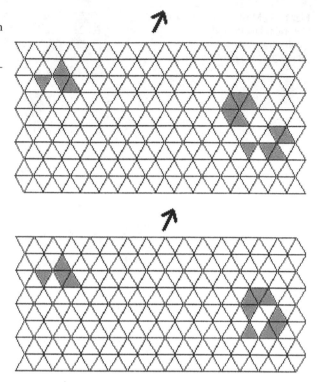

Fig. 17.11 After 800 generations, the two guns from Fig. 17.10 will have produced the output shown. Motion is in the direction given by the *arrows*. The gun at the *left* yields period 18 gliders, one every 80 generations, and the gun at the *right* produces a period 80 glider every 160 generations

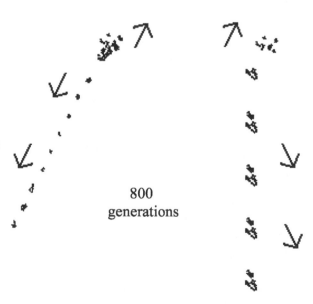

800 generations

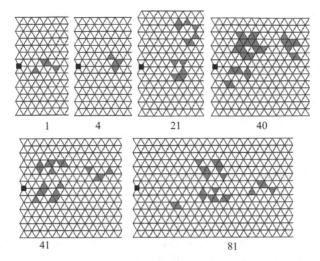

Fig. 17.12 Here we depict the "large" 2,7,8/3 glider. Perhaps "flamboyant" would be a better description, for this glider spews out much debris as it moves along. It has a period of 80 and its exact motion can be traced by observing its position relative to the *black dot*. Note that the debris tossed behind does not interfere with the 81st generation, where the entire process repeats 12 cells to the right. By carefully positioning two of these gliders, one can (without too much effort) construct a situation where the debris from both gliders interacts in a manner that produces another glider. This was the method used to discover the two guns illustrated in Figs. 17.10 and 17.11

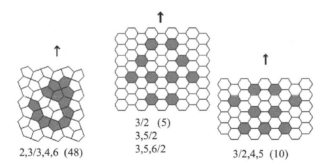

Fig. 17.13 GL rules are supported in pentagonal and hexagonal grids. The pentagonal grid (*left*) is called the "Cairo Tiling", supposedly named after some paving tiles in that city. There are many different topologically distinct pentagonal grids; the Cairo tiling is but one. At the *right* are gliders for the hexagonal rules 3/2 and 3/2,4,5. The 3/2 glider also works for 3,5/2, 3,5,6/2 and 3,6/2. All four of these rules are GoL rules. The rule 3/2,4,5 is unfortunately disqualified (barely) as a GoL rule because very large random blobs will grow without bounds. The periods of each glider are given in *parentheses*

The hexagonal grid supports the GoL rule 3/2, along with GoL rules 3,5/2, 3,5,6/2 and 3,6/2, which all behave in a manner very similar to 3/2. The glider for these three rules is shown in Fig. 17.13. It is possible that no other distinct hexagonal GoL rules exist, because with only 6 touching neighbors, the set of interesting rules is quite limited. Moreover the fertility portion of the rule must start with "2" and rules of the form */2,3 are unstable. Thus, any other hexagonal GoL rules must be of the form */2,4; */2,4,5; etc. (i.e. only seven fertility combinations).

A valid GoL rule has also been found for at least one pentagonal grid (Fig. 17.13). Since there are several topologically unique pentagonal tessellations (see [5]), probably other pentagonal gliders will be found, especially when all the variants of the pentagonal grid are investigated.

17.3 Three Dimensional Game of Life Rules

In 1987, the first GoL rules in three dimensions were discovered. The initially found gliders and their rules are depicted in Fig. 17.14. It turns out that Conway's 2D rule 2,3/3 is in many ways "contained" in the 3D GoL rule 5,6,7/6 [1, 7]. During the ensuing years, several other 3D GoL rules were found (Figs. 17.15 and 17.16) by searching for gliders in rules that stabilize. Most of these gliders were unveiled by employing random but symmetric small initial configurations. The large number of live cells in these 3D gliders implies that they are uncommon random occurrences in their respective GoL rules; hence it is highly improbable that the plethora of

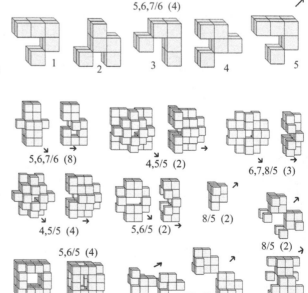

Fig. 17.14 The first three-dimensional GoL rules were found in 1987; these are the original gliders that were discovered. The rule 5,6,7/6 is analogous to the 2D rule 2,3/3 [1]. Note the similarity between this glider and Conway's original 2D version

Fig. 17.15 Several more 3D GoL rules were discovered between 1990–1994. They are illustrated here. The 8/5 gliders were originally investigated under the rule 6,7,8/5

Fig. 17.16 By 2004, computational speed had greatly increased, so another effort was made to find 3D gliders under GoL rules; these latest discoveries are illustrated here

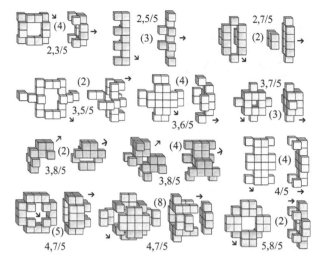

Fig. 17.17 Some work has been done with the 3D grid of "dense packed spheres". Two gliders have been discovered for the rule 3/3, which almost qualifies as a GoL rule

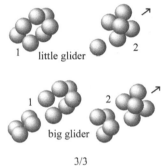

interesting forms (e.g. glider guns) such as those for 2D rule 2,3/3 exist in three dimensions.

The 3D grid of dense packed spheres has also been investigated somewhat; here each sphere touches exactly 12 neighbors. What is pleasing about this configuration is that each neighbor is identical in the manner that it touches the subject cell, unlike the square and cubic grids, where some neighbors touch on their sides and others at their corners. The gliders for spherical rule 3/3 are shown in Fig. 17.17. This rule is a "borderline" GoL rule, as random finite configurations appear to stabilize, but infinite ones apparently do not.

17.4 Conclusion

The original Game of Life, seminally defined by John Horton Conway's rule, has been aggressively investigated. Yet not surprisingly, when we look beyond Conway's configurations, we find that games of life abound — in both three dimensions and in non-square two dimensional grids. There has even been some work in four

Fig. 17.18 Some work (not much) has been done in four dimensions. Here is an example of a glider for the GoL rule 11,12/12,13. Many more 4D gliders exist

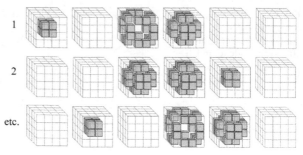

4D glider for GL rule 11,12/12,13

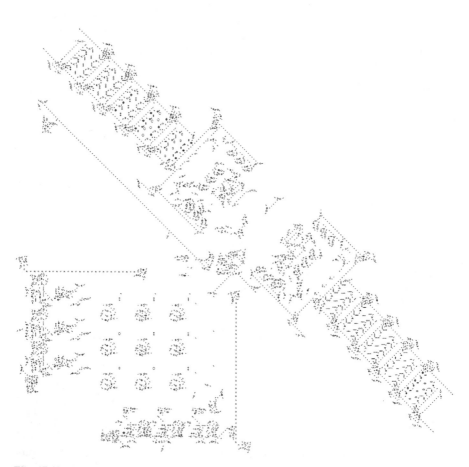

Fig. 17.19 The discovery of the glider in 2,3/3, along with the development of several glider guns, has made possible the construction of many extremely complex forms. Here we see a Turing machine, developed in 2001 by Paul Rendell. Figure 17.20 enlarges a small portion of this structure

Fig. 17.20 We have enlarged a tiny portion at the *upper left* of the Turing machine shown in Fig. 17.19. One can see the complex interplay of gliders, glider guns, and various other stabilizing forms

dimensions, where each cell has 80 touching neighbors. In Fig. 17.18 we illustrate a glider for the GoL rule 11,12/12,13; undoubtedly many other 4D GoL rules exist. Nevertheless, Conway's 2D rule 2,3/3 will likely forever be the "gold standard" for complex "life forms" that can be created — consider for example the Turing machine configuration of Figs. 17.19 and 17.20. But if we are willing to stretch our definition of a GoL rule somewhat, consider simple one dimensional cellular automata with four possible states. It will be a long time before all 10^{38} possible rules have been investigated!

References

1. Bays, C.: Candidates for the Game of Life in three dimensions. Complex Syst. **1**, 373–400 (1987)
2. Bays, C.: Patterns for simple cellular automata in a universe of dense packed spheres. Complex Syst. **1**, 853–875 (1987)
3. Bays, C.: Further notes on the Game of Three Dimensional Life. Complex Syst. **8**, 67–73 (1994)
4. Bays, C.: Cellular automata in the triangular tessellation. Complex Syst. **8**, 127–150 (1994)
5. Bays, C.: A note on the Game of Life in hexagonal and pentagonal tessellations. Complex Syst. **15**, 245–252 (2005)
6. Bays, C.: The discovery of glider guns in a Game of Life for the triangular tessellation. J. Cell. Autom. **2**(4), 345–350 (2007)
7. Dewdney, A.K.: The game Life acquires some successors in three dimensions. Sci. Am. **286**(2), 16–22 (Feb. 1987)
8. Wolfram S.: A New Kind of Science. Wolfram Media, Champaign (2002)

Chapter 18
The Game of Life Rules on Penrose Tilings: Still Life and Oscillators

Nick Owens and Susan Stepney

John Horton Conway's *Game of Life* [1, 4] is a simple two-dimensional, two state cellular automaton (CA), remarkable for its complex behaviour [1, 13]. That behaviour is known to be very sensitive to a change in the CA *rules*. Here we continue our investigations [7, 10, 11] into its sensitivity to changes in the *lattice*, by the use of an aperiodic Penrose tiling lattice [5, 12].

Section 18.1 describes Penrose tilings; Sect. 18.2 generalises the concepts of neighbourhood and outer totalistic CA rules (which include the Game of Life) to aperiodic lattices, and introduces a naming convention for Penrose Life oscillators. Section 18.3 presents various Penrose lattice still life configurations; Sects. 18.4–18.7 present various oscillators with periods from 2 to 15. Section 18.8 presents an algorithm to detect oscillators, and a means to classify them based on their underlying neighbourhood graph.

18.1 Penrose Tiling

18.1.1 Kites and Darts, and Rhombs

Grünbaum & Shephard [6, Chap. 10] provide a good introduction to aperiodic tilings, including Penrose tilings. The two variants of Penrose tiling we consider here are 'kites and darts', and 'rhombs'.

The kite and dart tile pair are shown in Fig. 18.1a; a large patch of kite and dart tiling is shown in Fig. 18.11. The thick and thin rhomb tile pair are shown in Fig. 18.1b; a large patch of rhomb tiling is shown in Fig. 18.13.

18.1.2 Matching Rules

The relationship of the rhomb tiles to the kite and dart tiles is shown in Fig. 18.2.

A. Adamatzky (ed.), *Game of Life Cellular Automata*,
DOI 10.1007/978-1-84996-217-9_18, © Springer-Verlag London Limited 2010

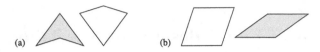

Fig. 18.1 Penrose tiles: (**a**) the dart (*grey*) and kite (*white*) tiles: the long and short sides are in the ratio $\phi : 1$, where the golden ratio $\phi = (1 + \sqrt{5})/2 = 2\cos(\pi/5)$; (**b**) the thick (*white*) and thin (*grey*) rhomb tiles

Fig. 18.2 Relationship between rhomb tiles and kites and darts: (**a**) a thick rhomb comprises a dart and two half-kites (matching rules, later, forbid a dart and full kite from being joined in this way); (**b**) a thin rhomb comprises two half-kites

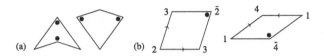

Fig. 18.3 Matching rules: (**a**) kite and dart vertex markings; (**b**) rhomb vertex marking and edge orientations plus vertex angle numbering, where interior angles are $\pi/5$ times the vertex angle number (note that vertices labelled 2, and labelled 4, come in two kinds, due to the matching rules: these are distinguished by *overbars*)

To avoid a kite and dart being joined to form a rhombus (Fig. 18.2a), which would allow a periodic tiling, there are additional 'matching rules' that force the tiling to be aperiodic: as well as edges of the same length being put together, certain vertices (given by the dots in Fig. 18.3a) must also be matched [5, 6].

To avoid rhomb tiles being used to form a periodic tiling, and force a periodic tiling, there are again additional 'matching rules': as well as edges of the same length being put together, the edge orientations (given by the arrows and dots in Fig. 18.3b) must also be matched [2].

18.1.3 Valid Vertex Configurations

There are many ways to put the tiles together, even with the restriction of the matching rules. However, in a true Penrose tiling (one that can be extended to infinity), not all of these configurations can exist.

There are only seven valid ways to surround any vertex in a kite and dart tiling [5] (Fig. 18.4).

There are only eight valid vertices in a rhomb tiling [2] (Fig. 18.5). The names of these vertices come from the names of the corresponding kite and dart vertices from which they can be derived [2]. Each vertex can be associated with a list of vertex angle numbers (after [14, Fig. 6.8], augmented here with overbars, Fig. 18.3b),

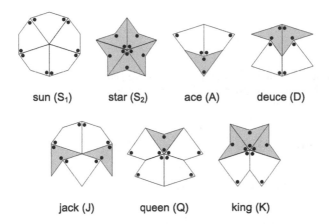

sun (S₁)   star (S₂)   ace (A)   deuce (D)

jack (J)   queen (Q)   king (K)

Fig. 18.4 The seven valid vertex configurations of a kite and dart tiling [5]

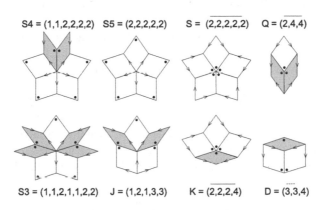

S4 = (1,1,2,2,2,2)   S5 = (2,2,2,2,2)   S = $\overline{(2,2,2,2,2)}$ Q = $\overline{(2,4,4)}$

S3 = (1,1,2,1,1,2,2)   J = (1,2,1,3,3)   K = $\overline{(2,2,2,4)}$ D = $\overline{(3,3,4)}$

Fig. 18.5 The eight valid vertex configurations of a rhomb tiling [2]

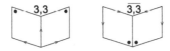

Fig. 18.6 Disambiguating the 3,3 vertices: the two distinct ways a 3,3 vertex can appear in a valid rhomb vertex configuration (in the *J* and *D*, see Fig. 18.5). This is a *context dependent* marking [16]

corresponding to the vertex angles of the tiles forming the central vertex. These are useful for determining how to complete forced vertices (see [10]). Note that there are two distinct occurrences of the 3,3 vertex configurations (in the J and D); see Fig. 18.6.

If a patch of tiling exhibits any other vertex configuration, it is not a true Penrose tiling: it will not be possible to extend it to infinity. We use these valid vertex configurations to analyse valid neighbourhood configurations later.

Fig. 18.7 Regular grid neighbourhoods: (**a**) the Moore neighbourhood, the eight cells with which the central cell shares a vertex; (**b**) the von Neumann neighbourhood, the four cells with which the central cell shares an edge

18.2 The Game of Life on a Penrose Tiling

18.2.1 Regular Game of Life Rules

Classic cellular automata are defined on regular lattices. The update rule depends on the state of each cell and its neighbourhood (the surrounding cells),[1] and the structure of that neighbourhood is invariant: all places in the lattice look the same, and the update rule can be applied uniformly across the lattice.

In general, the update rule depends on the particular state of each separate neighbour. For *outer totalistic* CA rules, like the Game of Life (GoL), the next state of a cell depends only on its current state, and the total number of neighbourhood cells in certain states.

In GoL, the neighbourhood of each cell comprises the regular Moore neighbourhood (Fig. 18.7a), the eight cells with which it shares a vertex. Each cell has two states, 'dead' and 'alive'. If a cell is alive at time t, then it stays alive if and only if it has two or three live neighbours (otherwise it dies of 'loneliness' or 'overcrowding'). If a cell is dead at time t, then it becomes alive (is 'born') if and only if it has exactly three live neighbours.

For aperiodic lattices such as a Penrose tiling, the detailed structure of the neighbourhood varies at different locations in the lattice. Outer totalistic rules can be given an interpretation in these aperiodic tiling neighbourhoods.

18.2.2 Generalising the Neighbourhood and the Rules

Generalised von Neumann Neighbourhood

The (Penrose) Game of Life rules use a (generalised) Moore neighbourhood. It is also convenient to define a generalised von Neumann neighbourhood, which we use later in the analysis of certain simple still lifes and oscillators.

---

[1] The standard definition of CA 'neighbourhood' includes both the surrounding cells and the updating cell. Here we use slightly different terminology: by 'neighbourhood' we mean *only* the surrounding cells.

Fig. 18.8 The generalised von Neumann neighbourhoods of a kite and dart Penrose tiling

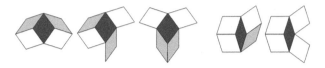

Fig. 18.9 The generalised von Neumann neighbourhoods of a rhomb Penrose tiling

Define the generalised von Neumann neighbourhood of a cell in a Penrose tiling to be all the cells with which it shares an edge (or, equivalently, two distinct vertices). Hence the size of the neighbourhood equals the number of edges of the central cell: four. Figures 18.8 and 18.9 show the distinct generalised von Neumann neighbourhoods which form valid vertices (established by exhaustive consideration of the valid vertex configurations, Figs. 18.4 and 18.5). Rotations and mirror images of these neighbourhoods are not considered to be distinct. de Bruijn [3] identifies the same rhomb neighbourhoods (but considers mirror images separately), and shows that a valid Penrose rhomb tiling can be constructed by considering just these neighbourhoods, without the need to use the rhomb matching rules of Fig. 18.3.

In the rectangular lattice none of the four von Neumann neighbourhood cells themselves share an edge. So if A is a neighbour of B, and B is a neighbour of C, then A is *not* a neighbour of C: neighbouring von Neumann neighbourhoods do not overlap (recall that we do not treat the central site as a member of the neighbourhood here). In the Penrose lattice, this is no longer the case: cells in a generalised von Neumann neighbourhood can share an edge, so neighbouring generalised von Neumann neighbourhoods can overlap. This may affect the communication paths through the Penrose CA.

Generalised Moore Neighbourhood

Define the generalised Moore neighbourhood of a cell in a Penrose tiling to be all the cells with which it shares a vertex.

Not only do cells have irregular shaped neighbourhoods, with the generalised Moore neighbourhood, not all cells have the same number of neighbours. The range of neighbourhood sizes and configurations is limited. Figure 18.10 shows the eight valid neighbourhood configurations in a kite and dart tiling: there are no other valid ways to surround a kite or a dart (established by exhaustive consideration of the valid kite and dart vertex configurations, Fig. 18.4). So there is one neighbourhood configuration of size 8 around a kite, and two around a dart; three of size 9 around a kite, and one around a dart; and one of size 10, around a dart ([7] incorrectly states that kite and dart tilings have neighbourhoods of size 8 and 9 only). Figure 18.11

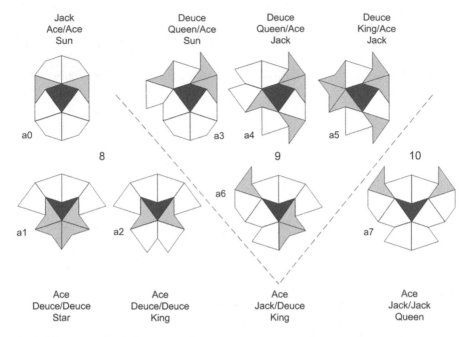

Jack
Ace/Ace
Sun

Deuce
Queen/Ace
Sun

Deuce
Queen/Ace
Jack

Deuce
King/Ace
Jack

a0

a3   a4

a5

8

9

10

a6

a1   a2

a7

Ace
Deuce/Deuce
Star

Ace
Deuce/Deuce
King

Ace
Jack/Deuce
King

Ace
Jack/Jack
Queen

Fig. 18.10 The generalised Moore neighbourhoods on a kite and dart Penrose tiling, with neighbourhood sizes, and the types of each vertex. Note that an ace vertex appears in every neighbourhood

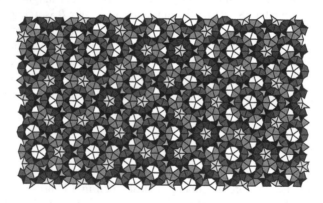

Fig. 18.11 A kite and dart tiling shaded by neighbourhood type. The neighbourhood shading is uniformly distributed between white and black such that a_0 is white and a_7 black

shows an area of kite and dart tilings with colouring to highlight the size of cells' neighbourhoods.

Similarly, Fig. 18.12 shows the valid neighbourhood configurations in a rhomb tiling (established by exhaustive consideration of the valid rhomb vertex configurations, Fig. 18.5). There is a larger range of distinct neighbourhood configurations for rhomb tilings. Figure 18.13 show an area of rhomb tilings with colouring to highlight the size of cells' neighbourhoods.

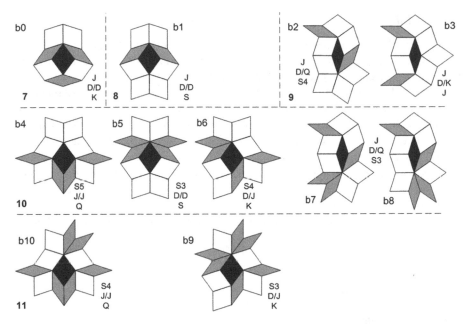

Fig. 18.12 The generalised Moore neighbourhoods on a rhomb Penrose tiling, with neighbourhood sizes, and the types of each vertex. Note that $b7$ and $b8$ have the same types of vertices, but in different orientations: we call $b7$ the "ortho" form, and $b8$ the "para" form. Not that the 3 angle in the fat rhombs is always a J or D vertex. Note that one of the 1 angles in the thin rhomb is always a J vertex

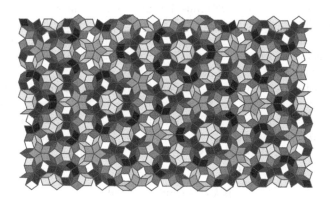

Fig. 18.13 A rhomb tiling shaded by neighbourhood type. The neighbourhood shading is uniformly distributed between white and black such that b_0 is white and b_{10} black. Socolar & Steinhardt [16] show that the number of thick to thin rhombs in a Penrose tiling is in the golden ratio $\phi : 1$

As can be seen from Figs. 18.11 and 18.13, not all sizes of neighbourhoods appear with the same frequency. [11] gives the distribution of neighbourhood sizes in a kite and dart tiling and in a rhomb tiling.

Penrose Life Rules

Using our definition of the generalised Moore neighbourhood, the definition of the
Game of Life as given in Sect. 18.2, in terms of the number of live and dead neighbours, can be used unchanged on a Penrose lattice.

Some early investigations are reported in [7]; further investigations are reported
in [10, 11]. We find that the Game of Life rules on Penrose tilings still has complex
behaviour. Both kinds of tiling need to be investigated, because we find that the
Game of Life on the rhomb tiling is statistically significantly different from that
on the kite and dart tiling: it has longer lifetimes to quiescence (the final periodic
activity state), lower ash densities (density of 'live' cells at quiescence), higher soup
growth (eventual extent of an initial random patch), significantly fewer oscillators
in the ash, and lower ash period.

Here we continue our Penrose Life investigations, into the variety of oscillators
supported by the tilings.

18.2.3 Identifying Oscillators

An Identification Code

GoL *patterns* are defined in terms of their 'live' cells. An *oscillator* is a pattern that
recurs after a given number of generations, called the *period* of the oscillator. The
Life Lexicon [15] defines the *rotor* to be "the cells of an oscillator that change state"
and the *stator* to be "the cells of an oscillator that are always on". It also states that
it is "easy to see that any rotor cell must be adjacent to another rotor cell".[2]

Following the regular Game of Life, we give some of the oscillators fanciful
names, loosely based on their static appearance at one or more timesteps, or their
dynamic appearance as they oscillate. We generally name a Penrose oscillator after
the isomorphic regular Life oscillator, if one exists.

In regular Life, oscillators are considered to be the same only if they have the
same 2D pattern of cells, up to a rotation. But with the Penrose grid, there are patterns that can look superficially different (using a kite rather than a dart, or a thick
rather than a thin rhomb), but have the same underlying structure. To help in their
identification we give all oscillators a short *code*, for example, r-p2-8-6. The code
has the following four-part form:

$$[l|kd|r\text{-}]p\text{nn-xx}[\text{-yy}] \tag{18.1}$$

The first part, $l|kd|r$, identifies the *tiling*: whether we are talking about a regular
life tiling l, a kite and dart tiling kd, or a rhomb tiling r. If the code is being used to

---

[2]*Proof* Clearly a rotor cell must be adjacent to some other cell in the oscillator, else it would be
dead. Consider an "off" rotor cell adjacent to only stator cells. Since it is a rotor cell, at some point
it changes state to "on", at which time it has three "on" neighbours. Since these neighbours are
stator cells by assumption, it would always have three "on" neighbours, so could no turn "off", and
so could not be a rotor cell. We have a contradiction, so such a rotor cell cannot exist.

refer to a pattern on all tilings, or if the tiling is clear from context, this part may be omitted.

The second part, *p*nn, identifies the *period* of the oscillator. For example, *p*1 means a still life.

The third part, xx, identifies the total number of cells involved in the oscillator. These are the cells that are "on", at some timestep. (For still lifes, this is just the number of cells.)

The fourth part, yy, is the minimum number of cells involved in the oscillator. It is determined by looking at the number of cells "on" in each timestep, and taking the minimum of these. (For still lifes, this is the same as the total, and so is omitted.) This follow the classification used in regular Life catalogues such as [9].

So, for example: p1-4 refers to a still life with four cells, that exists on all tilings (it may be either a *block* or a *tub*, see later); r-p2-8-6 refers to a rhomb period 2 oscillator, with a total of 8 cells live over its period, and a minimum of 6 live cells in one timestep (a *marcher*, see later).

Note that the code is not sufficient to uniquely identify an oscillator. Two clearly different oscillators may share a code; in particular, there are many *p*1 still lifes that have the same number of cells, but very different shapes. The third part of the code gives the number of nodes in the underlying "oscillator graph" (Sect. 18.8.2), which exposes this structure (the underlying topology of cell connections). We use this graph in addition to the code to identify "essentially similar", or *isomorphic*, oscillators: these have the same code *and* the same graph.

This scheme also allows us to say that oscillators on different tilings are nevertheless isomorphic. So, some regular Life oscillators can also exist on Penrose tilings, and some oscillators can exist on both forms of Penrose tiling.

Variant Forms

Despite the Penrose tiling being aperiodic, it has much underlying structure. Some isomorphic oscillators are due to this structure, and can be systematically constructed from regularities in the underlying tiling.

The underlying structure is generally captured in terms of "empires", or tiles forced to be in certain positions given the existence of particular other tiles (see, for example, [6, Fig. 10.6.6]). A *forced vertex* is one that can be completed in only one way to give a valid vertex. Completing the tiling around a forced vertex may result in new forced vertices; that a vertex is forced may be a result of constraints imposed by several surrounding vertices.

One class of variants is given by the different ways that a completed kite and dart Star vertex can be extended, described here. Consider a patch of tiling comprising a Star S_2 (Fig. 18.4), formed from five inward-pointing darts. Each of the Star's inward pointing vertices is forced to be an ace. Filling in these vertices results in what we call the "Complete Star", or S_C (Fig. 18.14).

The Complete Star S_C has no further forced vertices. However, there are still constraints on how further tiles can be attached to produce valid tilings. There are precisely three different ways to extend S_C, shown in Fig. 18.15. The different extensions can support variant forms of a given oscillator: three forms of the

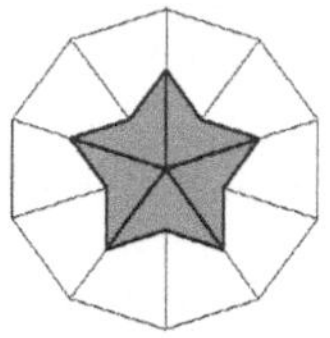

Fig. 18.14 The "Complete star" (S_C), found by extending all the "forced" vertices of the Star

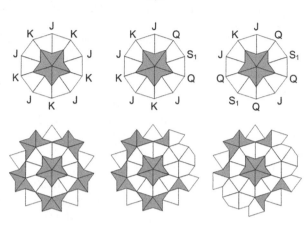

Fig. 18.15 The three ways to extend the S_C tiling. The *top row* labels the perimeter vertices with the vertices used to extend the tiling; the *bottom row* shows the resulting tilings

kd-p1-15 ring (Fig. 18.36), the symmetric and asymmetric forms of kd-p2-12-9, the *fast shuffler* (Fig. 18.48), the symmetric and asymmetric forms of kd-p4-14-6, the *bat* (Fig. 18.71), and three forms of kd-p15-40-8, the *dancer* (Fig. 18.71).

18.3 Still Life

A *still life* is a pattern that remains constant: it can be thought of as a period 1 ($p1$) oscillator, or an oscillator that has no rotor component. (Strictly, a still life is a minimal such pattern, where no subset of its cells can be removed and leave a still life.)

Here we give a preliminary catalogue of Penrose still lifes. These have been discovered by a combination of systematic construction and random search: systematic construction of the small still lifes possible around certain vertices and neighbourhoods; an examination of ash contents from multiple runs of the GoL rules on random initial conditions (300,000 runs over a range of initial conditions, for each tiling [11]); and constructions of large still lifes that extend the structure of smaller ones. Because of the diversity of tiling patterns, there is no guarantee that the catalogue is exhaustive, particularly for larger numbers of tiles. Some of these still lifes clearly have commonalities, and some we classify as isomorphic. We provide a definition of oscillator isomorphism in Sect. 18.8.2.

18.3.1 Blocks and Tubs

The smallest still lifes in regular Life have four cells (see [9] for an enumeration of all small still lifes); they are the *tub* (or *diamond*) and the *block* still lifes

Fig. 18.16 The regular Life
tub and block still lifes, l-p1-4

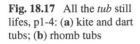

Fig. 18.17 All the *tub* still
lifes, p1-4: (**a**) kite and dart
tubs; (**b**) rhomb tubs

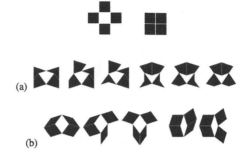

(Fig. 18.16). A tub is four "on" cells forming a chain around an "off" cell; the "on" cells are precisely the von Neumann neighbourhood of the surrounded "off" cell. A block is four "on" cells sharing a vertex.

These have isomorphic forms in Penrose Life, occurring in a variety of forms due to the more variable grid.

Tubs

The Penrose Life isomorphic forms of the tub are formed from the generalised von Neumann neighbourhoods (Figs. 18.8 and 18.9). All the resulting *tubs* are shown in Fig. 18.17.

Kite and Dart Blocks

We can discover the isomorphic forms of blocks by exhaustive examination of the seven valid kite and dart vertices (Fig. 18.4).

Of these seven valid vertex configurations, one (the ace) has three cells meeting at the vertex. These three cells comprise a *small block* still life, Fig. 18.18a. There are no three cell still lifes in regular Life.

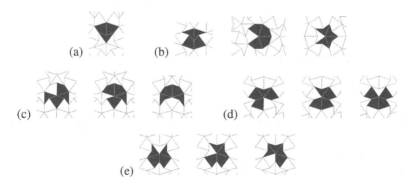

Fig. 18.18 All the kite and dart block still lifes, kd-p1-3 and kd-p1-4, classified according to the valid vertex configurations: (**a**) the three cell ace *small block*; (**b**) the deuce *block*; the sun *block*; the star *block*; (**c**) the three jack *blocks*; (**d**) the three queen *blocks*; (**e**) the three king *blocks*

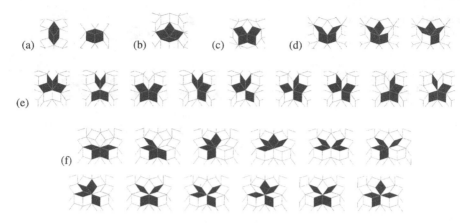

Fig. 18.19 The rhomb block still lifes, r-p1-3 and r-p1-4: (**a**) the three cell Q *small block* and D *small block*; (**b**) the K *block*; (**c**) the S *block*, equivalent to the S5 *block*; (**d**) the three J *blocks*; (**e**) the nine S4 *blocks*; (**f**) a selection of the several S3 *blocks*

The deuce has four cells meeting at the vertex. These four cells comprise a block still life, Fig. 18.18bi.

All the other valid vertices have five cells meeting at the vertex; any four of these chosen to be "live" form a still life. Two of these vertices, the sun and the star, are symmetric, so have one block form, Fig. 18.18bii and biii. The others (the jack, queen, and king) each have three block variants (up to a reflection), Fig. 18.18c, d, and e.

Rhomb Blocks

We can discover the isomorphic forms of blocks by exhaustive examination of the valid rhomb vertices (Fig. 18.5).

Of the eight valid vertex configurations, two (the Q and the D) have three cells meeting at the vertex. These three cells comprise a *small block* still life, Fig. 18.19a.

One valid vertex (the K) has four cells meeting at the vertex. These four cells comprise a block, Fig. 18.19b.

Three valid vertices (the S5, the S, and the J) have five cells meeting at the vertex; any four of these chosen to be "live" form a still life. Two of these, the S5 and the S, are symmetric, so have one block form each (or one between them if vertex orientation is ignored), Fig. 18.19c. The J has three block variants (up to a reflection), Fig. 18.19d.

One valid vertex (the S4) has six cells meeting at the vertex; again, any four of these chosen to be "live" form a still life, Fig. 18.19e.

One valid vertex (the S3) has seven cells meeting at the vertex. There are many ways of choosing four of these to be "live" to form a still life; a selection is shown in Fig. 18.19f.

Fig. 18.20 The regular Life five cell boat still life, l-p1-5

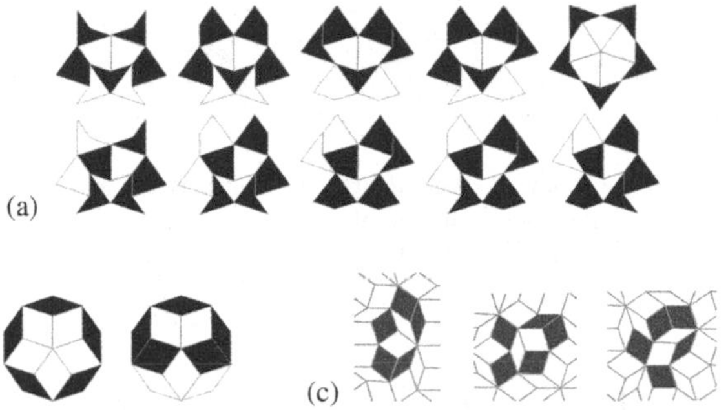

Fig. 18.21 Some five cell still lifes, p1-5: (**a**) kite and dart, from examining the ace and sun vertices; (**b**) rhomb, from examining the S vertex; (**c**) rhomb, found in the ash. We call the left pattern in (**b**) an S-chain

(a)

(b)   (c)

Fig. 18.22 All the regular Life six cell still lifes, l-p1-6: the *snake*, the *ship*, the *barge*, the *beehive*, and the *carrier*

Fig. 18.23 Some six cell Penrose snakes, p1-6: (**a**) kite and dart; (**b**) rhomb

(a)   (b)

18.3.2 Five and More Cell Still Lifes

There is a single regular Life five cell still life [9], the *boat* (Fig. 18.20). There is no Penrose life isomorphic to the boat: it is not possible to add a fifth stable cell to any of the tubs without giving an adjacent dead cell three live neighbours. All the identified Penrose five cell still lifes form chains (Fig. 18.21). (We call rings of cells, where every node has precisely two neighbours, *chains*.)

There are five regular Life six cell still lifes [9], the *snake*, the *ship*, the *barge*, the *beehive*, and the *carrier* (Fig. 18.22). The identified Penrose six cell still lifes form snakes (Fig. 18.23), chains (Fig. 18.24), or are disconnected (Fig. 18.25).

There are four regular Life seven cell still lifes [9], the *long snake*, the *fishhook*, the *long boat*, and the *loaf* (Fig. 18.26). The identified Penrose seven cell still lifes form snakes or rings (Fig. 18.27) or are disconnected (Fig. 18.28).

There are nine regular Life eight cell still lifes [9], the *shillelagh*, the *hook with tail*, the *very long snake*, the *canoe*, the *long ship*, the *long barge*, the *pond*, the *tub with tail*, and the *cigar* (Fig. 18.29). The identified Penrose eight cell still lifes form chains, rings, and snakes (Fig. 18.30).

The number of regular still lifes increases sharply with size (from 10 with nine cells, 25 with 10 cells, 46 with 11 cells, on up to 112,243 with 20 cells [9]). We have no reason to believe that the Penrose still lifes do not similarly increase in numbers, however our ash searches have not revealed these. Our hand constructions are based

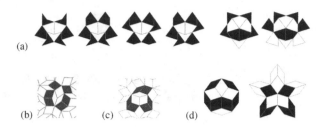

(a)

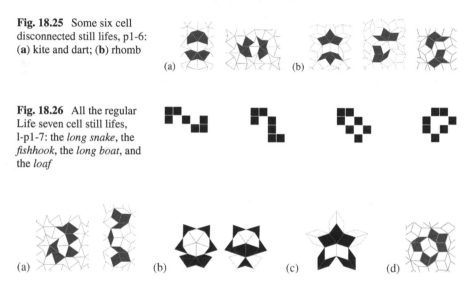

(b)   (c)   (d)

Fig. 18.24 Some six cell Penrose still life chains, p1-6: (**a**) kite and dart, from examining the ace and sun vertices; (**b**) rhomb, found in the ash: note that it forms a chain around a Q vertex; (**c**) rhomb, constructed by analogy to (**b**), based on a D vertex; (**d**) rhomb, from examining the S and S5 vertices

Fig. 18.25 Some six cell disconnected still lifes, p1-6: (**a**) kite and dart; (**b**) rhomb

(a)   (b)

Fig. 18.26 All the regular Life seven cell still lifes, l-p1-7: the *long snake*, the *fishhook*, the *long boat*, and the *loaf*

(a)   (b)   (c)   (d)

Fig. 18.27 Some seven cell still lifes, p1-7: (**a**) kite and dart *7-snake* [7], and a rhomb *7-snake*; (**b**) kite and dart rings, from examining the sun vertex; (**c**) rhomb ring, from examining the S5 vertex; (**d**) rhomb ring

Fig. 18.28 Some seven cell disconnected still lifes, p1-7: (**a**) kite and dart; (**b**) rhomb

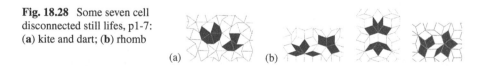

(a)   (b)

on patterns already seen: such still life rings and chains are shown in Figs. 18.31–18.33. A systematic cataloguing search would need to consider all vertices, and all valid extensions of those vertices; the large number of block still lifes alone (Figs. 18.18 and 18.19) indicates this would be a significantly larger job than for regular Life.

Fig. 18.29 All the regular
Life eight cell still lifes,
l-p1-8: the *shillelagh*, the
hook with tail, the *very long
snake*, the *canoe*, the *long
ship*, the *long barge*, the *tub
with tail*, the *pond*, and the
cigar

Fig. 18.30 Some eight cell still lifes, p1-8: (**a**) kite and dart chain; (**b**) rhomb chain, and ring;
(**c**) another rhomb chain; (**d**) rhomb *8-snake*

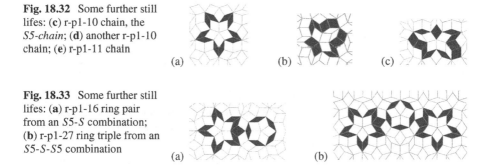

Fig. 18.31 Some nine cell still lifes, p1-9: (**a**) kite and dart chain; (**b**) rhomb chain; (**c**) rhomb
*9-snake*s

Fig. 18.32 Some further still
lifes: (**c**) r-p1-10 chain, the
S5-chain; (**d**) another r-p1-10
chain; (**e**) r-p1-11 chain

(a)   (b)   (c)

Fig. 18.33 Some further still
lifes: (**a**) r-p1-16 ring pair
from an *S5-S* combination;
(**b**) r-p1-27 ring triple from an
S5-S-S5 combination

(a)   (b)

18.3.3 Large Rings

Dart Rings

Large rings are not possible in regular Life, as a ring requires a "corner", which
results in a dead site with three live neighbours. It was noted in [7] that large
ring-shaped kite and dart still lifes can be formed. Some examples are shown in
Fig. 18.34.

Arbitrarily large dart rings can be constructed, in the following way. Pick a dart
that is not part of a Star S_2 vertex. Complete the "string" of darts that is formed

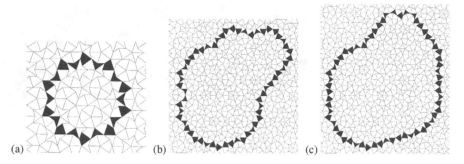

Fig. 18.34 Some large kite and dart still life chains: (**a**) kd-p1-20; (**b**) kd-p1-57; (**c**) kd-p1-61

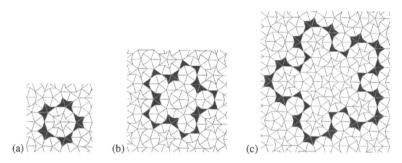

Fig. 18.35 Dart still life rings: (**a**) kd-p1-15; (**b**) kd-p1-25; (**c**) kd-p1-55

Fig. 18.36 The kd-p1-15 ring variant forms, from the three different extensions of the "Complete Star" S_C (Fig. 18.15)

from the darts in its generalised Moore neighbourhood (this is always possible, see Fig. 18.10). This string is a still life ring. Figure 18.35 shows example constructions.

Note that the dart ring in Fig. 18.35a occurs on the symmetric extension of the Complete Star S_C (Fig. 18.15). There are variants of this ring, on the other two extensions, shown in Fig. 18.36.

Rhomb Rings

Large still life chains can be formed in the rhomb tiling; some examples are shown in Fig. 18.37.

Arbitrarily large chains can be constructed, in the following way. Pick a thick rhomb that is not part of an S or $S5$ vertex. Complete the "ribbon" of thick rhombs that is formed from the two thick rhombs adjacent to its edges (all thick rhombs have precisely two such thick rhomb neighbours, see Fig. 18.9). Figure 18.38 shows several such ribbons of thick rhombs.

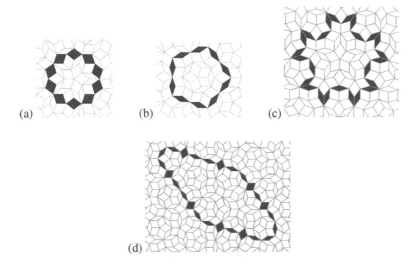

Fig. 18.37 Some large rhomb still life chains: (**a**) r-p1-10, all thick rhombs; (**b**) r-p1-10, all thin rhombs; (**c**) r-p1-25, all thin rhombs; (**d**) r-p1-25, thick and thin rhombs. Note that our later classification in terms of oscillator graphs identifies the chains in (**c**) and (**d**) as isomorphic still lifes

Fig. 18.38 Some "ribbons" of thick rhombs

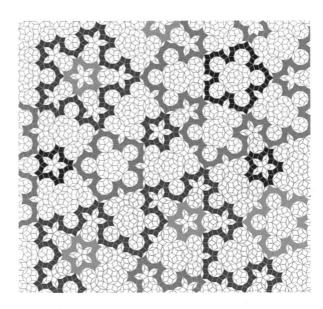

Each thick rhomb ribbon is edged on the inside and outside by a thin rhomb ribbon; choose one of these. Delete any S-chains (Fig. 18.21bi). What remains is a thin rhomb still life chain. Figures 18.39 and 18.40 show two examples of such constructions, plus some variant forms.

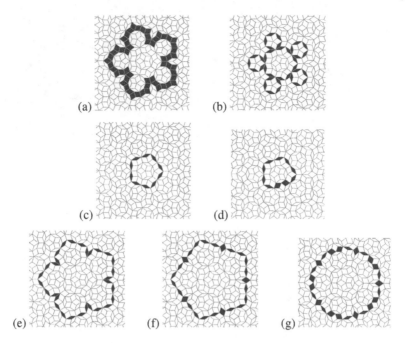

Fig. 18.39 Constructing still life chains: (**a**) a thick rhomb ribbon; (**b**) the internal thin rhomb ribbon; (**c**) the internal ribbon with the *S*-chains removed: r-p1-10, a still life chain; (**d**) a variant r-p1-10, constructed by "folding in" a pair of thin rhombs to become thick rhombs; (**e**) the external thin rhomb ribbon: r-p1-30, a still life chain; (**f**) a modification r-p1-25, constructed by replacing folding the thin rhomb "elbow" pair into a single thick rhomb, five times (this has similarities to the relationship between the regular Life *pond*, Fig. 18.29, and *loaf*, Fig. 18.26, with two squares folded into one); (**g**) a variant r-p1-25, constructed by "folding in" five pairs of thin rhombs to pairs of thick rhombs. A modification of this "folding" construction can be applied to convert chains into *p*2 oscillators (see Fig. 18.61)

18.3.4 Large Snakes

Long linear kite and dart still lifes can be formed; some examples are shown in Fig. 18.41. By combining the rhomb chain construction process with the linear still life termination on a K vertex (Fig. 18.5), long linear rhomb still lifes can also be constructed, see Fig. 18.42.

There is potential for using long snakes when making larger "machines". Disrupting a snake at some site, such as one end, causes it to "disintegrate" from that site, propagating the disruption along the snake, somewhat like a (messy) "fuse". Or if some activity were to break a long chain, leaving a terminator on one end, then a circular fuse would burn, with activity returning to the original point some time later. By choosing the length of the chain, this could provide a form of timer.

Fig. 18.40 Constructing still life chains. *Top row*: (**a**) a thick rhomb ribbon; (**b**) the external thin rhomb ribbon; (**c**) the external ribbon with the *S*-chains removed: r-p1-60, a still life chain; (**d**) r-p1-50, a modification of (**c**); (**e**) the internal thin rhomb ribbon: r-p1-60, a still life chain

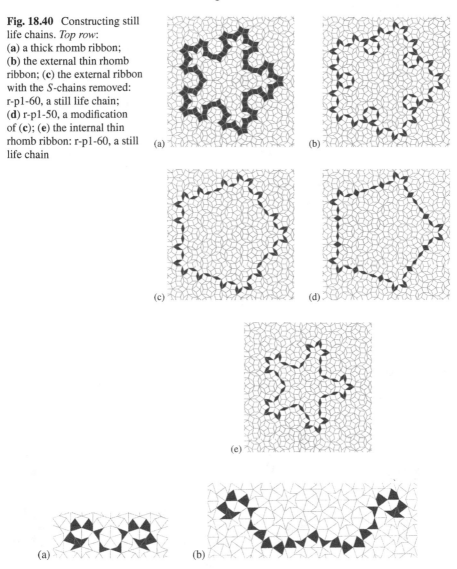

Fig. 18.41 Kite and dart still life snakes: (**a**) kd-p1-14: a *14-snake*; (**b**) kd-p1-25: a *25-snake*

18.4 Period 2 Oscillators

18.4.1 Blinkers and Plinkers

A regular Life "blinker" is a period 2 oscillator comprising a line of three live cells. If in generation 0 it is a horizontal line, then in generation 1 it is a vertical line sharing the same central cell (Fig. 18.43a).

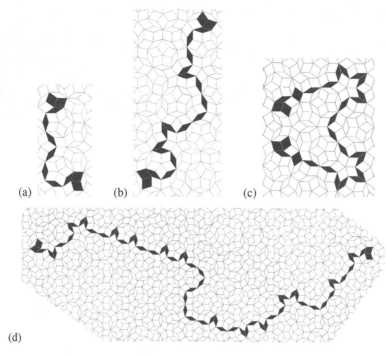

Fig. 18.42 Rhomb still life snakes, constructed from a combination of thin rhomb chains and K vertex terminators: (**a**) r-p1-11: an *11-snake*; (**b**) r-p1-20: a *20-snake*; (**c**) r-p1-28: a *28-snake*; (**d**) r-p1-61: a *61-snake*

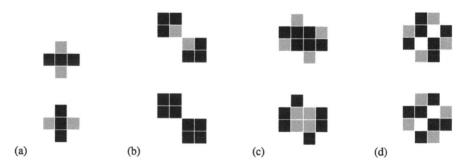

Fig. 18.43 The smallest regular life *p2* oscillators: (**a**) the *blinker*, 1-p2-5-3; (**b**) the *beacon*, 1-p2-8-6; (**c**) the *toad*, 1-p2-10-6; (**d**) the *clock*, 1-p2-10-6

There are isomorphic three cell, *p2 plinkers*[3] in Penrose life. A plinker can exist at any cell in the tiling.[4]

---

[3] These particular oscillators were dubbed "plinkers" in [7], and we continue that usage here, as an exception to our naming convention.

[4] *Proof* Consider any cell, which will be the 'central' cell of the plinker. Consider a pair of opposite (non-adjacent) edges of this central cell. Consider the two cells adjacent to this pair of edges. Make

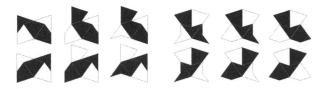

Fig. 18.44 The six distinct kite and dart plinkers: kd-p2-5-3

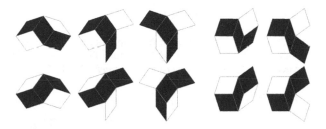

Fig. 18.45 The five distinct rhomb plinkers: r-p2-5-3

The set of generalised von Neumann neighbourhoods (Figs. 18.8 and 18.9) can be used to enumerate the complete set of distinct plinkers. There are six plinkers for kites and darts (Fig. 18.44, as noted in [7]), and five for rhombs (Fig. 18.45). Each plinker has 3 cells alive at any time, and a total of 5 distinct cells used.

18.4.2 Other p2 Oscillators

All six varieties of the kite and dart *plinker* (Fig. 18.44) were discovered while exploring the behaviour of the rules in [7]. In our subsequent searches new larger $p2$ kite and dart oscillators (Figs. 18.46–18.49) have been discovered.

18.5 Period 3 Oscillators

Four essentially different $p3$ kite and dart oscillators have been discovered: the *breather* with four isomorphic forms (Fig. 18.63), the *wagger* (Fig. 18.64), a disconnected oscillator with two isomorphic forms (Fig. 18.65), and a more irregular oscillator (Fig. 18.66).

The rhomb tiling also exhibits its own zoo of oscillators. In addition to the rhomb plinkers (Fig. 18.45), there are further $p2$ rhomb oscillators (Figs. 18.50–18.60). Large $p2$ rhomb oscillators can be constructed from rhomb chain still lifes (Figs. 18.61–18.62).

---

these two cells and the central cell alive, and all other cells dead. The result is a plinker, oscillating between the chosen pair of cells, and the pair of cells adjacent to the other two edges of the central cell.

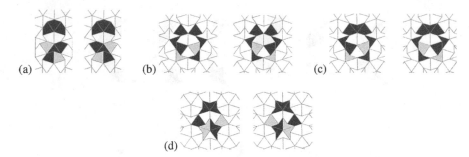

Fig. 18.46 Further $p2$ kite and dart oscillators: (**a**) kd-p2-8-6, a disconnected oscillator, constructed from the still life of Fig. 18.25ai by converting the lower 3-block to a plinker; (**b**) kd-p2-8-6, the *marcher*; (**c**) kd-p2-9-7, the *crowned marcher*; (**d**) kd-p2-9-6

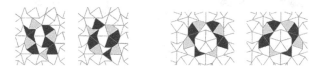

Fig. 18.47 kd-p2-10-6: the *hollow clock*, two isomorphic $p2$ kite and dart oscillators related to, but not the same as, the regular Life $p2$ oscillator the *clock* (Fig. 18.43d)

Fig. 18.48 kd-p2-12-9: the *fast shuffler*, which exists in a symmetric and asymmetric variant form, from the different extensions of the "Complete Star" S_C (Fig. 18.15)

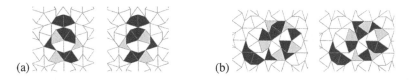

Fig. 18.49 Further $p2$ kite and dart oscillators: (**a**) kd-p2-10-7; (**b**) kd-p2-15-10

Fig. 18.50 $p2$ rhomb oscillators: (**a**) r-p2-8-6, a disconnected oscillator, constructed from the still life of Fig. 18.25bi by converting the lower 3-block to a plinker (the same construction can be applied to the still life of Fig. 18.28bi, but the resulting 4-block and plinker are each individually stable, so the result does not count as a disconnected oscillator); (**b**) r-p2-8-6, the *marcher*; (**c**) r-p2-9-7, the *crowned marcher*

Fig. 18.51 r-p2-10-6, two
isomorphic forms of the
hollow clock

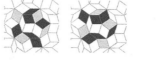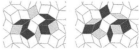

Fig. 18.52 r-p2-10-8

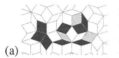

(a)   (b)

Fig. 18.53 (a) r p2 11 8; (b) r p2 11 9

Fig. 18.54 r-p2-12-9: the *big
beacon*, named by analogy to
the regular Life *p2 beacon*
(Fig. 18.43b), but here with
three components instead of
two

Fig. 18.55 r-p2-12-10:
(a) a ring oscillator; (b) two
isomorphic oscillators. The
way that (a) is shown to be
different from (b), and that
the two forms of (b) are
found to be isomorphic, is
described later

(a)

(b)

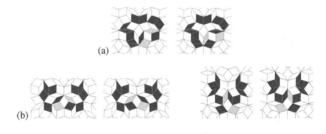

Fig. 18.56 r-p2-13-11:
(a) note that this is the
r-p2-10-8 with three extra
cells added; (b) note the
relationship to r-p1-11 of
Fig. 18.31d

(a)   (b)

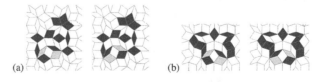

Fig. 18.57 r-p2-14-12

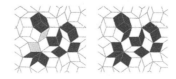

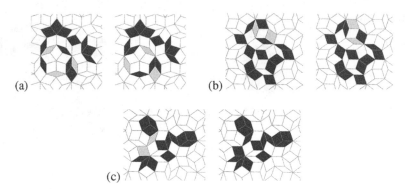

Fig. 18.58 (**a**) r-p2-15-12; (**b**) r-p2-15-13; (**c**) r-p2-15-13

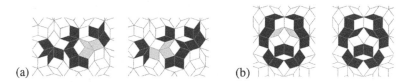

Fig. 18.59 (**a**) r-p2-16-13; (**b**) r-p2-16-14

Fig. 18.60 r-p2-17-14

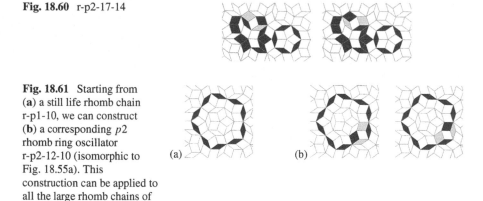

Fig. 18.61 Starting from
(**a**) a still life rhomb chain
r-p1-10, we can construct
(**b**) a corresponding p2
rhomb ring oscillator
r-p2-12-10 (isomorphic to
Fig. 18.55a). This
construction can be applied to
all the large rhomb chains of
Fig. 18.39

Fewer period three rhomb oscillators have been found. There is a rhomb oscillator isomorphic to the kite and dart p3 *breather* (Fig. 18.67), and a further p3 oscillator (Fig. 18.68).

18.6 Period 4 Oscillators

Several kite and dart p4 oscillators are shown in Figs. 18.69–18.72.

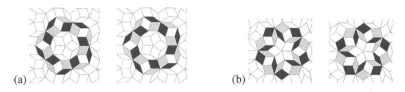

Fig. 18.62 The construction of Fig. 18.61 can be applied to more than one site in a chain: (**a**) a maximal *p*2 construction applied to Fig. 18.61a; (**b**) a maximal construction applied to Fig. 18.37a. These are both r-p2-20-10; many variants exist

Fig. 18.63 kd-p3-14-8: (**a**) the kite and dart *breather*; (**b**) an isomorphic variant breather; (**c**) a further isomorphic variant breather; (**d**) a fourth isomorphic variant breather

(a)

(b)

(c)

(d)

Fig. 18.64 kd-p3-15-10: the *wagger*

Fig. 18.65 kd-p3-16-13: (**a**) a disconnected oscillator; (**b**) a variant form

(a)

(b)

Fig. 18.66 kd-p3-17-8

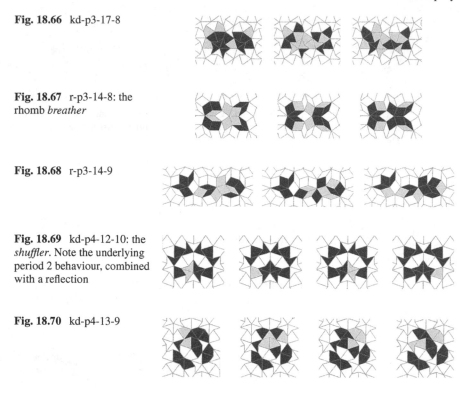

Fig. 18.67 r-p3-14-8: the
rhomb *breather*

Fig. 18.68 r-p3-14-9

Fig. 18.69 kd-p4-12-10: the
shuffler. Note the underlying
period 2 behaviour, combined
with a reflection

Fig. 18.70 kd-p4-13-9

The 6 cell kite and dart *p4 bat* (Fig. 18.71a) was discovered in [7]. Here
we also see variants: a *p4 asymmetric bat*, a variant form located on a different Complete Star extension (Fig. 18.71b), and a further isomorphic *bat*, not located on a Complete Star vertex extension (Fig. 18.71c). There is also a *bat-to-
bat* oscillator (Fig. 18.71d), which is two bat oscillators that actually touch at one
point.

Period 4 rhomb oscillators are shown in Figs. 18.73–18.77.

18.7 Higher Period Oscillators

18.7.1 Kite and Dart High Period Oscillators

One of the interesting discoveries in [7] was the existence of spatially small, but
relatively long period, oscillators in the kite and dart tiling, relative to the regular
lattice. An *p8* oscillator (Fig. 18.81), and the *p15 dancer* (Fig. 18.84), were discovered while exploring the behaviour of the rules.

In subsequent searches new kite and dart long period oscillators have been
discovered: the *p5 ninja* (Fig. 18.78) and *drummer* (Fig. 18.79), the *p6 tick-*

Fig. 18.71 kd-p4-14-6: (**a**) the *bat*, from [7]; (**b**) the variant *asymmetric bat*, from a different extension of the "Complete Star" S_C (Fig. 18.15); (**c**) an isomorphic oscillator, not located on a "Complete Star" S_C; (**d**) the *bat-to-bat*, where the left and right halves are each an asymmetric bat oscillator, but the bats touch at step two, kd-p4-28-12

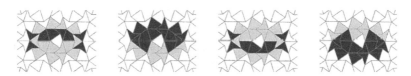

Fig. 18.72 kd-p4-22-8: the *hedgehog*. Note the underlying period 2 behaviour, combined with a reflection

Fig. 18.73 r-p4-12-9: the *clown*

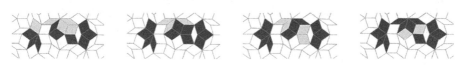

Fig. 18.74 r-p4-14-9

Fig. 18.75 r-p4-15-8: the *goldfish*. Note the underlying period 2 behaviour, combined with a reflection

Fig. 18.76 r-p4-15-10. Note the underlying period 2 behaviour, combined with a reflection

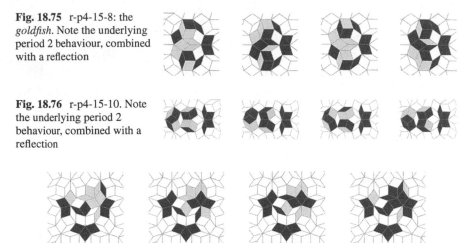

Fig. 18.77 r-p4-17-10: the *pirate*

Fig. 18.78 kd-p5-15-8: the *ninja*. Note the constant pattern undergoing a $4\pi/5$ rotation per timestep

ler (Fig. 18.80), the *p9 moustaches* (Fig. 18.82), and the *p11 malformed bat* (Fig. 18.83). Many of these exist in multiple isomorphic forms; we no longer display variant forms, for reasons of space, but show only the most "symmetric" variant found.

Note that when a period is composite, the oscillator may exhibit subperiods (for example, the *p4 shuffler* has a *p2* behaviour that is then reflected; the *p15 dancer* has a *p3* behaviour that undergoes a five-fold rotation), or not (the *p6 tickler* and the *p9 moustaches* have no obvious sub-periodic behaviours).

18.7.2 Rhomb High Period Oscillators

Higher period rhomb oscillators are shown in Figs. 18.85–18.96. One of these, the *ninja* (Fig. 18.86) is isomorphic to the kite and dart ninja (Fig. 18.78).

These high period complex rhomb oscillators help demonstrate that the rhomb-based CA, whilst having statistically significantly different statistical behaviour from the kite and dart-based CA under Game of Life rules [11], also exhibits complex and interesting behaviour.

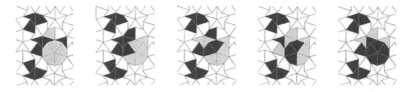

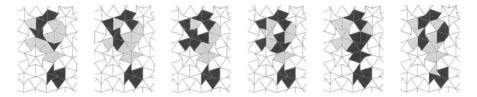

Fig. 18.79   kd-p5-16-10, a disconnected oscillator: the *drummer*

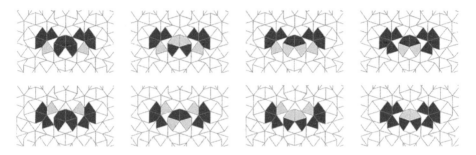

Fig. 18.80   kd-p6-20-10: the *tickler*

Fig. 18.81   kd-p8-12-8, from [7]

Fig. 18.82   kd-p9-36-10: the *moustaches*

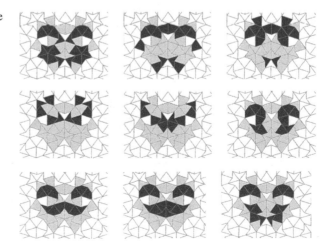

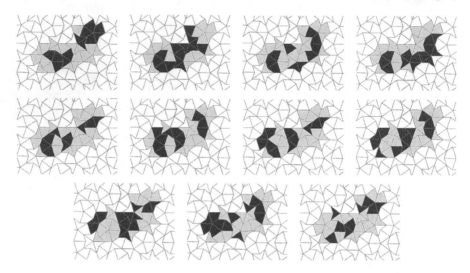

Fig. 18.83 kd-p11-31-9: the *malformed bat*. It has similarities to the *p4 bat* [7]

Fig. 18.84 kd-p15-40-8: the
dancer, from [7]. Note the
underlying period 3
behaviour, combined with a
5-fold rotation. The dancer
exists in three variant forms
(not shown), from the three
different extensions of the
"Complete Star" S_C
(Fig. 18.15)

Fig. 18.85 r-p5-14-7,
a disconnected *p*5 rhomb
oscillator: the *juggler*

Fig. 18.86 r-p5-15-8: the *ninja*

Fig. 18.87 r-p5-16-8: the *jumper*

Fig. 18.88 r-p5-18-11, another disconnected *p*5 rhomb oscillator

Fig. 18.89 r-p5-18-14, a further disconnected *p*5 rhomb oscillator

Fig. 18.90 r-p6-13-9

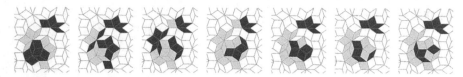

Fig. 18.91 r-p7-19-7: the *hattipper*

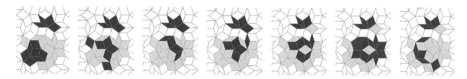

Fig. 18.92 r-p7-23-8, related to the *hattipper*

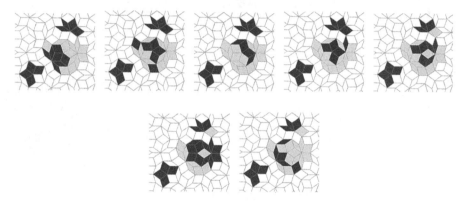

Fig. 18.93 A disconnected relation of the *hattipper*

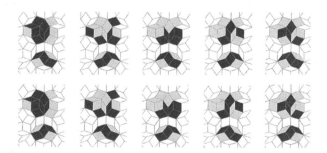

Fig. 18.94 A disconnected $p10$ rhomb oscillator, the *long juggler*. Note the combination of a period 5 behaviour (upper cells of the oscillator, identical to the upper portion of the $p5$ juggler) and a period 2 plinker (lower cells). Compare Figs. 18.85 (which shows the underlying $p5$ *juggler* and 18.50a (which shows a similar construction of a $p2n$ oscillator from an underlying pn oscillator)

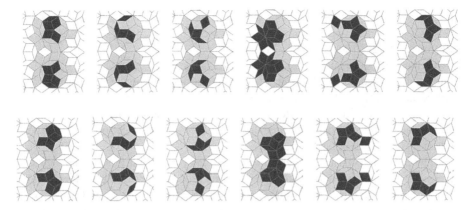

Fig. 18.95   r-p12-36-8: the *reflected bouncer*

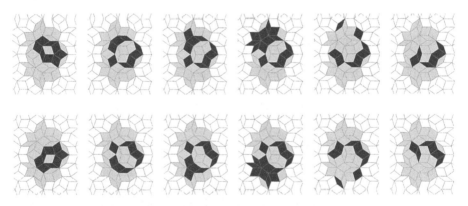

Fig. 18.96   r-p12-33-6: *fireworks*. Note the underlying period 6 behaviour, combined with a reflection

18.8   Oscillator Analysis

Although some of the oscillators presented earlier were constructed manually, many (particularly the more irregular and long period oscillators) were discovered by random search, by examining the ash left from a set of CA runs starting from random initial conditions. The majority of such runs end with mostly small, low period oscillators (blocks, small rings, and plinkers, and, in the kite and dart case, the occasional bat or dancer; note that random search will be biassed against finding oscillators with few or no states leading to them). In order to analyse the large number of runs needed to find the rarer oscillators, it is necessary to have algorithms to detect the oscillators, and to identify and classify them. The irregular Penrose tiling makes this more difficult than for regular Life.

For identification, we have the code name, but this does not uniquely identify oscillators (although collisions are rare). We would like to classify to oscillators to identify those that are isomorphic (for example, the plinkers in Figs. 18.44

and 18.45, or the bats in Fig. 18.71), and those that are truly different (for example, the p1-4 tubs in Fig. 18.17 versus the p1-4 blocks in Figs. 18.18 and 18.19, or the different r-p2-12-10 oscillators in Fig. 18.55).

We have the general requirement that the algorithms should be efficient. The computational overheads of our lazy tiling algorithm to define the Penrose grid (described in [10]), and of running the CA rules, are reasonable; the more efficiently we can detect oscillators the better we can explore oscillator space.

Oscillator analysis requires three steps: (i) quiescence detection; (ii) oscillator detection; (iii) oscillator classification.

Quiescence detection is straightforward: we detect when the behaviour of the CA becomes periodic (this would not work if ever a propagating structure like a regular Life glider were to form, but this has never been the case so far). Periodic behaviour is detected by a comparison between each new state of the CA to every old state.[5] Detection and classification algorithms are described in the following sections.

18.8.1 Oscillator Detection

Given a large patch of tiling containing many potential oscillators, we need a way to identify each individual oscillator, separate from the rest.

For the purposes of detection, define an oscillator O with period p as a set of n pairs, each pair being a cell and its sequence of p states:

$$O = \{(c_0, S_p(c_0)), (c_1, S_p(c_1)), \ldots, (c_n, S_p(c_n))\} \tag{18.2}$$

where $S_p(c) = \sigma_0(c), \sigma_1(c), \ldots, \sigma_{p-1}(c)$ and $\sigma_t(c)$ is the state of cell c at time t; $\sigma_t(c) = \blacksquare$ if the cell is alive at time t and $\sigma_t(c) = \square$ if the cell is dead at time t. We say $c \in O$ to mean cell c is one of the cells in the pairs contained in O. For all cells $c \in O$, $\sigma_0(c) = \sigma_p(c)$.

For example, consider a particular plinker O_π (one of those in Fig. 18.44 or 18.45). It has five cells, the central one on all the time, and two pairs on and off in antiphase. Labelling the central cell c_0, one pair of cells c_1 and c_2, and the other pair c_3 and c_4, we have

$$O_\pi = \{(c_0, \blacksquare\blacksquare), (c_1, \blacksquare\square), (c_2, \blacksquare\square), (c_3, \square\blacksquare), (c_4, \square\blacksquare)\} \tag{18.3}$$

Define $N(c)$, the oscillator neighbourhood of cell c, to be the set of cells in the oscillator, $n \in O$, that are in the generalised Moore neighbourhood of c. For the

---

[5]We have investigated other methods of period detection. Application of Floyd's algorithm [8, p. 7] uses two versions of the CA: the first is updated normally, the second is updated twice for every update of the first; when they become equal the automaton has cycled. This method offers $O(1)$ space, and $O(n)$ time, where $n = t_q + p$ is the time to quiescence plus the period of the oscillation. Our method requires $O(n)$ space and $O(n^2)$ time (requiring a triangle number of comparisons, so $(n^2 + n)/2$). However in practice our method is much faster as the size of the tiling greatly dominates the lifetimes involved here. The costs of running a second CA, even implemented just as a second state in each cell, is quite a performance hindrance.

plinker O_π, we have $N(c_0) = \{c_0, c_1, c_2, c_3, c_4\}$ and $N(c_1) = \{c_0, c_1, c_3, c_4\}$. Note that the neighbourhood relationship is symmetric: $c_1 \in N(c_2) \Leftrightarrow c_2 \in N(c_1)$.

Define the live and dead neighbourhoods, $N_t^\blacksquare(c)$ and $N_t^\square(c)$, by:

$$N_t^s(c) = \{n \in N(c) \mid \sigma_t(n) = s\} \tag{18.4}$$

where $s \in \{\square, \blacksquare\}$. So $N_t^\square(c)$ is the set of the oscillator's dead cells in the neighbourhood of c at time t, and $N_t^\blacksquare(c)$ is the set of the oscillator's live cells. Note that $N(c) = N_t^\square(c) \cup N_t^\blacksquare(c)$, for any t.

For the plinker O_π, the live and dead neighbourhoods are:

$$N_0^\square(c_0) = \{c_3, c_4\}; \qquad N_0^\blacksquare(c_0) = \{c_0, c_1, c_2\} \tag{18.5}$$

$$N_1^\square(c_0) = \{c_1, c_2\}; \qquad N_1^\blacksquare(c_0) = \{c_0, c_3, c_4\} \tag{18.6}$$

$$N(c_0) = N_0^\square(c_0) \cup N_0^\blacksquare(c_0) = N_1^\square(c_0) \cup N_1^\blacksquare(c_0) = \{c_0, c_1, c_2, c_3, c_4\} \tag{18.7}$$

We take a minimalist view of an oscillator: a cell is included in an oscillator if its removal would destroy the oscillator. So when considering the cells of an oscillator alone, they may have incomplete neighbourhoods, but sufficient neighbourhoods to allow correct oscillation. When the oscillator is considered as part of a larger tiling there may be cells bordering the oscillator that must always be dead for the oscillator to exist; these always dead bordering cells are not considered part of the oscillator, since, as will become clear, they are not actively influencing the oscillator. This approach to oscillator detection provides a clean platform to perform the subsequent oscillator classification.

The oscillator detection is performed on the automaton ash in two stages: (A) an assignment of cells to potential oscillators; (B) a removal of non-necessary cells from the potential oscillators.

Detection Stage A

For each CA ash state, a breadth-first search starting from each live cell is performed, to construct the set of all oscillators, Ω, as described in the algorithm in Fig. 18.97.

Examination of the two conditions under which dead cells are added to an oscillator (the "or" condition of step 9 of Fig. 18.97) reveals why a second removal stage (B) is necessary in oscillator detection to remove certain dead cells.

If the dead cell n has exactly three live neighbours, $|N_t^\blacksquare(n)| = 3$, then n will become alive on the next automaton iteration, and so is added to O. It is added immediately, to guard against the following situation: all three neighbours that caused n's birth die in the same iteration, no more neighbours of n are born, and all now dead neighbours of n fail to satisfy the conditions under which dead cells are added to an oscillator. So n would not be associated with the correct oscillator if it were not for the $|N_t^\blacksquare(n)| = 3$ condition. This situation arises in the $p12$ fireworks (Fig. 18.96).

If the dead cell n has more than three live neighbours, $|N_t^\blacksquare(n)| > 3$, then the cells of the oscillator are exerting an influence on the dead cell, forcing it to stay

```
 1:  for t = 0 to p − 1 {each automaton state in the ash} do
 2:      while there is a live cell c that has not been processed yet do
 3:          if c ∉ Ω {c is not contained in any oscillator} then
 4:              {create a new oscillator O containing just c and its current state}
                 O := {(c, σ_t(c))}; Ω := Ω ∪ {O}
 5:          else {c is in an oscillator, with (c, S) ∈ O}
 6:              S := S + σ_t(c) {update c's state list with c's current state}
 7:          end if
 8:          for each n ∈ N(c) {each of cell c's neighbourhood cells} do
 9:              if σ_t(n) = ■ {n is alive}
                 or σ_t(n) = □ and |N_t^■(n)| ≥ 3
                     {n is dead and has three or more live cells in its neighbourhood} then
10:                  O := O ∪ {(n, σ_0(n), . . . , σ_t(n))} {add n to O}
11:              end if
12:          end for
13:          if any of the cells n added to O are already a member of another oscillator R then
14:              O := O ∪ R; Ω := Ω − {R} {combine O and R}
15:          end if
16:          continue recursively processing all neighbourhood cells n added to O
17:      end while
18:  end for
```

Fig. 18.97 Detection stage A algorithm

dead (from 'overcrowding'). Adding it to O allows detection of oscillators with disjoint sections that interact via a forced dead cell. However, adding it may also add cells that are forced dead on the interior of the oscillator, or on the boundary but which may also connect coincidentally to non-interacting static or periodic structures. It is these cells that must be removed, potentially resulting in a splitting of the candidate oscillator. But exactly which cells must be removed is not known until the completion of stage A, and so a second removal stage B is needed.

Removal Stage B

The objective of the second stage is to remove cells from the oscillator that are not required by the oscillator. Given an oscillator, we test any two sub-structures connected by only a dead cell (for example, cell e in Fig. 18.99). We test for the existence of a time step in which one of the structures would cause this connecting dead cell to become live, were it not for presence of the other structure. If there is such a time step, then the dead cell is necessary to the oscillator; if there is no such time step, the dead cell should be removed, and the oscillator split in two.

We perform this removal stage by labelling oscillator cells and considering oscillator neighbourhood graphs.

Define $G(C)$ to be the (possibly disconnected) graph corresponding to a set of cells C, which has a node for every $c \in C$ and an undirected edge between every $c_n, c_m \in C$ if $c_n \in N(c_m)$. For an oscillator O we take $G(O)$ to be the oscillator graph of all the cells in O. For example, the plinker O_π has the oscillator graph shown in Fig. 18.98.

Define a γ-cell to be a cell c in an oscillator O that remains dead for every timestep i of the oscillation: $\forall t \bullet \sigma_t(c) = \square$.

Fig. 18.98 Oscillator graph
$G(O_\pi)$

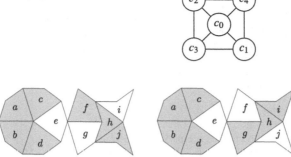

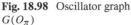

Fig. 18.99 A potential oscillator P found by detection stage A. It contains a $p2$ plinker with a close stable static structure, and the "on" cells are shown for its two timesteps. Cell e is always off and is the only γ-cell

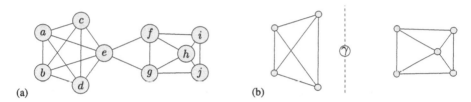

Fig. 18.100 (a) Oscillator graph $G(P)$ marked with cell labels from Fig. 18.99; (b) oscillator graph $G(\hat{\gamma}(P))$; the γ-cell is not in $G(\hat{\gamma}(P))$ but is drawn for clarity; the *dotted line* delineates the two sub-graphs. There are two disconnected sub-graphs, so $|G_\kappa(\hat{\gamma}(P))| = 2$

Define $\gamma(O)$ to be the set of all γ-cells in O: $\gamma(O) = \{c \in O \mid \forall t \bullet \sigma_t(c) = \square\}$. Define the complement, $\hat{\gamma}(O)$, to be the set of all cells in O that are not γ-cells (the set of all cells that are alive for at least one timestep): $\hat{\gamma}(O) = \{c \in O - \gamma(O)\}$. Note that $\gamma(O)$ and $\hat{\gamma}(O)$ partition the cells in O.

So $G(\hat{\gamma}(O))$ is the (potentially disconnected) graph of all cells in O that are not γ-cells (the graph of all cells that are alive for at least one timestep). $G(\hat{\gamma}(O))$ will be a disconnected graph if there are sub-graphs in the oscillator that are disconnected but for a γ-cell.

Define $G_\kappa(\hat{\gamma}(O))$ to be the set of the disconnected sub-graphs in $G(\hat{\gamma}(O))$. So the size of the set $G_\kappa(\hat{\gamma}(O))$ is the number of sub-graphs in $G(\hat{\gamma}(O))$. There are no neighbouring nodes in different sub-graphs: if $g_1, g_2 \in G_\kappa(\hat{\gamma}(O))$ then $g_1 \neq g_2 \Leftrightarrow \forall c_1 \in g_1, c_2 \in g_2 \bullet c_2 \notin N(c_1)$.

Define an internal γ-cell to be a $c_\gamma \in \gamma(O)$ that is connected to only one sub-graph in $G_\kappa(\hat{\gamma}(O))$, and so does not connect otherwise disconnected sub-graphs in $G(\hat{\gamma}(O))$.

For example, consider the potential oscillator P (Fig. 18.99) after detection stage A; Fig. 18.100 shows the oscillator graph $G(P)$. P contains a $p2$ plinker next to a $p1$ still life, positioned such that they share a common neighbour, a γ-cell. The plinker does not require the still life for its $p2$ oscillation, and the still life does

```
 1:  O := O − {internal γ-cells}
 2:  {any γ-cells still in O connect sub-graphs}
 3:  for each γ-cell cγ ∈ γ(O) {check for its survival requirement} do
 4:      for each sub-graph g ∈ Gκ(γ̂(O)) do
 5:          for each timestep t do
 6:              if |Nₜ■(cγ) ∩ {c ∈ g}| = 3 {cγ has three live neighbours at t in g} then
 7:                  mark cγ for survival
 8:              end if
 9:          end for
10:      end for
11:  end for
12:  O := O − {all γ-cells not marked for survival}
13:  {O now contains only those γ-cells it needs}
14:  for each disconnected sub-graph g ∈ Gκ(O)
15:      construct a new oscillator from g
16:  end for
```

Fig. 18.101 Removal stage B algorithm, for potential oscillator O

not require the plinker for its $p1$ oscillation. From the oscillator graph in Fig. 18.100 it is clear that removal of the γ-cell would disconnect these two independent structures. This is the aim of stage B: to remove unwanted γ-cells whilst leaving the needed ones.

There are several examples earlier of oscillators that need γ-cell survival, including disconnected still lifes (Figs. 18.25 and 18.28), the juggler (Fig. 18.85), and the long juggler (Fig. 18.94). In all of these cases, the γ-cell has exactly three live neighbours in at least one of the subgraphs for at least one timestep. It is stopped from coming alive in the following timestep by live neighbours in another subgraph in this timestep. If this other component were not present (if the oscillator were split into parts), then this cell would become live, and the assumed structure would not be preserved. Hence each of these oscillators is a true disconnected oscillator, and not two separate oscillators in close proximity.

Removal stage B works as follows. Internal γ-cells are removed.[6] Each remaining γ-cell connects two otherwise disconnected structures, and is checked for a timestep in which one of the disconnected structures contains exactly three live neighbours of the γ-cell. If there is such a timestep, then the γ-cell survives, otherwise it is removed. (Since by definition the γ-cell is always dead, then if the γ-cell is connected to a sub-graph containing three live neighbours, then there must be some other sub-graph disconnected from the first with at least one live neighbour of the γ-cell that prevents the γ-cell from coming alive. Hence that γ-cell in needed.) The algorithm for removing unnecessary γ-cells and constructing the true oscillators from a potential oscillator O is given in Fig. 18.101.

---

[6]Note that internal γ-cells can also be necessary for the survival of an oscillator, in that there may be three live cells from one part of the oscillator that would cause the γ-cell to come alive, were it not for more live cells from another, but still connected, part. The "holes" in the oscillator might need to be "narrow". However, we ignore this point here because we are using the oscillator graph simply to classify oscillators, not to discover suitable patches of tilings to support them.

Fig. 18.102 The live
neighbours of the single
γ-cell $e \in \gamma(P)$ at $t = 0$
(**a**) and $t = 1$ (**b**). *Black nodes*
are live cells, *grey nodes* are
dead cells. The edges between
nodes that are not connected
to the c_γ are not shown.
There is no timestep at which
one of the two sub-graphs has
three live neighbours of the
γ-cell e, and so it does not
survive. This removal leaves
two disconnected sub-graphs,
and hence two resulting
oscillators

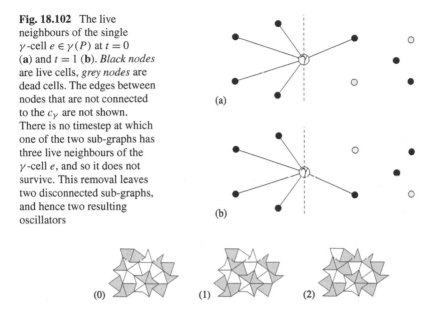

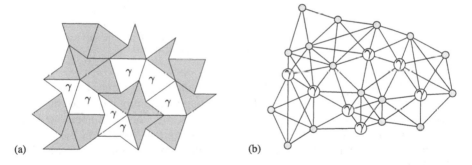

Fig. 18.103 The example kd-p3-16-13 oscillator (see also Fig. 18.65a), at $t = 0, 1, 2$. Every cell depicted is included in the oscillator after detection stage A

Fig. 18.104 The example kd-p3-16-13 oscillator: (**a**) the marked γ-cells; (**b**) graph with marked γ-cells

The γ-cell in potential oscillator P does not meet these survival conditions: see Fig. 18.102.

An oscillator that does require γ-cells is kd-p3-16-13 (Fig. 18.103). After detection stage A it has seven γ-cells (Fig. 18.104). Three of these are internal γ-cells, and so are removed (Fig. 18.105a). The remaining four γ-cells connect otherwise disconnected sub-graphs (Fig. 18.105b), and must be checked for survival. The checking of the survival conditions is shown in Fig. 18.106: all four γ-cells meet the survival condition.

The fact that all four γ-cells survive, each connecting the same two sub-graphs, presents interesting issues. For example, one might wish to eliminate three of the

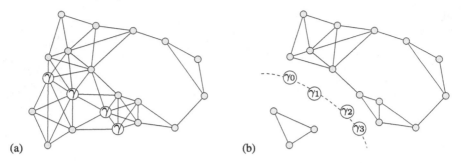

(a)   (b)

Fig. 18.105 The example kd-p3-16-13 oscillator: (**a**) the graph with internal γ-cells removed; (**b**) the disconnected sub-graphs: there are four γ-cells that require checking

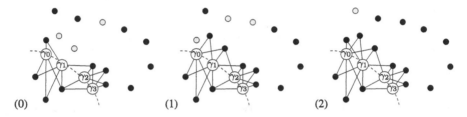

(0)   (1)   (2)

Fig. 18.106 The example kd-p3-16-13 oscillator, depicting γ-cell survival conditions at $t = 0, 1, 2$. Live nodes are *black*, dead nodes are *grey*, edges between non-gamma cell nodes are not shown. For all time steps the survival condition is fulfilled for γ_0 and γ_1, each with three live neighbours in the lower sub-graph. For all timesteps the survival condition is fulfilled for γ_2 and γ_3, each with three live neighbours in the upper sub-graph. Note that at $t = 2$ the survival conditions are satisfied for γ_0 with three live neighbours in the sub-graph also

γ-cells from the description, as only one cell is required to connect the two sub-graphs. Further, one could question whether a tiling variant might produce the same oscillator sub-graphs with a different number of γ-cells. For the moment we leave the full set of retained γ-cells in the oscillator description. We discuss these issues further in the subsection "Oscillators with γ-Cells: Macroscopic Isomorphism" of Sect. 18.8.2.

18.8.2 Oscillator Classification

Oscillator Graph Isomorphism

We use the oscillator graph $G(O)$ introduced in the subsection "Removal Stage B" of Sect. 18.8.1 in addition to the identification code defined in the subsection "An Identification Code" of Sect. 18.2.3 as the basis for a classification to group the zoo of oscillators into isomorphism classes: those with identical neighbourhood graphs *and* identical codes.

Define two neighbourhood graphs to be identical if they are isomorphic. If $C(O)$ denotes the cells of oscillator O, then two oscillators O_a, O_b are isomorphic if there exists a mapping $m : C(O_a) \to C(O_b)$ such that any two cells $c_1, c_2 \in O_a$ are neighbours, $c_1 \in N(c_2)$, in $G(O_a)$ if and only if $m(c_1)$ and $m(c_2)$ are neighbours, $m(c_1) \in N(m(c_2))$, in $G(O_b)$. If two oscillator graphs are isomorphic, we write $G(O_a) \sim G(O_b)$.

An oscillator graph isomorphism checking algorithm can use the underlying structure of the neighbourhood graphs, and the limited ways that they can be extended, to make the checks efficient.

The oscillator graph defines the topology of cells and neighbourhoods involved in the oscillator (including any γ-cells), but does not define which cells are live at each timestep of the oscillator. Note that the oscillator graph alone is not sufficient to uniquely identify an oscillator, although exceptions are rare. Two different oscillators may share a graph; in particular, the $p2$ *fast shuffler* (Fig. 18.48) and the $p4$ *shuffler* (Fig. 18.69) share a graph, Fig. 18.118c.

We have not (at least thus far) found a case where two clearly different oscillators share the same graph *and* the same code, and so we use this combination to identify isomorphic oscillators. If such cases are subsequently discovered, a further disambiguation marking would be needed.

Oscillators with γ-Cells: Macroscopic Isomorphism

As noted at the end of "Removal Stage B" in Sect. 18.8.1, there is a potential issue with the classification of disconnected still lifes. Two oscillators may have isomorphic sub-graphs in $G_\kappa(\hat{\gamma}(O))$ that are connected by differing numbers of γ-cells: should these be considered the same, of different, oscillators?

We define an extra layer of classification, which allows us to say that such oscillators are *macroscopically* isomorphic, but may be *microscopically* distinct. A macroscopic or microscopic classification may be appropriate in different contexts.

To define macroscopic isomorphism we first define the macroscopic oscillator graph of a disconnected oscillator O. Consider the disconnected subgraphs of γ-cells in O, given by $G_\kappa(\gamma(O))$. Each $g_\gamma \in G_\kappa(\gamma(O))$ is a connected graph representing γ-cells in O. Define $M(O)$, the *macroscopic graph* of O, to have a node for every $g_i \in G_\kappa(\hat{\gamma}(O))$ and an edge for every $g_\gamma \in G_\kappa(\gamma(O))$ connecting the relevant g_i (connecting the sub-graphs that the corresponding *gamma*-cells connect). Note that $M(O)$ discards the specifics of the γ-cells, and just asserts that there are γ-cells which connect two otherwise disconnected sub-graphs containing live nodes. Figure 18.107 shows the construction of the macroscopic graph for an oscillator with multiple γ-subgraphs in $G_\kappa(\gamma(O))$.

Define two oscillators O_a and O_b to be *macroscopically isomorphic* if their macroscopic graphs are isomorphic, $M(O_a) \sim M(O_b)$, and each pair of sub-graphs represented by their corresponding nodes are isomorphic.

Classifying the Still Lifes

Figures 18.108–18.115 show the oscillator graphs for various still lifes. (Note that in a still life oscillator graph with no γ-nodes, every node must have precisely 2 or 3 neighbours.)

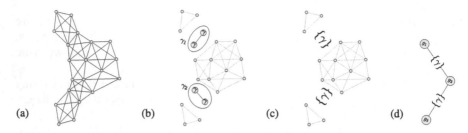

Fig. 18.107 Constructing the macroscopic graph for D, the kd-p5-16-10 *drummer* oscillator (Fig. 18.79): (**a**) the oscillator graph $G(D)$; (**b**) the oscillator sub-graphs $G_\kappa(\hat{\gamma}(D))$ and the γ-cell sub-graphs $\gamma_1, \gamma_2 \in G_\kappa(\gamma(D))$; (**c**) the oscillator sub-graphs $G_\kappa(\hat{\gamma}(D))$, and markers corresponding to γ_1 and γ_2; (**d**) the macroscopic graph $M(D)$

The blocks have fully connected oscillator graphs (Fig. 18.108a and c). The graphs serve to distinguish the 4 cell tubs from the 4 cell blocks (Fig. 18.108b and c). This is the case for kite and dart and for rhomb still lifes (and indeed for the regular Life case), justifying the common terminology.

All the discovered five cell chains in Fig. 18.21 have the same oscillator graph (Fig. 18.109b). The regular Life five cell boat has a different oscillator graph (Fig. 18.109a), and so is a distinct structure.

The regular Life snake and Penrose 6-snakes have the same oscillator graph (Fig. 18.110a). The six cell chains have a circular oscillator graph (Fig. 18.110d), as does the regular Life beehive.

The discovered six cell disconnected still lifes exhibit several different kinds of microscopic oscillator graphs (Fig. 18.111a–c). These graphs contain extra nodes corresponding to γ-cells: cells that are always "dead" but that are necessary to the oscillator; note that each γ-cell is connected to precisely three nodes in at least one of the "live" subgraphs. Also note that the graphs in Fig. 18.111b and c have isomorphic subgraphs (isomorphic patterns of "on" cells) but different numbers of γ-cells. They all have the same macroscopic graph (Fig. 18.111d).

The regular Life long snake and Penrose 7-snakes have the same oscillator graph (Fig. 18.112a). All the discovered seven cell rings have the same oscillator graph (Fig. 18.112d), which does not have a chain structure. Rather, it has a main ring of six nodes, with the seventh node providing a third neighbour for two nodes in the main ring. No regular Life still life has this graph; there is a regular Life pattern with a seven node chain: the loaf (Fig. 18.26d).

There are no seven cell regular Life disconnected still lifes. The discovered seven cell Penrose disconnected still lifes exhibit several different kinds of microscopic oscillator graphs (Fig. 18.113a–d). There are two different macroscopic graphs (Fig. 18.113e and f).

The discovered eight node Penrose still lifes exhibit three different kinds of oscillator graphs: snakes, chains, and rings (Fig. 18.114).

A similar structure to the seven node ring (Fig. 18.112d), but at points all around the main ring, is seen in the dart ring oscillator graphs (Fig. 18.115).

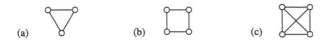

Fig. 18.108 Three and four node graphs, corresponding to the still lifes p1-3 and p1-4: (**a**) three cell Penrose ace block, Fig. 18.18a, and Q block and D block, Fig. 18.19a; (**b**) regular Life tub, Fig. 18.16, and Penrose tubs, Fig. 18.17; (**c**) regular Life block, Fig. 18.16, and the remaining four cell Penrose blocks in Fig. 18.18, and in Fig. 18.19

Fig. 18.109 Five cell graphs, corresponding to the still lifes p1-5: (**a**) regular Life boat, Fig. 18.20; (**b**) five cell kite and dart and rhomb chains, Fig. 18.21

Fig. 18.110 Six node graphs, corresponding to the still lifes p1-6: (**a**) regular Life snake, Fig. 18.22a and the Penrose 6-snakes, Fig. 18.23; (**b**) the ship; (**c**) the barge; (**d**) the beehive, Fig. 18.22d, and the six cell Penrose chains, Fig. 18.24

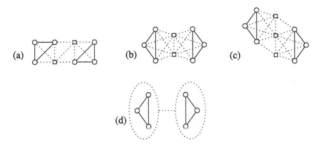

Fig. 18.111 Six node microscopic graphs for disconnected still lifes p1-6: (**a**) the regular Life carrier in Fig. 18.22; (**b**) the still lifes in Fig. 18.25a and bi; (**c**) the still lifes in Fig. 18.25bii and iii. In the microscopic graphs, edges from γ-cell nodes are shown *dotted*, and γ-cells are shown as *squares*, to help clarify the microscopic structure; (**d**) the macroscopic graph for all three forms of still life

Fig. 18.112 Seven node graphs, corresponding to the still lifes p1-7: (**a**) regular Life long snake and fishhook, Fig. 18.26, and the Penrose 7-snakes, Fig. 18.27a; (**b**) the long boat; (**c**) the loaf; (**d**) seven cell Penrose rings, Fig. 18.27b–d

Fig. 18.113 Seven node
microscopic graphs for
disconnected still lifes p1-7:
(**a**) kd-p1-7, Fig. 18.28a;
(**b**) r-p1-7, Fig. 18.28bi;
(**c**) r-p1-7, Fig. 18.28bii;
(**d**) r-p1-7, Fig. 18.28biii;
(**e**) macroscopic graph for
(**a**) and (**b**); (**f**) macroscopic
graph for (**c**) and (**d**)

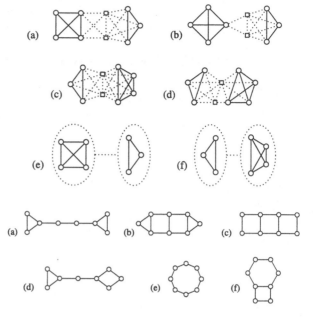

Fig. 18.114 Eight node
graphs, corresponding to the
still lifes p1-8: (**a**) regular
Life shillelagh, hook with
tail, very long snake, and
canoe in Fig. 18.29, and the
Penrose rhomb 8-snake in
Fig. 18.30d; (**b**) the long ship;
(**c**) the long barge; (**d**) the tub
with tail; (**e**) the pond, the
cigar, and the Penrose
8-chains in Fig. 18.30a, bi, c;
(**f**) the Penrose S5 vertex
variant in Fig. 18.30bii

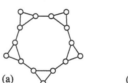

Fig. 18.115 The graph for
the Penrose dart still life rings
in (**a**) Figs. 18.35a and 18.36;
(**b**) Fig. 18.35b

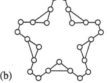

Classifying the Oscillators

Figure 18.98 shows the oscillator graph for all kite and dart and rhomb $p2$ plinkers, as well as the regular Life blinker.

Figure 18.116 shows the graphs for the 8 and 9 cell $p2$ oscillators. The regular Life *beacon* (Fig. 18.43b) and the kite and dart *marcher* (Fig. 18.46b) are both p2-8-6, but have different underlying graphs (Fig. 18.116a and b), so are different oscillators. However, the kite and dart marcher (Fig. 18.46b) and the rhomb marcher (Fig. 18.50b) have the same underlying graph, and so are isomorphic oscillators. Similarly, the p2-9-7 *crowned marcher* (Figs. 18.46c and 18.50c) are isomorphic.

Figure 18.117 shows the graphs for the p2-10-6 oscillators. The oscillators in Figs. 18.47 and 18.51 have isomorphic graphs (Fig. 18.117c), and are the isomor-

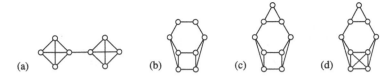

Fig. 18.116 Graphs for *p2* oscillators (**a**) l-p2-8-6, the *beacon*, Fig. 18.43b; (**b**) p2-8-6, the *marcher*, Figs. 18.46b and 18.50b; (**c**) p2-9-7, the *crowned marcher*, Figs. 18.46c and 18.50c; (**d**) kd-p2-9-6, Fig. 18.46d

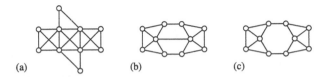

Fig. 18.117 Graphs for the p2-10-6 oscillators: (**a**) the *toad*, Fig. 18.43c; (**b**) the *clock*, Fig. 18.43d; (**c**) the *hollow clock*, Figs. 18.47 and 18.51

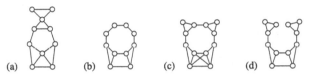

Fig. 18.118 Graphs for further *p2* oscillators: (**a**) kd-p2-10-7, Fig. 18.49a; (**b**) r-p2-10-8, Fig. 18.52; (**c**) kd-p2-12-9, isomorphic *fast shufflers*, Fig. 18.48; and also the graph for the longer period kd-p4-12-10, the *shuffler*, Fig. 18.69; (**d**) r-p2-12-10, isomorphic oscillators from Fig. 18.55b

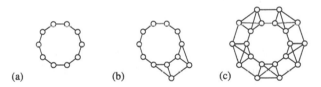

Fig. 18.119 Graphs for *p2* oscillators constructed from chains: (**a**) r-p1-10 chain, Fig. 18.61a; (**b**) r-p2-12-10, Figs. 18.55a and 18.61b; (**c**) r-p2-20-10, Fig. 18.62

phic *hollow clock*s. They have similarities to, but are not identical to, the graph of the regular Life *clock* (Fig. 18.117b).

Figure 18.118 shows graphs for further *p2* oscillators. Note that kd-p2-12-9, the *fast shuffler*, Fig. 18.48 has an isomorphic oscillator graph to the longer period kd-p4-12-10 *shuffler*, Fig. 18.69. Here the code is also needed to distinguish them.

Figure 18.119 shows the graphs for the r-p1-10 chain, and the associated oscillators constructed from it. It demonstrates that the r-p2-12-10 oscillators in Figs. 18.55a and 18.61b are isomorphic to each other, but distinct from the r-p2-12-10 oscillators in Fig. 18.55b, which have a different oscillator graph (Fig. 18.118d).

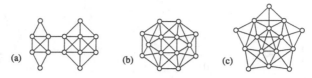

Fig. 18.120 The graph for (**a**) p3-14-8, the *breather* and all variants, Figs. 18.63 and 18.67; (**b**) kd-p4-14-6, the *bat* and all variants, Fig. 18.71a–c; (**c**) p5-15-8, the *ninja*, Figs. 18.78 and 18.86

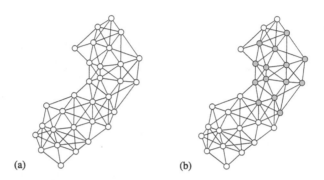

Fig. 18.121 (**a**) The oscillator graph for kd-p11-31-9, the *malformed bat*, Fig. 18.83; (**b**) the *bat* graph as a subgraph of the *malformed bat* graph

Figure 18.120 shows the oscillator graphs for the breather, the bat, and the ninja. These further demonstrate how we are able to identify both variant forms (the various kite and dart breathers, and the various bats), and how we are able to identify forms across tilings (the kite and dart breather and the rhomb breather, the kite and dart ninja and the rhomb ninja): they have isomorphic oscillator graphs.

Note that kite and dart ninja exists on the a0 neighbourhood (Fig. 18.10) rotated around the Sun vertex (Fig. 18.4), and the rhomb ninja exists on the b1 neighbourhood (Fig. 18.12) rotated about the S vertex (Fig. 18.5). The a0 and b1 neighbourhood graphs are isomorphic.

Figure 18.121a shows the oscillator graph for the malformed bat, so named because a part of it has similarities to the bat oscillator. It has the bat oscillator graph as a subgraph (Fig. 18.121b). Having another oscillator as a subgraph is not sufficient to mirror its behaviour, because the graph does not capture which cells are alive at any one time, and the other cells of the oscillator will have some influence on the subgraph cells. Also, due to the structure of the tiling, large oscillators will tend to have subgraphs corresponding to smaller oscillators. Nevertheless, oscillators that do appear visually related will probably have this subgraph structure; for example, the graph of the r-p7-19-7 *hattipper* (Fig. 18.91) appears as a subgraph of the related r-p7-23-8 oscillator (Fig. 18.92).

18.9 Summary and Conclusions

We have continued our investigations of the Game of Life on Penrose tilings, producing a preliminary catalogue of still lifes and oscillators, including comparisons with similar oscillators on the regular Life tiling. We have catalogued many still life and periodic oscillators, on both kite and dart and rhomb tilings. We have also demonstrated that arbitrarily large snakes and chains can exist. We have found no propagating features analogous to gliders, but the existence of "fuses" burning along chains makes us optimistic that such structures may be discovered.

We have introduced an identification code to help refer to different oscillators. However, it does not completely distinguish different oscillators (although collisions are rare, except among the still lifes). There are visually variant forms due to changes in the tiling (for example, the variant oscillators on different S_C extensions); these have the same code. But a few oscillators that are obviously of a different form also share the same code. We further distinguish these by considering the underlying oscillator graph, which captures the topology of oscillator cells neighbourhoods.

The oscillator graph by itself is also insufficient to completely distinguish different oscillators (although collisions are again rare). The combination of code and oscillator graph has, so far, proved sufficient to unify isomorphic forms whilst discriminating different oscillators.

Isomorphic forms can exist on one, two, or all three tilings. For example, several variants of the bats and of dancers exist on the kite and dart tiling; marchers, breathers, and ninjas exist on both kite and dart and rhomb tilings; tubs, blocks, snakes, and blinkers exist on all three tilings. If an oscillator exists on both the regular and one of the Penrose tilings, then it tends to exist on the other Penrose tiling, too. The only exception in this catalogue is that we have not been able to find a kd-p1-8 snake, despite there being a regular Life form (the *very long snake*) and a rhomb form.

Our classification of disconnected oscillators requires the inclusion of γ-cells: cells that are always dead, but whose removal would destroy the oscillator. We define a further *macroscopic* oscillator graph, to provide a form of isomorphism between disconnected oscillators with different numbers of γ-cells.

Further work includes extending the catalogue. Regular Life has very many small oscillators; we suspect that our preliminary catalogue of Penrose oscillators has identified only a small proportion of them; a more systematic search is needed. Additionally, larger and longer period oscillators are still to be found. Here an evolutionary search might be more appropriate.

Oscillators themselves, although demonstrating that Penrose life is complex, are not the ultimate goal. That would be to use Penrose life to implement larger computations in a way analogous to regular Life. We have made a start by suggesting that "fuses" can be used to propagate information, or to implement timers. Many more such components are needed: a starting place is to look at the structures and interactions of small oscillators.

Acknowledgements   We thank Adam Nellis for helpful comments on an earlier version.

References

1. Berlekamp, E.R., Conway, J.H., Guy, R.K.: Winning Ways for Your Mathematical Plays, vol. 2: Games in Particular. Academic Press, San Diego (1982)
2. de Bruijn, N.G.: Algebraic theory of Penrose's non-periodic tilings of the plane I and II. Indag. Math. (Proc.) **84**, 39–66 (1981)
3. de Bruijn, N.G.: Remarks on Penrose tilings. In: Graham, R.L., Nesetrilm, J. (eds.) The Mathematics of P. Erdös, vol. 2, pp. 264–283. Springer, Berlin (1996)
4. Gardner, M.: Mathematical games: The fantastic combinations of John Conway's new solitaire game "life". Sci. Am. **223**(4), 120–123 (1970)
5. Gardner, M.: Mathematical games: Extraordinary non-periodic tiling that enriches the theory of tiles. Sci. Am. **236**(1), 110–121 (1977)
6. Grünbaum, B., Shephard, G.C.: Tilings and Patterns. Freeman, New York (1987)
7. Hill, M., Stepney, S., Wan, F.: Penrose Life: ash and oscillators. In: Capcarrere, M.S., Freitas, A.A., Bentley, P.J., Johnson, C.G., Timmis, J. (eds.) Advances in Artificial Life: ECAL 2005, Canterbury, UK, September 2005. LNAI, vol. 3630, pp. 471–480. Springer, Berlin (2005)
8. Knuth, D.E.: The Art of Computer Programming: Seminumerical Algorithms, vol. 2, 3rd edn. Addison–Wesley, Reading (1998)
9. Niemiec, M.D.: Life page. http://home.interserv.com/~mniemiec/lifepage.htm (1998)
10. Owens, N., Stepney, S.: Investigations of Game of Life cellular automata rules on Penrose tilings: lifetime and ash statistics. In: Automata 2008, Bristol, UK, June 2008, pp. 1–34. Luniver Press, Beckington (2008)
11. Owens, N., Stepney, S.: Investigations of the Game of Life cellular automata rules on Penrose tilings: lifetime, ash and oscillator statistics. J. Cell. Autom. **5**(3), 207–255 (2010)
12. Penrose, R.: Pentaplexity. Eureka **39**, 16–32 (1978)
13. Rendell, P.: Turing universality of the Game of Life. In: Adamatzky, A. (ed.) Collision-Based Computing. Springer, Berlin (2002)
14. Senechal, M.: Quasicrystals and Geometry. Cambridge University Press, Cambridge (1995)
15. Silver, S.: Life lexicon, release 25. http://www.argentum.freeserve.co.uk/lex.htm (2006)
16. Socolar, J.E.S., Steinhardt, P.J.: Quasicrystals. II. Unit-cell configurations. Phys. Rev. B **34**, 617–647 (1986)

Chapter 19
A Spherical XOR Gate Implemented in the Game of Life

Jeffrey Ventrella

Are there uniquely spherical cellular automata machines? Might there be computational processes that come about more naturally on spheres than they would in the plane? This chapter describes an exploration of geodesic grids as environments for cellular automata (CA) and specifically addresses the movements of Game of Life (GoL) gliders whose interactions are affected by the positive curvature of spheres. 2D CA are typically arranged on regular planar grids with periodic boundary conditions — equivalent to the topology of a torus. This chapter instead considers the dynamics of CA on spheres. The unavoidable discontinuities that arise from mapping a 2D grid onto the sphere are accepted as integral components of the environment. A novel XOR gate built on GoL is demonstrated, utilizing the double-crossing of glider paths following geodesic great circles.

19.1 Introduction

Cellular automata (CA) are used for modeling many kinds of biological processes. Some CA rules can give rise to complex dynamics, self-replicating patterns, and the primary components of Boolean logic. CA-based models sometimes accompany discussion on the origins of life [4]. CA are typically modeled on 1D, 2D, or 3D grids. To avoid boundary artifacts, it is common to create periodic boundary conditions — so that opposing boundaries wrap around. In the case of a 2D cellular automaton on a rectangular grid, like Conway's Game of Life (GoL) [1], this periodic boundary creates the topological equivalent of a torus. A 2D grid can be mapped onto a torus without introducing cell neighborhood discontinuities. This is not the case when mapping a grid onto the sphere.

For any two points on a sphere, if they are set to motion and follow a straight path, staying on the surface of the sphere, they will come back to their original starting points. But more importantly: the two paths will intersect — twice. This is because all lines drawn on a sphere are actually geodesic arcs that close to form great circles. All great circles intersect all other great circles. Euclid's parallel postulate

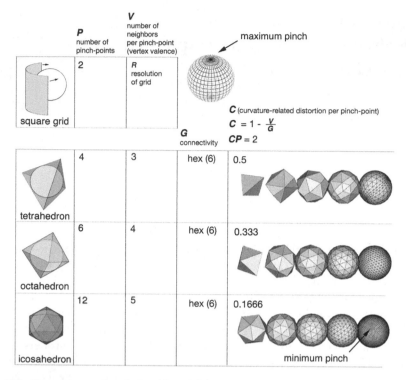

Fig. 19.1 Comparisons of geodesic grids and distortions to grid topology at pinch-points

works fine when you are drawing figures in the sand, but not if you are considering large trajectories on a sphere, where curvature matters.

CA implemented in non-Euclidean spaces, such as hyperbolic surfaces, have been studied [5, 8, 9], and the sphere has been studied with multiple resolution by [3], however, there does not appear to be substantial research in exploring glider dynamics on the sphere. This chapter specifically addresses the movements of information quanta (gliders), and explores structures from GoL implemented on a sphere, to find out if there are properties of the sphere that might be exploited for computation.

19.1.1 Grid Distortion

Figure 19.1 shows some examples of how the curvature of the sphere introduces distortions in the topology of a grid, concentrated at "pinch-points". The top example shows a square grid stretched onto the sphere, resulting in two pinch-points located at the north and south poles — the entire top and bottom edges of the grid are each reduced to a single point. The other examples show the generation of triangular geodesic grids based on the three triangle-faced Platonic solids. Grid dis-

Fig. 19.2 Six rectangular grids corresponding to the faces of the cube projected onto a sphere

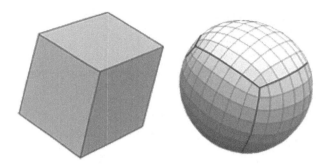

tortion from spherical curvature is distributed to the vertices, corresponding to the number of pinch-points P. Grid connectivity G is 6 (based on the hexagonal grid). The number of neighbors per pinch point V is equal to polyhedral vertex valence. The value C represents the amount of distortion per pinch-point. It normalizes to 2 when multiplied by P in all three examples. One implication from this comparison is that for implementing geodesic CA, there is a trade-off between fewer pinch-points with greater distortion per pinch-point vs. more pinch-points with less distortion per pinch-point. This is related to the concept of balance in tiling [2]. Observe also that the only possible regular spherical tilings (having no pinch-points) are represented by the five Platonic solids. But these have far too few tiles (i.e., cells) for useful glider dynamics to emerge.

The three polyhedral geodesics shown are triangular. Their dual polyhedra consist of mostly hexagons (with fewer-sided polygons at the pinch-point locations). These mostly-hexagonal grids can be seen as the cells of the CA — this helps to visualize neighborhood connectivity. Implementation of GoL on a sphere requires a rectangular grid, and so a cube is used as the base polyhedron. Figure 19.2 shows six rectangular grids comprising the faces of a cube projected onto the sphere. This mapping results in eight pinch-points, corresponding to the vertices of the cube. Although it has more pinch-points than the single-grid mapping at the top of the illustration, it is more regular and well-behaved: the pinch-points present less grid distortion, and the grid overall affords the multiple crossings of great circles (see Fig. 19.4). The topology of the 9-cell Moore neighborhood is preserved across the twelve edges of the cube.

19.2 Spheres and Computation

Are there uniquely spherical CA computational machines? Logic gates have been simulated using glider guns and other still-life's and short-period structures which, when arranged in just the right way, can simulate the primary elements of computation, such as AND, NOT, and OR gates. A Turing Machine was built using patterns of GoL [6]. Imagine a machine with logic gates that not only relies on periodicity (as a torus affords) but also on a uniquely spherical kind of periodicity. It must be pointed out that a "spherical universal Turing machine" is not possible

Fig. 19.3 GoL implemented on six grids such that dynamics is continuous across the twelve cubical edges (*left*); and, projection onto the sphere (*right*)

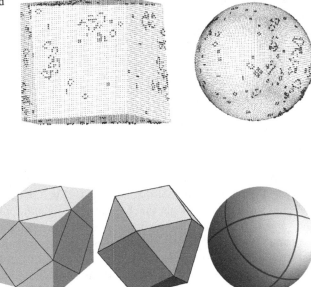

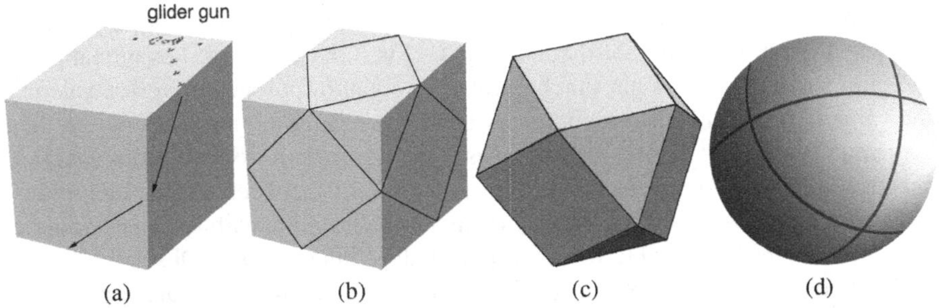

glider gun

(a)   (b)   (c)   (d)

Fig. 19.4 (**a**) Glider gun signal crossing midpoints of cube edges; (**b**) overlapping glider streams; (**c**) cuboctahedron; (**d**) there are four cuboctahedron great circles — each with six crossings

because a universal Turing machine requires an unbounded tape. But the finiteness of a geodesic grid does not exclude computation. It can support models of computational processes in nature and human technology, which are typically bounded, and embodied in finite physical structures.

Figure 19.3 illustrates GoL implemented on a unit cube and then projected onto a sphere. The six grids are sewn together so that the dynamics can pass across the twelve edges uninterrupted. Transformation from cube to sphere does not introduce any changes in grid connectivity.

Let's consider some of the properties of this cubical geometry. In Fig. 19.4a, we see what happens if Gosper's glider gun is performing its job on one of the face of a cube. The path of gliders generated by the gun moves diagonally, and therefore visits all the faces of the cube (and unless it is dealt with on another face, it will come back around and destroy the gun!). Imagine alternatively a continual stream of gliders (no gun) and a gun in another location on the cube sending a perpendicular signal of gliders that interact with this stream. Or imagine four continuous glider streams with no guns, as illustrated in Fig. 19.4b. This wasn't tested, but it is suspected that there are spherical implications to what emerges. These signal paths correspond to the edges of the cuboctahedron (Fig. 19.4c), which can be realized as four intersecting great circles (Fig. 19.4d).

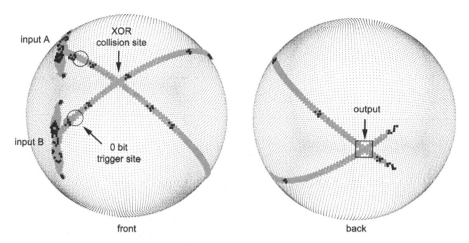

front   back

Fig. 19.5 A XOR gate in the Game of Life implemented on a sphere takes advantage of crossings of geodesic great circles which provides a single output location

19.2.1 A Spherical XOR Gate

To explore the effect of spherical curvature on computation, an XOR gate was implemented based on the geometry shown in Fig. 19.4. It employs similar techniques as demonstrated by Rennard [7]. This is illustrated in Fig. 19.5, showing GoL with 60×60 cells per cube-face (21,600 total cells). On the left is a view of the two inputs, and on the right is a view from the opposite side, showing the output. Two period 30 Gosper guns are used to generate the glider streams for inputs A and B. These streams are terminated by two eaters on the opposite side of the sphere.

Glider streams represent the one's and zero's of a binary signal, and the absence of a glider in the stream represents zero. A zero value is created by destroying a glider in the stream soon after leaving the gun. The site for this action is shown in Fig. 19.5 as the "0 bit trigger site". The input clock value $c =$ time step t mod 30 (the period of the gun). A new value is fed to the input stream every $c = 0$. If the input signal calls for a zero, the cell at the trigger site is set to true ("on", or "1") and causes the glider to dissolve within a few time steps. If the input signal calls for a 1, the glider is allowed to pass through. If gliders pass through from both inputs A and B, they collide and annihilate each other at the XOR collision site.

The binary output value is determined by reading the state of a specifically-chosen cell at the output site when $C = 10$. This cell is on only when a glider is passing over it, and so the output is 1. This cell detects the passage of a glider from either stream, as shown. With this implementation, there are two critical crossover points, corresponding to the intersection points of two great circles. The first one is responsible for the XOR collision, and the second one is responsible for reading the output.

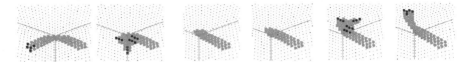

Fig. 19.6 GoL gliders passing over a cubical vertex

19.2.2 Exploiting Grid Discontinuity

The above experiment specifically avoids the eight pinch-points of the cube. As Fig. 19.6 shows, if a glider passes over one of these points, it may dissolve or get converted into two blinkers and a six-cell block (which normally changes to a bee-hive, but because of the abnormal neighborhood at the pinch-point, it remains as a still-life).

In this illustration, each glider is offset vertically by one cell. The first and last examples in the illustration show gliders escaping the outlaw nature of the pinch-point and veering off, following the grain of the grid.

A CA machine engineered to live on a geodesic grid could take advantage of these regions as termination points for gliders, or other computational purposes. In GoL these would normally be considered defects littering the otherwise regular grid. But consider a generalization of GoL allowing for the existence of isolated pinch-points at known locations. It is well-known that material defects can be catalysts for complex and characteristic growth, such as with crystals. Evolutionary simulations for studying emergent complex behavior might discover emergent CA structures that exploit these regions.

19.3 Conclusion

This chapter addresses the question of whether spherical curvature (manifested in the crossings of great circles) provides for unique opportunities for computational CA. The fact that GoL gliders follow straight paths means that when they are traveling on a sphere, as long as they do not pass over any pinch-points, they will follow along great circles, and experience collisions in different ways than if they were implemented on a torus. The fact that all great circles intersect at two points is the basis for the XOR gate shown here.

This chapter considers geodesic grids as valid environments for studying CA dynamics. Furthermore, grid discontinuities manifested as pinch-points are considered as properties of the environment that can be exploited for various purposes — including computation. Do there exist specialized, bounded GoL machines which can be built more efficiently using the properties of geodesic grids? The XOR gate described in this chapter indicates this possibility.

References

1. Berlekamp, E.R., Conway, J.H., Guy, R.K.: What Is Life? Winning Ways for Your Mathematical Plays, vol. 2: Games in Particular, Chap. 25. Academic Press, London (1982)
2. Grunbaum, B., Shephard, G.C.: Tilings and Patterns. Freeman, New York (1987) (balanced tilings explained in pp. 129–134)
3. Kiester, R.A., Sahr, K.: Planar and Spherical Hierarchical, Multi-resolution Cellular Automata. Computers, Environment and Urban Systems. Elsevier, Amsterdam (2008)
4. Langon, C.: Life at the Edge of Chaos. Artificial Life. Addison–Wesley, Reading (1992)
5. Margenstern, M.: Cellular Automata in Hyperbolic Spaces. Old City Publishing, Philadelphia, London, Paris (2007)
6. Rendell, P.: Turing universality of the Game of Life. In: Adamatzky, A. (ed.) Collision-Based Computing. Springer, Berlin (2002)
7. Rennard, J.P.: Implementation of logical functions in the Game of Life. In: Adamatzky, A. (ed.) Collision-Based Computing. Springer, Berlin (2002)
8. Yukita, S.: Dynamics of cellular automata on groups. IEICE Trans. Inf. Syst. **E82-D** (10), 1316–1323 (1999)
9. Yukita, S.: Cellular automata in non-euclidean spaces. In: Proceedings of the 7th WSEAS International Conference on Mathematical Methods and Computational Techniques in Electrical Engineering, pp. 200–207. World Scientific and Engineering Academy and Society, Stevens Point (2005)

Part V
Complexity

Chapter 20
Emergent Complexity in Conway's Game of Life

Nick Gotts

It is shown that both small, finite patterns and random infinite very low density ("sparse") arrays of the Game of Life can produce emergent structures and processes of great complexity, through ramifying feedback networks and cross-scale interactions. The implications are discussed: it is proposed that analogous networks and interactions may have been precursors to natural selection in the real world.

20.1 Introduction

This chapter explores the emergence of complex structures and processes in Conway's Game of Life (henceforth GoL). Two very different kinds of initial position are explored. First, initial positions in which there are only a small number of state 1 ("on") cells (50, in the positions explored in detail); second, initial positions with an infinite (or at least very large) number of state 1 cells, randomly and very sparsely distributed.

Studying emergent properties in mathematically well-defined systems such as CA may be particularly useful in constructing a typology of emergence: we may be better able to say exactly what "emerges" in a particular case than we generally can in the case of real-world systems. Casti [9] defines emergence as: "system behaviour that comes out of the interaction of many participants." This can be taken to cover any interaction of lower-level entities that produces qualitatively different or novel kinds of behaviour by higher-level entities. It is useful to distinguish synchronic emergence, where interactions at one level simply maintain ongoing behaviour at a higher one, from diachronic emergence, where lower-level interactions produce new levels of organisation over time. Good examples of the former are provided by the ways cells interact to maintain the functioning of an organ such as the heart [38]; of the latter, by the "major transitions in evolution" described by [27]. This chapter is primarily concerned with diachronic emergence in what may be among the simplest systems to display the phenomenon, although the diachronic emergence discussed rests upon an underlying layer of synchronic emergence.

A. Adamatzky (ed.), *Game of Life Cellular Automata*,
DOI 10.1007/978-1-84996-217-9_20, © Springer-Verlag London Limited 2010

A "feedback network" is a collection of components (variables or structures) linked by a network of interactions including cyclical paths, so at least some components influence themselves indirectly. In what is called here a "ramifying feedback network", some interactions produce new components, and hence new interactions. Ramifying feedback networks may be important to the emergence of complexity in both biological and socio-technical domains. Chemical evolution involving increasingly numerous and complex chemical species and reactions appears a necessary precursor to the appearance of self-replicating entities. Kauffman [23] argues that systems of polymer catalysts would pass a "critical complexity threshold" as the number of different polymers increases, past which subsystems would form where production of each member is catalysed by other subsystem members. In technological development, the identification of "components" is harder, but [13] argues that artefact-activity pairs (e.g. a bicycle and the activity of riding it) are the evolving entities. These components interact by coevolution (bicycles changed once roads were tarred) and transfer of materials and subcomponents; and although [13] does not mention the point, such interactions can give rise to new artefact-activity pairs — as improvements in steam engines and rail tracks, both developed for mining, made possible steam locomotives. Institutional systems may also provide examples: interactions between evolving institutions may reveal conflict or ambiguity, prompting the development of new institutional "components" and hence, ramification of feedback networks.

Real-world feedback networks often contain multiple layers. Complex systems are often "lumpy" [22]: despite the association of complexity and self-organisation with scale-free structures and behaviour [3], the most complex entities (organisms, ecosystems, societies) include diverse kinds of entities and processes, at multiple scales, each involving a characteristic set of processes that can to some extent be understood without considering larger or smaller scales. However, this autonomy of scales is incomplete, and events that transgress it ("cross-scale interactions") are dynamically important.

The biologically characteristic parts of an animal include organs, tissues, cells, organelles and macromolecules, each with its own characteristic range of scales, and interactions with similar entities. In social systems, we can similarly distinguish individuals, households, settlements, economic and political units of various kinds, and civilizations. The different types of entity may not fit into a single hierarchy, particular types may cover wide and sometimes overlapping size ranges, and a scale-free size distribution may cover several orders of magnitude as claimed for cities [8] and firms [1]. Nonetheless, the kind of lumpiness described, and the relative autonomy of processes at different scales may be a necessary attribute of highly complex entities [34].

Recent studies of cross-scale interactions range over plasma physics [10], oceanography [14], and the study of sepsis [36] and insect societies [35]. A model of wildfire spread is presented as an example of the production of catastrophic events by cross-scale interactions and system feedbacks in [29], with additional examples from desertification, epidemics, and structural failures in engineering. However, the most extensive work concerns ecological and social-ecological systems. The paper [30] proposes a model of "cross-scale resilience" — ecological resilience being the amount of disturbance required to shift an ecosystem from one set of mutually reinforcing structures and processes to another.

Ramifying feedback networks and cross-scale interactions, then, are important in many systems involving nonlinear dynamics and spatial heterogeneity. How simple can a system be, in terms of local elements and interactions, global structure, and initial conditions, while generating such dynamics? Once in existence, do cross-scale interactions tend to generate further complexity, and in particular additional kinds of cross-scale interactions? This chapter shows that ramifying feedback networks and cross-scale interactions can arise in remarkably simple systems, and that in some cases, they do produce additional kinds of cross-scale interaction, in a process that continues as far as it has been computationally possible to follow it. The example sets of systems, which are necessarily explored in some detail, have both features in common with the natural systems mentioned above, and marked differences from them.

20.2 Preliminaries

This section introduces some concepts and terminology that will be used in the course of the chapter.

A configuration or position is any GoL array, with the states of its cells specified. A pattern is a finite set of state 1 cells in an otherwise "empty" (state 0) array. We start by distinguishing three successively more inclusive classes of pattern:

- Still-lifes are stable patterns: all state 1 cells remain in that state; no more are created.
- Oscillators are the same at step $t + n$ as at step t, for some n. Still-lifes are oscillators of period 1.
- Repeaters are patterns for which step $t + n$ is a translation of step t, for some n. Oscillators are repeaters for which the translation is the identity transformation. Repeaters that move are spaceships.

If the state of one cell in a configuration is switched, the influence of the change cannot possibly spread faster than one link (between neighbouring cells) per step in any direction: the speed of light, or c — and in a position with a finite number of state 1 cells, cannot maintain this indefinitely. Among moving repeaters, gliders travel at the maximum possible diagonal speed, $c/4$, while the three smallest orthogonal spaceships (known as LWSS, MWSS and HWSS) travel at the maximum orthogonal speed, $c/2$.

Indefinite growth patterns increase their count of state 1 cells without bound. The smallest, and the only ones ever seen to emerge from random configurations, incorporate the pattern shown in Fig. 20.1, top left, the switch engine. After 48 steps (top center) this produces a glide-reflected copy of itself, plus additional debris; after 96 steps (top right) the switch engine has moved 8 cells diagonally, for a speed of $c/12$, and produced a larger debris clump. The debris soon overtakes and destroys a lone switch engine, but there are numerous ways of stabilizing the pattern: for

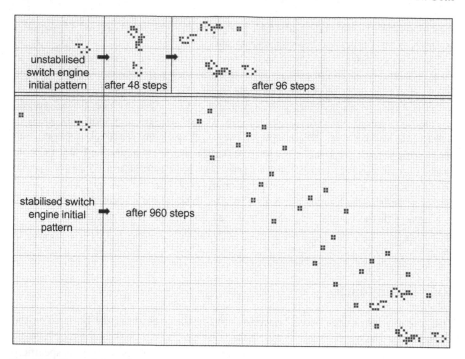

Fig. 20.1 The switch engine

example, a block placed as shown at bottom left in Fig. 20.3 interacts with the debris to produce an endless trail of blocks as the switch engine progresses (bottom right). (This pattern contains 12 state 1 cells; [20] report 10-cell patterns that produce a stabilised switch engine, and that 10 is the minimum number of state 1 cells needed for indefinite growth.)

The current study shows that certain small and apparently quite simple patterns grow in structural and dynamic complexity over large numbers of steps. Clearly, the complexity must be inherent in the combination of the GoL transition rule, the initial pattern, and the otherwise empty array; it is cross-scale interaction that makes it manifest. The type of complexity generated is markedly different from that produced from simple initial patterns by "type 3" cellular automata and even from any previously demonstrated for "type 4" cellular automata [39], such as GoL and the one-dimensional CA 110 [11]; and as a dynamic rather than a computational property, it is not an automatic consequence of computational universality. In the GoL patterns of Fig. 20.1, we can already distinguish levels of structure. The lowest consists of individual cells. The next level, or set of levels, consists of clumps of state 1 cells. A d-cluster is a maximal set of state 1 cells such that each pair of cells in the set is linked by a neighbor-to-neighbor path that never goes through more than d successive state 0 cells. The term cluster on its own refers to a set of cells constituting a d-cluster for some d. 1-clusters are of particular significance, as the state 1 cells in a 1-cluster cannot share any neighbors with state 1 cells outside it,

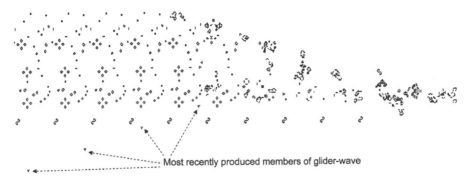

Most recently produced members of glider-wave

Fig. 20.2 Gosper2, the second discovered orbit of R. William Gosper's "puffer train"

so the states of the 1-cluster's cells, and their state 0 neighbors, can be determined for one step without considering any other cells. Any configuration can be assigned a hierarchical cluster structure, with 0-clusters grouped into 1-clusters, 2-clusters, ..., n-clusters. Dynamically, however, what is most important is how the 1-clusters composing a pattern merge, divide and disappear over time: the 1-cluster dynamics of the pattern. The 1-cluster dynamics of the patterns considered in detail here have intricate patterns of nested components, isolated for longer or shorter periods.

The stabilized switch engine of Fig. 20.1 is a puffer train (puffers form a subcategory of indefinite growth patterns). A puffer has a head which moves indefinitely across the array, leaving a trail of oscillators. It may also produce streams and/or waves of mobile repeaters: its limbs. (In a stream, all the mobile repeaters follow the same path, while in a wave, successive mobile repeaters follow different but parallel paths; streams produced by a puffer travel parallel or anti-parallel to the head's movement, while waves do not.)

The trails left by the stabilised switch engines of Figs. 20.2 and 20.3 are typical in having an initial irregular part (the tail), followed by an indefinitely lengthening regular body. The switch engine's head, however, is unusual among GoL puffers in consisting of a single 1-cluster through most of its cycle, and being "indivisible": no proper subpart of it can traverse the array alone. Other such puffers have considerably larger heads, while all other small puffers have heads with multiple separate but interacting parts and a three-level structure: individual cells, 1-clusters, and the head as a whole.

The first puffer discovered in the exploration of GoL (by R. William Gosper) was the puffer train [15], which can begin with a minimum of 22 state 1 cells. The head includes two LWSSs and an additional cluster, moves orthogonally at $c/2$ and in its original form, goes through a 140-step cycle producing oscillators but no waves or streams. Many puffers have alternative stabilizations, or orbits: that is, adding a small cluster to the original pattern, either behind or ahead of the head can cause a different body and in some cases limbs to form. If the additional cluster is located ahead of the head, the puffer can start in one orbit and switch when the additional cluster is encountered — and such switches may be repeated. These effects are

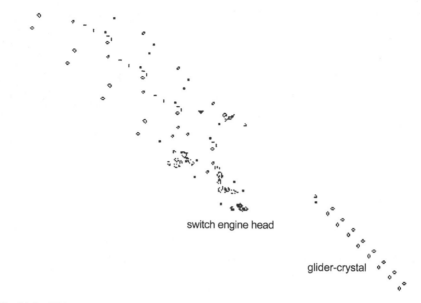

switch engine head

glider-crystal

Fig. 20.3 Glider-stream switch engine constructing glider crystal

reminiscent of the way some macromolecules can be shifted between alternative configurations, by interaction with small signaling molecules.

The switch engine has just two known orbits. The one shown in Fig. 20.1 is the block-laying switch engine. The other is the glider-stream switch engine (see Fig. 20.3), which produces a stream of gliders ahead of itself (the head moves at 1/3 of a glider's speed). A block placed at various points relative to the 8-cell switch engine head can bring about this stabilization, but it can also arise from crashing a glider into the front of a block-laying switch engine. The puffer train has several orbits. The one of importance here (Fig. 20.2) requires at minimum an initial 25 state 1 cells. It produces a wave of gliders, moving at 135 degrees to the puffer's head. This pattern will be referred to as gosper2. It settles into a 100-step cycle, producing a body with a spatial periodicity of 50 cells. There are many points during the 100-step cycle at which, if all the state 1 cells in a column perpendicular to the direction of growth are removed, it reverts to the 140-step cycle of the original puffer train, which appears more stable.

20.3 The Finite Case: Diverging Puffer Pairs

20.3.1 Pairs of Patterns

Combining any pair of patterns in a two-dimensional cellular automaton produces a Z^2 collection of larger patterns, each member being defined by the pair's relative placement. Interesting periodicities arise within such sets of patterns in GoL. We identify a particular member of the Z^2 collection by the relationship between the two sub-patterns' northwest cells — those at the northwest corners of their minimal containing rectangles. Placements causing the two components to overlap require special treatment: here, we ignore such cases. Clearly, if each of the two sub-patterns would on its own remain forever within a finite block of cells, placing them far enough apart will prevent interaction. Even where this is not so a complete analysis may be possible: although a glider can be placed at any distance from a block and still interact with it eventually, all members of the Z^2 collection of possibilities divide into a few classes, depending on how (if at all) the glider eventually hits the block.

More complex but still fully analyzable classes of examples involve a glider-stream switch engine and a small cluster such as a block or glider. The forward glider-stream can interact with such a cluster to construct a trail of six-cell still-lifes called beehives ahead of the switch engine. This trail is glider-crystal (Fig. 20.3), and once the interaction between glider-stream and glider-crystal enters a regular cycle, it takes 22 gliders to add two pairs of beehives, extending the crystal toward the switch engine's head. When the latter overtakes the crystal, interactions between head and crystal send glider-waves to either side. Depending on the phase of crystal building and of the heading at right angles to the glider-stream switch engine, may also be produced.

In other cases again, complete analysis may be impossible. The rest of Sect. 3 explores one such class of patterns (there are many others), formed by two gosper2s diverging at right angles.

20.3.2 Diverging Puffer Pairs, Expanding Feedback Loops, and Ramifying Feedback Networks

Consider a pair of gosper2 puffers, so placed in an otherwise empty array that they do not overlap, they grow at right angles, and neither will destroy its head by crashing into the other. These constraints still permit an infinite set of distinct ways of placing the two gosper2s; this set of patterns divides into families and subfamilies, which differ in the complexity of the feedback networks they generate. One subfamily is singled out for detailed attention.

A gosper2 can be grown from 25 cells, by adding a blinker or a preblock (three state 1 cells in an orthogonal "L", which becomes a block in one step) to the 22 required for the original puffer train. The patterns investigated here all start with two mirror-image copies of the same 25-cell pattern, as shown in Fig. 20.4. The two puffers travel east and south (left-to-right and top-to-bottom of the array), and are oriented so members of the backward wave of gliders from each travel toward the other's body.

Successive members of the backward glider-wave from a gosper2 follow paths 50 half-diagonals apart (moving 1 cell southeast is a shift of two half-diagonals southeast, while moving 1 cell east or south is a one half-diagonal shift southeast). For

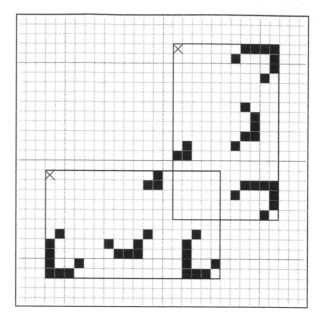

Fig. 20.4 Initial pattern of g2-g2-s0sw13. *Crosses* mark "northwest cells" of the two gosper2s constituting the pattern

the gosper2 headed east, the $n+1$th member of the wave is 75 cells east and 25 cells north of the nth, and follows a path 50 half-diagonals southeast of it. Shifting one puffer relative to the other along a northeast-southwest line leaves the relationship between the diagonals along which members of the glider-waves travel unchanged, while shifting along a north-south or east-west line, unless by a multiple of 50 cells, does not. This makes it convenient to use non-orthogonal coordinates, specifying the spatial relationship of the two gosper2s in terms of how far north or south, and how far southwest or northeast, the northwest cell of one is from that of the other. For the pattern g2-g2-s0sw13 (Fig. 20.4), the northwest cell of the south-headed gosper2 is 13 cells south and 13 cells west of that of the east-headed gosper2, but in the coordinate system used here, that is 0 south and 13 southwest.

Consider the set of patterns g2-g2-s$j$sw$k$, where j and k can take any integer values. For any j, sufficiently low values of k mean that either the two patterns will overlap, or the head of one gosper2 will crash into some part of the other; our interest is in the patterns where k is high enough to avoid these situations. Holding k constant, as the value of j modulo 50 is varied from 0 to 25, the spatial relationship between members of the two glider-waves as they either collide or pass through each other changes; as j modulo 50 is varied from 26 to 49, however, the spatial relationships that occurred as it ranged from 1 to 24 are repeated, but in reverse order and with the roles of the two waves reversed. Indeed, any g2-g2 pattern where $j > 25$ mod 50 is an exact enantiomorph of one where $j < 25$ mod 50. Hence we need consider only the patterns where $0 \leq j \leq 25$ mod 50, which we divide into families 0 through 25. For families 0 through 6, members of the two waves collide; in some of these families the collisions generate further gliders that then collide with the body of one of the gosper2s, but none of these collisions generate any further gliders

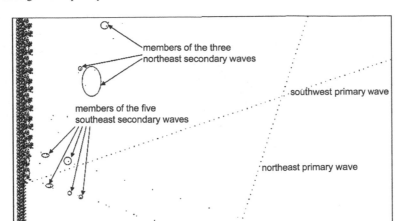

Fig. 20.5 Pattern g2-g2-s24sw13: closeup of interaction between southwest primary wave and south-headed gosper2 body

(or other spaceships). For families 7 through 25 the waves pass through each other, and the component gliders hit the other gosper2's body. For most of these families, members of the wave from each gosper2 hit the other's body without generating any secondary spaceships; the exceptions are families 7, 15, 17, 18, 19 and 24. In family 19, each collision with the body of the south-headed gosper2 generates a single secondary northeast-headed glider which hits the body of the east-headed gosper2 but generates nothing more. The other four families are more complicated, and divide into subfamilies according to whether secondary glider-waves interact with the northeast primary wave produced by the south-headed gosper2. This is determined by the value mod 25 of k in g2-g2-s$j$sw$k$ (g2-g2-sj+50sw$k$ is in the same subfamily as g2-g2-s$j$sw$k$).

In family 24, the most complex, the collisions of successive southwest primary wave gliders with the south-headed gosper2 interact, creating secondary southeast and northeast compound waves. A compound wave consists of two or more simple waves; all the gliders in a simple wave (like the primary waves) are produced in exactly the same way, and will (in g2-g2 patterns) follow paths that are multiples of 50 half-diagonals apart, and be separated by multiples of 25 cells in both orthogonal directions. In the simplest subfamilies, where the southeast secondary wave does not interact with the northeast primary, such as that represented by g2-g2-s24sw13 (Fig. 20.5), the northeast secondary wave consists of three simple waves (which collide with the east-headed gosper2 without creating any tertiary waves), the southeast secondary of five.

For many subfamilies of family 24, all members of the southeast waves either pass through the northeast primary wave without collision, or mutually annihilate with members of that wave. For some, collisions with northeast primary gliders create small remnants that the southeast primary avoids, while for three (subfamilies 4, 7 and 22), remnants are created with which members of the southeast primary

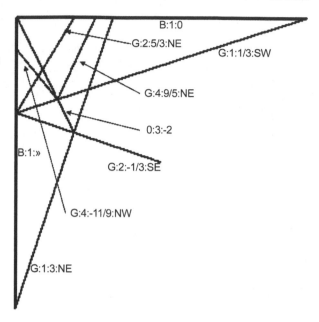

Fig. 20.6 Pattern g-g-s24sw29, after 2^{24} steps, showing all the linear structures that develop in this pattern

wave collide. Of these, only subfamily 22, which produces patterns that increase in manifest complexity fastest, is described in detail here, but Fig. 20.6 shows a member of subfamily 4, g2-g2-s24sw29, as it appears after 2^{24} steps (at the right scale, it would appear the same after any greater number).

Each line is labeled with a type (B, G or O — for gosper2 body, glider-wave and oscillator ray respectively), a generation number, a slope (distance traveled from south to north (positive) or north to south (negative) when moving a unit distance west to east along the line — for vertical lines, the symbol "≫" is used), and in the case of glider-waves, the gliders' direction of travel. The gosper2 bodies and primary glider-waves are given generation number 1, indicating production directly by the gosper2 heads; other waves, and the oscillator rays, are all products of interaction between pairs of lines with lower generation numbers, and are given a number one greater than the maximum of those of these lines. Thus the two lines labeled "G:2:5/3:NE" and "G:2:-1/3:SE" are the secondary waves created when the primary southwest wave (G:1:1/3:SW≫) hits the south-headed gosper2 body (B:1:≫); the line labeled "O:3:-2" is the ray of oscillators created by collisions between members of G:1:3:NE and G:2:-1/3:SE (which are both modified by the collisions); G:4:9/5:NE and G:4:-11/9:NW are glider-waves created when G:1:1/3:SW interacts with O:3:-2 (again, both interacting lines are themselves modified). The moving intersection between any two lines itself follows a straight-line path away from where the gosper2s started as the entire pattern grows; at each such moving intersection, one or more specific types of interaction may occur repeatedly — or the two lines may simply pass through each other without interacting, as in the case of G:2:5/3:NE and G:4:-11/9:NW here. If any of the types of interaction between the waves produce oscillators, the path of the moving intersection becomes visible

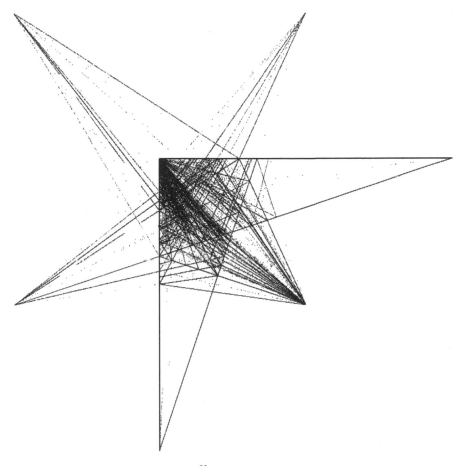

Fig. 20.7 Pattern g2-g2-s24sw72, after 2^{32} steps (entire pattern shown)

as an oscillator ray. In g2-g2-s24sw29, the range of interactions occurring at every moving intersection soon (i.e. within some tens of thousands of steps) becomes fixed.

Pictures of members of subfamily 22 are less easy to interpret because the number of lines increases continually (at least, as far as the patterns' development has been followed), and later lines are sparser than earlier ones, so it may be unclear what lines there are. Figure 20.7 shows a member of this family, g2-g2-s24sw72, after 2^{32} steps. The tips of the eastern and southern points of the irregular six-pointed star are the gosper2 heads. All the lines leading to the other points of the star are glider-waves, with the gliders moving outwards. Other members of the subfamily appear similar in general, but different in detail. Figure 20.8 shows part of the same pattern after 2^{27} steps, in which multiple oscillator rays and glider-waves of different densities, some with long gaps, are clearer. The patterns in this subfamily, and some other g2-g2 patterns, can surely be characterized as ramifying feedback networks:

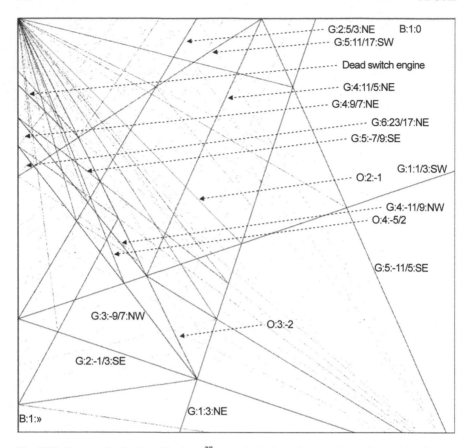

Fig. 20.8 Pattern g2-g2-s24sw72, after 2^{27} steps (only the main area of complex activity shown)

multiple, interacting (and in this case expanding) feedback loops in which new types of local interactions (collisions between spaceships, or between spaceships and os-cillators) are continually generated. Some of the early stages in the development of the ramifying feedback network in g2-g2-s24sw72 (and other members of its sub-family) can be understood from Fig. 20.8:

- The lines labeled O:3:-2, and G:3:-9/7:NW are created by the interaction between members of G:1:3:NE and G:2:-1/3:SE. These interactions can create oscillators, and send gliders in all four directions.
- Without the modifications these collisions cause, G:1:3:NE would pass through G:1:1/3:SW, but the modifications cause interactions, leaving O:2:-1 as evidence, and removing gliders from both waves.
- Both O:3:-2, and G:3:-9/7:NW also interact with and modify G:1:1/3:SW. Each remnant in O:3:-2 affects a part of G:1:1/3:SW that was northeast of the collision creating that remnant when it occurred, while the gliders in G:3:-9/7:NW created in that same collision affect members of G:1:1/3:SW that were created earlier, and were due north of the collision when it occurred. Thus the collision between

G:1:3:NE and G:2:1/3:SE indirectly modifies G:1:1/3:SW at two different points, which become increasingly distant with time.

- The interaction between G:1:1/3:SW and O:3:-2 also creates G:4:11/5:NE, which collides with B:1:0 creating G:5:-11/5:SE, which in turn collides with G:1:1/3:SW, modifying a part of it that has not yet been through any of the other interactions mentioned.

Where a linear macrostructure such as a wave or ray undergoes interactions with several other such macrostructures, all these interactions occur simultaneously, involving different parts of the macrostructure. The relationships between these interactions can become confusing because of the different scales involved: any given part of G:1:1/3:SW encounters G:5:-11/5:SW before it encounters O:3:-2; but any particular member of G:5:-11/5:SW results from a collision between an earlier member of G:1:1/3:SW, and a member of O:3:-2.

Figure 20.8 shows macrostructures made both of oscillators (the gosper2 bodies and the oscillator rays), and of spaceships (the glider-waves): the former simply grow at one end, while the latter both grow and move. Cutting across the oscillator/spaceship distinction is one between structures produced by a single growth point (the gosper2 bodies and primary waves), and those produced by successive collisions between clusters belonging to pre-existing macrostructures and taking place at a moving intersection (the oscillator rays and the remaining glider-waves). As already noted (in relation to G:1:1/3:SW and B:1:≫), there are cases where successive collisions of this kind interact with each other. Such a sequence may persist until a specific event disrupts it, establishing an alternative pattern, which itself persists until disrupted. Removing a single glider from G:1:1/3:SW in g2-g2-s24sw13 can produce such a switch into an alternative sequence, in which every six gliders in G:1:1/3:SW produce just two gliders in G:2:-1/3:SE and two in G:2:5/3:NE, instead of four in G:2:-1/3:SE and four in G:2:5/3:NE. Removing another glider reverses the switch. Such switching is a form of cross-scale interaction: a local change that can — if the conditions are right — have an effect on a scale which is non-locally determined. It creates meso-structures, distinct from both fixed-size micro-structures and ever-growing macro-structures.

Why does the feedback network continue ramifying in g2-g2-s24sw72, but not in g2-g2-s24sw29? In g2-g2-s24-sw72 the interaction between G:2:-1/3:SE and G:1:3:NE produces larger and more varied oscillator clusters in O:3:-2 than in g2-g2-s24sw29, and new gliders going in all four directions (no new gliders are produced in g2-g2-s24sw29); and further interactions involving these products and G:1:1/3:SW lead to a considerable variety of glider-clusters in this wave, and oscillator clusters in O:3:-2 and B:1:≫, in turn leading to additional types of collision. Given sufficient variety of collisions, at least some will produce novel oscillator and glider-clusters and hence the opportunity for further novel collisions; there may thus be quite a sharp division between diverging puffer pair patterns that produce enough initial variety to initiate this positive feedback loop, and those that do not.

20.3.3 Emergence of Additional Puffers

As noted, linear macro-features of the g2-g2-s24 family of patterns are produced
by micro-level mechanisms interacting in two different ways: through the action of
growth points (gosper2 heads) existing in the starting pattern, and by the repeated
interaction of components of other macro-features. Here, the focus is on linear features arising in a third way: by the production of new puffers — specifically, block-
laying switch engines. In this kind of cross-scale interaction a sequence of local
interactions widely distributed across two-dimensional space gives rise to a new
(meso-scale) linear structure.

The block-laying switch engines appear from the moving intersection between
B:1:$\gg$ (the south-headed gosper2's body) and G:3:-9/7:NW, the glider-wave produced by collisions between G:2:-1/3:SE and G:1:3:NE. This moving intersection
travels down B:1:$\gg$ at a speed of $c/14$. For a block-laying switch engine to be produced, a section of B:1:$\gg$ must have been "prepared" in the right way. Full details of
the process are too lengthy for this chapter, but Figs. 20.8 and 20.9 illustrate an outline account. The most important point is that a glider belonging to G:5:11/17:SW
(the preadjuster mentioned below) plays an essential part.

Gliders beginning as parts of primary glider-waves at four separate points on
B:1:0, and three on B:1:$\gg$, contribute to the final result.

- Gliders going southwest from around point D on B:1:0 in Fig. 20.9 reach B:1:$\gg$
  at D$'$.
- To contribute to production of a switch engine the wave must, when it reaches D$'$,
  include the sequence 01101(111) $*$ 0111 (the key sequence), where 1 indicates a
  glider, 0 a gap where a glider has been removed, and ($\langle$subsequence$\rangle$)$*$ a subsequence that can be repeated any non-zero number of times.
- The third 0 must represent a glider removed in a collision at point Y (leaving a
  cluster of oscillators there), with a glider headed northwest from B$''$ (see below).
- The other two gliders may have been removed in any of several ways.
- The start of the key sequence institutes a shift from the most stable form of interaction between B:1:$\gg$ and an unaltered G:1:1/3:SW to a second, less stable
  interaction. The final missing glider in the key sequence reverses this shift, and
  the gliders sent southeast as part of G:2:-1/3:SE around this time will interact with
  gliders from E$'$ to send gliders back northwest (as members of G:3:-9/7:NW).
- These can interact again with B:1:$\gg$ around D$'$ to make a southeast-headed
  switch engine.

For this to occur, however, a small change (one small still-life — a tub — being
turned into another — a boat) must be made to B:1:$\gg$ by a southeast headed glider
arriving on exactly the right diagonal from a second collision at point Y. Producing
this adjuster involves interactions between stretches of B:1:0 around points A, B and
C in Fig. 20.9, and stretches of B:1:$\gg$ around C$'$ and E$'$:

- The interaction between B:1:$\gg$ and G:1:1/3:SW around A$'$ produces gliders in
  G:2:-1/3:SE that interact with G:1:3:NE gliders from C$'$ to produce a collection
  of oscillators at A$''$ (belonging to O:3:-2).

Fig. 20.9 Construction of
block-laying switch engines
in g2-g2-s24sw22 subfamily
of patterns

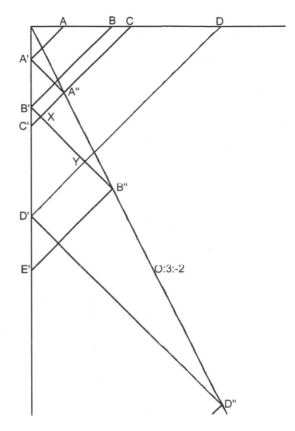

- Gliders in G:1:1/3:SW generated around C hit these oscillators, sending gliders back as part of G:4:11/5:NE.
- These collide with B:1:0 to produce gliders belonging to G:5:11/17:SW, which again encounter oscillators at A″ producing a glider (the preadjuster) that proceeds to point X.
- Here it may meet a glider produced in a similar journey from B, involving collisions at B′ and B″. The latter sends five gliders northwest as part of G:3:-9/7:NW. It is one of these that must cause the third gap in the key sequence, leaving a remnant at Y; another must meet the preadjuster at X, sending another glider back to Y to collide with the remnant there, producing the southwest headed adjuster.

This sequence of collisions was neither designed nor foreseen; the ramifying feedback network produces many such sequences, and it "just happens" that this one produces a switch engine — although this must be a logical consequence of the rules of GoL and the initial pattern used.

   If successive positive integers are assigned to the members of G:1:1/3:SW (call the integer assigned to a member its index), those contributing to the construction of a switch engine as described must have an index with certain residues in relation to specific moduli. In particular, a glider producing the remnant at Y and hence leaving the final gap in the key sequence is constrained to have a specific index residue modulo 119 ($= 7 \times 17$) which is fixed for a particular member of the subfamily, and varies in a systematic way across subfamily members. Since members of every simple wave of gliders are necessarily separated from each other by multiples of fifty half-diagonals (because of the regularities in the gosper2 body and primary glider-wave), the possible tracks of the members of G:1:1/3:SW that take part in the collision leaving the final 0 in a key sequence are separated by multiples of 5,950 half-diagonals, and hence the positions down B:1:$\gg$ from which switch engines can grow are separated by multiples of 5,950 cells.

Empirically, there is also a bias in the values modulo 3 of the starting positions of switch engines. Any two members of the g2-g2-s24sw22 subfamily of initial patterns can be mapped onto each other by shifting the relative position of the two gosper2s in the initial pattern by a multiple of 50 cells in the north-south direction, and then by a multiple of 25 cells in the northeast-southwest direction. Three groups within the subfamily can be defined by the residue modulo 3 of the north-south distance between the northwest cells of the two gosper2s, minus the east-west distance between the same two cells (which is the value of j in g2-g2-s$j$sw$k$). For each group, the north-south distance between B:1:0 and the starting positions of switch engines shows a strong bias toward one residue modulo 3: 108 of the 114 switch engines found have one of the three residues (standard 3-residue switch engines), 6 have the residue reached by adding 1 to this (standard-plus-1 3-residue), while none have yet been found in the standard-plus-2 3-residue class. This bias has not been fully accounted for, but appears to stem from the interactions that create O:3:-2, which consists of several types of remnant, each a cluster of oscillators. For any particular type of remnant, only every third member of G:1:3:NE can take part in the collision that creates it. Both of these leave a large remnant in O:3:-2 that will remove at least two gliders from G:1:1/3:SW. These interactions will give a modulo 3 bias to the index of the gliders, and particular sequences of gliders and gaps, remaining in G:1:1/3:SW after its encounter with O:3:-2, and hence a modulo 3 bias to the north-south positions at which particular sequences reach B:1:$\gg$.

There are also apparent differences in the ease with which switch engines form in the three groups of patterns defined above, according to a systematic study of 72 patterns followed for at least 2^{30} steps, and 144 more followed for at least 2^{28}. Again, this may be due to the way O:3:-2 removes gliders from G:1:1/3:SW.

20.3.4 Interacting Secondary Puffers: Emergence of a Quasi-population

Once a switch engine is formed, it grows southeast, the head moving both south and east at $c/12$. As long as its growth continues, the head is thus south of the moving intersection between B:1:$\gg$ and G:3:-9/7:NW (which moves at $c/14$), and south of any subsequently generated switch engine (heads of all growing switch engines form a straight line with a slope of $-1/7$). The switch engine head would eventually

Fig. 20.10 Switch engine
survival in g2-g2-s24sw22
subfamily of patterns

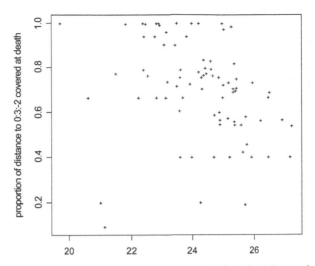

log (base 2) of switch engine's distance south of east-haeded gosper2

crash into the O:3:-2 remnant that produced the gliders that created it — hence, a switch engine's maximum lifespan is determined by its time of origin — but it is often destroyed before this can happen.

Figure 20.10 shows, for all the switch engines followed from "birth" to "death" in all members of the g2-g2-s24sw22 subfamily investigated, how far they grew before a fatal collision. The horizontal scale (logarithmic for clarity) shows the distance south of the tail of B:1:≫ at which a switch engine was formed; the vertical scale shows the proportion of the distance to O:3:-2 the switch engine grew before dying. Several points may be noted:

- The proportion of switch engines alive, at the last point at which the pattern was observed, grows toward the right of the plot, so the apparent decline in the average proportion of the distance to O:3:-2 covered before death may be partly an artifact. However, this would not explain the larger proportion of later-born switch engines killed before reaching 2/3 of the way to O:3:-2.
- A number of switch engines reached O:3:-2.
- At least two switch engines died extremely close to 0.2 (1/5) of the way to O:3:-2, several extremely close to 0.4 (2/5) of the way, or extremely close to 0.667 (2/3) of the way. These are hypothesized to have been killed by collision with members of pre-existing waves of gliders or lines of oscillators.
- Most of the remaining switch engines are believed to have been killed by an indirect process beginning when a glider hits the switch engine body. Several glider-waves hit a growing switch engine from the side (Fig. 20.11): some gliders pass through without hitting any blocks, but others do hit them; these may generate further gliders going in any direction. Those going southeast will overtake the head, and may then collide with something and either form an obstacle to the switch engine's progress, or produce gliders traveling northwest, which may collide with the switch engine head or with a later southeast glider. Hence, lo-

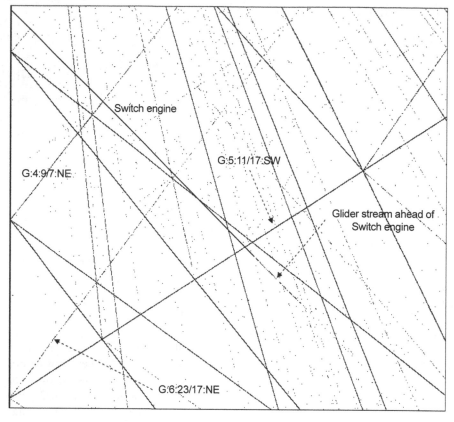

Fig. 20.11 Pattern g2-g2-n126sw297 after 2^{30}, showing growing switch engine's interaction with its environment

cal interactions between the switch engine's body and its environment can lead indirectly to its "premature death".

Even if the deaths at 1/5, 2/5 and 2/3 of the way to O:3:-2 are excluded, Fig. 20.10 shows apparent clusters of deaths: one to the right and between 0.5 and 0.6, a second spread more evenly across and between 0.7 and 0.85, and possibly a third above 0.9. The last two groupings might merge when further data is collected, but there are likely explanations for there to be two groupings at least. Taking the second and third groupings together, the densest wave a switch engine crosses is G:5:11/17:SW; it does so when it has gone 0.4 (2/5) of the way to O:3:-2. From then on, interactions between the two will generate an irregular glider-stream ahead of the switch engine. The leaders of this will reach O:3:-2 when the switch engine has completed 0.6 (3/5) of its journey, and northwest-headed gliders created by the reaction between the stream and the O:3:-2 remnants it hits cannot meet the head before it has completed 0.7 (7/10) of its journey. Hence, the number of switch engines dying should rise sharply at this point, then tail off (since if an earlier interaction kills the switch engine, a later one cannot). The peak in numbers dying between 0.5 and

0.6 probably results from the switch engine's interaction with two less dense waves. G:4:9/7:NE first hits the switch engine right near its root in B:1:≫, and resulting northwest-headed gliders can start meeting the switch engine halfway to O:3:-2. However, later interaction with G:6:23/17:NE can produce lumps on the switch engine body that block most of the southeast gliders produced by G:5:9/7:NE, while interaction with G:6:23/17:NE produces fewer southeast gliders itself.

How long a switch engine lives will, in general, depend on a large number of local interactions, most of them taking place along its body, which acts as a kind of "catalytic surface" in the two-dimensional array, linking local interactions in a complex causal network: another form of cross-scale interaction. Thus the exact sequence of gliders hitting a switch engine determines how long it lasts, and hence its effect on the wider pattern — but omitting some gliders will make no non-local difference, while others will be crucial, as has been suggested is characteristic of cross-scale interaction.

Additionally, switch engine bodies act as a kind of "semi-permeable membrane", thinning the glider-waves that cross them, and altering their composition — sometimes producing altogether novel glider-clusters. This is yet another form of cross-scale interaction, as the switch engine alters the relative frequency of different types of glider-cluster in a wave (eventually, the switch engine will die and the wave will, later, move past its southeast end, so their interaction can be regarded as a meso-scale event — non-local but limited in size and duration). Block-laying switch engines produce a body with a spatial periodicity of 48 half-diagonals. Of the 48 ways a single glider can approach the body from the side, only eight allow the glider to pass through, one produces an onward glider shifted by two half-diagonals toward the head, and one produces gliders going in all three other directions, but none going onward. For glider-clusters the situation is more complicated, but usually fewer travel onward than were originally in the cluster. The changes to the wave will affect lines it subsequently intersects, so the switch engine's influence will extend through the pattern's network of lines. This effect on its environment will continue even after its death, although any individual switch engine should become less influential as the main areas of activity move south and east.

Switch engines with starting positions on B:1:≫ which are not a multiple of 48 cells apart can interact differently with glider-waves, if the latter also show certain types of spatial periodicity. Specifically, G:5:11/17:SW, has a modulo 300 periodicity in the northwest-southeast direction: any two clusters of gliders of the same type (i.e., produced by homologous processes) will be separated by a multiple of 300 half-diagonals in that direction. Thus for any such type of cluster, and a given switch engine, there are eight possible interactions, and switch engines belonging to the three different 3-residue classes will have non-overlapping sets of interactions. Hence the effects of the interaction with this wave will be different for the three classes.

If a block-laying switch engine's head hits an oscillator or oncoming glider, the outcome depends on the obstacle's exact position and the switch engine's phase. The switch engine may be killed; the disturbance may be transitory; or the switch engine can change phase, so that once production of a regular trail of blocks resumes, it will have shifted in the northwest-southeast direction (for a southeast headed switch engine) by 16 or 32 half-diagonals modulo 48. Finally, conversion into a glider-stream switch engine has been observed in one g2-g2 pattern, and in others of similar types. The glider-stream switch engine itself was quite soon killed, but another outcome is possible, although not yet observed: if the glider-stream hits an obstacle in the right way, it may build up glider crystal which the head may subsequently overtake.

Switch engines will influence each other, particularly if more than one is or has recently been growing, as should become more common over time (across the 142 patterns followed as far as 2^{30} steps, the total number of switch engines produced is 0 after 2^{23} steps, 1 after 2^{24}, 1 after 2^{25}, 2 after 2^{26}, 13 after 2^{27}, 23 after 2^{28}, 44 after 2^{29}, and 85 after 2^{30}, indicating that the rate of switch engine production is roughly constant at least up to 2^{30} steps). Consider two switch engines starting their growth close in time. As soon as the second begins to grow, it will interpose itself between G:4:9/7:NE and the first. Conversely, the first switch engine will be interposed between the second and G:5:11/17:SW (the densest wave that interacts with switch engines), once the second has completed 2/5 of its journey. Such interposition is likely to be protective on average, since interactions with the interposing switch engine will tend to reduce the number of gliders hitting the other, and hence the number in the stream of gliders overtaking the head. However, interaction with the first switch engine may produce gliders that would not otherwise have existed, and in some cases these may generate southeast gliders when hitting the second.

The effect of interposition will vary with the exact distance between the starting positions of the two switch engines. The north-south distance between the starting positions of two switch engines produced within the same pattern may have any even residue modulo 48; although because of the modulo 3 bias in starting positions noted earlier, the residues 0, 6, 12, 18, 24, 30, 36 and 42 are (over the cases known) most common. Any particular residue (modulo 48) difference in starting position corresponds to a distinct way in which the trails of blocks making up the bodies of the switch engines line up with each other in a southwest-northeast direction. If (and only if) the residue modulo 48 is 0 or 24, any glider passing through one switch engine body will pass through the other. Only for residues 0, 2, 22, 24, 26 and 46 will there be any path a glider can take through both switch engine bodies without a collision. Hence, the difference in residue modulo 48 may make a big difference in the effect each switch engine has in altering the interactions the other has with glider-waves. Pairs with distances of residue 0 or 24 modulo 48 are likely to offer each other most protection, as any glider that passes through one without interaction will also pass through the other in the same way, whereas such gliders may cause the production of southeast gliders in the case of other residues.

The fact that switch engines can cross G:5:11/17:SW means they can inhibit production of new switch engines, because the preadjuster glider belongs to G:5:11/17:SW. Standard 3-residue and standard-plus-1 3-residue switch engines will block 3/4 of preadjusters, while standard-plus-2 residue switch engines (if any exist), and most pairs of switch engines that are not aligned in the northwest/southeast direction modulo 24 will block all preadjusters. If the starting position of a switch engine is r cells south of B:1:0, some or all switch engines may be prevented from forming that would have had starting positions between $17r/5$ and $17r/3$ cells south of B:1:0. Restricting attention to relationships between standard 3-residue switch engines, a switch engine that grows all the way to O:3:-2, with a starting position r cells south of B:1:0, where r has residue m modulo 24, will only permit the formation of possible switch engines between $17r/5$ and $17r/3$ cells south of B:1:0 if they have a starting position $m - 6$ mod 24 cells south of B:1:0 (for patterns where j in g2-g2-s$j$sw$k$ is 0 mod 4); and only if they have a starting position $m + 6$ mod 24 cells south of B:1:0 (for patterns where j in g2-g2-s$j$sw$k$ is 2 mod 4).

Consider two sequences of standard 3-residue switch engines arising in similar patterns and growing across G:5:11/17:SW, sequence A having north-south gaps in their starting positions of residue 0 modulo 24, sequence B including a mix of gaps of 0, 6, 12 and 18 modulo 24. Sequence A will increase the chance of several successive residue-3 switch engines being separated by residue 0 gaps modulo 24 at a later point, by blocking most of the preadjusters that could give rise to switch engines in different residue classes modulo 24. Sequence B, on the other hand, will block all preadjusters so long as any two with different residues modulo 24 are across G:5:11/17:SW. This will produce a stretch of B:1:$\gg$ with no switch engines at all — but that in turn will increase the likelihood that there will be a period when all preadjusters get through and create another sequence of switch engines including many gaps that are non-zero modulo 24.

Moreover, if the effect in at least some patterns is to lead to sequences of switch engines in which most adjacent pairs have a gap of 0 modulo 24 (the interactions discussed above between switch engines that simultaneously interact with the same glider-waves make this plausible), a slightly stronger form of quasi-hereditary effect could arise. The sequence of gaps between the starting positions of successive standard 3-residue switch engines would consist mostly of 0s, with an occasional 6, 12 or 18. Ignoring the 0s, consider the sequence of non-zero residues that remains. Each member of that sequence now represents a switch from one batch of standard 3-residue switch engines with gaps of residue 0 modulo 24 to another, with three different types of switch possible. Any switch or subsequence of switches might then have a tendency to produce a copy of itself further down the sequence. For example, suppose a batch of switch engines with starting position residues of 0 modulo 24 is followed by a batch with residues of 12 modulo 24. If the pattern is of the $j = 0$ mod 4 type, then while only members of the first batch lie across G:5:11/17:SW, preadjusters giving rise to switch engines with starting position residue 0, 6 or 12 mod 24 will be blocked, so that only switch engines with residue 18 can be created on a specific stretch of B:1:$\gg$ further down. (Strictly, this is true only for standard 3-residue switch engines; non-standard 3-residue switch engines with residues 2 or 34 could be created, but as noted, these are much rarer.) There is then likely to be a period when residue 0 and residue 12 switch engines both lie across G:5:11/17:SW, blocking all preadjusters. Once the last residue 0 engines no longer lie across G:5:11/17:SW, preadjusters able to create switch engines with starting positions of residue 6 can get through. Hence, once these preadjusters have done their work, a batch of residue 18 engines will be followed by a residue 6 batch, replicating the 12 in the non-zero gap-residue sequence. Parallel arguments can be constructed for patterns of the $j = 2 \mod 4$ type, for a non-zero gap-residue of 6 or 18, and for sequences of two or more non-zero gap-residues.

This line of reasoning is tenuous, but raises intriguing possibilities. Given the ways switch engines can influence each other's survival by modulating the effect of glider-waves discussed above, different non-zero gap-residues are likely to give the batches of engines they separate different expected lifetimes — and hence, differing abilities to influence the production of future engines, and non-zero gap residues. Even the possibility that partially heritable differences in ability to self-replicate might emerge from the ramifying feedback networks and cross-scale interactions in g2-g2 patterns is surely of considerable interest. There are also ways in which switch engines could promote the formation of additional switch engines by manufacturing additional adjusters. Those so far discovered (but not yet observed in growing patterns) involve at least two transformations of block-laying switch engines into the glider-stream variety, and the interaction of at least one of these with a length of glider-crystal. They could, like the inhibitory mechanism described above, affect the relative frequency of occurrence of gaps of different sizes between the starting positions of adjacent switch engines.

As far as available computational resources have allowed g2-g2-s24sw22 subfamily patterns to be followed, their apparent complexity continues to grow. There is no proof this will continue, but conversely, new complexity-generating mechanisms could arise — for example, additional switch-engines, or other forms of puffer. The long-term fate of these patterns, therefore, remains unknown.

20.4 The Infinite Case: Sparse Life

20.4.1 Random Configurations

Recall that a d-cluster is a maximal set of state 1 cells such that each pair of cells in the set is linked by a neighbor-to-neighbor path that never goes through more than d successive state 0 cells, while a cluster is a set of state 1 cells which constitutes a d-cluster for some d. The number of state 1 cells in a cluster is its size. The definitions of still life, oscillator, repeater and spaceship apply to clusters just as to patterns.

A quiet cluster consists of a set of repeaters, no two of which would ever overlap the same 1-cluster in such an array. The repeaters may move relative to one another, but no two will ever interact. An indefinite growth cluster would increase its number of cells without limit if it were a pattern; this definition implies that it would never become quiet. It will be said that two indefinite growth patterns (and by extension clusters) P1, P2 are of the same kind if and only if there are integers m and n such that if P1 is advanced m steps, and then both are advanced synchronously by any number of steps, it will always be possible to change the states of $\leq n$ cells in the P1 configuration so that an automorphism of the array can map it onto that containing P2. Intuitively, the patterns grow indefinitely, but the difference between them does not.

The layout arising after t steps from an infinite random GoL array, with initial state 1 probability p for each cell, will be called $L_{p,t}$. Specifying t and p determines the expected frequency of every possible arrangement of cell-states in the array (how many copies there will be, on average, per unit area). For $L_{p,0}$ the frequency of a specific cell-state arrangement will depend only on p and the number of cells required to be in each state. All possible arrangements will appear at non-zero frequencies, including self-reproducing clusters (believed to exist in GoL — see [6] surrounded by arbitrarily large empty areas (so long as $0 < p < 1$). For all values of p and t, we can in principle calculate exact cell-state arrangement frequencies for any size and shape of patch of cells, but the effort required appears to grow very rapidly with t. However, we can say that, as effects cannot propagate faster than one link per step, arbitrarily large empty areas will survive at any finite t. Copies of all possible types of repeater, and of all possible kinds of self-reproducing cluster, will also survive at any finite t, since copies of all types of cluster will occur at $t = 0$, surrounded by arbitrarily large empty areas. However, it may be that for some or all such p values, the density at which such clusters occur in the array tends to zero. Any given repeater has a characteristic velocity with which it moves across the array (which is zero if it is an oscillator). Repeaters with different velocities may begin arbitrarily far away from each other, but on course for an eventual collision. Copies of any cluster which is the outcome of such a collision of two or more repeaters will therefore exist at any finite t. In $L_{p,0}$, all possible finite arrangements of cell-states will appear infinitely often. However, some of these have no possible predecessor [6], and will not appear at any $t > 0$.

20.4.2 Arbitrarily Sparse Random Configurations

For any infinite binary CA, there is a p below which there will only be finite 0-clusters in random arrays. The critical value of p will be $\geq 1/(z - 1)$, where z is the number of non-self neighbours a cell has [32]. We can extend the result to any value of d by noting that adding links to an infinite but locally finite vertex-transitive graph so that each node is joined directly to all and only those previously no more than x links away, for some x, gives us another such vertex-transitive graph. There will be a density below which there are only finite 0-clusters in random configurations on the network formed in this way, and removing the additional links will give a random configuration without infinite $(n - 1)$-clusters on the original network [33]. We can therefore begin analysing arbitrarily sparse random arrays of a CA by thinking about the frequency with which different clusters will occur. Very sparse CA configurations give rise to a logarithmic time scale of eras expressed in terms of p, or its reciprocal, which we designate N. "In era 1", for example, means "after $\approx N$ steps"; "in era 3" means "after $\approx N^3$ steps". Powers of p are used in referring to probabilities and densities, powers of N in referring to distances, sizes, and durations; $\approx N$ should be read as "around N". The assertion that $F(p) \sim x^a$, for some function F defined over positive values of x, means that $(F(x)/x^a + x^a/F(x))/(x^a + x^{-a}) \to 0$ as $x^a + x^{-a} \to \infty$. This covers cases where x is either greater than or less than 1. More simply, $F(x) \ll x^a$ means that $F(x)/x^a \to 0$ as $x^a + x^{-a} \to \infty$, and $F(x) \gg x^a$ means that $x^a/F(x) \to 0$ as $x^a + x^{-a} \to \infty$.

In considering the history of an infinite random configuration from the beginning, any class of events which occurs within a fixed number of steps can be regarded as short term. Long term, for infinite arrays, refers to what happens in the temporal limit. Events which occur on a timescale specifiable in terms of powers of N are medium term. The medium term covers all the ground between the short and long terms, but this study does not consider any medium term events beyond era 3. Local, global and intermediate levels of spatial structure can be defined analogously, and we can again use powers of N to describe distances on the intermediate level. The term "local cluster" will be used below. A local cluster is one with diameter $< N^a$ where a can be any positive value sufficiently near zero that a lower value would make no difference to the analysis; a can be taken to be $1/4$ here.

Consider cluster-types defined so that two clusters are of the same type if and only if a translation maps one onto the other. The exact density at which such a cluster-type will be found in $L_{p,0}$ depends on how many cells must be in state 0 for a cluster of that type to exist, as well as how many must be in state 1. However, with p very low, p^s where s is the number of state 1 cells in the cluster, gives a very good approximation so long as s and d are not too large. Only $\approx p$ of the state 1 cells will be $< N^{1/2}$ links from any others, only $\approx p^2$ of them will have two others that near, and so on. All translation-defined cluster-types of the same size will be almost equal in frequency in $L_{p,0}$ for small s and d, while there will be large differences in frequency (growing without bound as $p \to 0$) between cluster-types of different sizes. The effect of this is to make the behaviour of initially small clusters very important for the dynamics of sparse random configurations. It will be useful to define the cumulative image of a set of 1-clusters existing at time t_x, over a period t_x to t_y. At t_x, it consists of the cells in the cluster set. At t_{x+1}, it consists of those cells, plus any neighbouring cells that have now entered state 1, so long as these cells, along with those original cells that are still in state 1, again make up a set of 1-clusters. At each subsequent step, we repeat the procedure of adding neighbouring cells that have just entered state 1, and checking that these plus the survivors from the previous step still make up a set of 1-clusters. So long as they do so, the cumulative image of the original set of 1-clusters remains well-defined.

20.4.3 Large Sparse Random GoL Arrays: The Short Term

The first 17,410 steps

In this subsection, when a quantity or ratio is said to approach or tend to some value, it is to be understood that this occurs as $p \to 0$. At $t = 1$, all clusters of size ≤ 2 will vanish (and some larger ones). By $t = 2$, all initial size 3 3-clusters will have vanished except for blinkers and pre-blocks (which become blocks at $t = 1$). The same will be true of a proportion of size 3 1-clusters approaching 1 (some initial size 3 1-clusters will be parts of larger 2- or 3-clusters, and some of these will interact with other parts of these super-clusters). The ratio of blocks to blinkers tends to 2:1 and the density of state 1 cells tends to $22p^3$. From $t = 2$ onward, the layout will, for a number of steps $\to \infty$ as $p \to 0$, contain mostly these blocks and blinkers, with the proportion of other cells approaching 0 as p does. (Collectively blocks and blinkers are blonks; those derived from 3 cells at $t = 0$ are original blonks; and "original" is also applied to other clusters derived from minimal size clusters at $t = 0$.) It is important to note that however low p is set, so long as it is non-zero, examples of 1-clusters of size greater than 3, and indeed examples of any type of finite cluster, will exist in an infinite array. Some of these larger clusters will, in the medium term, become dynamically important. By $t = 11$, all initial size 4 clusters which have not interacted with other clusters (approaching all of them as $p \to 0$) reach a stable form or enter a 2-cycle, producing additional oscillators scattered thinly among the original blonks: more blonks, plus beehives, traffic lights, ponds and tubs in ratios approaching 58:12:4:1 (reflecting the numbers of cluster-types leading to each). Size 5 clusters give rise to two new kinds of oscillator. More significantly, some are or become gliders — and others are or become r-pentominos, each of which produces six gliders, plus a clump of oscillators. These two forms are produced in a ratio approaching 7:9. Some of the r-pentomino predecessors take 1,105 steps to reach quiescence. All clusters with between six and nine cells reach quiescence by $t = 17,410$, apart from some that consist of two sub-clusters, one of which produces a glider that later hits the other (a survey of all clusters up to nine cells was undertaken by Paul B. Callahan). In all these cases, the two sub-clusters both reach quiescence before $t = 17,410$.

Minimal Clusters and Their Histories

In the short term, the minimum number of cells that can give rise to any type of cluster is the principle determinant of how common that type of cluster will be. In analysing very sparse GoL fields we are interested in the most common clusters with particular dynamic properties, and therefore in those with the smallest initial cluster sizes. We have seen the smallest persistent GoL clusters (three cells), and the smallest spaceships and clusters with indefinite growth in diameter (five cells). Paul B. Callahan found three 10-cell indefinite growth clusters that increase their number of cells linearly with time [20], and others have been found since. These patterns grow into switch engines: most into block-laying switch engines, but at least one into a glider-stream switch engine. The number of cells in a GoL pattern cannot indefinitely grow at more than a quadratic rate; its diameter (as for any CA) can at most grow linearly. The contribution made to overall density by any type of original cluster will at most grow in proportion to t^2 (interaction between clusters cannot overcome this limit, although it might produce new quadratic growth clusters). The linear limit on diameter growth means that an original cluster's distance from a given cell, or a straight strip of cells (such as a glider's prospective path or a switch engine's trail) cannot shrink faster than proportionally to t. The size of the cumulative image may grow at the same rate as the current cell count or may grow faster, unless the latter grows quadratically. There are known patterns of 16 cells with a cumulative image that grows quadratically: these are pairs of switch engines that interact in such a way as to fire gliders at right angles to the movement of the head, so that each successive glider follows a previously untraveled path. The author recently found the smallest known patterns with indefinite superlinear (in fact, quadratic) cell count growth. These have 26 cells, initially consisting of one such 16-cell pattern, plus ten cells that give rise to a glider-stream switch-engine; their interaction produces an ever-growing set of block-laying switch engines.

So long as we focus attention on the short-term, and hence on local events, the overwhelming majority of state 1 cells will belong to original blonks. Almost all local clusters with current size ≥ 12 (the size of the largest oscillator starting from four cells) will be original r-pentominos.

20.4.4 Large Sparse Random GoL Arrays: The Medium Term

Era 1 and Beyond

After $\approx N$ steps, original gliders will have travelled a distance of $\approx N$ links across the array, if they have not collided with anything. The cumulative image each has produced will have a size of $\approx N$ cells. Original r-pentominos will have a diameter (maximum distance between state 1 cells) of $\approx N$ links, original switch engines a size of $\approx N$ cells, and original clusters with quadratic growth in number of cells a size of $\approx N^2$ cells. Comparable numbers of current cells will be descended from original local clusters of sizes 9 and 10, each making a contribution to the global density of state 1 cells (a global density contribution) of $\approx p^9$ (assuming that there are no clusters of size 10 with superlinear cell count growth). Of the original gliders and those generated by r-pentominos, $\approx p^2$ will have collided with something; in all but $\approx p$ of these cases, this will be an original blonk. In (say) era 4/3, $\approx N^{1/3}$ times as many cells will belong to local clusters of original size 10 as to those of original size 9 (again assuming there are no size 10 clusters with superlinear cell count growth). The maximum possible global density contribution from original indefinite growth local clusters in any era is p^{10-2E}, where E is the era. This would only be reached if there are 10-cell indefinite quadratic growth cluster types — which seems highly implausible, but an exhaustive check has not been made. The known 10-cell indefinite linear growth cluster types will contribute p^{10-E} in early eras.

Interactions Between Initially Distant Clusters

Consider the future of an original glider in a very sparse, random, infinite GoL field. The expected distance to the first original blonk so placed that the glider would interact with it is $\approx N^3$ links. If nothing else interfered, the expected time to such a collision would therefore be $\approx N^3$ steps. After this amount of time, the density contribution of other original local clusters could not exceed $\approx p^4$, even if there are 10-cell quadratic growth patterns. However, interactions between initially distant clusters must be considered.

For $\approx p^2$ of the original gliders and those produced by r-pentominos, the nearest original blonk along the line of travel will be $< N$ links away, and in general $\approx p^E$ of them will have one at $< N^{3-E}$ links for any $E < 3$. While original blonks remain much commoner than anything else, the rough proportion of original gliders that will have collided with one can be calculated. In the first $\lfloor N \rfloor$ time steps (where $\lfloor x \rfloor$ is the largest integer no greater than x), $\approx p^2$ of the original gliders will have hit a blonk ($\approx p^3$ of them will have hit something else instead). The same calculations apply to the gliders emitted by an original r-pentomino (a single glider going in one direction, and fleets of two and three gliders in two others). The result of a glider or fleet of gliders hitting a blonk depends on precisely how the collision occurs. Two of the 12 glider/blinker collisions, and some of those between r-pentomino derived fleets and a blinker or block, produce further gliders (plus clusters of oscillators). Of the two glider/blinker collisions, one ("glider/blinker 6") produces five new gliders (two travelling onward in the same direction as the incoming glider, one in the opposite direction, and two perpendicular to the incomer). The "onward" pair both follow the same path, 52 half-diagonals from the path of the incomer. The "perpendicular" pair both travel in the same direction, their paths 53 half-diagonals apart. If one of the fleets produced by a first generation collision between a glider and a blinker itself hits something, then with probability near 1 this will be an original blonk, as long as the global density remains dominated by these. This collision may itself produce further fleets and another collision — on the same assumption, almost certain to be with either an original blonk, or the oscillators left by the first collision. This second possibility depends on exact details of the collisions, but the path of any backward glider from the second collision must at least pass near to the debris from the first, and prior to era 3, that debris is almost certain to be closer to the second collision than any original blonk in the path of any gliders produced. Similar but more complicated possibilities arise when gliders stemming from an original r-pentomino are considered.

A collision sequence involving just one original five cell cluster (a glider or r-pentomino), and a set of (zero or more) original blonks, is a standard collision sequence. (The null sequence is included for convenience.) The number of possible standard collision sequences grows rapidly with the number of original blonks involved; the probability of any given sequence occurring falls correspondingly, but these probabilities are asymptotically independent of p as $p \to 0$. Prior to era 3, and still assuming that original blonks make the predominant contribution to the global density, the proportion of original gliders and r-pentominos taking part in standard collision sequences involving b original blonks (order b standard collision sequences) by era E ($\approx N^E$ steps) is $\approx p^{b(3-E)}$. Factors due to variation in the number of output gliders different collisions produce can be neglected, as p is taken to be much smaller than any finite probability that occurs in the analysis. The density of points in the array near which the last collision in an order b standard collision sequence has at some time occurred (the cumulative occurrence density of such events) will therefore be $\approx p^{5+b(3-E)}$. Non-standard collision sequences, beginning with an original cluster of six or more cells, or involving the $\approx p$ chance of a fleet stemming from an original glider or r-pentomino encountering something other than an original blonk (or the debris from an earlier collision in the sequence), will have a cumulative occurrence density $\ll p^6$.

The switch engines growing from original 10-cell clusters have roughly the same expected distance to an obstacle in their path — $\approx N^3$ links — as a glider. Unlike glider fleets, however, switch engines leave persistent trails of oscillators, which may be struck by gliders travelling perpendicular to the trail. The proportion of original gliders involved in such collisions will be insignificant, but the proportion of original switch engines will not. At characteristic diameter and size $\approx N^{5/2}$, in era 5/2, the rate at which original switch engines are struck by gliders will reach $\approx p^{5/2}$ per step, so that one such collision could be expected in $\approx N^{5/2}$ steps. This is long before the switch engine would be expected to run into an obstacle, so long as the expected distance to such an obstacle is $\gg N^{5/2}$. As a switch engine grows past a particular point, the nearest glider moving toward it will usually be N^5 links away, but for $\approx p^{5/2}$ cases it will be $\approx N^{5/2}$ links away (or in a small proportion of those cases, less), and will hit the switch engine's trail after $\approx N^{5/2}$ steps. The glider-stream switch engine's growth can be halted by a single glider hitting the forward glider stream, creating an obstruction in the switch engine's path. The block-laying switch engine cannot be killed by a single glider, or fleet from an r-pentomino, but may be vulnerable to multiple collisions of this kind. If so, both kinds of original switch engines will "die" when they reach $\approx N^{5/2}$ cells. Moreover, it is conceivable that some sequence of collisions with the trail could "ignite" it and burn it away completely. There are orthogonally-moving counterparts to the switch engines, and at least one has a trail known to be vulnerable to this kind of reaction (in this case, the reaction ends when it reaches the head, and the growth of the trail resumes). Whether or not they "die", switch engines will almost certainly "reproduce" (catalyse the production of more switch engines) at size $\approx N^{5/2}$ cells (it is very likely that interactions between successive collisions will in some cases generate new switch engines at right angles to the original).

Original local clusters with indefinite superlinear growth in their cumulative image (whether or not this also occurs in their current cell count), will interact with their environment at an earlier era than the switch engines. In the case of quadratic growth of the cumulative image, interaction with the environment will become significant in era 3/2, when the growing cluster will have impinged on the neighbourhood of $\approx N^3$ cells, and could therefore expect to encounter an original blonk. Such a pattern would then be impinging on $\approx N^{3/2}$ previously unvisited cells per step, and would therefore expect to encounter an original blonk once in $\approx N^{3/2}$ steps. There are some quadratic growth patterns (known as spacefillers) that would be "killed" by any single encounter of this kind. Others, such as the 26 cell pattern mentioned above, appear less vulnerable, but none have yet been proved able to survive unlimited numbers of encounters with isolated blonks; doing so would require showing that the damage caused by any single encounter would always be self-limiting. No such encounter could lead to a faster than quadratic growth rate, however, so their overall density contribution could not be increased past p^{10-2E}.

Standard Collision Sequences and Auto-catalysis

Successive collisions with blonks can reduce any fleet to a single glider, and there is a five collision sequence starting from a single glider, in which the fifth collision reproduces the glider fleets and oscillators produced by an r-pentomino. Any cluster which can be produced by a collision sequence starting from an original glider or r-pentomino, and which itself produces any gliders, can therefore start a collision sequence leading to a copy of itself. Gliders, r-pentominos, and the glider fleets and other clusters produced by such collision sequences could all be considered to replicate themselves in sparse GoL fields, using the original blonks as raw material. There are at least three possible objections to this description of affairs. First, none of these patterns replicate entirely autonomously: all require interaction with objects in a non-uniform environment. However, the same is true of real organisms; it is replication in an otherwise uniform environment [25, 37] that is unbiological. Nevertheless, in the CA literature, the term "self-replication" has generally been used to refer to patterns that, placed in an otherwise uniform or periodic array, will in time give rise to multiple copies of themselves, without any finite limit. Second, it could be said that this form of replication is trivial. The term trivial self-replication is generally applied to the phenomenon found in Fredkin's parity-rule CA [15], where any pattern whatever develops into multiple copies of itself. In the current case, this is not so: only certain clusters can lead to copies of themselves being produced, and different ones do so (and produce each other), with different probabilities. However, the mechanism does not involve a universal computer, as in [37], nor even a "genome" which is both copied and read [25]. The term auto-catalysis will therefore be used for the type of "self-replication" just described. Third, no fleet or stationary cluster is known to be able to catalyse its own production in this way with sufficient efficiency to increase its global density of occurrence. Such a process could not in any case become self-sustaining until era 3, and other processes may radically alter the environment before then.

As a prelude to describing these processes, it will now be shown that glider fleets of any finite size can arise through standard collision sequences. The glider/blinker 6 collision, as noted, produces a two-glider fleet with a 53 half-diagonal sideways separation. Either of these gliders, colliding with a second original blinker in the right way, produces two forward gliders, travelling behind and one half-diagonal to the side of the other member of the fleet. The first two of this new fleet (the third neither helps nor hinders) can react successively with a block and another blinker to give another three-glider fleet, of which two again follow paths a half-diagonal apart, while the third is some 70 half-diagonals to the side. The half-diagonal pair can then go through alternating block and blinker collisions repeatedly, adding one glider to the fleet every two collisions (see Fig. 20.12). The extra gliders (represented

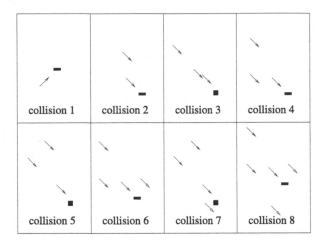

Fig. 20.12 Building arbitrarily large fleets: the lucky rake

by open-headed arrows in the figure) are regularly spaced, 82 half-diagonals apart. They can, with the right sequence of collisions, produce further half-diagonal pairs, giving a further kind of auto-catalysis, of sub-fleets within a fleet.

Self-organized Construction of Indefinite Growth Clusters

If, contrary to fact, one of the collisions between a glider and a blonk produced a cluster with indefinite linear growth in current cell number and cumulative image (like the two forms of switch engine), this process could have a significant effect on the global density by era 5/2. By this era, $\approx p^{1/2}$ of the original gliders would have taken part in such a collision, meaning that overall such a collision would have occurred near $\approx p^{11/2}$ of the cells in the field. Almost all such collisions would have generated a structure containing $\approx N^{5/2}$ cells, for a total density contribution of around $\approx p^3$ — the same order as for the original blonks. If the cluster were linear in shape as well as growth rate (like all known clusters with indefinite linear growth in both current cell number and cumulative image), an original fleet (allow-ing "fleet" to cover the case of a single glider) would then be roughly as likely to hit such a resulting cluster as to hit an original blonk: the expected distance to ei-ther kind of obstacle would be $\approx N^3$. If a cluster with indefinite quadratic growth could be produced in this way, the same density contribution would be reached in era 5/3, provided collisions of the cluster produced with original blonks did not halt its quadratic growth. By this era, $\approx p^{4/3}$ of the gliders would have taken part in such a collision, and almost all would have produced a structure of $\approx N^{10/3}$ cells. However, an original fleet would not be likely to hit such a cluster until era 5/2. A cluster must spread across a fleet's path to be hit by it; the number of glider paths a cluster obstructs can at most grow linearly. It is known that there is no standard collision sequence involving fewer than three original blonks that gives rise to an indefinite growth cluster. The author found a standard collision sequence involving 49 original blonks (an order 49 sequence), beginning from an r-pentomino, that produces a block-laying switch engine. (Incidentally, we take a standard collision sequence to have ended when an indefinite growth cluster is produced: any subsequent collisions come under the heading of interaction between that cluster and its environment.) For a standard collision sequence producing a linear growth pattern and involving collisions with b original blonks, a first calculation suggests a density contribution of $\approx p^3$ in era $(3b + 2)/(b + 1)$. In this era, $\approx p^{b/(b+1)}$ of original gliders and r-pentominos would have produced standard collision sequences of at least order b. Some fraction of these ($\gg p$, and asymptotically independent of p as $p \to 0$) will be the desired sequence. The pattern concerned would therefore have begun to grow near $\approx p^{(6b+5)/(b+1)}$ of the cells in the array. Each one would reach a size of $\approx N^{(3b+2)/(b+1)}$ during the era $(3b + 2)/(b + 1)$, for a total density contribution of $\approx p^3$. This calculation neglects the fact that, at characteristic diameter $\approx N^{5/2}$, growing switch engines will begin to be struck by gliders at intervals comparable to the time they have taken to grow to that size. The possible consequences of this fact are considered in detail in the next subsection: it makes it impossible to prove, from what is currently known, that the density contribution of emergent indefinite growth clusters — those produced by a standard collision sequence, or any other sequences of collisions starting from a set of quiet clusters — will ever exceed, or even reach, $\approx p^3$. However, it can be shown that if such clusters never do make a density contribution $\gg p^3$, other interesting emergent effects will occur.

Before turning to these matters, which require consideration of the secondary interactions between emergent structures and their environment, an outline proof will be given that standard collision sequences can lead to any collision between glider fleets obeying a simple constraint on the spacing of individual gliders within each fleet. This in turn indicates that an endless variety of indefinite growth clusters could be produced by standard collision sequences, probably including clusters capable of universal computation, and of self-replication. First, the class of glider fleet collision concerned must be specified. The relative placement of any two gliders heading in the same direction can be described in terms of the separation of their paths (expressed in half-diagonals), and of which (if either) leads the other, and by how many steps. A glider occupies a cell in a new half-diagonal perpendicular to its path every second step. For any two gliders in any one of the colliding fleets, either their paths must differ by at least 12 half-diagonals, or one must be at least 15 steps ahead of the other. A fleet obeying this constraint is well-spaced. Note that this does not rule out one glider following directly behind the other, nor two being equally far advanced along parallel paths: the two spacing requirements are alternatives (and it is probable they could be further relaxed). Note also that there are fleets, hence fleet collisions, which cannot be produced by the methods to be described. They cannot, for example, produce two gliders on the same path, one 14 steps behind the other, although two such gliders can travel without mutual interference.

Now consider a rectangular chunk of a GoL array, and mark out an orthogonal cross, with arms stretching from the centre to the four sides of the rectangle, two cells wide. The collisions we are concerned with begin with these arms empty. The quadrants between them can be designated SE, SW, NE and NW. Each quadrant is either unoccupied, or contains a well-spaced fleet moving toward the centre of the cross. Any gliders in the NW quadrant are thus heading SE, any in the NE quadrant are heading SW, any in the SE quadrant NW, and any in the SW quadrant NE. The rectangle is at the centre of a larger rectangle, empty apart from the glider fleets, and sufficiently large to ensure that nothing else is close enough to affect the collision for any period we like to specify. Any glider fleet collision beginning from such an arrangement is a well-spaced glider collision, and any such collision can arise from a standard collision sequence.

The proof cannot be given here in full detail (see [18]), but consists of three parts.

Part 1   Standard collision sequences can construct any fleet moving SE — or equally in any other direction — in which each glider is on a path at least 82 half-diagonals different from any other, and no glider is both on a path to the SW of another, and (even a single step) to the west of that other; a glider on a path SW of another will then necessarily be south of the other. (Again, it could equally be specified that no glider is on a path NE of another, and even a single step north of that other.) Such a fleet is a spaced-out fleet.

First, consider the evenly-spaced gliders running on parallel paths produced by the lucky rake collision sequence. These are 82 half-diagonals apart, and each is 1,279 steps behind its predecessor. If the gliders are travelling SE, and the paths of the generating pair of gliders are shifted 82 half-diagonals to the NE on each cycle, each glider of the rake will be 1,115 steps west and 1,443 steps north of its predecessor. Removing the generating pair at any point (which can be done by a single block) will leave a spaced-out fleet.

The lucky rake sequence can be modified by introducing subsequences of $3n + 1$ collisions (for any positive integer n) which have the effect of moving the generating pair $65n$ half-diagonals NE, and delaying their subsequent collisions by $1,170n$ steps, measured in the direction of movement (SE). No additional gliders are added to the rake by such a subsequence. Indeed, the first of the $3n + 1$ collisions removes the last glider added to the rake; if this is not done, it interferes with the process of moving the generating pair. The effect is to divide the rake into parts. The first member of each part after the first is separated from the last member of the preceding part by $164 + 65n$ half-diagonals and $2,558 + 1,170n$ forward steps.

If any such sequence of collisions occurs, and the generating pair is then removed, the result is a spaced-out fleet. Gliders in the rake can then be removed (by one blonk per glider), to leave the number of gliders required in the desired spaced-out fleet. By choosing how the rake is constructed and which gliders are then removed, any adjacent pair of those remaining can have either an odd or an even number of forward steps separating them, and this number of steps can be as large as desired. At the same time, the number z of half-diagonals between their paths can be such that $z \equiv y \bmod x$ for any integers $x > 1$ and $0 \le y < x$.

In particular, an odd or even number of steps difference can be combined with any number of half-diagonals path difference considered mod 52. The glider/blinker 6 collision can then be used to reduce these path differences by multiples of 52 half-diagonals. Glider/blinker 6 produces a pair of gliders going in the same direction as the incoming glider, both on the same path, 52 half-diagonals (in either direction) from the incoming glider's path. It also produces a clump of oscillators, and three additional gliders, one going NW, the remaining two travelling perpendicular to the incoming glider. If the forward gliders' path is shifted 52 SW relative to the incoming glider, the two perpendicular gliders also travel SW; conversely, if the forward gliders' path is shifted 52 NE, the two perpendicular gliders also travel NE. Care must be taken to avoid the additional gliders and the oscillators produced by a 52-shift interacting with other gliders in the developing fleet. In this sideways shifting phase, that can be ensured by first shifting the leading glider 52 SW (for reasons explained below), then shifting the other gliders twice or more 52 SW in turn, starting with the second and moving backward through the fleet, until the path difference between first and second is as desired (any number over 82). Then each glider in turn starting from the third is shifted, until the path difference between second and third is as desired. The process continues in this fashion until all path differences have been adjusted. If the differences in the forward direction were sufficiently large, the result will be a variation on a spaced-out fleet, in which each glider is replaced by a pair of gliders on the same path, but each pair is separated from the next by at least 82 half-diagonals, and the rear member of each pair is no further west than the front member of the next. The initial shift of the leading glider ensures that the entire fleet consists entirely of such pairs at the end of this phase.

Differences in the forward direction can then be adjusted as required. The glider/blinker 6 collision is used again. This time the rearmost glider is dealt with first, and collisions are performed in pairs, shifting the glider path 52 one way (SW), then back again NE (the minimum path separation of 82 may be violated after the first collision, but is restored after the next). The fact that the forward gliders from the collision are paired makes it possible to choose either a short delay 52-shift or a long delay 52-shift in each collision. The short delay 52-shift retards a glider pair by 374 steps, the long delay 52-shift by 2,240 steps. Since the highest common factor of these numbers is 2, any desired multiple of 2 steps alteration in the relative forward position of two glider pairs can be achieved by adjusting the number of shifts of each type made. By extension, the same can be achieved for any fleet produced by the method described. At the end of the process, only the rear member of each pair will in fact be retained; once a pair has been shifted back and forth as required, the front member can be removed using a block. The next pair can then go through its back-and-forth shifts, and so long as the rear member of the pair does not end up west of its immediate predecessor, there will be no effect on any of its predecessors: all the lateral gliders sent NE will be generated to the east of the immediate predecessor, and hence of all its predecessors. At any stage in this procedure, the single gliders at the rear will form a spaced-out fleet, while after every pair of shifts, the front of the fleet consists of pairs separated from each other in similar fashion.

Part 2 Any well-spaced fleet moving SE can be produced from a spaced-out fleet by a succession of collisions with isolated blocks and blinkers. Thus any well-spaced fleet moving SE can be produced by a standard collision sequence. The key to this phase is the design of a six-glider adjustment subfleet (Fig. 20.13) which has the following properties:

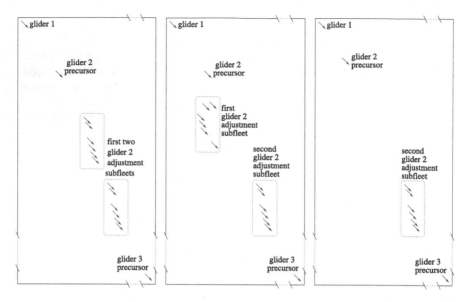

Fig. 20.13 Use of an adjustment subfleet

- When correctly positioned south of its "target" glider the adjustment subfleet can, on encountering a blinker in the right position, generate a two-glider fleet travelling NE. The eastern member of this pair collides with the target. The result of this collision is a single glider going NW. The western member of the pair collides with this, producing a new SE glider shifted two half-diagonals NE and 2,788 steps backward relative to where the original would have been. The collisions are sufficiently "tidy" that the adjustment will succeed so long as the glider's adjusted position is in the front rank of what was a well-spaced fleet before it was added, and remains well-spaced when it takes up that position. The adjustment subfleet can be shifted north or south by multiples of two cells without affecting the resulting change in the target's position after adjustment.
- While not itself spaced-out, the adjustment subfleet can be produced from a spaced-out fleet, using only the glider/blinker 6 collision, and without causing disturbance to any following fleet that is at least 82 half-diagonals NE of it, and has no glider east of the final position of any member of the subfleet.
- The adjustment subfleet can then be shifted laterally to paths further to the SW (and backwards relative to an unshifted glider going in the same direction), using only glider/blinker 6 collisions. The lateral gliders produced by these collisions always go SW and so do not disturb the following fleet. Given any desired well-spaced fleet, it is possible to construct a spaced-out fleet that can give rise to it by working backwards.

Number the gliders in the desired fleet, beginning at the rear and, if two or more are equally far forward, proceeding from SW to NE along the diagonal. Each will have a precursor in the spaced-out fleet (glider 1 being its own precursor).

Glider 2's precursor will be the second rearmost member of the spaced-out fleet, placed $2n$ half-diagonals SE of glider 2's desired position relative to glider 1, and $2,788n$ steps forward from that position, for some n sufficient to make its relationship to glider 1 compatible with being part of the same spaced-out fleet.

The next members of the spaced-out fleet will be $6n$ gliders, making up n copies of the adjustment subfleet's spaced-out predecessor.

Following these will be the precursor of glider 3, then its adjustment subfleets' spaced-out predecessors, and so forth. Figure 20.13 gives a schematic picture of events in the spaced-out fleet near the start of the adjustment procedure. Specifically, it shows the rear part of the fleet at the very start of phase 2 (left), after the first adjustment subfleet has been moved into position (centre), and after the first adjustment has occurred (right). As noted above, the adjustment subfleet can, just prior to the adjustment, be placed arbitrarily far south of the target. It can therefore be constructed arbitrarily far to the SE of the target, so this construction can always take place to the east of the part of the well-spaced fleet that has already been constructed, as required. Similarly, the precursor of the mth member of the desired well-spaced fleet can always be placed on a path SW and a position east of the $m - 1$th adjustment fleet's predecessor gliders, by placing it sufficient multiples of the adjustment distance away from its final position.

Part 3   Any collision of well-spaced fleets beginning with the fleets in separate quadrants as described above can be produced from a well-spaced SE fleet by a collision with a single blinker. Thus any well-spaced fleet collision can be produced by a standard collision sequence. The key to this phase is a set of closely related collisions between a well-spaced fleet of nine gliders and a blinker. Assume the fleet to be travelling SE.

- The first two members of the fleet transform the blinker into a block, and create a glider travelling NW.
- The third glider collides with the NW glider, creating a pond. The fourth and fifth turn this into a boat. The distance between the second and third gliders determines how far NW of the block this boat is.
- The sixth glider collides with the boat, resulting in a new glider going NE (its path determined by how far NW the boat was), and nothing else. Once the boat has been created, the amount of delay can be freely varied by changing the timing of this glider.
- The last three gliders turn the block back into a blinker.

The result is a blinker shifted somewhat from the original position, and a glider travelling at right angles to the incoming fleet. The process can be repeated any number of times, with the relative paths of successive gliders, and their timing, above a certain minimum delay, freely variable. The final cycle can be truncated, with the block either left behind, or removed using a single glider. Once a NE fleet has been created in this way, it can be adjusted using additional members of the SE well-spaced fleet. The adjustment subfleet of phase 2 produced two gliders that made a specific adjustment. The amount of longitudinal adjustment can be varied

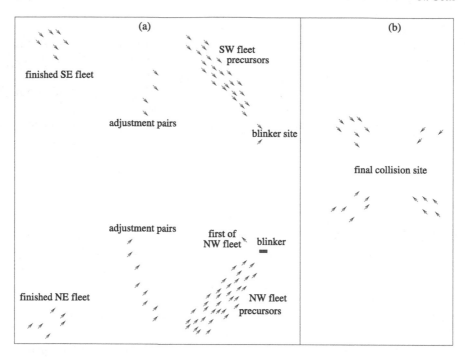

Fig. 20.14 Two well-spaced fleets used to create a collision between four. (**a**) Creating four fleets from two. (**b**) Approaching the final collision

by multiples of eight steps by altering the spacing of the two adjusting gliders, and in the current phase, we are able to assume that the desired spacing has already been achieved. Any well-spaced NE-travelling fleet can therefore be created from a well-spaced SE-travelling fleet (approximately 11 times larger), and a blinker. Of course, this SE-travelling fleet may be part of a larger well-spaced fleet, the rest of which could meet the NE-travelling fleet in a well-spaced collision. As the first acts in this collision, two pairs of gliders can create two blinkers, after which part of the SE-travelling fleet can be used to create a well-spaced SW-travelling fleet, part of the NE-travelling fleet to create a well-spaced NW-travelling fleet. Figure 20.14 shows, schematically, the start of the process of creating these last two fleets from subfleets of the first two (a), and the approach to the final collision (b). Since fleet timings can be adjusted freely, we can conclude that any well-spaced fleet collision can be created by a standard collision sequence.

The advantage of this proof is that it makes it possible to draw on the work of the GoL community, who have expended considerable effort in showing that a very wide range of clusters can be constructed in glider collisions. So long as the collisions concerned are well-spaced, or can be adapted into well-spaced collisions, they can be used to show that the clusters produced can be constructed by standard collision sequences. This, as will be seen, advances the analysis of medium-term events in infinite sparse random GoL arrays. It can be shown that, among others, the following clusters can be constructed by standard collision sequences:

- The three smallest orthogonally-moving spaceships: LWSS, MWSS and HWSS (in fact, there is a 2-collision standard collision sequence that creates an LWSS).
- The glider-stream switch engine.
- The minimal known pattern with quadratic growth of its cumulative image, mentioned above.
- The minimal known pattern with quadratic growth in its number of cells, mentioned above.
- The g2-g2 patterns explored in Sect. 20.3.
- The space rake, the first to be discovered of a class of indefinite growth patterns with a moving head that emits spaceships (in this case gliders) travelling at an angle to its own path, without leaving a permanent trail of oscillators.
- The glider gun, an indefinite growth pattern that emits a stream of gliders, one every 30 steps. (Unlike the glider-stream switch engine, the only growth in this pattern is the increasing number of gliders.)
- The eater, a small pattern that can be created in a two-glider collision, and can absorb the glider gun's stream of gliders.

The significance of the last three items is that they can be used to construct many more complex mechanisms. Among these are infinite families of more complex guns that fire either gliders, the orthogonal spaceships referred to, or structures such as the space rake itself (in which case the whole pattern shows quadratic growth in cell count); and infinite families of puffer trains, with moving heads and various products including oscillators, spaceships, guns, and other puffer trains. Furthermore, Conway's argument that a universal computer can be constructed within GoL [6] uses only glider guns, eaters and blocks (there are several two-glider collisions that make a block and nothing else). The suggested approach to construction appears to provide sufficient freedom in the placing of components to make it highly plausible that a well-spaced glider collision could produce such a mechanism, but is insufficiently detailed for this to be beyond reasonable doubt. Conway [6] regards it as obvious that some glider collision would do so, and this forms the basis for his claim that self-replicating computers (universal if desired), can be realised as GoL patterns:

> Eaters and guns can be made by crashing suitable fleets of gliders, so it's possible to build a computer simply by crashing some enormously large initial pattern of gliders. In this way one computer can give birth to another.

However, Conway (from whom the idea of using a pair of parallel gliders to adjust the path of a third stems) does not prove that the glider collisions necessary to construct arbitrary automata out of glider guns and eaters can be produced by the mechanisms he describes. Nor does Poundstone [31], although he goes into more detail. Indeed, as illustrated, his mechanism appears to require glider fleets travelling in opposite directions to pass through each other, without ensuring that this would be possible for the fleets required. Also, it is certainly not shown that, as he states:

> Life computers can produce any desired stream of gliders.

The methods described by Conway and Poundstone, like those described above, cannot for example produce two gliders on the same path, one 14 steps behind the other. The current author has no doubt the general approach described by Conway and Poundstone would work; but is unwilling to claim that universal computers or self-replicating patterns could definitely be produced by well-spaced glider collisions on the basis of the published descriptions of that approach.

Finally, all persistent clusters beginning with four cells, and all the glider-containing quiet clusters that can be produced starting from six cells, can be produced in well-spaced glider collisions. Standard collision sequences can thus construct anything that can be produced by the most commonly occurring non-standard collision sequences: all those involving just one original six cell cluster plus original blonks, or one original five cell cluster, one original four cell cluster, and original blonks. Anything which can be constructed by a standard collision sequence (which includes anything constructable in a well-spaced glider collision) will be so constructed in an infinite random GoL array of any density. Thus the emergence of arbitrarily complex structures from a single glider and a set of initially distant blocks and blinkers certainly occurs in infinite random GoL arrays. In a sparse random array, any specific structure will be constructed in this way at a rate determined by the shortest standard collision sequence that produces it (which is unlikely to be the immensely long sequence that would result from the process outlined above). It is worth noting that the collection of clusters that form each of these structures would not, except in a vanishingly small proportion of cases, form the whole of a cluster either at $t = 0$, $t = 2$, or indeed at any time $\ll N^{3/2}$ steps before the final collision: other original blonks would until then in almost all cases be closer to each of the components than their distance from each other.

Emergent Indefinite Growth Clusters and Their Environment

It has already been argued that almost all original indefinite growth clusters will begin to react with their environment after characteristic numbers of steps: after $\approx N^{5/2}$ steps (in era 5/2) for clusters with linear growth in diameter and both current and cumulative cell count, and after $\approx N^{3/2}$ steps (in era 3/2) for those with quadratic growth in cumulative cell count, whatever their growth in current cell count. When we consider emergent indefinite growth clusters, approximately the same numbers of steps can be expected to pass between such clusters' emergence and their first interaction with other clusters, so long as some additional conditions remain true:

1. The expected distance from a randomly selected cell to the nearest state 1 cell is not $\ll N^{3/2}$. A sufficient but not necessary condition for this to be the case is that the global density of state 1 cells is not $\gg p^3$.
2. The expected distance from a randomly selected cell to the nearest 3-cluster which includes a different set of cells from those it included on both of the two preceding steps, is not $\ll N^{5/2}$. A sufficient but not necessary condition here is that the global density of such clusters is not $\gg p^5$.

In era 0 (and after $t = 0$), these conditions hold. The first expected distance is $\approx N^{3/2}$, the expected distance to an original blonk; the second is $\approx N^{5/2}$, the expected distance to the nearest glider (or in the first 1,100-odd steps, developing r-pentomino). As long as condition 2 holds, there can be nothing moving towards most cells at a distance of $\gg N^{5/2}$ links (although the converse may not be true). Only clusters emerging from standard collision sequences themselves can possibly stop these conditions being true before era 3. First, note that no event can possibly have effects as much as $\approx N^3$ steps away before that time. Events with a cumulative occurrence density of $\approx p^6$ or less cannot therefore affect any point within $\approx N^{5/2}$ steps of the vast majority of cells: the expected distance to the nearest such event at any time before era 3 will be $\gg N^3$ links. This rules out all events involving original clusters of six or more cells. Since an original glider or r-pentomino comes into contact with only a fixed number of new cells on each step, and the chance of any one of these containing anything derived from an original local cluster of four or more cells during the time the glider or r-pentomino is close to it is $\approx p^4$, events involving such meetings will also have a cumulative occurrence density of $\ll p^6$ at any time before era 3. Only standard collision sequences which generate an indefinite growth cluster could undermine conditions 1 and 2 in any specified era, prior to era 3, for arbitrarily low values of p. The cumulative occurrence density of the final collisions of an order b standard collision sequence is $p^{5+b(3-E)}$, where E is the era. To make a global density contribution $\gg p^5$, necessary if either condition is to be undermined, any such class of completion events must therefore generate a cluster of $\gg N^{b(3-E)}$ cells. Since any collision which does not generate an indefinite growth cluster must by definition have a maximum cell count, there will be values of N (and hence p), for which it does not do this at any specified era. As the set of standard collision sequences of a given length is finite, this means that if we want to be sure no growth-limited cluster from a standard collision sequence undermines conditions 1 and 2 above before any particular era prior to era 3, sufficiently low values of p will always ensure this. If a class of clusters can be constructed by a standard collision sequence of order b, and cannot be constructed by a sequence of any lower order, then so long as condition 1 above remains true, its rate of production per step will depend primarily upon the product of the global densities of original blonks, and the cumulative occurrence density of the penultimate collision in the sequence: $\approx p^3 \cdot p^{5+(b-1)(3-E)}$ or $\approx p^{8+(b-1)(3-E)}$. The probability of the first collision between the result of the penultimate collision and some other cluster being the right one may be low, but will not be dependent on p. (For non-standard collision sequences, the upper limit on rate of production per step is $\approx p^9$, as such a sequence must end in a collision between two types of structure which have global densities with a product of at most that magnitude.) All kinds of indefinite growth clusters can then expect to remain unaffected by the environment for at least $\approx N^{3/2}$ steps. The global density of any kind of indefinite growth cluster constructable by an order b standard collision sequence (as opposed to its global density contribution — the amount clusters of that kind add to the overall density of state 1 cells) will therefore be at least $\approx p^{13/2+(b-1)(3-E)}$, which will tend toward $\approx p^{13/2}$ as $E \to 3$. Expected distance to such a cluster from a randomly chosen cell will tend toward $\approx N^{13/4}$. We can therefore say that for low enough p and for any kind of indefinite growth cluster constructable by a standard collision sequence, one of the following must be true:

1. At some time before era 3, indefinite growth clusters emerging from collision sequences will have been causally concerned in generating almost all state 1 cells in the array and raising the global density to $\gg N^3$.
2. At some time before era 3, the global density of indefinite growth clusters of the specified kind that have been produced by a standard collision sequence will exceed the density of such clusters in the original array by any specified factor. Also, the expected distance to such a cluster from a randomly chosen cell will be less than the corresponding distance to a cluster of the same kind present in the original array, or produced with the involvement of non-standard collision sequences.

The second alternative has to be phrased carefully: some indefinite growth clusters might be produced only or in the greatest density by secondary interactions between those produced by standard collision sequences and objects in their environment: either other such indefinite growth clusters, or repeaters. Another possibility is that quadratic growth clusters produced by non-standard collision sequences and not themselves constructable by standard collision sequences might include sufficient copies of a kind of indefinite linear growth cluster to contribute more to the global density of that kind of cluster than standard collision sequences. Standard collision sequences can produce anything constructable by the commonest subclass of non-standard collision sequences, as noted above. If this were not so, such sequences might also produce a smaller expected distance to some kinds of indefinite growth cluster. The expected distance to the nearest cluster produced by any other type of non-standard collision sequence, however, would be $\gg N^{7/2}$ at any time before era 3. Concerning standard collision sequences ending in the production of an indefinite growth cluster with no more than linear growth in cumulative image size, such as the 49 collision sequence, we can say a little more. So long as conditions 1 and 2 both hold, these clusters can expect to remain unaffected by the environment for $\approx N^{5/2}$ steps. The global density of a kind of indefinite growth cluster produced by such a sequence will therefore be at least $\approx p^{11/2+(b-1)(3-E)}$, tending towards $\approx p^{11/2}$ as $E \to 3$. Therefore one of the following must hold if condition 1 does so:

1. At some time before era 3, indefinite growth clusters emerging from standard collision sequences will have been causally concerned in generating the closest 3-cluster to almost all cells, which includes a different set of cells from those it included on both of the two preceding steps.
2. At some time before era 3, the expected distance to an example of any such kind of indefinite growth cluster (which tends to $\approx N^{11/4}$ as $E \to 3$) will be less than the expected distance to any indefinite growth cluster not derived from a standard collision sequence.

We can also be certain that at some time before era 3, almost all cells that began in state 0 and with empty neighbourhoods, but are now in state 1 (this excludes the original blonks, but does not exclude traffic lights descended from size 4 clusters), will have been produced by emergent indefinite growth clusters. We know that patterns with quadratic growth can be constructed by standard collision sequences. If the global density does not rise $\gg p^3$ these collision sequences will occur, and will approach a completion rate of p^8 per step as $E \rightarrow 3$. Almost all the quadratic growth patterns will last for at least $\approx N^{3/2}$ steps. On the pessimistic assumption that they all disappear at an age of $\approx N^{3/2}$ steps, there will at any time be at least one such pattern for every $\approx N^{13/2}$ cells, most of them containing $\approx N^3$ cells. Their global density contribution will therefore tend toward $\approx p^{7/2}$ as $E \rightarrow 3$, outnumbering all descendants of original local clusters of more then 3 cells which have not taken part in producing emergent indefinite growth clusters. The discovery of the 49 collision sequence producing a block-laying switch engine sets a medium-term time limit by which either emergent indefinite growth clusters will outnumber original indefinite growth clusters, or state 1 cells produced by emergent indefinite growth clusters will form the vast majority of those in the array. The rate of production of block-laying switch engines by this mechanism, unless the global density of state 1 cells becomes $\gg p^3$, will be $\approx p^{8+48(3-E)}$. Under the same proviso, each will exist for at least $\approx N^{3/2}$ steps, so the number in existence at any era will be at least $\approx p^{13/2} \cdot p^{48(3-E)}$. This will reach $\approx p^{10}$, the density of original indefinite growth clusters, in era $281/96$. In any later era, emergent structures will therefore dominate the array in one of these two senses, although we cannot say which.

20.5 Discussion

Patterns in any binary CA with the same neighbourhood as GoL have a 1-cluster dynamics analogous to GoL patterns, but usually far simpler. For example, if at least 4 neighbours are needed to shift a cell into state 1, 1-cluster mergers soon cease; for "3–4 Life", in which a cell is in state 1 at t if and only if either 3 or 4 of its neighbours (not including itself) were in state 1 at $t-1$, any random blob of sufficient size quickly consolidates into a single large 1-cluster and expands apparently forever, possibly throwing off small spaceships which move faster than blob expansion and thus escape. These CA thus do not support the temporally extended merging, splitting and vanishing dynamics of 1-clusters that GoL does. Even within GoL, almost all moderate-sized and dense random blobs (i.e. where the initial density of state 1 cells is anywhere near 50%) support this only for thousands or tens of thousands of steps. Prolonged interactions between large and varied collections of physically distinct "objects" of different sizes, may be a precondition for the appearance of complexity; certainly their existence in the physical world, at both astronomical and molecular levels, appears to have supported life's emergence. Note that GoL puffers already have at least three distinct structural levels: individual cells, 1-clusters, and repeated sequences of splitting and merging among 1-clusters. The extent to which small patterns with distinctive dynamic properties, such as LWSSs and switch engines, can combine with each other and with other small patterns to produce larger patterns with more sophisticated dynamic properties is one root of GoL's ability to generate complexity — rather as the ability of small molecules with distinctive chemical properties to combine into macromolecules underlies the complexity of organic chemistry and ultimately, life.

Computational universality is not necessary for the type of complexity-emergence described here — all the phenomena described could emerge in large but finite GoL arrays, which are not universal, having limited storage capacity. Nor is computational universality sufficient for their emergence, except in the weak sense that any universal computational system could be used to simulate GoL. Such a simulation in a one-dimensional universal CA such as CA 110 would require extremely complex initial patterns to simulate the GoL patterns described here, and the simulation would slow drastically as the simulated pattern grew. It is not obvious that rule 110 (or any other one-dimensional CA) can support anything like the ramifying feedback networks and resulting cross-scale interactions described for GoL in this chapter, except by simulating GoL, with the resulting slowdown. On the other hand, it is possible that two dimensions are not sufficient for an efficiently parallelized artificial biosphere to arise within a CA. (It could also be the case that no CA will suffice; that biosphere origin depends, for example, on quantum mechanical features of reality.) It could be argued, for example, that knots, and non-planar communication networks, play crucial parts in cell biochemistry and networks of neurons respectively. However, GoL itself could certainly support non-planar communication networks, with signals passing through each other; and all knots can be represented in a plane if there is a way of showing which strand passes over the other at a crossing point.

Initially, patterns in the g2-g2 class consist of two small subpatterns, each of which turns into a head, body and glider-wave. The heads thereafter remain the same size, while the bodies and waves grow linearly with time; we can thus distinguish micro or local objects and processes, of fixed size, from macro or global objects and processes, with apparently unending growth. Within the micro-scale there are at least three functionally important subscales: individual cells, 1-clusters, and groups of 1-clusters which interact in a regular cycle, or else are produced close together in space and time, like the clusters of oscillators or gliders constituting most of the oscillator rays and compound glider-waves discussed. Within the macro-scale we can distinguish subscales for the pattern as a whole (the area covered expands quadratically, but the number of state 1 cells linearly, as far as any pattern has been followed), and the ever-growing linear features. All these grow at a constant rate, but the rate constants differ between lines, as do the average density of the lines in terms of state 1 cells per unit length. We have noted distinctions between rays and waves, and between lines produced from a single growth point and those produced by interactions between other lines. These interactions produce secondary microfeatures, the points of intersection, each of which moves at a constant velocity. The lines also divide a pattern into subareas: the puffer bodies separate off the southeast quadrant; rays divide the inner part of this quadrant into segments, and waves subdivide each segment into parts which shift and grow as the waves move south and east, but also get subdivided as additional waves form. In those patterns which produce ramifying feedback networks, later-formed rays and lines are generally less dense, so the areal division of the south-east quadrant has a continually developing hierarchical aspect: the edges of an area act as barriers to and modulators of the spaceship-clusters moving through them, and the denser barriers generally make most difference. Lines running in semi-cardinal directions, and to a much lesser extent orthogonal lines, can also act as quasi-catalytic sites, as collisions with them can generate spaceships which travel along the line until they meet an obstruction or spaceship from another collision.

While both micro and macro scales have complex structure including subscales, there are phenomena in g2-g2-s24-sw22 subfamily patterns that grow over time but not forever, and can be described as meso-scale. Most of the kinds of cross-scale interaction noted in these patterns involve meso-scale phenomena. The exception is the existence of the primary expanding feedback loops themselves. Here, the global structure of the pattern (the pair of diverging puffers sending glider-waves toward each other), combined with the exact spatial relationship between the two that define the subfamily, permit early members of G:1:1/3:SW to initiate chains of local interactions that remove later wave members, and add new types of glider-cluster to it. Once this type of cross-scale interaction initiates the formation of a ramifying feedback network, however, further types of cross-scale interaction either produce meso-scale phenomena, or involve these phenomena in further interaction with micro- or macro-scale phenomena, or both. These include:

- The reversible switches in the form of interaction occurring between G:1:1/3:SW and B:1:$\gg$ that can occur when a member of the former is removed. It is these that give the secondary southeast wave sufficient variety to trigger the ramification of the feedback network.
- The production of block-laying switch-engines through a specific sequence of local interactions.
- The subsequent death of such switch engines, or their transformation into glider-stream switch engines, by collision with a glider or oscillator.
- The influence of such switch engines, once produced, on the "population dynamics" of particular cluster types, as they act as semi-permeable and cluster-transforming membranes; and as catalytic sites, bringing about the interaction of glider-clusters that would otherwise remain functionally unconnected.
- The way specific sequences of glider-clusters hitting a switch engine, along with any that have interacted with the O:3:-2 remnant, can determine the switch engine's fate, via the production of a compound stream of gliders that overtakes the head of the switch engine — and hence the fate, or indeed existence, of other switch engines.

We have already noted the involvement of both ramifying feedback networks and cross-scale interactions in real-world complex systems. Clearly, the specific entities and processes involved in the apparent growth in complexity in the GoL patterns described here are unlike those in any real-world domain, and specifically any involved in the origin of life. However, there is at least one current proposal for the origin of life where pre-existing groups of meso-scale structures, themselves formed through what can be seen as cross-scale interactions and influencing each others' formation and persistence, play a crucial role. A major problem in understanding life's beginnings lies in the apparent need for self-replicating molecules or sets of molecules, and semi-permeable cell walls or membranes, to arise simultaneously. This is, specifically, a problem for the idea that life arose in a "primordial soup" of interacting prebiotic chemicals, within which molecules could move freely and so encounter each other: if reaction products are free to diffuse away, it is unclear how (in the absence of highly evolved cell membranes), prebiotic chemicals could be sufficiently concentrated to support the emergence of self-replicating polymers, or how selection could get started. (The "soup" metaphor has influenced artificial life research into the origins of life [16, 28] although the latter confines the soup within a "continuously stirred flow reactor".) Among alternatives to the primordial soup idea is that proposed in [24, 26]: that life arose around hydrothermal vents on the ocean floor, where the geothermally-driven flow of "exudate" rich in metal sulphides and other reactive molecules, into a cooler and more acid early ocean, could both create a continuously growing network of metal-sulphide lined compartments (which would themselves influence the flow of exudate, although these authors do not say so) and provide chemical building blocks for the formation of self-replicating sets of organic polymers. The formation of compartments requires the "cooperation" of processes on the molecular and geological scales, and would in turn permit the concentration of particular chemical species, while their surfaces could catalyse the production of more complex from simpler molecules. Free-living cells could have arisen by the later construction of lipid membranes, independently for archaebacteria and eubacteria. In this connection, the role of switch engine bodies as both semi-permeable barriers and sites of "catalysis" is of particular interest.

More generally, the way in which ramifying feedback networks give rise to an ever-growing variety of novel local interactions in the patterns examined here, and eventually to some which are capable of producing non-local dynamic effects and of further expanding the range of processes occurring, is surely relevant to the study of a broad range of systems showing increasing complexity: in GoL we have a system with a completely transparent "basic physics", remote from real physics, in which deterministic processes operating on quite simple initial conditions can generate structures and processes which show cross-scale but scale-dependent complexity reminiscent of that found in many real-world domains. Perhaps, contrary to proponents of various "anthropic principles" [4], complexity sufficient to serve as the basis for natural selection, and so for much greater increases in complexity culminating in the evolution of intelligent observers, would arise in a very wide variety of logically possible worlds.

The latter part of the chapter has shown that the medium term dynamics of infinite sparse random GoL arrays are dominated, in ways that can be made precise, by emergent structures, and that some simple forms of auto-catalysis will occur in producing these structures. More complex forms of self-replication almost certainly occur, although their cumulative occurrence density in the array may be limited by other emergent processes. Specifically, it has been demonstrated that a well-defined class of processes, standard collision sequences (which will necessarily occur in infinite sparse random GoL arrays), have the potential to construct an infinite variety of structures which would increase in cell count indefinitely if given the space to do so, from the interactions of initially distant structures without this dynamic property. It has been argued that the structures emerging from these processes will have a crucial influence on the dynamics of the arrays studied. For example, this collection of structures will in time either produce almost all the state 1 cells in the array, or include structures with a wide variety of computational and dynamic properties — including self-reproducing universal computers if we accept Conway's argument that these exist in GoL.

So far as infinite sparse random GoL itself is concerned, considerable further progress may be possible. It is certainly feasible to check whether the two switch engines are indeed the only 10-cell indefinite growth patterns, to build up a catalogue of kinds of indefinite growth clusters that can be produced in standard collision sequences (this could include developing more detailed proofs of the constructability of self-replicating universal computers), and to expand the class of glider collisions shown to be constructable by standard collision sequences. It may be possible to find the minimal standard collision sequence that produces indefinite growth clusters, to prove that anything constructable by broader classes of non-standard collision sequences is also constructable by a standard collision sequence, and to discover more about interactions between emergent indefinite growth clusters and their environment.

We can extend some of the findings concerning infinite arrays to toroidal sparse GoL arrays which are very large in both "north-south" and "east-west" directions: if we first choose some value of p, we can then select dimensions for the toroidal array which will, for example, make it likely that there will be a very large number of gliders and r-pentominos in the original configuration (and even larger numbers of blonks and original clusters of size 4), but no sets of six or more state 1 cells close together. We can push the probability that this will be so as close to certainty as we like by decreasing p and increasing the size of the torus at the same time. In such arrays, standard collision sequences will ensure the emergence of local indefinite growth clusters in an array initially completely bereft of them, and the effects of these will dominate the medium term dynamics of the field. In the long term, of course, any finite array will become periodic.

So far as other CA are concerned, a little work has been done both on the elementary one-dimensional CA (for example, the smallest indefinite growth pattern for that generally called ECA 120 [7] grows proportionally to $t^{1/2}$, which does not appear to have been noted before), and on CA closely related to GoL. Of these, the most interesting may be "HighLife" [5], which differs from GoL only in that a cell switches from 0 to 1 if and only if either three or six of its non-self neighbours are in state 1. Many GoL patterns "carry over" to HighLife, but there are two major differences. First, no standard collision sequence can get going: the glider carries over, but its collisions with blonks produce no new gliders, and there is nothing corresponding to the r-pentomino. Second, there is a six-cell cluster with indefinite growth, of a kind which is difficult (though possible) to produce in GoL. This rapidly becomes an 11-cell self-replicating pattern, which produces copies of itself along a diagonal line. In time, any given number of copies will be exceeded, but, at ever-increasing intervals, the number of copies returns to two, then rises to new heights. The effect is to "embed" copies of the one-dimensional CA known as ECA 18 into the HighLife field; the mean number of cells increases at a rate proportional to $t^{(\ln 3/\ln 2)-1}$. A glider striking this growing structure in the right way can cause it to send off a branch at right angles, but calculations suggest this process will not occur even once in the expected lifetime of a replicator (after $\approx N^3$ steps, the line of replicators will hit a blonk at each end, and be eaten away). HighLife raises a wider point. Both these CA show complex behaviour, but the two are very different, and how the differences manifest themselves depends on the range of initial conditions tested. Also, the features displayed in very sparse random fields are the outcome of interplay between patterns whose behaviour depends on details of the transition rule, and can be radically changed by small changes in the rule. The relative sizes of the smallest clusters with crucial dynamic properties are important, as are the interactions between these minimal clusters. The approach taken here, like the work of [12, 21] and others, casts doubt on claims that CA fall into a few "universality classes" within which the details of rules do not matter much. Whether this is so appears to depend on how you look at them. GoL appears to be a highly unusual CA in at least two ways: the critical or near-critical configurations produced by running finite arrays to periodicity, explored by [2] and others; and the unusual facility with which complex structures and mechanisms can be built up within it. Applying the approach described here to a range of CA may reveal whether its ability to produce emergent complexity in very sparse arrays is similarly unusual. Whether there are theoretically significant connections between its known properties requires further investigation.

Studying "complex" CA such as GoL makes it possible, as shown here, to isolate and characterise some simple forms of emergence. At the lowest level, even small structures such as blocks and gliders can be said to "emerge" synchronically from the interactions of the lowest-level "participants", the individual cells. At a higher level, in the large, sparse random GoL arrays studied, systems or parts of systems populated only by objects of limited size and simple behaviour give rise, through these objects' interactions, to diachronically emergent structures with qualitatively different and more complex behaviour. Most of the possible patterns of initial objects give rise to nothing new, and those which do differ from the rest only in quantitative aspects of the spatial relationships between the objects. At a third level, it has been shown that the diachronically emergent structures have an important influence on the dynamics of the system as a whole. CA, and GoL in particular, offer the opportunity to study emergence in a context where the lowest-level participants have very simple behaviour relative to that of participants in social, biological or other real-world systems. These lowest-level participants also interact with a fixed set of partners — although the objects forming the basis of the higher-level types of emergence explored here do not. In the approach taken here, the lowest-level and higher participants also exist in very large numbers, so the systems studied stand at one end of a range of possible types in that way too. In seeking to understand a phenomenon as ubiquitous but as hard to pin down as emergence, extreme examples may be particularly useful. In real-world systems, neither the local rules of interaction, nor the global structure tying the local interactions together, nor the initial conditions, are ever likely to be as easily described or as readily manipulated as in the systems studied here. In studying relatively simple artificial systems such as CA we can vary each of these factors independently (substitute HighLife rules for GoL, toroidal for infinite arrays, high or intermediate values of p for low ones) and compare the resulting systems with regard to emergent phenomena. In this way, we can hope to understand more about when and how different types of emergence arise.

Acknowledgements This chapter is based on articles that originally appeared in the International Journal of Systems Science [17], and Artificial Life [19]. Work for the chapter was partially funded by the Scottish Government Rural and Environment Research and Analysis Directorate. It would not have been possible without free software written by Johan Bontes, Tom Rokicki and Andrew Trevorrow. Tom Rokicki also most generously allowed me to use two of his computers for extended periods. The contributions of Paul B. Callahan and the late Robert Norman are gratefully acknowledged, as are those of numerous members of the community of Game of Life enthusiasts, including David Bell, Jon Bennett, Johan Bontes, Dave Buckingham, Tim Coe, John Conway, Charles Corderman, Noam Elkies, Achim Flammenkamp, Bill Gosper, Alan Hensel, Dean Hickerson, Dan Hoey, Heinrich Koenig, the late Dietrich Leithner, Mark Niemiec, Rich Schroeppel, Steven Silver, and Allan Wechsler.

References

1. Axtell, R.: U.S. firm sizes are Zipf distributed. Science **293**, 1818–1820 (2001)
2. Bak, P., Chen, K., Creutz, M.: Self-organized criticality in the "Game of Life". Nature **342**, 780–781 (1989)
3. Barabási, A.-L.: Linked: The New Science of Networks. Perseus, Cambridge (2002)
4. Barrow, J.D., Tipler, F.J.: The Anthropic Cosmological Principle. Oxford University Press, Oxford (1986)
5. Bell, D.I.: Highlife. an interesting variant of Life. Available from http://www.tip.net. au/~dbell/ (1994)
6. Berlekamp, E., Conway, J.H., Guy, R.: Winning Ways, vol. 2. Academic Press, San Diego (1982)
7. Braga, G., Catteneo, G., Flocchini, P., Quaranta Vogliotti, Q.: Pattern growth in elementary cellular automata. Theor. Comput. Sci. **145**, 1–26 (1995)
8. Carroll, G.R.: National city-size distributions. Prog. Hum. Geogr. **6**, 1–43 (1982)
9. Casti, J.L.: Would-Be Worlds: How Simulation Is Changing the Frontiers of Science. Wiley, New York (1997)
10. Chang, T., Tam, S.W.Y., Wu, C.-C., Consolini, G.: Complexity, forced and/or self-organised criticality, and topological phase transitions in space plasmas. Space Sci. Rev. **107**, 425–445 (2003)
11. Cook, M.: Universality in elementary cellular automata. Complex Syst. **15**(1), 1–40 (2004)
12. Dhar, A., Lakdawala, P., Mandal, G.: Role of initial conditions in the classification of the rule-space of cellular-automata dynamics. Phys. Rev. E **51**(4, Pt. A), 3032–3037 (1995)
13. Fleck, J.: Artefact activity: the coevolution of artefacts, knowledge and organization in technological innovation. In: Ziman, J. (ed.) Technological Innovation as an Evolutionary Process, pp. 248–266. Cambridge University Press, Cambridge (2000)
14. Fu, L.-L.: Interaction of mesoscale variability with large-scale waves in the Argentine basin. J. Phys. Oceanogr. **37**, 787–797 (2007)
15. Gardner, M.: Wheels, Life and Other Mathematical Amusements. Freeman, New York (1983)
16. Gönerup, O., Crutchfield, J.P.: Hierarchical self-organization in the finitary process soup. Santa Fe Institute working paper 06-03-008 (2006)

17. Gotts, N.M.: Emergent phenomena in large sparse random arrays of Conway's "Game of Life". Int. J. Syst. Sci. **31**(7), 873–894 (2000)
18. Gotts, N.M.: Self-organised construction in sparse random arrays of Conway's Game of Life. In: Griffeath, D., Moore, C. (eds.) New Constructions in Cellular Automata. Santa Fe Institute. Studies in the Sciences of Complexity, pp. 1–53. Oxford University Press, Oxford (2003)
19. Gotts, N.M.: Ramifying feedback networks, cross-scale interactions, and emergent quasi individuals in Conway's Game of Life. Artif. Life **15**(3), 351–375 (2009)
20. Gotts, N.M., Callahan, P.B.: Emergent structures in sparse fields of Conway's "Game of Life". In: Adami, C., Belew, R.K., Kitano, H., Taylor, C. (eds.) Artificial Life VI: Proceedings of the Sixth International Conference on Artificial Life, pp. 104–113. MIT, Cambridge (1998)
21. Hanson, J.E., Crutchfield, J.P.: Computational mechanics of cellular automata: an example. Physica D **103**(1–4), 169–189 (1997)
22. Holling, C.S., Peterson, F., Marples, P., Sendzimir, J., Redford, K., Gunderson, L., Lambert, D.: Self-organization in ecosystems: lumpy geometries, periodicities and morphologies. In: Walker, B.H., Steffen, W.L. (eds.) Global Change and Terrestrial Ecosystems, pp. 346–384. Cambridge University Press, Cambridge (1996)
23. Kauffman, S.A.: The Origins of Order: Self-organization and Selection in Evolution. Oxford University Press, Oxford (1993)
24. Koonin, E.V., Martin, W.: On the origin of genomes and cells within inorganic compartments. Trends Genet. **21**, 649–654 (2005)
25. Langton, C.G.: Self-reproduction in cellular automata. Physica D **10**, 134–144 (1984)
26. Martin, W., Russell, M.J.: On the origins of cells: a hypothesis for the evolutionary transitions from abiotic geochemistry to chemoautotrophic prokaryotes, and from prokaryotes to nucleated cells. Philos. Trans. R. Soc. Lond. A: Biol. Sci. **358**, 59–85 (2002)
27. Maynard Smith, J., Szathmáry, E.: The Major Transitions in Evolution. Freeman, New York (1995)
28. Pargellis, A.N.: The evolution of self-replicating computer organisms. Physica D **98**, 111–127 (1996)
29. Peters, D.P.C., Pielke, R.A. Sr., Bestelmeyer, B.T., Allen, C.D., Munson-McGee, S., Havstad, K.M.: Cross-scale interactions, nonlinearities, and forecasting catastrophic events. Proc. Natl. Acad. Sci. USA **101**(42), 15130–15135 (2004). www.pnas.org_cgi_doi_10.1073_pnas.0403822101
30. Peterson, G.D., Allen, C.R., Holling, C.S.: Ecological resilience, biodiversity and scale. Ecosyst. **1**, 6–18 (1998)
31. Poundstone, W.: The Recursive Universe. Morrow, New York (1985)
32. Shante, V.K.S., Kirkpatrick S.: An introduction to percolation theory. Adv. Phys. **20**, 325–357 (1971)
33. Silver, S.: Personal communication (1998)
34. Simon, H.A.: The Sciences of the Artificial, 3rd edn. MIT, Cambridge (1996)
35. Theraulaz, G., Bonabeau, E.: A brief history of stigmergy. Artif. Life **5**(2), 97–116 (1999)
36. Tjardes, T., Neugebauer, E.: Sepsis research in the next millennium: concentrate on the software rather than the hardware. Shock **17**(1), 1–8 (2002)
37. von Neumann, J.: The Theory of Self-reproducing Automata. University of Illinois, Urbana (1966)
38. Winfree, A.T.: When Time Breaks Down: The Three-Dimensional Dynamics of Electrochemical Waves and Cardiac Arrhythmias. Princeton University Press, Princeton (1987)
39. Wolfram, S.: Universality and complexity in cellular automata. Physica D **10**, 1–35 (1984)

Chapter 21
Macroscopic Spatial Complexity of the Game of Life Cellular Automaton: A Simple Data Analysis

A.R. Hernández-Montoya, H.F. Coronel-Brizio, and M.E. Rodríguez-Achach

In this chapter we present a simple data analysis of an ensemble of 20 time series, generated by averaging the spatial positions of the living cells for each state of the Game of Life Cellular Automaton (GoL). We show that at the macroscopic level[1] described by these time series, complexity properties of GoL are also presented and the following emergent properties, typical of data extracted complex systems such as financial or economical come out: variations of the generated time series following an asymptotic power law distribution, large fluctuations tending to be followed by large fluctuations, and small fluctuations tending to be followed by small ones, and fast decay of linear correlations, however, the correlations associated to their absolute variations exhibit a long range memory. Finally, a Detrended Fluctuation Analysis (DFA) of the generated time series, indicates that the GoL spatial macro states described by the time series are not either completely ordered or random, in a measurable and very interesting way.

21.1 Introduction

Cellular Automata (CA) are discrete in time and space dynamical systems, where space consists of a geometrical regular n-dimensional lattice and time advances in discrete steps. Each cell can be in one of a possible numerical finite set of states and makes transitions synchronously in time between these states, in accordance with a local updating rule called the CA rule. The application of the CA rule also depends on a fixed neighbourhood, which captures the topology of the interactions between

---

[1] When we study a system with a very large number of particles or components, in order to describe a micro-state of the system, we need to know all the states of each one of its components at certain time. In real life complex systems, this is possible only in principle. To specify a macro-state of the system, we need to know at certain time, usually a scalar, global quantity obtained from averaging a quantity that depends of the components of the system, in fact this is done in Statistical Physics. In our context, a micro-state at certain time of the Game of Life will be two-dimensional array of size $N \times N$ of 0's and 1's and a macro-state any scalar constructed from the micro-states, for example the sum of all 1's, the total number of 0's, their difference, etc.

A. Adamatzky (ed.), *Game of Life Cellular Automata*,
DOI 10.1007/978-1-84996-217-9_21, © Springer-Verlag London Limited 2010

438   A.R. Hernández-Montoya et al.

the cells, and applies to every cell in the lattice to update its current state [1, 6, 14, 21, 22, 26, 39, 41, 43, 44].

Although the origins of cellular automata are relatively recent, they can be traced back to the ambitious studies of Ulam and von Neumann in 1950 [40], who attempted to study biological processes such as self-replication. They formulated an abstract, discrete model of a universal self-reproductive automaton. Some important advances following von Neumann's line of research and its implementation were obtained [15, 33, 36] culminating with the new Sciences of Artificial Life [28, 29] and Neural Networks [25]. Other important results were John Conway's invention of the Game of Life Cellular Automaton [9, 19] and Wolfram's Mechanical Statistical Studies of CA and classification of CA Rules in accordance with their dynamical behaviour [42].

Nowadays, cellular automata are important both theoretically and for their applications. Currently cellular automata are being used to model and to study physical, biological and social systems and in related applications such as diffusion processes, phase transitions, lattice gases, cancer spreading, epidemic modeling, randomness theory, artificial life, traffic engineering problems and modeling of speculative markets, among many others (see [6, 7, 11, 14, 26, 27, 29, 30, 34, 37, 39, 43–45] and references therein). From a more fundamental point of view, cellular automata are possibly the simplest type of complex system. A very brief mention of some basic terms related to Game of Life Cellular Automaton and to Complexity Sciences are presented below.

21.1.1 Game of Life Cellular Automaton

Following Wolfram [42], in accordance with the dynamical behaviour of the Cellular Automata, their rules can be classified as:

- Class I: The Cellular Automaton evolves to a homogeneous final state.
- Class II: In the long term, Cellular Automaton reach a periodic or stable final state.
- Class III: Leads the Cellular Automaton to have a chaotic behaviour.
- Class IV: Cellular Automaton evolves to states that form complex, self-organized and localized structures. It is proposed that such Cellular Automata have the property of Universal Computation, i.e. they are equivalent to a Turing Machine.

It is clear that Class IV rules are the most interesting and important.

The Game of Life (GoL), as it was already mentioned, was invented by John Conway, and is a class IV, binary cellular automaton, that evolves using a deterministic updating rule in a two-dimensional rectangular lattice with a Moore neighbourhood.

In a binary Cellular Automaton, all cells can take only two possible states and a Moore neighbourhood includes the eight nearest cells surrounded the central cell, i.e. the one to be updated, forming a 3×3 squared array of cells in the simplest case.

Due to the simplicity of its updating rule (which allows computer enthusiasts an easy implementation) and the amazing richness of the patterns generated by its dynamical evolution, GoL is surely the most studied and popular of all deterministic two state cellular automata. Let us review in plain language its updating rule:

- A living cell with two or three living neighbours survives, otherwise it dies.
- A dead cell with three living neighbours becomes alive, otherwise it stays dead.

It is remarkable that the above microscopic rules, as simple as they are, entail the rich dynamics of GoL, such as its self-organizing structures. Even in a macroscopic level, complexity is maintained, as it is shown in this work. In our opinion, due to its simplicity, GoL has been underestimated; but it is precisely due to this simplicity that GoL can be used as a source of complexity to explore and study under controlled conditions, self-organizing processes and emergence properties. As an example of this, in reference [24] we show that using GoL as in the present work, is possible to generate the most important statistical properties of financial data, called stylized facts [16] by the economists.

21.1.2 Complex Systems Science

Nowadays, although no formal and unique definition of complexity or complex system exists, we have a minimally descriptive one, and it is generally agreed that a complex system must have the following properties:

- Complex systems are constructed from a large number of components, called agents.
- Agents interact between themselves in a "non-trivial way", usually through local, nonlinear interactions.
- Global macroscopic non-trivially predictable behavior arises from the microscopic integrated dynamics of individual agents and their mutual interactions: the so-called "emergent properties" of the system.

Some other properties such as non-stationarity, adaptation, evolution, etc., may be included in the above definition, usually depending on the field of science to which the particular complex system under study belongs. Complex Systems Sciences, with origins in several areas in Physics (non-equilibrium Statistical Mechanics), Mathematics (Chaos, Cellular Automata Theory, Stochastic Processes) and Computing Sciences (Computational Complexity and Information Theories), have evolved to cover a wider and more general class of problems, including applications in Biology, Economics and Social sciences. They have become a truly new, interdisciplinary and fast growing area of research.

21.1.3 Power Law Distribution and GoL's Complexity and Criticality

A power law distribution is considered as a signature of complexity, criticality and self-similarity phenomena. Diverse mechanisms of creation or generation of data samples with a power law distribution have been proposed [4, 12, 18, 20, 32, 38].

Among the most widely studied we can mention self-organized criticality (SOC) [4] proposed by Per Bak, Chao Tang and Kurst Wiesenfield (BTW). Its mechanism was illustrated by means of a cellular automaton based on a simplified sandpile model, whose dynamics makes it evolve towards a critical point, i.e. a state where the system is in transition between two phases, without the need of adjusting exterior conditions of critical parameters, as in the case of more familiar critical systems such as boiling (at critical temperature $T_c = 100°C$) or freezing water (at $T_c = 0°C$), both measured to 1 atm; or ferromagnetic transitions (paramagnetic-ferromagnetic phase transition to certain critical temperature T_c). For the case of GoL, the SOC model states that GoL's micro-states with a random uniformly distributed number of living cells are critical, and that is possible to reach them by choosing, initially, a stationary or periodical state (a state reached usually after GoL has evolved during a long time) [5], [3, Chap. 6] and perturbing it by swapping individual cells states, one by one (in analogy to the act of "dropping" individual grains of sand on a pile, as is done in the Sandpile Model [4]) to cause "avalanches" of GoL activity, that start when the perturbation is applied and end when GoL returns to a stationary or quiescent state. By using this mechanism, the distribution of sizes of GoL's "avalanches" follows asymptotically a power law distribution with exponent $\alpha \simeq 1.3$ [3, Chap. 6]. In this way, and following BTW theory, GoL increases its complexity until it reaches a self-organized critical state.[2]

Organization of This Work

This work is organized as follows: in the following section we explain the geometrical procedure applied on GoL micro-states to generate the sample data, denoted r_t, $t = 1, 2, \ldots, M$, where $M = 20,000$ is the number of generated events. Section 21.3 is divided in two subsections, in the first one, it is shown that $S(t)$, obtained by standardizing the logarithmic differences to unitary time lag of the original time series r_t, has the following statistical properties: non-stationarity, clustered volatility, a non-gaussian distribution and that is strongly leptokurtic, with tails decaying as a power law. In the next subsection it is shown that the autocorrelation functions of $S(t)$ and $|S(t)|$ decay to noise fluctuations after one time step in the first case and have a long memory in the second one. Also, we present a Detrended Fluctuation Analysis of r_t time series and jointly with the correlation analysis above explained, we explore some consequences of the obtained results with respect to the equilibrium between order and randomicity in GoL. Finally, in Sect. 21.4 the conclusions of this work are discussed.

21.2 Generating the Sample Data

The implementation of GoL uses the standard rules with periodic boundary conditions, an initial state with 20% of living cells random selected and uniformly dis-

---

[2]Although, we most mention at this point, that there are also researchers that have concluded that under that conditions GoL reaches a subcritical state. Whether GoL is SOC or not, is in our opinion still under debate. See [2, 8, 10, 23] for further details.

tributed in a lattice size of $N \times N = 3{,}000 \times 3{,}000$ cells, we have chosen a not very high initial density of living cells in order to give GoL enough room to evolve. After setting up a fixed, discrete cartesian coordinate system XY with origin in the center of the lattice and axis parallels to its horizontal and vertical sides or borders, and by using of a natural analogy with a real system of particles, we can consider to each cell as a point of unit mass when it is in a living state, and of zero mass when its state is "dead". Then we can compute the "Center of Mass" (CM) two-dimensional vector of living cells for each GoL micro-state, at time step t, denoted $\mathbf{R}(t)_{CM}$ and defined in the usual way:

$$\mathbf{R}(t)_{CM} := \big(X(t)_{CM}, Y(t)_{CM}\big) = \frac{1}{N} \sum_{k=1}^{N} \sum_{l=1}^{N} C_{kl}(t)\big(x_k(t), y_l(t)\big) \qquad (21.1)$$

where $C_{kl}(t)$ denotes the state (1 or 0) of the (k, l) cell at the coordinate $(x_k(t), y_l(t))$ with respect to XY at time step t, $t = 1, 2, 3, \ldots, M$.

As GoL evolves, the "Center of Mass" vector, $\mathbf{R}(t)_{CM}$, is calculated for each time step $t = 1, 2, 3, \ldots, M$ and its variations give us information about the way the distributions of living cells tends to concentrate and evolves on time. Note that due to the election of our coordinate system XY, with origin in the center of the lattice, the discrete point $(l, m) \neq (x_l, y_m)$, in fact both discrete frames are related by an operation of coordinates translation. Since we will focus our interest on the module of the vector $\mathbf{R}(t)_{CM}$ as explained below, this choice of the coordinate system is more natural and intuitive. Here, we should note that even if we are using periodic boundary conditions and initial states with 20% living cells uniformly and randomly distributed, there is no ambiguity in the calculations of $\mathbf{R}(t)_{CM}$ with respect to any plane coordinate system on the lattice, in particular with respect to our coordinate system XY as described above, i.e. periodic boundary conditions do not imply that our lattice (space) is curved.

Figure 21.1 shows the evolution of $\mathbf{R}(t)_{CM}$ for three realizations of the numerical experiment. It can be seen that two of them tend to die out or become periodic because GoL becomes quiescent after reaching about 14,000 time steps and that the third one continues fluctuating without showing stationary or periodical behaviour even after 20,000 time steps.

Figure 21.2 shows 10,000 generations of the evolution of the vector $\mathbf{R}(t)_{CM}$ in the x–y plane, in what appears to be a random pattern.

In fact, and in order to simplify the analysis, we will not focus our attention in the two-dimensional time series $\mathbf{R}(t)_{CM}$, but in the scalar one $r_t := |\mathbf{R}(t)_{CM}| := \sqrt{X(t)_{CM}^2 + Y(t)_{CM}^2}$, $i = 1, 2, 3, \ldots, M$, that describes how the distance from the state's center of mass to the origin evolves with time. $|\mathbf{R}(t)_{CM}|$ evolution can be seen in Fig. 21.3.

From same Fig. 21.3, it can be observed the presence of diverse trends in the r_t series, this fact is a clear evidence that the series r_t is not stationary and therefore the series of logarithmic differences d_t, of the r_t points, is constructed as usual: $d_t := \log r_{t+1} - \log r_t$, for $t = 1, 2, 3, \ldots, M$.

In the generation of the data, initial state configurations with a 20% density of random uniformly placed living cells were used. In this way, we obtained a sample

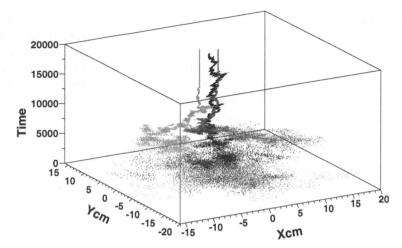

Fig. 21.1 Spacetime history of $\mathbf{R}(t)_{CM}$ for three different realizations of the computational experiment

Fig. 21.2 Two dimensional random walk obtained by plotting $\mathbf{R}_{CM}$ as a function of time

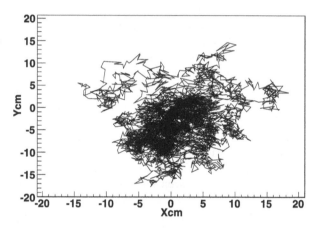

data of 20 random walks of 20,000 steps each. Because GoL evolves to a quiescent state, most of the variations in these time series begin to behave stationary or to die out after a different number of time steps. Therefore, a low cut off was applied to all the time series eliminating the last 15,000 events, which are most stationary or periodic, remaining a total of $M_c = 5,000$ events to be analyzed for each time series. See annotation of Fig. 21.4.

After the cut off was applied, the d_t series was standardized by subtracting its mean and dividing by its standard deviation, obtaining a new set of values S_t, $t = 1, 2, 3, \ldots, M - 1$, which we will denote in this work only as $S(t)$, with mean zero and standard deviation one. The standardization process will not change its distribution but allows us to compare and concatenate our diverse generated data samples. Figure 21.3 displays the evolution in time of r_t corresponding to the three typical generated random walks already shown in Fig. 21.1.

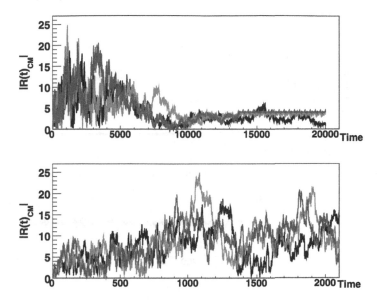

Fig. 21.3 Evolution of our observable $r_t = |\mathbf{R}_{CM}|$ (CM distance to origin) for each one of the three random walks shown in Fig. 21.1. We show two different time scales

21.3 Data Analyses

The three panels of Fig. 21.4 display the evolution in time of $S(t)$ for a typical realization of the computational experiment. The damping effect on the amplitude of the oscillations is a consequence of the natural dying out of GoL over time. However, there is an initial regime where $|S(t)|$ fluctuations do not appear to be "damped", this is shown in the lower panel of the same Fig. 21.4; this regime with its corresponding data sample is of the most interest to us. The first thing we observe is that large fluctuations tend to be followed by large fluctuations, of either sign, and small fluctuations tend to be followed by small ones. This property was observed by Mandelbrot [31] in the context of financial returns variations and is referred as "Volatility Clustering" and is a property non-presented in stationary stochastic processes [16].

Figure 21.5 compares the distribution of $S(t)$ fluctuations for all the generated data sample with those generated from a gaussian distribution with the same values for the mean and variance. From the same figure it becomes clear that fluctuations are not gaussian. This fact is more clearly appreciated in Fig. 21.6, where the $S(t)$ and the simulated standard gaussian processes are compared. The $S(t)$ distribution is strongly leptokurtic, with kurtosis $\simeq 55$ (about 18 times that of a normal distribution) and appears to have "Pareto tails".

In the next subsection it will be shown that the asymptotic form of the $S(t)$ distribution is consistent with a power law.

Fig. 21.4 Standardized logarithmic differences at time lag 1, denoted by $S(t)$ for a typical realization of our experiment. We show different time intervals for the same run. As expected, variations die out as time increases and GoL becomes quiescent. Note: We have selected a particular run that dies out faster than others, to justify the need of applying our cut off value of 5,000 time steps

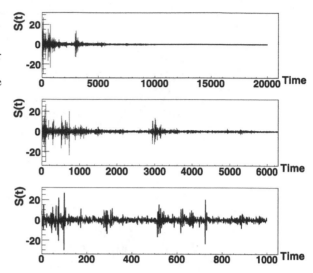

Fig. 21.5 (Color online) $S(t)$ evolution (*red*) in comparison with a Gaussian process (*black*) with the same mean and variance

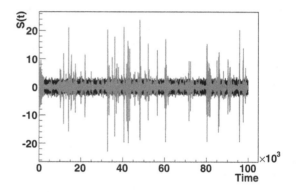

Fig. 21.6 Distribution of $S(t)$. As it can be seen, this distribution is not gaussian and it has fat tails

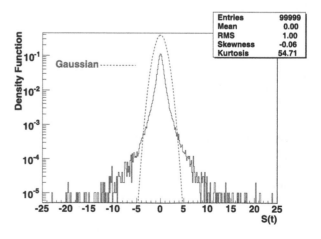

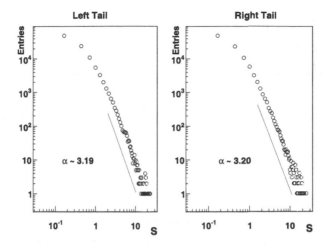

Fig. 21.7 (**a**) Log–log $S(t)$ left tail distribution (actually $-S(t)$). (**b**) Log–log $S(t)$ right tail distribution. Both decay as a power law with exponent larger than three. The exponent α is recovered from fitting a power law to the Cumulative Distribution Function of $S(t)$, which decays asymptotically following a power law with an exponent $\simeq \alpha = \beta + 1$

Table 21.1 Power law fit on $S(t)$ CDF parameters. γ is the fitted cut off parameter, i.e. the value where we cut off the $S(t)$ and $-S(t)$ distributions tails before performing the fit

Tail	Anderson–Darling statistic	γ	β
Positive tail	3.67	2.435	2.1959
Negative tail	3.15	2.380	2.1916

21.3.1 Power Law of S(t) Distribution

Distributions of $S(t)$ and $-S(t)$ in a log–log scale are displayed respectively in right and left panels of Fig. 21.7: power law fits were not performed directly on the empirical distribution of $S(t)$. Instead, in order to avoid fitting errors due to bin size sensitivity, power law fits on the positive and negative tails of the Cumulative Distribution Function of $S(t)$ were performed using the methodology described in [17]. The fitting procedure results are summarized in Table 21.1.

As it could be expected from the property of symmetry of the initial conditions in GoL, the results from both fits for the positive and negative tails are very similar. The $S(t)$ distributions decays as a power law with exponent $\alpha \simeq 3.2$ for both tails.

21.3.2 Autocorrelation Function

In order to study the stochastic independence of $S(t)$ fluctuations, we compute its linear Autocorrelation Function (ACF), by means of the following formula for stan-

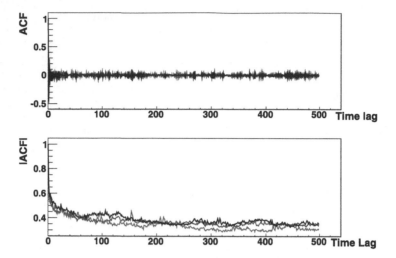

Fig. 21.8 *Upper figure*: $S(t)$ ACF decays in less than two time steps to noise level. *Lower figure*: ACF of $|S(t)|$ shows a long range memory and decays very slowly

dardised data:

$$ACF(t) = \frac{1}{(M_c - t)} \sum_{k=1}^{M_c - t} S(k)S(k+t); \quad t = 1, 2, 3, \ldots, M_c. \quad (21.2)$$

Where $M_c = 5{,}000$ is the number of points remaining after applying the cut, as explained previously.

Upper panel of Fig. 21.8 shows the typical behavior of the ACF for three different time series $S(t)$. In fact it can be seen that they decay quickly to zero after one time step. On the other hand, the lower panel of Fig. 21.8 displays the ACF of absolute variations $|S(t)|$, showing a completely different history. If $S(t)$ variations were independent, any function of $S(t)$ would also be independent, it can be seen that this is not the case: $|S(t)|$ ACF shows long range correlations in the three realizations. In fact in all of the 20 data samples, the same behavior is observed: $S(t)$ ACF decays very quickly to zero after only one time step. On the other hand, $|S(t)|$ ACF decays very slowly. The above facts indicates that it is not obtained a fully random states by mapping GoL's micro-states to a scalar r_t (this is really a ramdomization process). This property is a very common one of Complex Systems and constitutes, for the case of Economic Complex System one of the so called stylized facts [16]. We show in [24] that $|S(t)|$ ACF decays as a power law.

21.3.3 Detrended Fluctuation Analysis

The statistical technique called detrended fluctuation analysis, introduced by Peng and collaborators [35], has proved to be a valuable tool in signal analysis, specially when searching for long time correlations in non-stationary time series. In

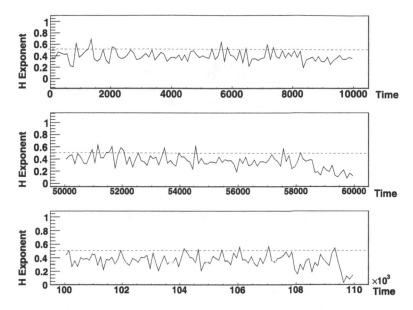

Fig. 21.9 H DFA scaling exponent evolution for a typical realization r_t with a moving time window of size 100. The three figures show different time intervals. Note: The *dashed line* corresponds to a value of $H = 0.5$

this method, a scaling exponent H is computed for the signal under analysis. The value of the scaling exponent H can be interpreted as (i) if $H = 0.5$, there is no correlation at all and the time series represents a random walk; (ii) if $0.5 < H < 1$, persistent long-range power law correlations are present in the time series, and (iii) if $0 < H < 0.5$, power law anti-correlations are present. We have calculated the scaling exponent, with a moving time window of 100 time steps for the time series obtained concatenating all the 20 $\{r_t\}$ time series. The results can be seen in Figs. 21.9 and 21.10.

It can be seen that the scaling exponent fluctuates around a value of $H = 0.38$, which is indicative of anti-correlations being present in all the r_t time series; however, the scaling exponent fluctuations also reach the random and the long range power law correlations regime. This fact confirms that GoL behaves as a system whose states are not ordered, neither fully random. In fact 90% of fluctuations are in the regime $H < 0.5$ and 10% in the regime $H > 0.5$ and including of course the random fluctuations regime $H = 0.5$.

21.4 Summary

As explained in Sect. 21.2, twenty non-stationary time series r_t carrying information about the spatial distribution of the living cells of a GoL were constructed, in analogy with the Center of Mass of a system of particles. These time series, have

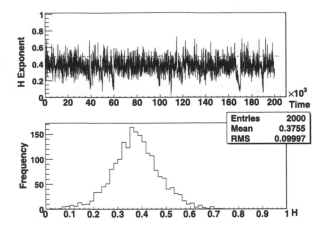

Fig. 21.10 *Upper figure*: H DFA scaling exponent with a moving time window of size 100 for all the samples. Again, the *dashed line* signals a value of $H = 0.5$. *Lower figure*: Its distribution

very interesting properties: their variations, calculated as the standardized logarithmic differences to time lag unitary, denoted $S(t)$, are non-gaussian and its larger variations tend to form clusters. Also, it was shown that the sample values from $S(t)$, have asymptotically a power law distribution with exponent $\alpha \simeq 3.2$ in left and right tails, implying that the variance of this fluctuations is finite because this value of $\alpha = \beta + 1 \simeq 3.2$ indicates that they are outside the stable Levy regime $0 < \beta < 2$. As a note, it is reminded here that micro-state fluctuations of GoL, as was already explained in Sect. 21.1.3, cause "avalanches" with sizes distributed as a power law with exponent $\mu = 1.3$.

A correlation analysis of $S(t)$ and r_t was also performed, and it was concluded that GoL also at a macroscopic level behaves trying to maximize complexity. This is appreciated because its spatial variations, as measured by r_t and $S(t)$ are not fully random (gaussian) neither completely ordered: the autocorrelation function of $S(t)$ behaves as if its fluctuation were stochastically independent and, at the same time, the autocorrelation function of $|S(t)|$ does not show this stochastic independence feature: it has a long memory. On the other hand, the DFA scaling exponent H of r_t, confirms this complex behaviour: we estimated a mean value of $H \simeq 0.3755$, but it fluctuates in such a way that it takes values corresponding to the three regimes described for H.

Given that from the construction of our computational setup, it would be expected only trivial fluctuations of r_t around the origin of the lattice, it is an interesting fact that GoL complexity behaves as a statistical physics system: at the macroscopic level described by the variable r_t, critical properties are observed. Although more studies still must be performed in this direction, this fact could be a consequence of, at the microscopic level, uniformly distributed micro-states, following [3], are critical.

Acknowledgements   In Memoriam of Dr. Augusto García, great human being and teacher. Authors wish to thank professor Andrew Adamatzky, UWE Bristol, for his kind invitation to collaborate in his project. Also, authors acknowledge support from the Sistema Nacional de Investigadores (CONACyT-México) and PROMEP (México) for financial support through grants No. 129141 and No. PTC-256 respectively. ARHM thanks S. Jiménez and A. Huerta for useful suggestions and valuable time. Analyses and plots of this work were made with ROOT [13].

References

1. Adamatzky, A.: Identification of Cellular Automata. Taylor & Francis, London (1994)
2. Alstrøm, P., Leão, J.: Self-organized criticality in the "Game of Life". Phys. Rev. E **49**, R2507–R2508 (1994)
3. Bak, P.: How Nature Works, The Science of Self-organized Criticalicity. Springer, Telos (1999)
4. Bak, P., Tang, C., Wiesenfeld, K.: Self-organized criticality: an explanation of $1/f$ noise. Phys. Rev. Lett. **59**, 381–384 (1987)
5. Bak, P., Chen, K., Creutz, M.: Self-organized criticality in the "Game of Life". Nature **342**, 780–782 (1989)
6. Bandini, S., Chopard, B., Tomassini, M.: Cellular automata. In: 5th International Conference on Cellular Automata for Research and Industry (ACRI) Proceedings, Geneva, Switzerland (2002)
7. Bartolozzi, M., Thomas, A.W.: Stochastic cellular automata model for stock market dynamics. Phys. Rev. E **69**, 046112 (2004)
8. Bennett, C., Bourzutschky, M.: "Life" not critical? Nature **350**, 468 (1991)
9. Berkelamp, E.R., Conway, J.H., Guy, R.K.: Winning Ways for Your Mathematical Plays, vol. 2. Academic Press, San Diego (1982)
10. Blok, H.J.: Life without bounds: does the Game of Life exhibit self-organized criticality in the thermodynamic limit? Master's thesis, University of British Columbia (1995). Available from http://www.zoology.ubc.ca/rikblok/lib/blok95b.html
11. Boon, J.P., Dab, D., Kapral, R., Lawniczak, A.: Lattice gas automata for reactive systems. Phys. Rep. **273**, 5 (1996)
12. Bouchaud, J.P.: Power laws in economics and finance: Some ideas from physics. Quant. Finance **1**, 105–112 (2001).
13. Brun, R., Rademakers, F.: ROOT — An object oriented data analysis framework. In: Proceedings AIHENP'96 Workshop, Lausanne (Sep. 1996). Nucl. Instrum. Methods Phys. Res. A **389**(1–2), 81–86 (1997). http://root.cern.ch
14. Chopard, B., Droz, M.: Cellular Automata Modeling of Physical Systems. Cambridge University Press, Cambridge (1998)
15. Codd, E.F.: Cellular Automata. Academic Press, New York (1968)
16. Cont, R.: Empirical properties of assets returns: stylized facts and statistical issues. Quant. Finance **1**, 223–236 (2001)
17. Coronel-Brizio, H.F., Hernández-Montoya, A.R.: On fitting the Pareto–Levy distribution to financial data: selecting a suitable fit's cut off parameter. Physica A **354**, 437–449 (2005)
18. Gabaix, X., Gopikrishnan, P., Plerou, P., Stanley, H.E.: A theory of power-law distribution in financial markets fluctuations. Nature **423**, 267–270 (2003)
19. Gardner, M.: Mathematical games: The fantastic combinations of John Conway's new solitaire game Life. Sci. Am. **220**(4), 120 (1970)
20. Goldenfeld, N.: Lectures on Phase Transitions and the Renormalization Group. Frontiers in Physics Series. Addison–Wesley, Reading (1992)
21. Griffeath, D., Moore, C.: New Constructions in Cellular Automata. Santa Fe Institute Studies in the Sciences of Complexity. Oxford University Press, London (2003)
22. Gutowitz, H. (ed.): Cellular Automata, Theory and Experiment. A Bradford Book. MIT, Cambridge, London (1991). Special issues of Physica D
23. Hemmingsson, J.: Consistent results on "life". Physica D **80**, 151–3 (1995)

24. Hernández Montoya, A.R., Coronel-Brizio, H., Politi, M., Rodríguez-Achach, M.E., Scalas, E., Stevens, A.: Stylized facts generated through cellular automata models. Case of study: the Game of Life. Phys. Rev. E (2010, submitted for publication)
25. Hopfield, J.J.: Neural networks and physical systems with emergent collective computational abilities. Proc. Natl. Acad. Sci. USA **79**, 2554–2558 (1982)
26. Illachinski, A.: Cellular Automata. A Discrete Universe. World Scientific, Singapore (2001)
27. Komosinski, M., Adamatzky, A.: Artificial Life Models in Software, 2nd edn. Springer, Berlin (2009)
28. Langton, C.G.: Studying artificial life with cellular automata. Physica D **22**, 120 (1986)
29. Langton, C.G.: Artificial Life: Proceedings of an Interdisciplinary Workshop on the Synthesis and Simulation of Living Systems. Santa Fe Institute Studies in the Sciences of Complexity. Addison–Wesley, Reading (1989). And others volumes of this series
30. Maerivoet, S., De Moor, B.: Cellular automata models of road traffic. Phys. Rep. **419**, 1–64 (2005)
31. Mandelbrot, B.B.: The variation of certain speculative prices. J. Bus. **36**, 394–419 (1963)
32. Mandelbrot, B.B.: The Fractal Geometry of Nature. Freeman, New York (1983)
33. Minsky, M.L.: Computation, Finite and Infinite Machines. Prentice–Hall, New York (1967)
34. Oono, Y., Yeung, C.: A cell dynamical system model of chemical turbulence. J. Stat. Phys. **48**, 593 (1987)
35. Peng, C.K., Buldyrev, S.V., Havlin, S., Simons, M., Stanley, H.E., Goldberger, A.L.: Mosaic organization of DNA nucleotides. Phys. Rev. E **49**, 1685–1689 (1994)
36. Pesavento, U.: An implementation of von Neumann's self reproducing machine. Artif. Life **2**, 337 (1995)
37. Qiu, G., Khandai, D., Sloot, P.M.A.: Understanding the complex dynamics of stock market through cellular automata. Phys. Rev. E **75**, 046116:21–28 (2007)
38. Stanley, H.E.: Introduction to Phase Transition and Critical Phenomena. Oxford University Press, London (1971)
39. Toffoli, T., Margolus, N.: Cellular Automata Machines. A New Environment for Modeling. MIT, Cambridge (1987)
40. von Neumann, J.: Theory of Self-reproducing Automata. A.W. Burks. University of Illinois, Urbana (1966)
41. Wolfram, S.: Statistical mechanics of cellular automata. Rev. Mod. Phys. **55**, 601–644 (1983). This and Ref. [42] can be found in [43]
42. Wolfram, S.: Universality and complexity in cellular automata. Physica D **10**, 1–35 (1984)
43. Wolfram, S.: Cellular Automata and Complexity: Collected Papers. Westview, Boulder (1994)
44. Wolfram, S.: A New Kind of Science. Wolfram Media, Champaign (2002)
45. Zhou, T., Zhou, P.L., Wang, B.H., Tang, Z.N., Liu, J.: Modeling stock market based on genetic cellular automata. Int. J. Mod. Phys. B **18**, 2697–2702 (2004)

Part VI
Physics

Part VIII

Chapter 22
The Enlightened Game of Life

Claudio Conti

The link between light and the development of complex behavior is as much subtle as evident. Examples include the moonlight triggered mass spawning of hard corals in the Great Barrier [10], or the *light-switch hypothesis* in evolutionary biology [13], which ascribes the Cambrian explosion [9] to the development of vision.

Developing simple mathematical models accounting for the interaction between a complex system and electromagnetic radiation, while stressing self-organization and collective dynamics, is an interesting and original enterprise. The basic idea is identifying the most limited set of ingredients including, on one hand, the electromagnetic origin of light (i.e., not limiting to ray-tracing and similar techniques) and, on the other hand, a minimal description of a complex system affected by illumination. Such an approach necessarily leads to extremely simplified and un-realistic theoretical representations, but these are expected to be the starting point for more complicated treatments for problems like DNA replication and accumulation under intense fields [2, 12], swarming [3], or nonlinear optics of complex soft-materials [4, 6, 11, 14, 16, 19]. Furthermore practical realizations of these models could be realized by using light-controlled chemical reactions [5, 15].

Here we consider the way the appearance of photosensitivity affects the dynamics, the emergent properties and the self-organization of a community of interacting agents, specifically, of cellular automata (CA). CA are historically the most fundamental paradigm of artificial life (see, e.g., [1]); in this work the renowned Conway's Game of Life [7, 18] is coupled to Maxwell's equations. This is the first example of photosensitive CA.

22.1 The Model

Our approach is based on two models: Maxwell equations for the electromagnetic (EM) field and the Game of Life (GoL) for the CA. The latter is represented by an ensemble of squares in a 2D box (or *cavity*) that can be occupied by a living cell (LC, symbol **1**), or not (symbol **0**); each cell has eight neighbors. The CA evolution is made by a series of temporal steps obeying the GoL rules: (i) if a LC has

A. Adamatzky (ed.), *Game of Life Cellular Automata*,
DOI 10.1007/978-1-84996-217-9_22, © Springer-Verlag London Limited 2010

0, 1, 4, 5, 6, 7 or 8 occupied neighbors, it dies (loneliness or overcrowding); (ii) if a LC has 2 or 3 occupied neighbors, it survives to the next step; (iii) if an unoccupied cell has 3 living neighbors, it becomes occupied (self-replication). In addition we assume the following rule: (iv) if a LC has collected enough energy from the EM field it survives. This is modeled by determining the EM energy $\mathscr{E}$ (see below) in the automaton square and calculating a quantity $\mathscr{P}$, which is the fraction of $\mathscr{E}$ that the CA is able to use for life-sustenance. $\mathscr{P}$ obeys the equation (one for each LC)

$$\frac{d\mathscr{P}}{dt} = -\frac{\mathscr{P}(t)}{T} + \frac{\eta}{T}\mathscr{E}(t), \tag{22.1}$$

where η is the efficiency ($\mathscr{P} = \eta\mathscr{E}$ in the steady state) and T is the dissipation rate, or memory time. Indeed we include a power consumption mechanism for the stored EM energy. We assume that (a) if a LC dies, it looses all its energy, (b) if $\mathscr{P}$ is greater than a threshold value $\mathscr{P}_{th}$, the LC survives independent of the number of living neighbors.

For a fixed efficiency η, the CA evolution depends on the available EM energy; however simple scaling arguments (as outlined below) show that one can use a single dimensionless parameter the *irradiance* J. If $J = 0$ the CA is "blind", as the standard GoL, conversely, as J increases the effect of the EM field grows.

22.1.1 Electromagnetic Field Equations

For a 2D cavity (with perfect mirrors as boundaries), with edge L, Maxwell equations are written in the transverse-electric (TE) polarization (i.e. only the fields E_x, H_y, H_z are not vanishing) as

$$\partial_z H_y - \partial_y H_z = \varepsilon_r \varepsilon_0 \partial_t E_x \qquad \partial_{y,z} E_x = -\mu_0 \partial_t H_{y,z}. \tag{22.2}$$

To each element of the CA is associated a square of material, whose electromagnetic response is determined by the relative dielectric permittivity ε_r. Since we are interested in the light-driven CA complex dynamics, we initially neglect any feedback mechanism of the CA on the field. This is the case of the "transparent CA", which corresponds to taking $\varepsilon_r \cong 1$ and neglect their light absorption. We will account for the nonlinear feedback of the CA on the field in a later section of this chapter.

Each CA element is mapped to a square with edge $L_{CA} \ll L$. The energy $\mathscr{E}$ in (22.1) is given by

$$\mathscr{E} = \int_{cell} \sigma E_x^2 \, dx \, dy \tag{22.3}$$

where σE_x the Ohmic current and σ is the conductivity. For the EM evolution we adopt the Finite Difference Time Domain approach [17] and take a monochromatic field with angular frequency ω; this is generated by an oscillating dipole placed in the middle of the cavity, which is switched on for a limited time-slot (10 optical cycles). The corresponding seeding current is sinusoidal with period $2\pi/\omega$ and amplitude J.

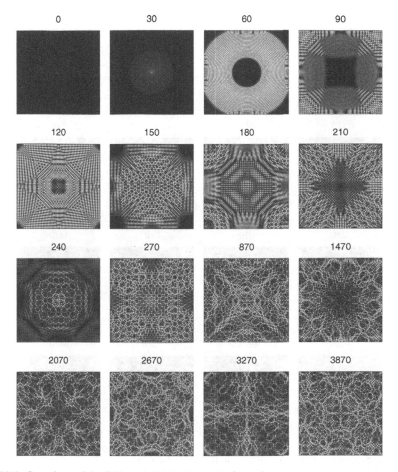

Fig. 22.1 Snapshots of the field evolution in the cavity for various t

22.1.2 Parameters

Straightforward rescaling of the relevant equations (22.1) and (22.2) shows that the dimensionless parameters ruling the dynamics are: ωT, the time constant (i.e. the memory) of each CA element expressed in units of the inverse angular frequency; ωt_{CA}, the time interval between each CA evolution stage; kL_{CA}, the spatial extension of each CA element in units of the inverse wavenumber k^{-1}, with $k = \omega c$ and c the vacuum light velocity; kL, the spatial extension of the cavity; $\kappa = \mathscr{P}_{th}/\mathscr{E}_\lambda$, the ratio between the threshold energy for the CA and a reference energy $\mathscr{E}_\lambda = \eta J^2/\sigma k^2$. Without loss of generality, we can fix σ and $\mathscr{P}_{th}$ to any value and change J (expressed in dimensionless units hereafter) to modulate the effect of the EM field on the CA dynamics. Here we choose units such that $\omega = 1$, $\omega t_{CA} = 1$, $kL_{CA} = 1$, $kL = 100$ and use T and J as control parameters.

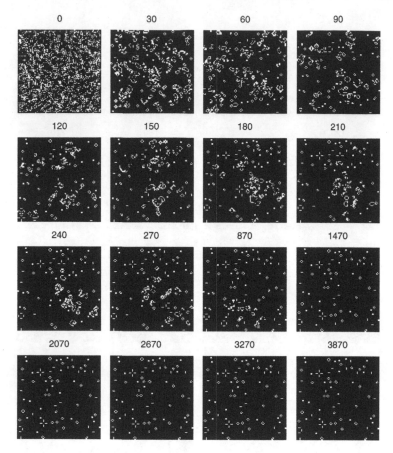

Fig. 22.2 Snapshots of the CA evolution in the cavity for various t at $J = 1$ ($T = 10$)

22.2 Field and CA Evolution

We consider the simultaneous EM-CA evolution by starting from a random configuration of 100×100 CA elements in the box. We show in Fig. 22.1 various snapshots at different t of the EM field that, being initially generated in the middle of the structure, progressively fills the cavity. We show (for $T = 10$) in Fig. 22.2 various snapshots of the CA with $J = 1$, in Fig. 22.3 for $J = 5$, and in Fig. 22.4 for $J = 50$. We show in Fig. 22.5 three snapshots of the EM field in the cavity with the corresponding CA distribution. In the early stages the CA is disordered, while a complex pattern appears at long times; this is largely affected by the degree of photo-sensitivity determined by the parameter J.

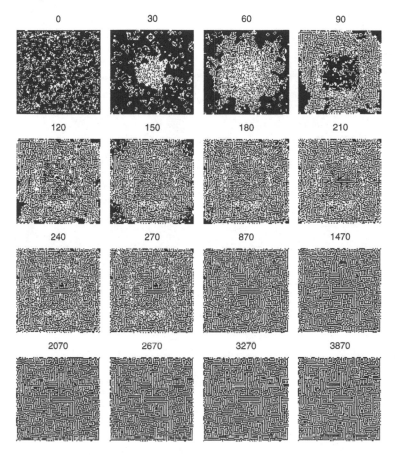

Fig. 22.3 Snapshots of the CA evolution in the cavity for various t at $J = 5$ ($T = 10$)

22.3 Stationary Properties of the CA

The various CA phases can be characterized by the number of LC; this is quantified by the relative life index (RLI) m, which is calculated by assigning an Ising spin σ with value "-1" to **0** and a "$+1$" to **1**.

The RLI is the average value of σ over all the CA. A configuration with many **0**s exhibits negative values of m ($m = -1$ for all **0**), while $m = 1$ for all LC. In the 3D plot of m vs T and J, three regions can be identified (Fig. 22.6). At very low irradiance (small efficiency or low EM intensity) the final population is organized as in the standard GoL (Fig. 22.6, right top panel): it is characterized by small-size unconnected communities of LC (*blind* phase). Two additional regimes are found while increasing J: (i) a *glassy* phase (where $m = 0$) with regular domains separated by various defects (Fig. 22.6, right middle panel); (ii) a region where the CA is frozen in a large disordered configuration with $m > 0$ (Fig. 22.6, right bottom panel). In the glassy phase (plateau in Fig. 22.6), the CA is not sensible to any increase of J. In this regime the EM field sustains a large amount of LC, but their number

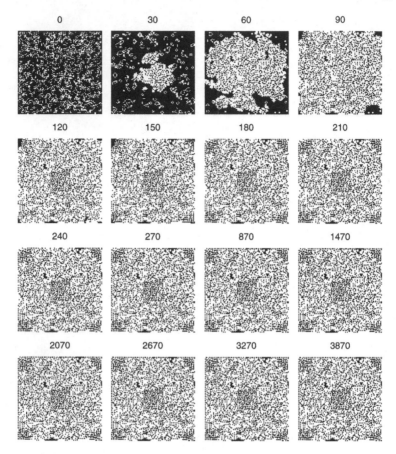

Fig. 22.4 Snapshots of the CA evolution in the cavity for various t at $J = 50$ ($T = 10$)

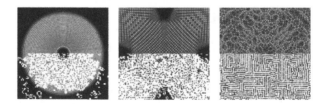

Fig. 22.5 Photosensitive cellular automata. Electromagnetic field distribution in the square cavity at different time instants (from *left* to *right*, $t = 550$; $1,210$; $33,000$). The contemporary CA distribution (a small *white box* for each LC) is superimposed to the field, only its bottom-half is shown (parameters $T = 0.1$, $J = 1,100$)

is frustrated by the internal self-organization. This is true as far as the region with $m > 0$ is entered, where an explosive growth of the LC with the irradiance (and the memory time) is found; this is the *evolved phase*. The existence of this transition is a result of the competition between the GoL rules and the effective employment of the EM energy for life-sustenance.

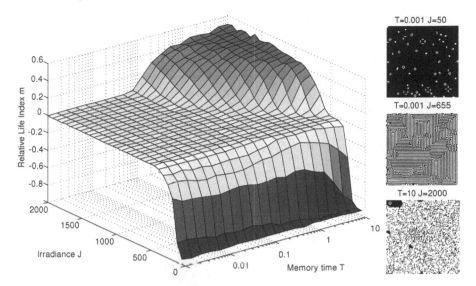

Fig. 22.6 Relative Life Index vs irradiance and memory. The panels on the *right* show three long-time ($t = 4,000$) CA configurations with the corresponding parameters

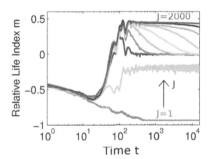

Fig. 22.7 Life index vs time for increasing irradiance (here $T = 1$)

22.4 Dynamics

In Fig. 22.7 we show the time evolution of m for increasing J at a fixed memory. Starting from the same random CA, different histories are determined by EM field. In the blind phase, the CA rapidly evolve to a small number of LC separated in isolated communities (see [18]). At sufficiently high fluence, the RLI overshoots and then decays to zero. This implies that the EM field favors the life of a large number of CA; however this is sustainable for long times only at very high irradiance (evolved phase); in the other regimes the population steadily decays to $m = 0$.

Figure 22.8 shows the autocorrelation function averaged over all the CA:

$$\phi(\tau) = \left\langle \int \sigma_i(t + \tau)\sigma_i(t)\, dt \right\rangle_{CA}, \qquad (22.4)$$

which is normalized such that $\phi(0) = 1$. When increasing the strength of the interaction with the EM field from the blind GoL ($J \cong 0$), the CA first display a disor-

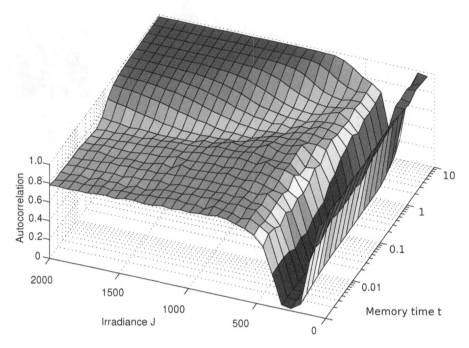

Fig. 22.8 Autocorrelation at $\tau = 4,000$

dered dynamical phase ($m \cong -1$ and $\phi \cong 0$), then a glassy region (where $m = 0$ and $\phi \cong 0.8$). In the evolved phase ($m \cong 1$), $\phi \cong 1$ denotes a frozen configuration.

22.5 Self-healing After a Catastrophic Event

We then consider the reaction of the photosensitive CA to "catastrophic events". We let the system evolve to a stationary state and then we "kill" all the cells occupying a square in the middle of the box and with size $L/2$ (see Fig. 22.9). In the blind phase the system does not react to this event, and the RLI is reduced. Conversely, in the glassy and in the evolved phases, the system rapidly restores the number of LC; at high fluences this is also accompanied by an overshoot of the RLI, which decays to zero in the glassy phase.

22.6 Topology and Self-organization

To characterize self-organization we count the number of edge-connected objects (or *communities*) in the large time ($t = 4,000$) CA configuration. Figure 22.10 shows a three-dimensional plot of the number of edge-connected regions versus the irradiance and the memory time. In the blind phase, one has a large number of unconnected very-small communities with $m \cong -1$. In the glassy phase, many connected

Fig. 22.9 Self-healing after a catastrophic event. Relative life index vs time for various irradiances J (not all values of J for the reported lines are shown) in the presence of the abrupt killing of the living cells in the middle area of the box (at $t = 2{,}000$, here $T = 1$)

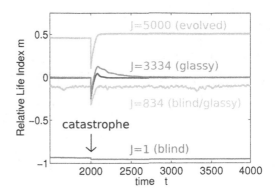

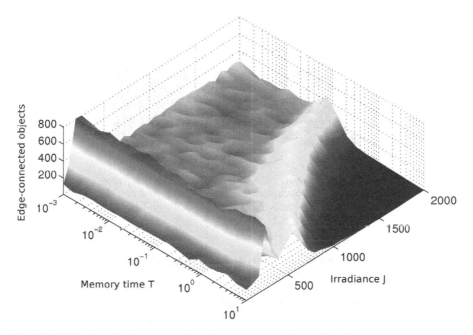

Fig. 22.10 Number of edge-connected regions vs memory time and irradiance

regions with $m = 0$ are found. In the evolved phase ($m > 0$), the CA is organized into a small number of large communities. Specific transition regions can be identified (peaks in Fig. 22.10, brighter lines in Fig. 22.11) and these are characterized by tiny ranges of the parameters with a huge number of small unconnected communities. Indeed, the transition from the glassy phase to the evolved one is driven by the breaking of the almost regular domains (see panels in Fig. 22.11). Correspondingly the number of communities first increases (in Fig. 22.11 they change from 189 to 492) and then rapidly decreases in correspondence to the formation of a large amorphous but connected CA at high irradiance.

Fig. 22.11 Colormap of the number of edge-connected regions; the three *dots* correspond to the CA panels on the *right*; connected regions are discriminated by different grey levels (colors online)

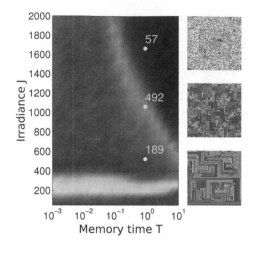

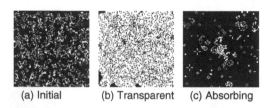

(a) Initial   (b) Transparent   (c) Absorbing

Fig. 22.12 (Color online) Photosensitive CA with genetic code. (**a**) Initial configuration with same number of blind (*red, darker*) and photosensitive (*white*) LC; (**b**) final configuration for the transparent CA; (**c**) as in (**b**) for the absorbing CA ($T = 0.1$, $J = 50$, $t = 1,000$)

22.7 Introducing a Genetic Code and Inheritance

One can argue if the photoreception ability can favor some evolution of the CA toward novel species. The simplest mechanism to be considered is that based on natural selection, such that one assumes that a "gene" responsible for photoreception is randomly distributed among the LC. Those LC not displaying such a gene are blind (they obey to the simple GoL rules); the others behave as described above and feel the presence of the EM field. When a new LC is born from the three neighbors [GoL rule (iii) above], it inherits the photoreceptive gene if this is present in two or three of the parents, otherwise it is blind. We find that, as far as no EM field is present, the photosensitive population (PLC) balances the blind one (BLC). When the EM field is introduced, the PLC rapidly supersede the BLC. Figure 22.12b shows the large-time state of the CA when starting from a balanced configuration (Fig. 22.12a). Letting n_p equal the number of PLC, and n_b equal that of the BLC, we show the ratio $g = (n_p - n_b)/(n_p + n_b)$ in Fig. 22.13 (left axis) vs time. In the transparent case, g rapidly reaches unity.

Fig. 22.13 *Left scale*:
fraction of PLC vs time
averaged over 100 initial
balanced random condition
for transparent (*dashed*) and
absorbing (*full line*) CA.
Right scale: field localization
length vs time (100 initial
random configuration) for the
absorbing CA. The insets
show the EM field
distributions (for the
absorbing case with the CA
superimposed) at two instants
($\otimes$ and $*$) unveiling the
self-organized dynamic wave
localization ($J = 50$, $T = 1$)

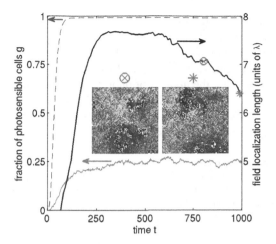

22.8 Energy Dissipating CA

The "gene" selection process implies that PLC are favored in the presence of an
external EM field. The situation however is different if one takes into account the
fact that the PLC absorb energy: as their number grows the life-sustaining field
is reduced and the selection process is frustrated. In Fig. 22.12 we compare the
transparent case with the absorbing one ($\sigma = 10^5 \, \mathrm{S\,m^{-1}}$ for the CA material). In the
presence of dissipation the fraction of photosensitive agents is reduced, however,
surprisingly enough, it stays constant with time after an initial build-up transient.

When considering the snapshots of the EM profile during the CA evolution, one
readily realizes that, at variance with transparent case (where the EM wave is de-
localized in the entire cavity, see Fig. 22.5), the field displays a certain degree of
localization. Indeed regions with high intensity appear, circumvented by various LC
(insets in Fig. 22.13). The effect can be quantified by calculating the EM localization length l_{EM} (see, e.g., [8]), reported in Fig. 22.13 (right axis). Notably, after a
transient over which the field fills the cavity (up to $t \cong 250$), l_{EM} starts to decrease
with time, while g stays constant. As the PLC dissipate energy, the CA is able to
localize light (insets in Fig. 22.13) in order to preserve the intensity level.

22.9 Conclusion

Within the proposed model of photosensitive (artificial) life, one finds that the development of photoreception largely affects not only the number of living automata
but also their organization. If the storage time is too small, the population cannot
grow; it wastes energy more quickly than the time needed to collect it. Conversely,
an explosive growth is found at the expense of large-scale self-organization, which
appears only after a critical degree of photosensibility has been developed. Self-
healing abilities after catastrophic events and dynamical hierarchies are triggered
by the EM radiation.

When introducing a genetic-like competition between photosensitive and blind CA, the former are favored by the irradiation. If the CA energy dissipation is included, the highly nonlinear EM-CA system results into a self-organized field localization effect, such that the EM localization length decreases with time in order to keep constant the number of photosensitive agents.

The proposed model shows that the competition between the internal rules of a complex system and the development of new abilities (as vision) nurtures abrupt evolutive steps and collective behavior.

Acknowledgements We acknowledge support from the INFM-CINECA initiative for parallel computing. The research leading to these results has received funding from the European Research Council under the European Community's Seventh Framework Program (FP7/2007-2013)/ERC grant agreement n.201766.

References

1. Bedau, M.A.: Trends Cogn. Sci. **7**, 505 (2003)
2. Braun, D., Libchaber, A.: Phys. Rev. Lett. **89**, 188103 (2002).
3. Camazine, S., Deneuborg, J.-L., Franks, N.R., Sneyd, J., Theraulaz, G., Bonabeau, E.: Self-organization in Biological Systems. Princeton University Press, Princeton (2001)
4. Conti, C., Ruocco, G., Trillo, S.: Phys. Rev. Lett. **95**, 183902 (2005)
5. de Lacy Costello, B., Toth, R., Stone, C., Adamatzky, A., Bull, L.: Phys. Rev. E **79**, 026114 (2009)
6. Duhr, S. Braun, D.: Appl. Phys. Lett. **86**, 131921 (2005).
7. Gardner, M.: Sci. Am. **223**, 120 (1970)
8. Gentilini, S., Fratalocchi, A., Angelani, A., Ruocco, G., Conti, C.: Opt. Lett. **34**, 130 (2009)
9. Gould, S.J.: Wonderful Life: The Burgess Shale and the Nature of History. Norton, New York (1991)
10. Levy, O., Appelbaum, L., Leggat, W., Gothlif, Y., Hayward, D.C., Miller, D.J.: Science **318**, 467 (2007)
11. Lumsdon, S., Kaler, E., Velev, O.: Langmuir **20**, 2108 (2004)
12. McCann, J., Dietrich, F., Rafferty, C.: Mutat. Res. **411**, 45 (1998)
13. Parker, A., In the Blink of an Eye. Simon and Schuster, London (2003)
14. Reece, P., Wright, E., Dholakia, K.: Phys. Rev. Lett. **98**, 203902 (2007)
15. Sendiña Nadal, I., Mihaliuk, E., Wang, J., Pérez-Muñuzuri, V., Showalter, K.: Phys. Rev. Lett. **86**, 1646 (2001)
16. Snoswell, D.R.E., Bower, C.L., Ivanov, P., Cryan, M.J., Rarity, J.G., Vincent, B.: New J. Phys. **8**, 267 (2006)
17. Taflove, A., Hagness, S.C.: Computational Electrodynamics: The Finite-Difference Time-Domain Method, 3rd edn. Artech House, London (2000)
18. Wolfram, S.: Rev. Mod. Phys. **55**, 601 (1983)
19. Yethiraj, A., van Blaaderen, A.: Nature **421**, 513 (2003).

Chapter 23
Towards a Quantum Game of Life

Adrian P. Flitney and Derek Abbott

Cellular automata provide a means of obtaining complex behaviour from a simple array of cells and a deterministic transition function. They supply a method of computation that dispenses with the need for manipulation of individual cells and they are computationally universal. Classical cellular automata have proved of great interest to computer scientists but the construction of quantum cellular automata pose particular difficulties. We present a version of John Conway's famous two-dimensional classical cellular automata *Life* that has some quantum-like features, including interference effects. Some basic structures in the new automata are given and comparisons are made with Conway's game.

Landauer based his research on a simple rule: information is physical. That is, information is registered by physical systems such as strands of DNA, neurons and transistors; in turn the ways in which systems such as cells, brains and computers can process information is governed by the laws of physics. Landauer's work showed that the apparently simple and unproblematic statement of the physical nature of information had profound consequences.

Seth Lloyd on Rolf Laundauer [21]

23.1 Introductory Concepts of Quantum Mechanics

A good introduction to all aspects of quantum computation is provided by a number of recent books. The bible remains the excellent work by Nielsen and Chuang [26]. In order for the non-specialist to follow this chapter we summarize some concepts from elementary quantum mechanics below.

In the so-called Dirac notation a quantum state labeled by ψ is the *ket* $|\psi\rangle$. The state $|\psi\rangle$ is a member of a complex vector space known as a *Hilbert space*. If $\{|\phi_i\rangle, i = 1, \ldots, N\}$ is an orthonormal basis for an N-dimensional Hilbert space, then $|\psi\rangle$ may written as the decomposition

$$|\psi\rangle = \sum_{i=1}^{N} c_i |\phi_i\rangle, \qquad (23.1)$$

where the c_i are complex numbers.

A. Adamatzky (ed.), *Game of Life Cellular Automata*,
DOI 10.1007/978-1-84996-217-9_23, © Springer-Verlag London Limited 2010

In classical computation, the *bit* is the fundamental unit of information, taking the values 0 or 1. The quantum bit or *qubit* is its quantum analog. The *computational basis* states of a qubit are $|0\rangle$ or $|1\rangle$. However, a qubit may also be in a *superposition* of states, a convex linear combination,

$$|\psi\rangle = \alpha|0\rangle + \beta|1\rangle, \tag{23.2}$$

subject to the normalization condition $|\alpha|^2 + |\beta|^2 = 1$. If we examine a qubit to determine if it is $|0\rangle$ or $|1\rangle$, that is, we take a measurement of $|\psi\rangle$ in the computational basis, the state $|0\rangle$ will be returned with probability $|\alpha|^2$ and $|1\rangle$ with probability $|\beta|^2$. The vector

$$\begin{bmatrix} \alpha \\ \beta \end{bmatrix}, \tag{23.3}$$

can be used to represent the quantum state (23.2). A superposition is not simply a classical ensemble of its component states, which would merely represent our lack of knowledge as to whether $|\psi\rangle$ is actually $|0\rangle$ or $|1\rangle$. Instead, each component of the superposition is simultaneously present. The state (23.2) is often referred to as a *coherent superposition* to emphasize the existence of coherence between the components. *Coherence* can be thought of as a measure of the "quantumness" of a state.

Multiple qubits each inhabit their own two-dimensional Hilbert space and can be written, for example, as $|0\rangle \otimes |1\rangle \equiv |01\rangle$. The Hermitian conjugate of a state $|\phi\rangle$ is known as the *bra* ϕ, or $\langle\phi|$. For example, the Hermitian conjugate of (23.2) is

$$\langle\psi| = \alpha^*\langle 0| + \beta^*\langle 1|, \tag{23.4}$$

or

$$[\alpha^* \quad \beta^*], \tag{23.5}$$

where $*$ refers to complex conjugation. The *bra-ket* $\langle\phi|\psi\rangle$ measures the overlap between two states. That is, $|\langle\phi|\psi\rangle|^2$ is the probability of a measurement[1] revealing $|\phi\rangle$ and $|\psi\rangle$ in the same state. The two states are orthogonal if this value is zero. If $\{|\phi_i\rangle,\ i = 1, \ldots, N\}$ is an orthonormal basis of an N-dimensional Hilbert space, then $|\langle\phi_i|\phi_j\rangle|^2 = \delta_{ij}$, where δ_{ij} is the usual Kronecker delta.

Two or more qubits may exist in an *entangled* state such as

$$|\psi\rangle = \frac{|0\rangle_A \otimes |0\rangle_B + |1\rangle_A \otimes |1\rangle_B}{\sqrt{2}} \equiv \frac{|0_A 0_B\rangle + |1_A 1_B\rangle}{\sqrt{2}}. \tag{23.6}$$

Such a state is not decomposable to a product state of individual qubits. Much of the peculiarity of quantum mechanics resides in such states, including Einstein's famous "spooky action at a distance" [9], but since they do not concern us in the present work they will not be discussed further. Suffice it to say that the measurement correlations in the quantum state (23.6) are stronger than can exist in *any* classical system [4].

---

[1] We are not going to be concerned with what constitutes a measurement in quantum mechanics, since there is no clear consensus on this. A common sense definition of a measurement will suffice.

Fig. 23.1 A classical NOT gate

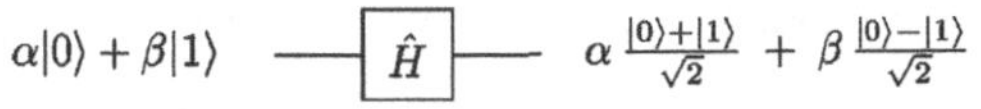

$$x \quad \longrightarrow \quad \bar{x}$$

Fig. 23.2 A Hadamard gate

$$\alpha|0\rangle + \beta|1\rangle \quad \boxed{\hat{H}} \quad \alpha \frac{|0\rangle+|1\rangle}{\sqrt{2}} + \beta \frac{|0\rangle-|1\rangle}{\sqrt{2}}$$

Classical computation is carried out by means of logic gates acting on bits that are transmitted down wires of some form. The NOT gate, transforming $0 \leftrightarrow 1$ (Fig. 23.1) is the only non-trivial gate on a single bit. The quantum NOT gate switches $|0\rangle \leftrightarrow |1\rangle$ and acts linearly on a superposition:

$$\alpha|0\rangle + \beta|1\rangle \rightarrow \alpha|1\rangle + \beta|0\rangle. \tag{23.7}$$

The action of the quantum NOT operator $\hat{X}$ can be represented by the matrix

$$\hat{X} \equiv \begin{bmatrix} 0 & 1 \\ 1 & 0 \end{bmatrix}, \tag{23.8}$$

where the hat over a symbol indicates that it is an operator. In quantum mechanics multiplying a state by an arbitrary phase $e^{i\alpha}$ has no physical effect since the probabilities of measurement are proportional to the square of the modulus. However, the relative phase between the components of a superposition is important as we shall see later. Thus, other important single qubit gates exist in quantum mechanics. The phase flip operator

$$\hat{Z} \equiv \begin{bmatrix} 1 & 0 \\ 0 & -1 \end{bmatrix}, \tag{23.9}$$

has the effect of reversing the relative phase between the $|0\rangle$ and $|1\rangle$ components of a superposition, while an arbitrary phase difference is applied by the phase operator

$$\hat{\mathscr{P}}(\alpha) \equiv \begin{bmatrix} 1 & 0 \\ 0 & e^{i\alpha} \end{bmatrix}. \tag{23.10}$$

The *Hadamard* gate

$$\hat{H} \equiv \frac{1}{\sqrt{2}} \begin{bmatrix} 1 & 1 \\ 1 & -1 \end{bmatrix}, \tag{23.11}$$

changes the computational basis states into a superposition half way between $|0\rangle$ and $|1\rangle$. That is,

$$\hat{H}|0\rangle = (|0\rangle + |1\rangle)/\sqrt{2},$$
$$\hat{H}|1\rangle = (|0\rangle - |1\rangle)/\sqrt{2}. \tag{23.12}$$

Note that all operators act linearly on a superposition. Figure 23.2 represents a quantum circuit that shows the action of the Hadamard gate. The "wires" represent any medium through which a qubit can propagate. Any single qubit operation can be represented by a 2×2 unitary matrix, where the overall phase is not physically relevant. An N-qubit operator can be represented by a $2^N \times 2^N$ unitary matrix.

Fig. 23.3 A controlled-NOT
gate, where the binary
operator ⊕ represents
addition modulo two

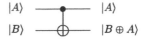

A particularly useful operator when discussing measurement is the projection operator onto the state $|\phi\rangle$:

$$\hat{P} = |\phi\rangle\langle\phi|. \tag{23.13}$$

It has the effect of projecting out of a state $|\psi\rangle$ only the component parallel with $|\phi\rangle$. For example, $\hat{P}_0 \equiv |0\rangle\langle0|$ applied to (23.2) results in the state $\alpha|0\rangle$. Multi-qubit projection operators, such as $|01\rangle\langle01|$, are also possible.

The two-bit gate NAND is known to be universal for classical computation. That is, combinations of single-bit gates (NOT gates) and NAND gates can perform any computation. In quantum computation all gates are necessarily reversible. Hence, there can be no quantum analogue of the NAND gate, or of the other two-bit gates, AND, OR, and XOR. However, the two-qubit controlled-not, or CNOT, gate together with the set of single-qubit gates forms a universal set for quantum computation [3]. The CNOT gate flips the second qubit if the first, or control qubit, is in the $|1\rangle$ state, as shown in Fig. 23.3.

23.2 Background and Motivation for Quantum *Life*

In this section some of the major results for one-dimensional classical cellular automata are presented and some of the problems of making quantizable versions of cellular automata are discussed.

23.2.1 Classical Cellular Automata

A cellular automaton (CA) consists of an infinite array of identical cells, the states of which are simultaneously updated in discrete time steps according to a deterministic rule. Formally, they consist of a quadruple (d, Q, N, f), where $d \in \mathbb{Z}^+$ is the dimensionality of the array, Q is a finite set of possible states for a cell, $N \subset \mathbb{Z}^d$ is a finite neighbourhood, and $f : Q^{|N|} \rightarrow Q$ is a local mapping that specifies the transition rule of the automaton. The simplest CAs are constructed from a one-dimensional array of cells taking binary values, with a nearest neighbour transition function, as indicated in Fig. 23.4. Such CAs were studied intensely by Wolfram [30] in a publication that lead to a resurgence of interest in the field. Wolfram classified CAs into four classes. The classes showed increasingly complex behaviour, culminating in class four automata that exhibited self-organization, that is, the appearance of order from a random initial state.

In general, information is lost during the evolution of a CA. Knowledge of the state at a given time is not sufficient to determine the complete history of the sys-

Fig. 23.4 A schematic of a one-dimensional, nearest neighbour, classical cellular automaton showing the updating of one cell in an infinite array

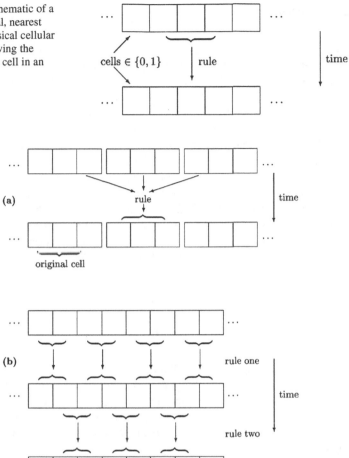

cells $\in \{0, 1\}$ rule

time

(a)   rule   time

original cell

(b)   rule one   time

rule two

Fig. 23.5 A schematic of a one-dimensional, nearest neighbour, classical (**a**) partitioned cellular automaton [24] and (**b**) block (or Margolus) partitioned cellular automata. In (**a**), each cell is initially duplicated across three cells and a new transition rule $f : Q^3 \rightarrow Q^3$ is used. In (**b**), a single step of the automata is carried out over two clock cycles, each with its own rule $f : Q^2 \rightarrow Q^2$

tem. However, reversible CAs are of particular importance, for example, in the modeling of reversible phenomena. Furthermore, it has been shown that there exists a one-dimensional reversible CA that is computationally universal [25]. Toffoli [29] demonstrated that any d-dimensional CA could be simulated by a $(d + 1)$-dimensional reversible CA and later Morita [24] found a method using partitioning (see Fig. 23.5) whereby any one-dimensional CA can be simulated by a reversible one-dimensional CA. There is an algorithm for deciding whether a one-dimensional CA is reversible [1], but in dimensions greater than one, the reversibility of a CA is, in general, undecidable [17].

23.2.2 Conway's Game of Life

John Conway's game of *Life* [11] (Life) is a well known two-dimensional CA where cells are arranged in a square grid and have binary values generally known as "dead" or "live." The status of the cells change in discrete time steps known as "generations." The new value depends upon the number of living neighbours, the general idea being that a cell dies if there is either overcrowding or isolation. In Life, a dead (or empty) cell becomes live if it has exactly three living neighbours, while a live cell survives if and only if it has two or three living neighbours. There are many different rules that can be applied for birth or survival of a cell and a number of these give rise to interesting properties such as still lives (stable patterns), oscillators (patterns that periodically repeat), spaceships or gliders (fixed shapes that move across the Life universe), glider guns, and so on [6, 12, 13]. Conway's original rule, Life, is one of the few rules with Moore neighbourhoods that are balanced. Much literature on the game of Life and its implications exists. For a discussion on the possibilities of this and other CAs the interested reader is referred to Ref. [32].

23.2.3 Quantum Cellular Automata

The idea of generalizing classical cellular automata to the quantum domain was already considered by Feynman [10]. Grössing and Zeilinger made the first serious attempts to consider quantum cellular automata (QCA) [14, 15], though their ideas are considerably different from modern approaches. QCAs are a natural model of quantum computation where the well developed theory of classical CAs might be exploited. Quantum computation using optical lattices [22] or with arrays of microtraps [7] are possible candidates for the experimental implementation of useful quantum computing. It is typical of such systems that the addressing of individual cells is more difficult than a global change made to the environment of all cells [5] and thus they become natural candidates for the construction of QCAs. An accessible discussion of QCAs is provided by Gruska [16] and a more recent review by Auon [2]. There is also a relationship between the heavily analyzed area of quantum walks [18] and QCAs [19, 20].

The simple idea of quantizing existing classical CAs by making the local translation rule unitary is problematic: the global rule on an infinite array of cells is rarely described by a well defined unitary operator. One must decide whether a given local unitary rule leads to "well-formed" unitary QCA [8] that properly transform probabilities by preserving their sum of squares to one. One construction method to achieve the necessary reversibility of a QCA is to partition the system into blocks of cells and apply blockwise unitary transformations. This is the quantum generalization to the scheme shown in Fig. 23.5(b) — indeed, all QCAs, even those with local irreversible rules, can be obtained in such a manner [27]. Formal rules for the realization of QCAs using a transition rule based on a quasi-local algebra on the

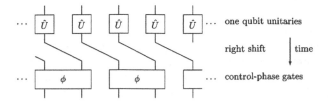

Fig. 23.6 A schematic of a one-dimensional nearest neighbour quantum cellular automaton according to the scheme of [27] (adapted from Fig. 10 of that publication). The right-shift may be replaced by a left-shift or no shift

lattice sites is described in Ref. [27]. In this formalism, a unitary operator for the time evolution is not necessary. The authors demonstrate that all nearest neighbour one-dimensional QCAs arise by a combination of a single qubit unitary, a possible left- or right-shift, and a control-phase gate,[2] as indicated in Fig. 23.6.

Reversible one-dimensional nearest neighbour classical CAs are a subset of the quantum ones. In the classical case, the single qubit unitary can only be the identity or a bit-flip, while the control-phase gate is absent. This leaves just six classical CAs, all of which are trivial.

23.3 Semi-quantum *Life*

We now present a plausible variant of the game of *Life* that reproduces some quantum features.

23.3.1 The Idea

Life is irreversible while, in the absence of a measurement, quantum mechanics is reversible. In particular, operators that represent measurable quantities must be unitary. A full quantum *Life* on an infinite array would be impossible given the known difficulties of constructing unitary QCAs [23].

Interesting behaviour is still obtained in a version of *Life* that has some quantum mechanical features. Cells are represented by classical sine-wave oscillators with a period equal to one generation, an amplitude between zero and one, and a variable phase. The amplitude of the oscillation represents the coefficient of the live state so that the square of the amplitude gives the probability of finding the cell in the live state when a measurement of the "health" of the cell is taken. If the initial state of the system contains at least one cell that is in a superposition of live and dead the

---

[2]A control-phase gate is a two-qubit gate that applies the phase operator (23.10) to the target qubit if the control qubit is 1.

neighbouring cells will be influenced according to the coefficients of these states, propagating the superposition to the surrounding region.

If the coefficients of the superpositions are restricted to positive real numbers, qualitatively new phenomena are not expected. By allowing the coefficients to be complex, that is, by allowing phase differences between the oscillators, qualitatively new phenomena such as interference effects, may arise. The interference effects seen are those due to an array of classical oscillators with phase shifts and are not fully quantum mechanical.

23.3.2 A First Model

To represent the state of a cell introduce the following notation:

$$|\psi\rangle = a|\text{live}\rangle + b|\text{dead}\rangle, \tag{23.14}$$

subject to the normalization condition

$$|a|^2 + |b|^2 = 1. \tag{23.15}$$

The probability of measuring the cell as live or dead is $|a|^2$ or $|b|^2$, respectively. If the values of a and b are restricted to non-negative real numbers, destructive interference does not occur. The model still differs from a classical probabilistic mixture, since here it is the amplitudes that are added and not the probabilities. In our model $|a|$ is the amplitude of the oscillator. Restricting a to non-negative real numbers corresponds to the oscillators all being in phase.

The birth, death and survival operators have the following effects:

$$\begin{aligned}
\hat{B}|\psi\rangle &= |\text{live}\rangle, \\
\hat{D}|\psi\rangle &= |\text{dead}\rangle, \\
\hat{S}|\psi\rangle &= |\psi\rangle.
\end{aligned} \tag{23.16}$$

A cell can be represented by the vector $\begin{bmatrix} a \\ b \end{bmatrix}$. The $\hat{B}$ and $\hat{D}$ operators are not unitary. Indeed they can be represented in matrix form by

$$\begin{aligned}
\hat{B} &\propto \begin{bmatrix} 1 & 1 \\ 0 & 0 \end{bmatrix}, \\
\hat{D} &\propto \begin{bmatrix} 0 & 0 \\ 1 & 1 \end{bmatrix},
\end{aligned} \tag{23.17}$$

where the proportionality constant is not relevant for our purposes. After applying $\hat{B}$ or $\hat{D}$ (or some mixture) the new state will require (re-)normalization so that the probabilities of being dead or live still sum to unity.

A new generation is obtained by determining the number of living neighbours each cell has and then applying the appropriate operator to that cell. The number of living neighbours in our model is the amplitude of the superposition of the oscillators representing the surrounding eight cells. This process is carried out on all cells effectively simultaneously. When the cells are permitted to take a superposition of states, the number of living neighbours need not be an integer. Thus a mixture of the $\hat{B}$, $\hat{D}$ and $\hat{S}$ operators may need to be applied. For consistency with Life the following conditions will be imposed upon the operators that produce the next generation:

- If there are an integer number of living neighbours the operator applied must be the same as that in Life.
- The operator that is applied to a cell must continuously change from one of the basic forms to another as the sum of the a coefficients from the neighbouring cells changes from one integer to another.
- The operators can only depend upon this sum and not on the individual coefficients. For example, survival will result if a cell has two neighbours with $a = 1$ or four with $a = \frac{1}{2}$, or any other combination summing to two.

If the sum of the a coefficients of the surrounding eight cells is

$$A = \sum_{i=1}^{8} a_i \tag{23.18}$$

then the following set of operators, depending upon the value of A, is the simplest that has the required properties

$$
\begin{aligned}
0 \le A \le 1; \quad & \hat{G}_0 = \hat{D}, \\
1 < A \le 2; \quad & \hat{G}_1 = (\sqrt{2}+1)(2-A)\hat{D} + (A-1)\hat{S}, \\
2 < A \le 3; \quad & \hat{G}_2 = (\sqrt{2}+1)(3-A)\hat{S} + (A-2)\hat{B}, \\
3 < A < 4; \quad & \hat{G}_3 = (\sqrt{2}+1)(4-A)\hat{B} + (A-3)\hat{D}, \\
A \ge 4; \quad & \hat{G}_4 = \hat{D}.
\end{aligned}
\tag{23.19}
$$

For integer values of A, the $\hat{G}$ operators are the same as the basic operators of Life, as required. For non-integer values in the range $(1, 4)$, the operators are a linear combination of the standard operators. The factors of $\sqrt{2}+1$ have been inserted to give more appropriate behaviour in the middle of each range. For example, consider the case where $A = 3 + 1/\sqrt{2}$, a value that may represent three neighbouring cells that are live and one the has a probability of one-half of being live. The operator in this case is

$$\hat{G} = \frac{1}{\sqrt{2}}\hat{B} + \frac{1}{\sqrt{2}}\hat{D}, \propto \frac{1}{\sqrt{2}}\begin{bmatrix} 1 & 1 \\ 1 & 1 \end{bmatrix}. \tag{23.20}$$

Applying this to either a cell in the live, $\begin{bmatrix} 1 \\ 0 \end{bmatrix}$ or dead, $\begin{bmatrix} 0 \\ 1 \end{bmatrix}$ states will produce the state

$$|\psi\rangle = \frac{1}{\sqrt{2}}|\text{live}\rangle + \frac{1}{\sqrt{2}}|\text{dead}\rangle \tag{23.21}$$

which represents a cell with a 50% probability of being live. That is, $\hat{G}$ is an equal combination of the birth and death operators, as might have been expected given the possibility that A represents an equal probability of three or four living neighbours.

Of course the same value of A may have been obtained by other combinations of neighbours that do not lie half way between three and four living neighbours, but one of our requirements is that the operators can only depend on the sum of the a coefficients of the neighbouring cells and not on how the sum was obtained.

In general the new state of a cell is obtained by calculating A, applying the appropriate operator $\hat{G}$:

$$\begin{bmatrix} a' \\ b' \end{bmatrix} = \hat{G} \begin{bmatrix} a \\ b \end{bmatrix}, \tag{23.22}$$

and then normalizing the resulting state so that $|a'|^2 + |b'|^2 = 1$. It is this process of normalization that means that multiplying the operator by a constant has no effect. Hence, for example, $\hat{G}_2$ for $A = 3$ has the same effect as $\hat{G}_3$ in the limit as $A \to 3$, despite differing by the constant factor $(\sqrt{2} + 1)$.

23.3.3 A Semi-quantum Model

To get qualitatively different behaviour from Life we need to introduce a phase associated with the coefficients, that is, a phase difference between the oscillators. We require the following features from this version of *Life*:

- It must smoothly approach the classical mixture of states as all the phases are taken to zero.
- Interference, that is, partial or complete cancellation between cells of different phases, must be possible.
- The overall phase of the *Life* universe must not be measurable, that is, multiplying all cells by $e^{i\phi}$ for some real ϕ will have no measurable consequences.
- The symmetry between $(\hat{B}, |\text{live}\rangle)$ and $(\hat{D}, |\text{dead}\rangle)$ that is a feature of Life should be retained. This means that if the state of all cells is reversed ($|\text{live}\rangle \longleftrightarrow |\text{dead}\rangle$) and the operation of the $\hat{B}$ and $\hat{D}$ operators is reversed the system will behave in the same manner.

In order to incorporate complex coefficients, while keeping the above properties, the basic operators are modified in the following way:

$$\begin{aligned}
\hat{B}|\text{dead}\rangle &= e^{i\phi}|\text{live}\rangle, \\
\hat{B}|\text{live}\rangle &= |\text{live}\rangle, \\
\hat{D}|\text{live}\rangle &= e^{i\phi}|\text{dead}\rangle, \\
\hat{D}|\text{dead}\rangle &= |\text{dead}\rangle, \\
\hat{S}|\psi\rangle &= |\psi\rangle,
\end{aligned} \tag{23.23}$$

where the superposition of the surrounding oscillators is

$$\alpha = \sum_{i=1}^{8} a_i = A e^{i\phi}, \tag{23.24}$$

A and ϕ being real positive numbers. That is, the birth and death operators are modified so that the new live or dead state has the phase of the sum of the surrounding cells. The operation of the $\hat{B}$ and $\hat{D}$ operators on the state $[\begin{smallmatrix} a \\ b \end{smallmatrix}]$ can be written as

$$\hat{B}\begin{bmatrix} a \\ b \end{bmatrix} = \begin{bmatrix} a + |b|e^{i\phi} \\ 0 \end{bmatrix},$$

$$\hat{D}\begin{bmatrix} a \\ b \end{bmatrix} = \begin{bmatrix} 0 \\ |a|e^{i\phi} + b \end{bmatrix}, \tag{23.25}$$

with $\hat{S}$ leaving the cell unchanged. The modulus of the sum of the neighbouring cells, A determines which operators apply, in the same way as before [see (23.19)]. The addition of the phase factors for the cells allows for interference effects since the coefficients of live cells may not always reinforce in taking the sum, $\alpha = \sum a_i$. A cell with $a = -1$ still has a unit probability of being measured in the live state but its effect on the sum will cancel that of a cell with $a = 1$. A phase for the dead cell is retained in order to maintain the live $\longleftrightarrow$ dead symmetry, however, it has no effect. Such an effect would conflict with the physical model presented earlier and would be inconsistent with Life, where the empty cells have no influence.

A useful notation to represent semi-quantum *Life* is to use an arrow whose length represents the amplitude of the a coefficient and whose angle with the horizontal is a measure of the phase of a. That is, the arrow represents the phaser of the oscillator at the beginning of the generation. For example

$$\longrightarrow = \begin{bmatrix} 1 \\ 0 \end{bmatrix},$$

$$\uparrow = e^{i\pi/2}\begin{bmatrix} 1/2 \\ \sqrt{3}/2 \end{bmatrix} = \begin{bmatrix} i/2 \\ i\sqrt{3}/2 \end{bmatrix}, \tag{23.26}$$

$$\nearrow = e^{i\pi/4}\begin{bmatrix} 1/\sqrt{2} \\ 1/\sqrt{2} \end{bmatrix} = \begin{bmatrix} (1+i)/2 \\ (1+i)/2 \end{bmatrix},$$

etc. In this picture, α is the vector sum of the arrows. This notation includes no information about the b coefficient. The magnitude of this coefficient can be determined from a and the normalization condition. The phase of the b coefficient has no effect on the evolution of the game state so it is not necessary to represent this.

23.3.4 Discussion

The above rules have been implemented in the computer algebra language *Mathematica* [31]. All the structures of Life can be recreated by making the phase of all the live cells equal. The interest lies in whether there are new effects in the semi-quantum model or whether existing effects can be reproduced in simpler or more generalized structures. The most important aspect not present in Life is interference. Two live cells can work against each other as indicated in Fig. 23.7 which shows an elementary example in a block still life with one cell out of phase with its

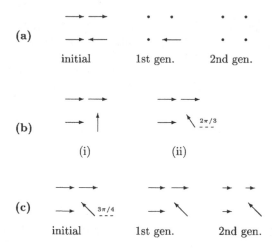

Fig. 23.7 (**a**) A simple example of destructive interference in semi-quantum *Life*: a block with one cell out of phase by π dies in two generations. (**b**) Blocks where the phase difference of the fourth cell is insufficient to cause complete destructive interference; each cell maintains a net of at least two living neighbours and so the patterns are stable. In the second of these, the fourth cell is at a critical angle. Any greater phase difference causes instability resulting in eventual death as seen in (**c**), which dies in the fourth generation

neighbours. In Life there are linear structures called wicks that die or "burn" at a constant rate [28]. The simplest such structure is a diagonal line of live cells as indicated in Fig. 23.8(a). In this, it is not possible to stabilize an end without introducing other effects. In the new model a line of cells of alternating phase ($\ldots \longrightarrow \longleftarrow \ldots$) is a generalization of this effect since it can be in any orientation and the ends can be stabilized easily. Figures 23.8(b)–(c) shows some examples. A line of alternating phase live cells can be used to create other structures such as the loop in Fig. 23.9. This is a generalization of the boat still life in Life that is of a fixed size and shape. The stability of the line of $\longrightarrow \longleftarrow$'s results from the fact that while each cell in the line has exactly two living neighbours, the cells above or below this line have a net of zero or one living neighbours due to the canceling effect of the opposite phases. No new births around the line will occur, unlike the case where all the cells are in phase.

Figure 23.10 shows a stable boundary that results from the appropriate adjustment of the phase differences between the cells. The angles have been chosen so that each cell in the line has between two and three living neighbours, while the empty cells above and below the line have either two or four living neighbours and so remain life-less. Such boundaries are known in Life but require a more complex structure.

Oscillators and spaceships cannot be made simpler than the minimal examples presented for Life. In Life interesting effects can be obtained by colliding gliders. In the semi-quantum model additional effects can be obtained from colliding gliders and "anti-gliders," where all the cells have a phase difference of π with those of the original glider. For example, a head-on collision between a glider and an anti-glider,

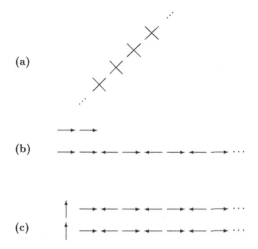

(a)

(b)

(c)

Fig. 23.8 (a) A wick (an extended structure that dies, or "burns," at a constant rate) in Life that burns at the speed of light (one cell per generation), in this case from both ends. ('*X*' represents a live cell.) It is impossible to stabilize one end without giving rise to other effects. (**b**) In semi-quantum *Life* an analogous wick can be in any orientation. The block on the left-hand end stabilizes that end; a block on both ends would give a stable line; the absence of the block would give a wick that burns from both ends. (**c**) Another example of a light-speed wick in semi-quantum *Life* showing one method of stabilizing the left-hand end

Fig. 23.9 An example of a stable loop made from cells of alternating phase. Above a certain minimum, such structures can be made of arbitrary size and shape compared with a fixed size and limited orientations in Conway's scheme

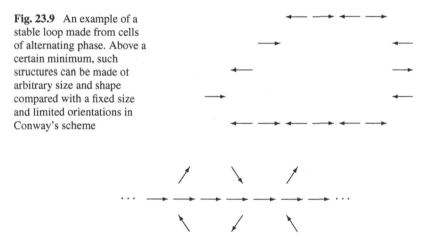

Fig. 23.10 A boundary utilizing appropriate phase differences to produce stability. The upper cells are out of phase by $\pm\pi/3$ and the lower by $\pm2\pi/3$ with the central line

as indicated in Fig. 23.11, causes annihilation, where as the same collision between two gliders leaves a block. However, there is no consistency with this effect since other glider–anti-glider collisions produce alternative effects, sometimes being the same as those from the collision of two gliders.

Fig. 23.11 A head on
collision between a glider and
its phase reversed counter
part, an anti-glider, produces
annihilation in six generations

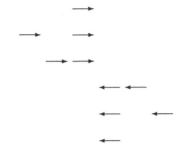

23.4 Conclusion

John Conway's game of *Life* is a two-dimensional cellular automaton where the
new state of a cell is determined by the sum of neighbouring states that are in one
particular state generally referred to as "live." A modification to this model is proposed where the cells may be in a superposition of the live and dead states with the
coefficient of the live state being represented by an oscillator having a phase and
amplitude. The equivalent of evaluating the number of living neighbours of a cell is
to take the superposition of the oscillators of the surrounding states. The amplitude
of this superposition will determine which operator(s) to apply to the central cell to
determine its new state, while the phase gives the phase of any new state produced.
Such a system show some quantum-like aspects such as interference.

Some of the results that can be obtained with this new scheme have been touched
on in this chapter. New effects and structures occur and some of the known effects
in Conway's *Life* can occur in a simpler manner. However, the scheme described
should not be taken to be a full quantum analogue of Conway's *Life* and does not
satisfy the definition of a QCA.

The field of quantum cellular automata is still in its infancy. The protocol of
Ref. [27] provides a construction method for the simplest QCAs. Exploration and
classification of these automata is an important unsolved task and may lead to developments in the quantum domain comparable to those in the classical field that
followed the exploration of classical CAs. Quantum cellular automata are a viable
candidate for achieving useful quantum computing.

Appendix

Listed below is a package written in the computer algebra language *Mathematica*
[31] for exploring semi-quantum *Life*. The package is written in a functional programming style. Commands are carried out by functions returning the desired value
or array. The functions are nested so that the "guts" of the calculations are carried out
at the deepest levels. The package is commented and usage statements are provided
for the main commands. Function names begin with capital letters, as is standard in
Mathematica, while variable names are lower case.

CountNeigh::usage = "CountNeigh[universe, x, y] returns the sum of the live coefficients of the cells surrounding (x,y)."

DisplayUni::usage = "DisplayUni[universe] displays a graphic of the universe with arrows representing living cells, the length being the magnitude and the angle with the horizontal being the phase.
DisplayUni[universe, m] displays the universe after m generations."

ExpandUni::usage = "ExpandUni[universe, n] returns an nxn universe with the old universe centred in it. n must be larger than the existing universe dimensions."

Generation::usage = "Generation[universe, x, y] returns a universe with the new value of the cell at (x,y)."

InsertBlinker::usage = "InsertBlinker[universe, x, y] returns a universe with the addition of a blinker (horizontal) starting at (x,y)."

InsertBlock::usage = "InsertBlock[universe, x, y] returns a universe with the addition of a block (2x2 square of live cells) with the lower left corner (x,y).
InsertBlock[universe, x1, y1, x2, y2] returns a universe with the addition of a group of live cells with lower left corner (x1,y1) and upper right corner (x2,y2)."

InsertGlider::usage = "InsertGlider[universe, x, y, dir] returns a universe with the addition of a glider with the lower left corner of the 3x3 square containing the glider at (x,y). dir gives the direction of the glider NE, NW, SW, SE."

InsertLine::usage = "InsertLine[universe, x, y, m] returns universe with the addition of a line of live cells starting at (x,y) of length |m|. The line is horizontal if m is positive, vertical if m is negative."

InsertString::usage = "InsertString[universe, x, y, m] returns a universe with the addition of a line of live cells of alternating phase (a 'string') of length |m|. The line is horizontal if m is positive, vertical if m is negative."

MakeUni::usage = "MakeUni[n] returns an empty universe of dimension nxn."

NextGeneration::usage = "NextGeneration[universe] returns the next generation of the universe.
NextGeneration[universe, m] returns the universe after m generations."

NormaliseCell::usage = "NormaliseCell[cell] returns the normalised value of the given cell. All cells are automatically normalised when producing the next generation by the function Generations."

Reflect::usage = "Reflect[universe, x, y] returns a universe with the cell at (x,y) phase Reflected. i.e., {1,0} -> {−1,0}.

Reflect[universe, x1, y1, x2, y2] returns a universe with the group of cells with bottom left corner at (x1,y1) and top right corner at (x2,y2) phase Reflected.
Reflect[universe, {{x1,y1}, {x2,y2}, ...}] returns a universe with all the points in the list Reflected."

Rotate::usage = "Rotate[universe, x, y, phi] returns a universe with the cell at (x,y) Rotated in phase by phi.
Rotate[universe, x1, y1, x2, y2, phi] returns a universe with the group of cells with bottom left corner at (x1,y1) and top right corner at (x2,y2) rotated in phase by phi.
Rotate[universe, {{x1,y1}, {x2,y2}, ...}, phi] returns a universe with all the points in the list Rotated by phi."

SetValue::usage = "SetValue[universe, x, y, value] returns a universe with the cell at (x,y) set to value (default=live).
SetValue[universe, {{x1,y1}, {x2,y2}, ...}, value] returns a universe with all the points in the list set to value (default=live).
SetValue[universe, {{x1,y1,val1}, {x2,y2,val2}, ...}] sets the cell {x1,y1} to val1, {x2,y2} to val2 etc."

TakeMeasurement::usage = "TakeMeasurement[universe] takes a measurement of the universe based on the quantum amplitudes, making each cell either {1,0} or {0,1}."

```
<<Graphics'PlotField'
dead = {0,1}
live = {1,0}


(*_____-*)
(* Basic functions for producing and displaying a universe *)

(* returns an empty nxn universe *)
MakeEmpty[n_Integer] := Table[ dead, {i,1,n}, {j,1,n} ]

(* take a 'measurement' - returns a universe where all cells are {1,0} *)
(* or {0,1} with the correct probabilities *)
TakeMeasurement[u_] :=
    Table[ If[ Random[] < Abs[ u[[i, j]][[1]] ]^2, {1,0}, {0,1} ],
        {i,1,Length[u]}, {j,1,Length[u]} ]

(* expand an existing universe by placing it in the centre of a *)
(* larger universe of dimension n *)
ExpandUni[u_, n_Integer] :=
    Block[ { i,j,nu, l=Length[u], ex=Floor[(n-Length[u])/2] },
        If[ n > l,
            nu = Table[
```

```
                    If[ (i > ex) && (j > ex) && (i <= l+ex) && (j <= l+ex),
                        u[[i−ex, j−ex]], dead ],
                    {i,1,n}, {j,1,n} ],
        u ];
    nu ]
```

(* display the Universe after m generations (default = 0) *)
```
DisplayUni[u_, m_Integer] := DisplayUni[ NextGeneration[u, m] ]
DisplayUni[u_] :=
    If[ !(u == MakeEmpty[ Length[u] ]),
        ListPlotVectorField[
            Flatten[
                Table[ { { (j−1)×10 + 5 − Re[ u[[i,j]][[1]] ] × 5,
                          (i−1)×10 + 5 - Im[ u[[i,j]][[1]] ] × 5 },
                         { Re[ u[[i,j]][[1]] ] × 10,
                          Im[ u[[i,j]][[1]] ] × 10 } },
                       {i,1,Length[u]}, {j,1,Length[u]} ],
                1 ],
            ScaleFactor -> None,
            Frame -> True,
            GridLines -> None
        ]
    ]
```

(* print the universe as an array of live coefficients *)
```
PrintUni[u_, m_Integer] := PrintUni[ NextGeneration[u, m] ]
PrintUni[u_] :=
    Do[
        Print[
            Table[ u[[i,j]][[1]] /.
                {0.+1. I -> I, 0. − 1. I -> −I, 1. + 0. I -> 1,
                 −1. + 0. I -> −1, 0. -> 0, 1. -> 1, −1. -> −1},
                {j,1,Length[u]} ]
        ],
        {i,Length[u],1,−1} ]
```

(*———————————————————————————————————*)
(* Functions for producing new generations *)

(* return the Universe after m (default = 1) generations *)
```
NextGeneration[u_, m_Integer] :=
    Block[ {k, nu=u}, Do[ nu = NextGeneration[nu], {k,1,m} ]; nu ]
NextGeneration[u_] :=
    Table[ Generation[u, i, j], {i,1,Length[u]}, {j,1,Length[u]} ]
```

```
(* generate the new value of the cell at position (x,y) *)
Generation[u_, x_, y_] :=
    Block[ {neigh=0, phi, A, cell=u[[x, y]], newcell={0,0}},
        neigh = CountNeigh[u, x, y];
        A = Abs[neigh];
        If[ !(neigh == 0), phi = Arg[neigh], phi=0 ];
        Which[
            (A <= 1) || (A >= 4), newcell = Death[cell, phi],
            (A > 1) && (A <= 2), newcell = (Sqrt[2] + 1)(A−1) cell +
                (2−A) Death[cell, phi],
            (A > 2) && (A <= 3), newcell = (Sqrt[2] + 1)(A−2) ×
                Birth[cell, phi] + (3−A) cell,
            (A > 3) && (A < 4), newcell = (Sqrt[2] + 1)(A−3) ×
                Death[cell, phi] + (4−A) Birth[cell, phi]
        ];
        NormaliseCell[newcell]
    ]


(* count the number of neighbours to cell (x,y) *)
CountNeigh[u_, x_, y_] :=
    Block[ {temp=0, n=Length[u]},
        If[ (x > 1) && (y > 1), temp += u[[x−1, y−1]][[1]] ];
        If[ (x > 1), temp += u[[x−1, y]][[1]] ];
        If[ (x > 1) && (y < n), temp += u[[x−1, y+1]][[1]] ];
        If[ (y > 1), temp += u[[x, y−1]][[1]] ];
        If[ (y < n), temp += u[[x, y+1]][[1]] ];
        If[ (x < n) && (y > 1), temp += u[[x+1, y−1]][[1]] ];
        If[ (x < n), temp += u[[x+1, y]][[1]] ];
        If[ (x < n) && (y < n), temp += u[[x+1, y+1]][[1]] ];
    temp ]


(* B, D operators *)
Birth[c_, phi_] := {c[[1]] + Exp[I phi] × Abs[ c[[2]] ], 0}
Death[c_, phi_] := {0, c[[2]] + Exp[I phi] × Abs[ c[[1]] ]}


(* normalise a cell so |a|^2 + |b|^2 = 1 *)
NormaliseCell[c_] :=
    Block[ {normfact=Sqrt[ Abs[c[[1]]]^2 + Abs[c[[2]]]^2] ], nc = dead},
        If[ !(normfact == 0),
            ( nc[[1]] = c[[1]]/normfact;
            nc[[2]] = c[[2]]/normfact )
        ];
        N[nc]
    ]
```

```
(*————————————————————————————*)
(* Functions for setting cell and manipulating cell values *)

(* set the (x,y) cell in u to value val (default=live) *)
SetValue[u_, x_Integer, y_Integer] := SetValue[u, x, y, live]
SetValue[u_, x_Integer, y_Integer, val_] :=
    Block[ {nu = u}, nu[[x, y]] = val; nu]
(* set the list of points to the specified value (default=live) *)
SetValue[u_, pts_List] := SetValue[u, pts, live]
SetValue[u_, pts_List, val_] :=
    Block[ {i, nu = u},
        Do[ nu[[ pts[[i]][[1]], pts[[i]][[2]] ]] = val,
        {i,1,Length[pts]} ];
    nu ]
(* set a list of points to the specified individual values *)
SetValue[u_, pts_List] :=
    Block[ {i, nu=u},
        Do[ nu = SetValue[ nu, pts[[i]][[1]], pts[[i]][[2]], pts[[i]][[3]] ],
        {i,Length[pts]} ];
    nu ]


(* Reflect the phase of the live coefficient of the cell at (x,y) *)
Reflect[u_, x_Integer, y_Integer] :=
    SetValue[ u, x, y, {−u[[x, y]][[1]], u[[x, y]][[2]]} ]


(* Reflect the phase of a group of cells with bottom left corner (x1,y1) *)
(* and top right corner (x2,y2). *)
Reflect[u_, x1_Integer, y1_Integer, x2_Integer, y2_Integer] :=
    Block[ {i,j, nu=u},
        Do[ nu = Reflect[ nu, i, j ], {i,x1,x2}, {j,y1,y2} ];
    nu ]


(* Reflect the list of points in 'pts' *)
Reflect[u_, pts_List] :=
    Block[ {i, nu=u},
        Do[ nu = Reflect[ nu, pts[[i]][[1]], pts[[i]][[2]] ],
        {i,1,Length[pts]} ];
    nu ]


(* Rotate the live coefficient of the cell at (x,y) by phi *)
Rotate[u_, x_Integer, y_Integer, phi_] := SetValue[ u, x, y,
    {u[[x, y]][[1]] * Exp[I phi], u[[x, y]][[2]]} ]


(* Rotates a group of cells with bottom left corner (x1,y1) and top right *)
(* corner (x2,y2), by phi *)
```

```
Rotate[u_, x1_Integer, y1_Integer, x2_Integer, y2_Integer, phi_] :=
    Block[ {i,j, nu=u},
        Do[ nu = Rotate[nu, i, j, phi], {i,x1,x2}, {j,y1,y2} ];
    nu ]
```

(* Rotate the list of points in the list 'pts' by phi *)
```
Rotate[u_, pts_List, phi_] :=
    Block[ {i, nu=u},
        Do[ nu = Rotate[ nu, pts[[i]][[1]], pts[[i]][[2]], phi ],
        {i,1,Length[pts]} ];
    nu ]
```

(*————————————————————————————*)
(* Some functions for inserting certain structures *)

(* blinker (period two oscillator – horizontal – starting at (x,y) *)
```
InsertBlinker[u_, x_Integer, y_Integer] :=
    Block[ {nu=u}, Do[ nu = SetValue[nu, x, y+i, live], {i,0,2} ]; nu ]
```


(* block still life, lower left corner at (x,y) *)
```
InsertBlock[u_, x_Integer, y_Integer] :=
    Block[ {nu=u},
        Do[ nu = SetValue[nu, i, j, live], {i,x,x+1}, {j,y,y+1} ];
    nu ]
```

(* set a block of cells to a value, (default = live) *)
```
InsertBlock[u_, x1_Integer, y1_Integer, x2_Integer, y2_Integer] :=
    InsertBlock[u, x1, y1, x2, y2, live]
InsertBlock[u_, x1_Integer, y1_Integer, x2_Integer, y2_Integer, val_] :=
    Block[ {nu=u},
        Do[ nu = SetValue[nu, i, j, live], {i,x1,x2}, {j,y1,y2} ];
    nu ]
```

(* Insert a glider, lower corner of 3x3 square containing glider at (x,y) *)
(* dir specifies the direction *)
```
InsertGlider[u_, x_Integer, y_Integer, dir_] :=
    Block[ {nu=u},
        Which[
            dir == "se") || (dir == "SE"),
                nu = SetValue[nu, x, y+1, live];
                nu = SetValue[nu, x−1, y+2, live];
                Do[ nu = SetValue[nu, x−2, y+i, live], {i,0,2} ],
            (dir == "sw") || (dir == "SW"),
```

```
                    nu = SetValue[nu, x−2, y+1, live];
                    nu = SetValue[nu, x−1, y+2, live];
                    Do[ nu = SetValue[nu, x−i, y, live], {i,0,2} ],
                (dir == "nw") || (dir == "NW"),
                    nu = SetValue[nu, x−1, y, live];
                    nu = SetValue[nu, x−2, y+1, live];
                    Do[ nu = SetValue[nu, x, y+i, live], {i,0,2} ],
                (dir == "ne") || (dir == "NE"),
                    nu = SetValue[nu, x−1, y, live];
                    nu = SetValue[nu, x, y+1, live];
                    Do[ nu = SetValue[nu, x−i, y+2, live], {i,0,2} ]
        ];
    nu ]


(* insert a line of live cells starting at (x,y) of length |m| *)
(* horizontal if m > 0, vertical if m < 0 *)
InsertLine[u_, x_Integer, y_Integer, m_Integer] :=
    Block[ {nu=u, i},
        If[m > 0,
            Do[nu = SetValue[nu, x, y + i, live], {i, 0, m−1}],
            Do[nu = SetValue[nu, x + i, y, live], {i, 0, −m−1}]
        ];
    nu ]


(* horizontal 'string' starting at (x,y) of length m, if m > 0 *)
(* or a vertical 'string' starting at (x,y) of length −m, if m < 0 *)
(* string starts with {1,0} *)
InsertString[u_, x_Integer, y_Integer, m_Integer] :=
    Block[ {nu=u, i},
        If[m > 0,
            Do[nu = SetValue[nu, x, y + i, {(-1)^i, 0}], {i, 0, m−1}],
            Do[nu = SetValue[nu, x + i, y, {(-1)^i, 0}], {i, 0, −m−1}]
        ];
    nu ]
```

References

1. Amoroso, S., Patt, Y.N.: Decision procedures for surjectivity and injectivity of parallel maps for tessellation structures. J. Comput. Syst. Sci. **6**, 448–464 (1972)
2. Auon, B., Tarifi, M.: Introduction to quantum cellular automata. Eprint: arXiv:quant-ph/0401123 (2004)
3. Barenco, A., Bennett, C.H., Cleve, R., DiVincenzo, D.P., Margolus, N., Shor, P., Sleator, J., Smolin J., Weinfurter, H.: Elementary gates for quantum computation. Phys. Rev. A **52**, 3457–3467 (1995)
4. Bell, J.S.: On the Einstein–Podolsky–Rosen paradox. Physics **1**, 195–200 (1964)

5. Benjamin, S.C.: Schemes for parallel quantum computation without local control of qubits. Phys. Rev. A **61**, 020301 (2000)
6. Berlekamp, E.R., Conway, J.H., Guy, R.K.: Winning Ways for Your Mathematical Plays, vol. 2. Academic Press, London (1982)
7. Dumke, R., Volk, M., Muether, T., Buchkremer, F.B.J., Birkl, G., Ertmer, W.: Microoptical realization of arrays of selectively addressable dipole traps: a scalable configuration for quantum computation with atomic qubits. Phys. Rev. Lett. **89**, 097903 (2002)
8. Dürr, C., Santha, M.: A decision procedure for well-formed unitary linear quantum cellular automata. SIAM J. Comput. **31**, 1076–1089 (2002)
9. Einstein, A., Podolsky, B., Rosen, N.: Can quantum-mechanical description of physical reality be considered complete? Phys. Rev. **47**, 777–780 (1935)
10. Feynman, R.P.: Simulating physics with computers. Int. J. Theor. Phys. **21**, 467–488 (1982)
11. Gardner, M.: Mathematical games: The fantastic combinations of John Conway's new solitaire game of "Life". Sci. Am. **223**(10), 120 (1970)
12. Gardner, M.: Mathematical games: On cellular automata, self-reproduction, the Garden of Eden and the game of "Life". Sci. Am. **224**(2), 116 (1971)
13. Gardner, M.: Wheels, Life and Other Mathematical Amusements. Freeman, New York (1983)
14. Grössing, G., Zeilinger, A.: Quantum cellular automata. Complex Syst. **2**, 197–208 (1988)
15. Grössing, G., Zeilinger, A.: Structures in quantum cellular automata. Physica B **151**, 366–370 (1988)
16. Gruska, J.: Quantum Computing. McGraw Hill, Maidenhead (1999)
17. Kari, J.: Reversibility of two-dimensional cellular automata is undecidable. Physica D **45**, 379–385 (1990)
18. Kempe, J.: Quantum random walks: an introductory overview. Contemp. Phys. **44**, 307–327 (2003)
19. Konno, N.: Quantum Walks and Quantum Cellular Automata. Lecture Notes in Computer Science. Springer, Berlin/Heidelberg (2008)
20. Konno, N., Mistuda, K., Soshi, T., Yoo, H.J.: Quantum walks and reversible cellular automata. Phys. Lett. A **330**, 408–417 (2004)
21. Lloyd, S.: Obituary: Rolf Laundauer. Nature **400**, 720 (1999)
22. Mandel, D., Greiner, M., Widera, A., Rom, T., Hänsch, T.W., Bloch, I.: Coherent transport of neutral atoms in spin-dependent optical lattice potentials. Phys. Rev. Lett. **91**, 010407 (2003)
23. Meyer, D.A.: From quantum cellular automata to quantum lattice gases. J. Stat. Phys. **85**, 551–574 (1996)
24. Morita, K.: Reversible simulation of one-dimensional irreversible cellular automata. Theor. Comput. Sci. **148**, 157–163 (1995)
25. Morita, K., Harao, M.: Computation universality of one-dimensional reversible (injective) cellular automata. Trans. IEICE **72**, 758–762 (1989)
26. Nielsen, M.A., Chuang, I.: Quantum Computation and Quantum Information. Cambridge University Press, Cambridge (2000)
27. Schumacher, B., Werner, R.F.: Reversible quantum cellular automata. Eprint: arXiv:quant-ph/0405174 (2004)
28. Silver, S.A.: http://www.bitstorm.org/gameoflife/lexicon
29. Toffoli, T.: Cellular automata mechanics. PhD thesis, The University of Michigan (1977)
30. Wolfram, S.: Statistical mechanics of cellular automata. Rev. Mod. Phys. **55**, 601–644 (1983)
31. Wolfram, S.: Mathematica: A System for Doing Mathematics by Computer. Addison–Wesley, Redwood City (1988)
32. Wolfram, S.: A New Kind of Science. Wolfram Media, Champaign (2002)

Part VII
Music

Chapter 24
Game of Life Music

Eduardo R. Miranda and Alexis Kirke

At the time when the first author was post-graduate student, in the evenings he used to entertain himself with the equipment in the electronic music studio at the University of York until dawn. It must have been around three o'clock in the morning of a rather cold winter night in the late 1980s, when he connected his Atari 1040ST computer to a synthesizer to test the first prototype of a system, which he was developing for his thesis. The system, named CAMUS (short for Cellular Automata Music), implemented a method that he invented to render music from the behaviour of the Game of Life (GoL) cellular automata (CA).

It took him a few hours to get the studio and software settings to work. When he finally managed to run the program, the music that it produced took him by surprise: it sounded remarkably interesting! It was an awesome experience, which marked the beginning of his enduring interest in making music with CA. Unwittingly, to the best of our knowledge, Miranda probably was a pioneer of GoL music. He soon learned that two other composers, Peter Beyls (in Belgium) and Dale Millen (in the USA), were also experimenting with CA at the time, but almost certainly not with GoL. Also, it would seem that Iannis Xenakis, a Greek composer then living in Paris, had used CA in the mid of the 1980s "to create complex temporal evolution of orchestral clusters" for the composition Horos [1, p. 122]. However, as reported in Hoffman's paper, this was poorly documented by the composer; it is not clear how the temporal orchestral clusters of Horos were generated by the automata.

The animated patterns produced by GoL are rather inspiring and we cannot help but think of ways to render musical forms from the automata rather than — or in addition to — visual patterns.

This chapter introduces the core of Miranda's original rendering method devised for the CAMUS system and subsequent methods developed in collaboration with Kirke and others.

A. Adamatzky (ed.), *Game of Life Cellular Automata*,
DOI 10.1007/978-1-84996-217-9_24, © Springer-Verlag London Limited 2010

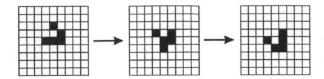

Fig. 24.1 GoL is a two-dimensional cellular automaton where each cell can be in one of two possible states: alive or dead. Given an initial configuration of "living" cells, the application of Conway's rules results in a sequence of patterns, two of which are shown

24.1 A Brief Introduction to GoL

A detailed introduction to GoL is beyond the scope of this chapter; please refer to the introductory chapters in this volume. Here we present only the very basics necessary to contextualize the chapter.

GoL is a two-dimensional CA invented by John Conway where the cells of a matrix of cells can be in one of two possible states: alive or dead (see example of propagating structure in Fig. 24.1). "Conway was fascinated by the way in which a combination of a few simple rules could produce patterns that would expand, change shape, or die out unpredictably. He wanted to find the simplest possible set of rules that would give such an interesting behaviour" [5, p. 44].

As time progresses, the state of each cell in the matrix is determined by the state of its eight nearest neighbouring cells, as follows:

- Birth: A cell that is dead at time t becomes alive at time $t + 1$ if exactly three of its neighbours are alive at time t.
- Death by overcrowding: A cell that is alive at time t will die at time $t + 1$ if four or more of its neighbours are alive at time t.
- Death by exposure: A cell that is alive at time t will die at time $t + 1$ if it has one or no live neighbours at time t.
- Survival: A cell that is alive at time t will remain alive at time $t + 1$ only if it has either two or three live neighbours at time t.

The rules above are applied simultaneously to all cells of the matrix. GoL is also characterised by a number of interesting initial configurations of "living" cells that are known to give rise to intriguing patterning behaviour. A few examples of well-known initial configurations are shown in Fig. 24.2.

24.2 Rending Musical Forms from GoL

The fundamental challenge of generating music with GoL is to design a suitable method to represent the patterns generated by the automata in terms of musical forms rather than — or in addition to — visual forms. How can one render the sequence of patterns generated by GoL (e.g., the ones shown in Fig. 24.1) in terms of music? In the next three sections we will introduce three mapping approaches for GoL which are summarized in Table 24.1.

Fig. 24.2 A few examples of initial GL configurations. Note that the "glider" configuration was used in the example in Fig. 24.1

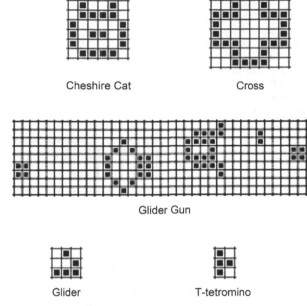

Cheshire Cat   Cross

Glider Gun

Glider   T-tetromino

Table 24.1 The three types of mappings to be introduced

Name	GoL dimensions	Coordinate system	Notes per cell
CAMUS	2	2-D Cartesian	3
CAMUS 3D	3	3-D Cartesian	4
Radial symmetry	2	2-D radial	1

24.3 CAMUS: Cartesian Representation of Note Sets

Miranda devised a musical representation scheme, whereby an ordered set of three notes is represented by a point on a two-dimensional Cartesian plane [4]. These three notes are ordered in terms of the distances between them. In musical jargon, the distance between two notes is referred to as an "interval" and the unit is the semitone.[1] Given a reference note, the abscissa of the Cartesian plane represents the interval between this reference and the second note of the

---

[1] This unit is based on the Western European tempered musical scale consisting of 12 notes separated by a semitone (or half of a tone). For example, on the piano keyboard shown in Fig. 24.3, the distance between C3 and the next white note, D3, is one tone; the distance between C3 and the next back note, C3 sharp, is half of a tone. The seventh white note from C3 is C4; C4 is called an octave of C3. The distance between a note and its octave is 12 semitones. Other cultures use different scales where the interval unit may be different; e.g., quarter of a tone.

Fig. 24.3 Musical representation using a two-dimensional Cartesian coordinate system to represent an ordered set of three notes. In this case the note *C3* is 19 semitones above *F1*, which is the reference note, and *G3* is seven semitones above *C3*

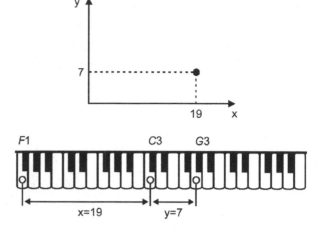

Fig. 24.4 As the automaton evolves, ordered sets of three notes are produced. These sets are defined in terms of the two-dimensional Cartesian coordinate system shown in Fig. 24.3

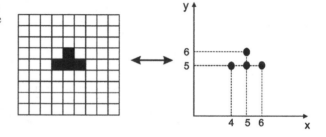

set. The ordinate represents the interval between the second note and the third (Fig. 24.3).

Given the representation shown in Fig. 24.3, a GoL matrix of cells can be mapped onto the Cartesian plane, where those cells that are alive become points of the its coordinate system (Fig. 24.4). Thus, as the automaton evolves in time, it generates points representing sets of three notes in terms of coordinates of a two-dimensional Cartesian plane.

To begin a composition process, a $[x \times y]$ GoL automaton is set up with a given initial configuration of "living" cells. At each cycle of the automata (a cycle corresponds to the application of the rules to the cells of the matrix), the cells are analysed column by column, starting with cell $(0, 0)$, continuing through to $(0, x)$, moving on to cell $(1, 0)$ through to cell $(1, x)$ and continuing in this manner until cell (x, y) has been checked. When the system arrives at a "living" cell, its co-ordinates are used to calculate the three notes, as shown in Fig. 24.3. A set of reference notes is specified beforehand by the user. Although the cell updates occur at each cycle in parallel, the system plays the "living" cells column by column, from top to bottom. The three notes of a set are not necessarily played simultaneously. Rather, they have their own starting and ending times, as explained below.

24.3.1 Temporal Morphology

The method for staggering the starting and ending times of the notes of a set represented by a Cartesian point (x, y) uses the states of its respective neighbouring cells in the GoL matrix. The system constructs a set of values from the states of the neighbouring cells, the value being equal to one if the cell is alive and zero if it is dead, as follows:

$$a = \text{cell}(x, y - 1)$$
$$b = \text{cell}(x, y + 1)$$
$$c = \text{cell}(x + 1, y)$$
$$d = \text{cell}(x - 1, y)$$
$$m = \text{cell}(x - 1, y - 1)$$
$$n = \text{cell}(x + 1, y + 1)$$
$$o = \text{cell}(x + 1, y - 1)$$
$$p = \text{cell}(x - 1, y + 1)$$

Then, the system forms four 4-bit words as follows: $abcd$, $dcba$, $mnop$ and $ponm$. Next, it performs the bit-wise inclusive OR operation, '|', to generate two four-bit words: Tgg and Dur:

$$Tgg = abcd|dcba$$
$$Dur = mnop|ponm$$

The system derives trigger information for the notes from Tgg, and duration information from Dur. With each relevant four-bit word, the system associates a code to represent time-forms where B denotes the bottom reference note, M the middle note, and U the upper one. The square brackets are used to indicate that the note events contained within that bracket occur simultaneously. The codes are as follows:

$$0000 = B[UM]$$
$$0001 = [UMB]$$
$$0010 = BUM$$
$$0011 = UMB$$
$$0101 = BMU$$
$$0110 = UBM$$
$$0111 = MBU$$
$$1001 = U[MB]$$
$$1011 = MUB$$
$$1111 = M[UB]$$

Fig. 24.5 Ten different time-forms combined in pairs define temporal morphologies for the set of three notes associated to a "living" GoL cell

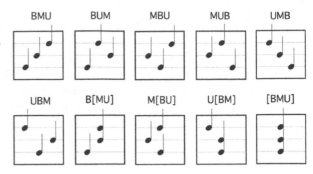

Fig. 24.6 The temporal morphology starting with *MBU* and ending with *B[MU]*

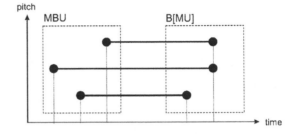

Fig. 24.7 A musical passage generated by a single cell with the temporal morphology portrayed in Fig 24.6

A visual representation of the time-forms assigned to the 4-bit words are shown in Fig. 24.5. Pairs of time-forms define a temporal morphology for the cells. For example, consider a temporal morphology starting with *MBU* and ending with

Fig. 24.8 The main steps of
the GoL music algorithm

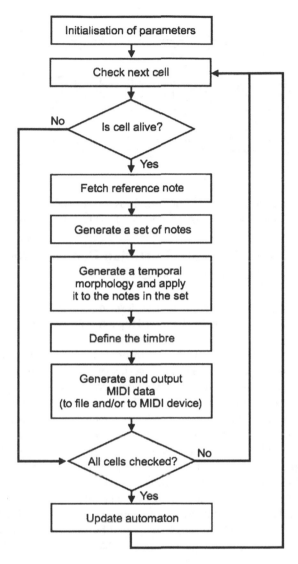

$B[MU]$ (Fig. 24.6). Figure 24.7 shows an instantiation of this morphology in musical notation.

The actual values in milliseconds for the trigger and duration parameters are calculated using a pseudo-random number generator. Finally, the music is written to a MIDI file and/or sent directly to a MIDI sampler or synthesiser to be played. Figure 24.8 illustrates the main steps of the algorithm in the form of a flowchart. An example of a musical passage generated by the system is shown in Fig. 24.9. Note the characteristic sequence of patterns of groups of three notes evolving in time.

Fig. 24.9 An example of a musical passage composed with CAMUS. In the first two bars, three patterns of groups of three notes, corresponding to GoL "living" cells are *circled*

24.4 CAMUS 3D: Cartesian Representation for Three Dimensional GoL

The Cartesian representation described above was extended in collaboration with Kenny McAlpine and Stuart Hoggar, who contributed to further develop CAMUS by introducing a number of variations to the original program, notably the use of three-dimensional GoL [3]. In order to achieve this, the three-dimensional Cartesian space was treated as a series of stacked two-dimensional spaces. Therefore, a three-dimensional GoL was defined as a series of two-dimensional GoL stacked parallel to the plane $x = 0$. Then each of the stacked planes has the form $x = a$, for some integer value, a. Thus, when it comes to assess the neighbouring cells of an arbitrary cell (a, b, c), the algorithm needs to restrict its attention only to the cells $(a, b + 1, c)$, $(a, b - 1, c)$, $(a, b, c + 1)$, $(a, b, c - 1)$, $(a, b + 1, c + 1)$, $(a, b + 1, c - 1)$, $(a, b - 1, c + 1)$ and $(a, b - 1, c - 1)$, because the focus is on the plane $x = a$. This means that each of the stacked two-dimensional GoL evolve independently; that is, none of the neighbouring cells can exert any influence on the cells in any of the other GoL. This configuration is illustrated in Fig. 24.10.

An example of music generated by the three-dimensional CAMUS is shown in Fig. 24.11. As on the case of Fig. 24.9, it is possible also here to identify the characteristic sequence of patterns of groups of four notes evolving in time.

24.5 Radial Representation for Two Dimensional GoL

In the original Conway's GoL rules, the state of a cell is determined by the state of the cells around it. If one rotates the neighbourhood of a cell around the centre of the matrix by 90, 180 or 270 degrees, the rule would still have the same effect on the cell's state. We believe that this is the reason that GoL generates such attractive and musically inspiring symmetric patterns. A number of researchers have discovered what are now well-known symmetrical life evolutions, and there is some discussion

Fig. 24.10 A three-
dimensional automaton
defined as a series of stacked
two-dimensional GoL. The
two-dimensional GoL in this
case are stacked parallel to
the plane $y = 0$. The *dark
grey* cell in the *middle* layer is
currently under examination.
Since we treat the y
co-ordinate as a constant,
only the *shaded* neighbouring
cells in the same plane are
also examined. Thus, each
two-dimensional game
evolves in isolation

Fig. 24.11 Example of musical passage generated by the three-dimensional CAMUS. In the first
two bars, two patterns of groups of four notes, corresponding to GoL "living" cells are *circled*

of the symmetrical tendencies of GoL [6]. And even for those evolutions, which
are non-symmetrical, the emergence of complex broken symmetries from simple
symmetrical rules is one element that appeals to our aesthetic sense during GoL
iterations.

The two mappings introduced so far have focused on a Cartesian approach to
mapping the cell content. In this third mapping, we devised a method to capture the
radial symmetry inherent in the aesthetic of the GoL — polar co-ordinates [2]. For

Fig. 24.12 In the polar
co-ordinates method to
capture the radial symmetry
inherent in the aesthetic of the
GoL, an origin is defined at
the centre of the GoL matrix,
and each live cell is mapped
into the musical domain
based on its (r, θ) co-ordinate

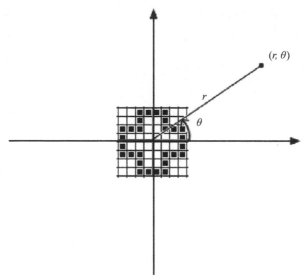

this approach, the origin is placed at the centre of the GoL matrix, and each live cell
is mapped into the musical domain based on its (r, θ) co-ordinate rather than its x or
(x, y) co-ordinate (Fig. 24.12). Once this radial mapping is chosen a decision needs
to be made as to what to map (r, θ) on to. The challenge here is: How to represent
music using the (r, θ) co-ordinate?

The following algorithm was developed to generate music using the (r, θ) co-
ordinate:

1. Choose a BPM (beats per minute) value. Initialise a variable *baseBeat* to 0,
   choose a fixed note duration D.
2. Run a generation of the GoL.
3. Iterate θ around the matrix from 0 to 2π in 127 steps. For each of the 127 values
   of θ, iterate r from the centre of the matrix to its edge, one cell at a time.
4. For each of the iterated values of r, examine the cell at (r, θ). If it is "alive"
   generate a MIDI note with pitch proportional to r, and of duration D. Locate the
   MIDI note at beat: $baseBeat + D + (16 * (\theta/(2\pi)))$.
5. After completing the nested iterations of (r, θ) over the whole matrix, update
   baseBeat to the beat of the last generated MIDI note.
6. Go back to (2) and repeat.

Let us consider the cell with (x, y) co-ordinate $(2, 2)$ in Fig. 24.12. By performing a transformation from Cartesian to Polar co-ordinates we obtain a quantized (r, θ) coordinate of $(3, \pi/2)$. This is halfway across the GoL matrix shown
in Fig. 24.12. Thus, the proportional MIDI pitch value will be 64 (halfway up the
MIDI pitch range of 0 to 127). Looking at step (1) above, if we assume a duration
D of 1 beat and a starting value of *baseBeat* of 0, then the start of the beat for the
MIDI note calculated by step 4 is: $0 + 16(\pi/2)/(2\pi) = 5$. So $(2, 2)$ will generate a
MIDI note of pitch 64 at beat 5.

Fig. 24.13 First two generations of a GoL starting with a circle of "alive" cells on the left (*white* = "alive")

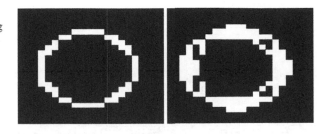

In order to illustrate the radial mapping in practice, let us assume a GoL matrix of size 20 cells by 20 cells is used. The matrix is initialised with a pattern of "living" cells, which is approximately a circle of radius of 7 squares. It is only approximately a circle since it is built using polar co-ordinates, and the grid quantization mapping does not allow it to be a perfect circle. The left hand picture in Fig. 24.13 shows the seed; note: in this case white cells are "alive", and black ones "dead". It can be seen that the next cycle of GoL — shown on the right hand side of Fig. 24.13 — does not have full radial symmetry either. However it still has significant amounts of radial symmetry that will be picked up by the generative polar co-ordinate process.

Figure 24.14 shows the resulting output of the two GoL matrices shown in Fig. 24.13. In this example, the pitches were scaled between MIDI values 60 and 90. A tempo of 120 beats per minute was used and the global duration D was set to 2 beats. Looking at the first generation in bars 1 to 8 (corresponding to the left hand side of Fig. 24.13) we can observe the melody falling and then rising again. In fact the melody should stay at a constant pitch for a perfect circle seed. But in reality one side of the "circle" is closer to the origin to the other side due to the generating pattern not being a perfect circle around the origin; that is, a quantization error for such a small circle in GoL. Furthermore, in bars 1 to 8, there would only be a single pitch at any time if we had a perfect circle. But the way the circle is generated in the seed causes it to have a thickness of 2 cells in certain places. From bar 9 onwards the second generation (on the right hand side of Fig. 24.13) is playing. Some of the "rise, fall, rise"-effect in pitch, seen in the first 8 bars, is still there. This is because the second generation still approximately holds the quantized circle shape. However, as can be seen in the right hand side of Fig. 24.13, the complexity of GoL is now starting to manifest in the score. A significant number of new cells have been brought to life, contributing to the harmony of the music. Furthermore, the quantization of the circle leads to θ generating slightly more complex rhythms.

Despite the quantization errors, these two generations show how the polar mapping is starting to capture the outward symmetrical complex movement of the circle seed. Thus better capturing some of the radially symmetric aesthetic of the visuals than a linear mapping would.

24.6 Concluding Remarks

Pioneers such as Xenakis, Beyls, Millen and others, rightly understood the suitability of cellular automata to model systems that change some feature with time.

Fig. 24.14 Music of the first 2 generations of the GoL shown in Fig. 24.13

Essentially, music is a time-based art form, where sequences of musical notes and rhythms form patterns of sonic structures organised in time. In this context, GoL is appealing because they produce sequences of coherent patterns, some of which can be very complex, and yet controlled by remarkably simple rules. In this chapter we have introduced three mapping approaches to composing music with GoL — a two-dimensional and three-dimensional Cartesian mapping, and a two-dimensional radial mapping. By no means do these exhaust the mapping possibilities, but they highlight some key aspects of GoL musical mappings, as shown earlier in Table 24.1.

The systems described in this chapter have been used to compose a number of fully-fledged pieces of music professionally. In relation to this, it should be noted that an inevitable problem with the great majority of systems that generate music automatically is that they do not possess knowledge about musical instruments. They often generate musical passages that would be technically impossible to be played on the respective instruments. These pieces tend to sound unconvincing, or "non-musical", when played on a MIDI synthesiser or sampler, because the music is not performed idiomatically; the clarinets do not sound "clarinetistically", the violins do no sound "violinistically", and so on. Better results can be achieved by amending the score manually in order to render the piece more realistic. The systems described in this chapter are not exceptions. For instance, the passages in Fig. 24.9 have been edited manually in order to alleviate the idiomatic problem. However, the edits in Fig. 24.9 were kept to a minimum and often boiled down to transposition of a note an octave upwards or downwards. Nevertheless, it is still possible to identify the characteristic sequence of patterns of groups of three notes evolving in time. Dynamics and articulation were also added manually. Note that Fig. 24.14 was not edited; it shows the raw output from the system.

The resulting pieces were performed in public concerts and some are available on CD. For instance, there are two compositions on the CD *Plural*, released in Brazil (389.726.04-7): tracks 1–3 contains the piece *Grain Streams* for piano and electronic sounds and track 14 contains the piece *Entre of Absurdo e o Mistério*, for orchestra. Please e-mail the authors should you wish to order a copy of this CD or obtaining recordings of other GoL-generated pieces.

References

1. Hoffman, P.: Towards and automated art: algorithmic processes in Xenakis' compositions. Contemp. Music Rev. **21**(2–3), 121–131 (2002)
2. Kirke, A., Miranda, E.R.: Capturing the aesthetic: radial mappings for cellular automata music. J. ITC Sangeet Res. Acad. **21**, 15–23 (2007)
3. McAlpine, K., Miranda, E.R., Hoggar, S.: Making music with algorithms: a case study system. Comput. Music J. **23**(2), 19–30 (1999)
4. Miranda, E.R.: Cellular automata music: an interdisciplinary project. Interface **22**(1), 3–21 (1993)
5. Wilson, G.: The life and times of cellular automata. News Sci. **October**, 44–47 (1988)
6. Wolfram, S.: Universality and complexity in cellular automata. Physica D **10**, 1–35 (1994)

Part VIII
Computation

Chapter 25
Universal Computation and Construction in GoL Cellular Automata

Adam P. Goucher

This chapter is concerned with the developments of universal computation and construction within Conway's Game of Life (GoL). I will begin by describing the history of the concepts and mechanisms for universal computation and construction in GoL, before explaining how a Universal Computer–Constructor (UCC) would operate in this automaton. Moreover, I shall present the design of a working UCC in the rule. It is both capable of computing any calculation (i.e. it is Turing-complete) and constructing most, if not all, of the constructible configurations within the rule. It cannot construct patterns which have no predecessor; neither can any machine in the rule (for obvious reasons). As such, it is more accurately a *general constructor*, rather than a universal constructor.

25.1 History

This section briefly outlines the previous work in GoL and other cellular automata. Some of the concepts introduced in this section will help you to understand the workings of the universal computer described later in this chapter.

25.1.1 Birth of Self-replication

In the late 1940s, John von Neumann invented cellular automata. He originally attempted to make a kinematic model of self-replication, but settled on a mathematical abstraction since it was much easier to formalise. He considered using partial differential equations to govern his system, [9, p. 876] due to their widespread use in physics. Conversely, the mathematician Stanislaw Ulam suggested he use a lattice grid instead; this gave von Neumann the inspiration to create his cellular automaton. It operates on an infinite, two-dimensional square grid, or *lattice*, where each *cell* can exist in one of 29 states. These 29 states fall into several subcategories:

A. Adamatzky (ed.), *Game of Life Cellular Automata*,
DOI 10.1007/978-1-84996-217-9_25, © Springer-Verlag London Limited 2010

- the *ground state*, U, exists as a passive background on which the rules operate (all but a finite configuration of cells are in this state);
- the *ordinary transmission states* are used to transfer information;
- the *special transmission states* are used to remove ordinary transmission states, and vice-versa;
- the *confluent state* can be used as a delay, AND gate or fanout device;
- the *sensitised states* are temporary states that quickly become quiescent cells.

Since this book is about Conway's Game of Life, rather than von Neumann's cellular automaton, I shall not describe the cellular automaton in great detail. Instead, I shall show you a self-replicator created in a (slightly different) rule, and outline its major components, see Fig. 25.1.

The configuration contains specialised organs to perform different functions. At the bottom of the configuration is a tape-reading device. The tape extends to the right for thousands of cells, far beyond the edge of Fig. 25.1. Situated directly above is the tape-writing organ, which produces an exact duplicate of the input tape. The remainder of the machine is dedicated to interpreting and executing the instructions. The instruction tape contains the entire template for describing the replicator.

Most replicators use the same construction arm to replicate the tape and construct the duplicate; this replicator is different in that respect. The upper-right corner of the configuration contains two separate arms: one for corpus construction and one for tape duplication. This means that the tape only has to be read once, making this a much faster machine. It also improves readability — you can immediately see the organs responsible for reading, writing and construction.

Unfortunately, replication in Conway's Game of Life is not as simple as that in von Neumann's cellular automaton. For instance, the mechanisms for signal propagation, logical operations and construction are not explicitly handled by the rules. Indeed, Conway himself was not certain of the universality of GoL before Gosper published the first infinite-growth pattern [5].

25.1.2 Early Attempts at GoL Self-replication

In October 1970, Martin Gardner published Conway's Game of Life in his amazingly successful *Mathematical Games* column. It wasn't long before the first pattern was found that grew indefinitely: Bill Gosper's perpetual glider gun.

The gun produces a glider every 30 generations. In Fig. 25.2, you can see two gliders travelling south-east. The gun continues to emit gliders at regular intervals. Since logical operations can be conducted using gliders, this allows any binary computation to be performed. An example of a small, useful configuration that uses glider guns is the binary adder.

Conway's design for a universal computer involved a 'tape' of instructions, much like the replicators in previous cellular automata. However, he chose to represent the tape as a single integer, rather than as a string of binary digits. The integer is specified by the diagonal distance of a block from the organ. Moving the block

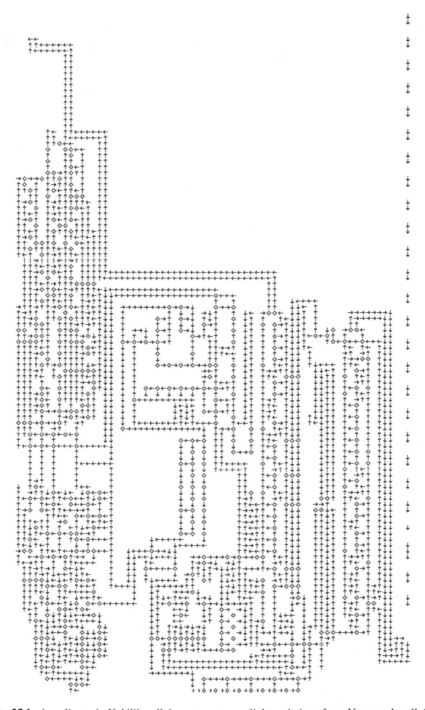

Fig. 25.1 A replicator in *Nobili's cellular automaton*, a slight variation of von Neumann's cellular automaton

Fig. 25.2 Gosper's period-30
glider gun

away from the organ corresponds to increasing the number by that distance, and
vice-versa.

The integer can be incremented or decremented by firing a salvo of gliders towards the block. This can 'pull' or 'push' the block by one cell diagonally. Moreover, a perpendicular glider can be fired to test whether the block is in the 'zero'
position. Dean Hickerson built a working model of the *sliding block memory*, based
on an elaborate configuration of glider guns [3]. Universal computers can be built
entirely out of these sliding-block registers. Computers like these are called *register
machines*, or *counter machines*.

Universal construction is much more difficult to implement. At that time, the
suggested method was to collide gliders to form basic elements, or *Buckingham bits*.
The entire machine could itself be composed solely of Buckingham bits, enabling
the machine to replicate. Buckingham bits are objects such as the pentadecathlon,
glider gun, block, eater and other vital components. They have relatively simple
glider syntheses, meaning that they are easy to construct. More than 30 years later,
Paul Chapman and Dave Greene demonstrated that a replicator can be built from a
more conservative set of components.

25.1.3 Stable Reflectors

A *stable reflector* is a passive device that can reflect gliders and return to its original
state. Such a device would be invaluable, since it requires no synchronisation. Since
all previous technology was based on either period-30 objects, such as the glider
gun, or period-46 objects, such as the 'twin bees shuttle', migrating to solely passive
components represented a giant leap in GoL engineering. In 1996, Paul Callahan
built the first-ever stable reflector [4]. As it was large, slow, and cumbersome, many
successive refinements were made. The smallest 90° reflector was made by Stephen
silverAG [1], as shown in Fig. 25.3.

This stable reflector is nearly an order of magnitude faster than its earliest counterpart, taking just 497 generations to recover. However, this is still much slower
than period-30 technology — a small price to pay, since stable technology radically
simplifies the design of a self-replicating machine. In fact, stable reflectors are so

Fig. 25.3 The smallest
known 90° stable reflector

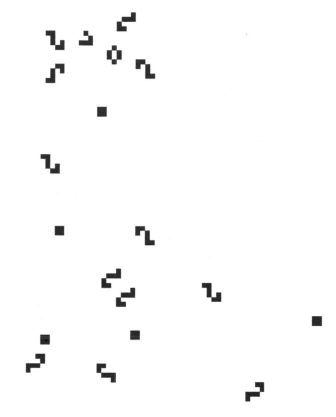

useful that Dave Greene has offered a prize of 50 USD [1] for the first 90° stable reflector that fits inside a 50 × 50 bounding box, and a further 50 USD for the first to fit inside a 35 × 35 box. Of course, you could potentially earn both prizes simultaneously, by discovering a 35 × 35 reflector before the other prize has been claimed.

Stable reflectors can be chained together with other aperiodic components to form more complex devices such as logic gates, fanout devices and registers. Registers operate in the same way as Hickerson's sliding-block memory, only are much larger due to the size of stable reflectors.

A variation of the stable reflector in Fig. 25.3 is the *edge-shooter*. The edge-shooter is slightly slower than silverAG's reflector, but has the additional property that the output glider is produced on a clear lane beyond the edge of the configuration. This is beneficial, as it allows multiple gliders to be placed on very close, even overlapping, paths. You will see a potential use for this ability in the subsequent sections of this chapter, where it is heavily utilised in the construction arm and sliding-block registers.

25.1.4 Chapman's Universal Computer

Paul Chapman implemented a slow, but powerful, computer in Life [2]. His original computer used period-30 technology, but he soon duplicated this feat using only stable components. The design that he used is called a *Minsky Register Machine*, or MRM [7]. Register machines differ from Turing machines in the way that they store information: Turing machines use an infinite tape, whereas register machines use a finite number of registers, each of which can store an unbounded integer. Registers are simpler to construct than tape interfaces, which is why the early designs for replication in GoL involved them. Although they are much easier to implement, they do have a downside: it takes exponential time to perform a calculation using registers; tapes only require polynomial time.

Chapman's machine contains 12 unbounded registers. The first of these registers contains the program for the machine, expressed as an integer. Imagine a very small program comprising only of 100 bits of information, the corresponding register would need to be 2^{100} (1.2×10^{30}) cells long. Just to give an idea of the magnitudes involved, if this machine was drawn at a scale of 1 pixel per millimetre, the resulting configuration would be slightly larger than the visible universe! A realistic estimate for the length of a self-replicating tape is 10^8 bits long, making the register unimaginably large. For this reason, my machine uses binary tapes for storing the program and data.

Despite the impracticality of using such a machine, it *does* still constitute a universal computer. In other words, it could perform any calculation, despite taking a colossal amount of time to do so. Registers are also useful for storing large integers, such as variables and memory addresses. Chapman's registers have two inputs:

- to increment the value in the register by one (INC);
- to attempt to decrement the value in the register (TDEC).

When the TDEC input is activated, the register returns a signal from one of two outputs:

- the decrement has failed, i.e. the register contained a zero (Z);
- the decrement has succeeded, i.e. the register contained a non-zero value (NZ).

A sufficient number of these simple registers, together with an appropriate microprogram, guarantees computation universality.

25.2 Components

The Universal Computer–Constructor implemented in GoL is highly integrated. This section describes the components of the UCC that co-operate to achieve universal computation and construction. The components include switches, latches, registers, tapes, and the construction arm.

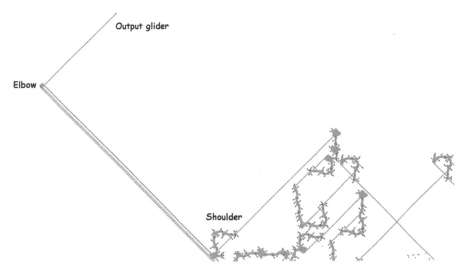

Fig. 25.4 Chapman and Greene's versatile construction arm

25.2.1 The Construction Arm

The construction arm is what makes the difference between a universal computer and a universal constructor. As opposed to the replicator in Fig. 25.1, my Universal Computer–Constructor uses the same construction arm for duplicating the tape and the corpus (body) of the machine. The arm itself was built by Paul Chapman and Dave Greene, and is composed of simple stable objects. These objects, which can easily be constructed, are known as *Spartan* objects.

The main components of the arm are the *shoulder* and *elbow*. The shoulder is a complicated configuration of reflectors and edge-shooters that together generate salvoes of gliders corresponding to different instructions. Whereas the shoulder is large and static, the elbow is small and dynamic. In fact, the elbow is the simplest stable object possible — a single block. This block is free to move along a single diagonal extending from the shoulder. Figure 25.4 shows the layout of the construction arm.

Remarkably, the arm uses just four instructions. Two of these are articulation commands, for pushing and pulling the elbow in the same fashion as sliding-block memory. The other two fire gliders perpendicular to the construction arm. The reason that two commands are needed is due to parity — the Life universe can be coloured like a chessboard, so unique commands are required to send gliders along the white and black diagonals.

The four commands are able to produce any salvo of gliders, with the restriction that the previous glider must have settled down before the next glider can arrive. Such a salvo is known as a *slow salvo*. This restriction does not prevent the replicator from being able to construct any Spartan configuration, including itself.

25.2.2 The Memory Tape

Differing from Conway's original design, I decided to make a binary memory tape. This would enable data to be stored much more efficiently, reducing replication time from exponential to polynomial. The binary tape consists of an infinite number of bits, each represented by the presence or absence of a block. A complex organ is required to interface with the memory tape, since it needs to be capable of reading, writing and articulation. In order to read both on and off bits, I included a regulation device. This waits for a pre-determined amount of time (related to the position in the tape) before checking whether the returned bit is on or off. A variable-length regulator is required, since it takes longer to read more distant bits from the tape. The time taken to read a binary digit from the data tape is equal to $64p + k$, where p is the position of the head, and k is a fixed constant. This is a consequence of the fact that it takes 32 generations for a glider to travel from one binary digit to the next, and another 32 generations to return. As such, it takes a quadratic amount of time to read the information in the entire tape, equal to

$$t = \sum_{p=0}^{l} 64p + k$$

generations, where l is the length of the tape, and k is a fixed constant. For large values of l, the linear terms will have a negligible effect on the overall reading time, so it can be approximated by $32l^2$. By comparison, a register machine takes $k \times 2^l$ generations to read the same amount of information.

Obviously, a single read of the tape is not sufficient to allow self-replication. John von Neumann addressed this issue, explaining that *two* reads of the tape are necessary: one for the execution of the instructions and one to copy those instructions. This requires the use of a *conditional jump* in the instruction set. The conditional jump is the simplest way of allowing the machine to make decisions.

The specification mentioned above refers to the data and marker tapes. The program tape differs in the following respects:

- the bits are separated by 4, not 8, diagonal cells;
- each bit is represented by an eater, as opposed to a block;
- the program tape is read-only, unlike the other tapes.

25.2.3 Registers

The idea of sliding-block memory registers has also been utilised in my machine. The registers are of a similar design to Chapman's, although the component-level architecture is different. As we have seen, sliding-block memory needs to be able to increment, decrement and test whether or not the block is in its initial (zero) position. The former two tasks can be accomplished using a glider salvo; the latter requires a more interactive approach. In order to test whether the block is in the zero

Fig. 25.5 An incident glider
approaches from the
south-west. If the block is in a
non-zero position, the glider
will continue unaffected. If
the glider is absorbed, another
glider diametrically opposite
is used to restore the block

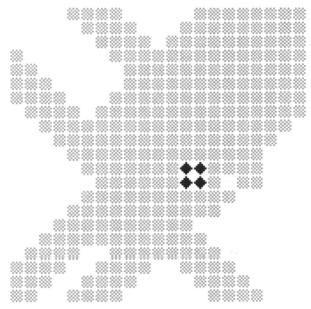

position, a glider is fired towards it. The glider will collide with it if, and only if, it
is in the zero position. If the detection glider is absorbed, it causes the block to be
repositioned. This means that another glider, anti-parallel to the first, must be fired
at the block to restore it. Otherwise, if the detection glider continues past, the block
is in a position corresponding to a (positive) non-zero integer. Figure 25.5 shows the
active site of the register.

25.2.4 Switches, Latches and Gates

Machines that use Boolean logic generally require some way of storing binary
data. In electronic computers, this is generally accomplished by a capacitor or R-S
flipflop. In my universal computer–constructor, Boolean data is stored in switches.
One type of switch would be sufficient, but I have used two different switches for
different purposes. In most cases, a basic demultiplexer is used. The data stored in
demultiplexers is cleared when the state is retrieved. The Boolean registers, or gates,
contain non-destructive memory. They each contain a *latch* and a *buffer*. The latch is
responsible for storing the data; the buffer acts as a safety feature to prevent certain
operations. Specifically, opening the latch when it is already open (or attempting to
close it when it is already closed) would result in the device exploding and damaging any circuitry in the vicinity. The buffer only allows the latch to be opened when
it is initially closed, and vice-versa.

Fig. 25.6 Layout of the
universal
computer–constructor

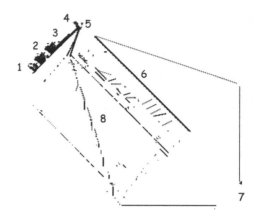

25.3 An Overview of the Design

So far, I have only described the individual components. This section explains how
the components interact to achieve computation universality. Figure 25.6 shows
some of the different sections of the computer:

1. Program tape
2. Other tapes
3. Registers
4. Construction arm
5. Latch header
6. Latches
7. Ground
8. Lookup table

25.3.1 The Layout

At the simplest level, the machine consists of two parts: the *control* section and the
devices. The devices are further subdivided into two groups: the *memory* devices
(tapes, registers and gates), and the single *output* device (the construction arm).
This table summarises the functions of the different components:

Section	Purpose
Control section	Orchestration
Output devices	Construction
Memory devices	Information storage

25.3.2 The Control Section

The control section is essentially a finite state machine, and is composed of two *lookup tables*. The first of these interfaces with the various devices. The other decides the next state of the control section. The behaviour can be formalised as a set of quadruples, (C, I, N, A), where:

- C is the current state of the machine (ranges from 1 to 147);
- I is the input from the devices (can be either *true* or *false*);
- N is the next state of the machine (ranges from 1 to 148, due to an extra *halt* state);
- A is the set of actions to perform (any subset of the 64 available actions, subject to the condition that exactly one action must return a value).

In total there are 294 quadruples, one for each possible pairing of C and I. Each quadruple is represented by a row of the lookup table. The data is stored as an array of reflectors. The first table is 64×294, with each column corresponding to a different action. In the second, much larger (148×294), table, each row has exactly one reflector — a logical necessity, since the machine is deterministic. Unlike the first table, the second table uses a form of unary encoding. Binary encoding could potentially reduce the overall area of the table, but would increase its complexity.

If you have encountered the state-transition table of the Turing machine, this description may seem familiar to you. Both the symbol to write and direction to move (two of the parameters in a Turing machine quintuple) are included, amongst other functions, in the A element of the UCC transition table.

25.3.3 The Operation Cycle

During each operation cycle, the control section waits for the input from the devices (I). Next, it performs one or more actions, as specified by A. Exactly one of these actions must return a boolean value. This boolean value becomes the I of the next cycle. Finally, the control section selects the next state of the machine (N).

The lookup involves a complicated system of activating and deactivating latches. There are 294 latches in total, one for each row of the lookup table. This series of latches is capped by two units: the *latch header* is upstream of the latches, whereas the *ground* is downstream. The latch header sends signals down the series of latches. The ground simply absorbs stray signals from the latch header. Its complexity is due to the fact that the signals may be on adjacent paths that cannot be consumed by standard eaters.

25.3.4 The Instruction Set

The instruction set is designed to be easy to program, rather than compact. The instruction set of Wang's W-machine [8] is a subset of the instruction set of my machine, proving that it is indeed Turing-complete. Unlike Chapman's MRM, this machine can solve any P-class problem in polynomial time.

The most difficult command to implement is the conditional jump instruction. It has the following code:

$$0001\, D\, Z\, 1\, M$$

The first four bits (0001) is the opcode for the conditional jump. Since it was the first command that I implemented, it has the opcode 0001. This is followed by a direction, D. D is set to 1 for a forward jump, or 0 for a backward jump. This is then followed by a series of zeros, Z. The number of zeros should be equal to the length of M. This is terminated by a 1, which separates Z and M. M is a binary integer representing the size (in bits) of the jump. As an example, consider the following string, which represents a backwards jump of length 42:

$$0001\, 0\, 0000000\, 1\, 101010$$

The remaining instructions are much simpler, each comprising of between 5 and 8 bits. The instructions are based on a binary tree structure, not unlike a Huffman tree. HALT corresponds to the sequence 00000, NOP corresponds to the sequence 00001 etc. The remaining commands are used to transfer data between registers, gates and tapes, and to perform basic arithmetical and logical operations on them. There are also four other commands of the form 0011xx, which correspond to the four different construction commands.

25.3.5 Proposed Method of Replication

There are several ways to program the tape-copying program and construction procedure into the machine. The simplest way is to use the approach that Tim Hutton used in his implementation of Codd's self-replicating machine [6]. The program tape contains the instructions to replicate the data tape twice and to build the duplicate; the data tape merely contains an exact copy of the program tape. The marker tape is reserved for storing internal values. This removes the need for separate program-copying and data-copying procedures.

The construction procedure could simply be represented as a series of 0011xx commands. Indeed, that would be the easiest to implement. However, the repetitive nature of the machine means that there will be lots of repeated commands. These could be compressed by utilising the powerful conditional jump instruction, which could potentially reduce the overall tape length, and thus replication time, by a few orders of magnitude.

25.4 Conclusion

In this chapter, you have seen the prior attempts at self-replication in GoL and other cellular automata, explored the design of a recent implementation of universal computation and construction, and learned how one would go about writing a self-replicating tape for the machine. Undoubtedly, much simpler methods of replication exist, i.e. ones that do not require the use of a universal computer. However, the computational abilities facilitate machine evolution, something that cannot be demonstrated with a simple replicator. This was John von Neumann's original motivation for exploring cellular automata — to demonstrate the mechanics of development itself.

References

1. http://www.argentum.freeserve.co.uk/reflect.htm
2. http://www.igblan.free-online.co.uk/igblan/ca/index.html
3. http://www.radicaleye.com/lifepage/patterns/sbm/sbm.html
4. http://www.radicaleye.com/lifepage/patterns/ssrefl/ssrefl.html
5. Gardner, M.: Wheels, life and other mathematical amusements (1983)
6. Hutton, T.: Codd's self-replicating computer (2010)
7. Minsky, M.: Recursive unsolvability of Post's problem of 'tag' and other topics in theory of turing machines. JSTOR (1961)
8. Wang, H.: A variant to Turing's theory of computing machines (1957)
9. Wolfram, S.: A New Kind of Science. Wolfram Media, Champaign (2001)

Chapter 26
A Simple Universal Turing Machine
for the Game of Life Turing Machine

Paul Rendell

In this chapter we present a simple universal Turing machine which is small enough to fit into the design limits of the Turing machine build in Conway's Game of Life by the author [7]. That limit is 8 symbols and 16 states. By way of comparison we also describe one of the smallest known universal Turing machines due to Rogozhin [9] which has 6 symbols and 4 states.

26.1 Introduction

The author constructed a Turing machine in Conway's Game of Life [7] see Fig. 26.1. The example constructed has a finite state machine of 3 states and uses 3 symbols. The program implemented doubles the length of a string of one of these symbols on the tape. Figure 26.5 shows this diagrammatically. In this chapter we examine the options and practicality of putting a universal Turing machine program in the Game of Life (GoL) Turing machine.

The Game of Life is a cellular automaton invented by J.H. Conway [2, 3]. It has an infinite universe divided into cells, each cell takes a state from a binary set and updates its state according to certain strict rules. All cells change their states simultaneously in discrete time. For Conway's Game of Life the cells have two states, call them live and dead, and the rules are based on the number of neighbouring cells, which are alive.

A universal Turing machine is a Turing machine which emulates another Turing machine. It has on its tape a description of that machine and its tape. The only thing fixed about a Turing machine is its finite state machine, if this is a universal Turing machine then this one machine can perform any calculation simply by emulating a more dedicated machine. With an infinite tape to store things on, the options are truly infinite. Turing first described his universal Turing machine in his 1936 paper [10].

We are looking for a simple Turing machine in the sense that it should be easy to understand as befits a demonstration model such as [7].

A. Adamatzky (ed.), *Game of Life Cellular Automata*,
DOI 10.1007/978-1-84996-217-9_26, © Springer-Verlag London Limited 2010

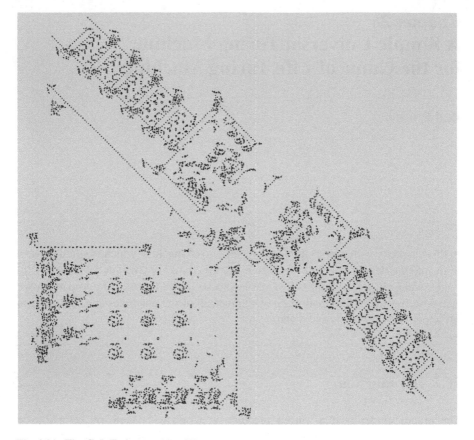

Fig. 26.1 The GoL Turing machine [7]

26.2 Turing Machines

In this section we explain how a Turing machine works using as an example the Turing machine mentioned above.

26.2.1 Turing Machine Operation

A Turing machine consists of a finite state machine which interacts with an infinite date storage medium. The data storage medium takes the form of a unbounded tape on which symbols can be written and read back via a moving read/write head. The symbols which can appear on the tape must be members of a finite alphabet.

The GoL Turing machine finite state machine is shown in Fig. 26.2. This shows the 3 × 3 array of memory cells each of which holds the actions required for the particular combination of state and symbol that corresponds to their position in the

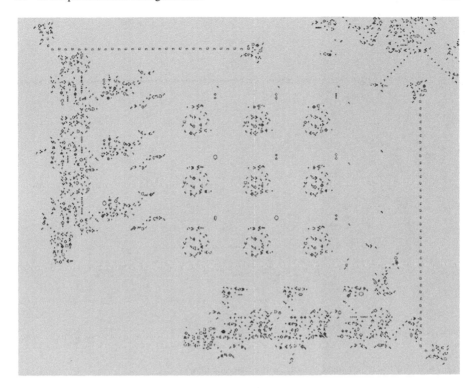

Fig. 26.2 The GoL Turing machine finite state machine

array. There are 2 stacks to represent the tape. Figure 26.3 shows part of the lower stack. Each stack cell can hold 3 gliders representing 3 binary digits. The two cells at the top left both hold one glider marked in grey.

The machine performs a regular cycle, reading a symbol off the tape, overwriting that symbol with another or the same one and them moving the read/write head one symbol width either right or left along the tape. This continues until the machine reaches a conclusion in its internal workings and halts.

The finite state machine is the program which determines from the symbol read and an internal 'state' which symbol to write and which way to move the read/write head. The internal 'state' is a member of a finite set of states.

The operation of the machine is completely determined by a table which gives for each combination of input symbol and internal state:

- The symbol to write.
- The new internal state.
- The direction to move the read/write head.
- Whether to halt or continue.

This Turing machine does not have a cell on its tape for the value under the read/write head. That value is inside the finite state machine after being popped from one stack.

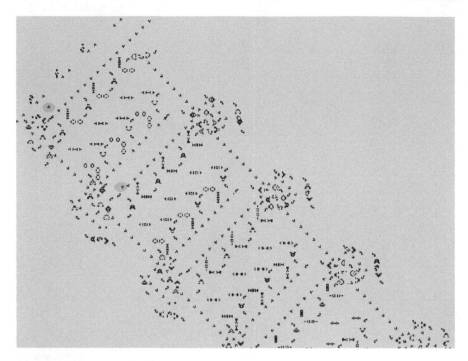

Fig. 26.3 The lower left stack of GoL Turing machine tape

A Turing machine has no other communication with the outside world than the contents of the tape. The contents of the tape before it starts represent its input and the contents on the tape when it has finished represents its output. It is possible that the machine never stops.

The Turing machine is as a mathematical tool to probe the limits of computability. Its power comes from its simplicity; speed and efficiency are of no consequence.

26.2.2 Example Turing Machine

We will use as an example the Turing machine used in the Game of Life pattern [7]. This machine doubles the length of a string of a particular symbol on the tape. The tape has alphabet $A = \{`0`, `1`, `2`\}$ and has 3 states $S = \{S0, S1, S2\}$. The tape looks like Fig. 26.4a when it starts. ⇑ marks the position of the read/write head

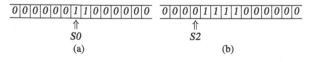

Fig. 26.4 Tapes of Turing machine used in GoL patterns [7]: (**a**) initial tape, (**b**) final tape

with the current state shown below. It will finish with twice as many '*1*' symbols as shown in Fig. 26.4b.

The operation of the machine can be shown clearly by means of a state transition diagram. The diagram for this example is shown in Fig. 26.5. Each state is represented by a hexagonal box. Either L or R (for left or right respectively) is written inside the box and below this is the state name (*S0, S1 or S2* in this case). Arrows from one state box to another represent state transitions. The symbol at the base of the arrow represents the symbol read from the tape which triggers this transition. The symbol half way along the arrow represents the symbol written to the tape during this transition and the direction to move after writing the symbol is found in the top part of the destination state box. An arrow may loop back to the same state indicating no change of state. If the symbol to write is the same as the symbol read, it is not shown on the arrow to reduce clutter in the diagram.

Each of the transitions in this diagram are coded as data in one of the cells of the GoL finite state machine. Table 26.1 shows each of the cells being read. The 8 spaceships moving along the row are picking up the contents of the cells, the spaceships marked in grey will be deleted. A deleted space ship stands for '1'. The spaceships are read from right to left in the order of arrival. The first on the right represents the direction of movement of the read/write head, which stack to push and which to pop, the next 3 represent the symbol to write during the push operation, and the last 4 are the next state. Table 26.2 shows a list of these transitions.

The machine starts in state *S0* with the read/write head over the first '*1*' symbol of the string to double. If there is a '*1*' symbol on the tape it will change state to *S1* and replace the symbol '*1*' with '*2*' symbol. The symbol '*2*' is a temporary mark replacing '*1*' symbols that have been processed. The read/write head is moved left as it enters state *S1*. State *S1* performs the job of finding a blank part of tape to use as the double of the symbol found. This will be to the left of the original string of '*1*'s. When a blank part of tape is found (symbol '*0*'), the machine changes to state *S0* replacing the '*0*' with a '*2*' and moving right. State *S0* now performs the task of finding the next '*1*' to the right of the current position. It will skip over any '*2*' symbols it finds and we expect at least one of these at this stage. If a '*1*' symbol is found he machine changes to state *S1* as before. This time we expect to skip some '*2*' symbols in state *S1* to find a blank part of tape.

This sequence will continue until all the '*1*' symbols in the string have been changed to '*2*' symbols, with one blank also changing to a '*2*' symbol for each of these. The machine will be in state *S0* and read the blank symbol '*0*' to the right of the last of the original '*1*' symbols. This will trigger a change to state *S2*. This state simply moves the read/write head left and changes all the '*2*' symbols into '*1*' symbols. It will stop when it reads a '*0*' symbol. The tape will then look like Fig. 26.4b.

There are a number of Turing machine simulators available on the Internet. The author used [1]. This simulator requires the definition of the finite state machine to be in the form of the list of state transitions. The list of the transitions for this example is shown in Table 26.2. This simulator treats halt as a state. It therefore performs the full state transition action into this state including moving the read/write head.

Fig. 26.5 The Turing
machine program used in the
GoL pattern [7]

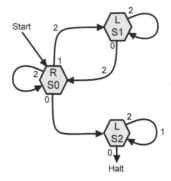

Table 26.1 The GoL Turing machine finite state machine program

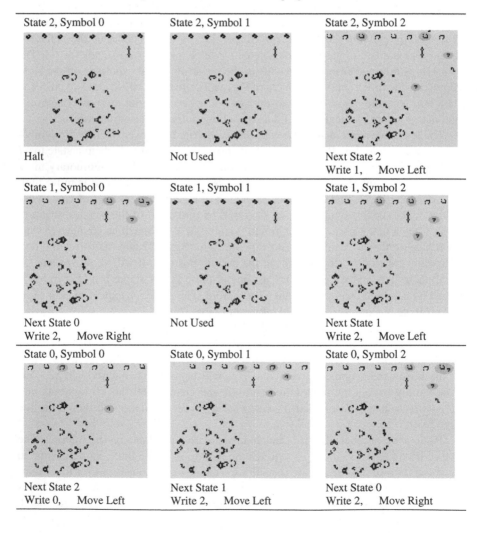

State 2, Symbol 0	State 2, Symbol 1	State 2, Symbol 2
Halt	Not Used	Next State 2 Write 1, Move Left

State 1, Symbol 0	State 1, Symbol 1	State 1, Symbol 2
Next State 0 Write 2, Move Right	Not Used	Next State 1 Write 2, Move Left

State 0, Symbol 0	State 0, Symbol 1	State 0, Symbol 2
Next State 2 Write 0, Move Left	Next State 1 Write 2, Move Left	Next State 0 Write 2, Move Right

Table 26.2 Symbol string
doublers transition list

State	Symbol	Next state	Next symbol	Direction
S0	0	S2	0	⇐
S0	1	S1	2	⇐
S0	2	S0	2	⇒
S1	0	S0	2	⇒
S1	2	S1	2	⇐
S2	0	Halt	0	⇐
S2	2	S2	1	⇐

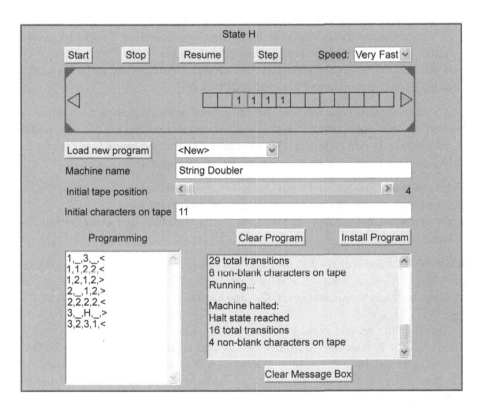

Fig. 26.6 Screen shot of the Turing machine in Fig. 26.5 being simulated by [1], this simulator numbers its states from 1 so for this they have been renumbered 1–3

It also has a special symbol '_' for a blank part of tape replacing the '0' used in this chapter. Figure 26.6 shows a screen shot of this simulator after completing the example program.

26.2.3 Universal Turing Machines

A universal Turing machine U is a Turing machine which takes as its input a description of another Turing machine T and a description of T's initial tape. U will leave on its tape a description of the output that T would have produced. Turing first described his universal Turing machine in his 1936 paper [10].

U is said to be universal because there exists a T which performs the equivalent calculation of any Turing machine A that meets the description of a Turing machine in Sect. 26.2.1.

There are two practical issues to overcome in showing that there exists a T equivalent to any A:

- U's tape must contain a description of T. It is awkward for U to have a description of T's tape when this is infinite in both directions. Therefore T will have a tape which is finite in one direction and infinite in the other.
- A can have any number of symbols in its alphabet. One technique we have adopted to keep U small is to have T's alphabet defined by U.

To cover the first point we note that for every A with a tape which is infinite in both directions there exists a T which is equivalent except that it has a tape which is only infinite in one direction. This can easily be arranged by folding T's tape in half. T will simulate A's tape by grouping 3 symbols together.

- One to hold a symbol on A's tape going to infinity on the left.
- One to hold a symbol on A's tape going to infinity on the right.
- One for space to hold a symbol marking the middle of A's tape at the fold on T's tape.

It is then a trivial matter for T to have two sets of states, both equivalent to A, one for each half of A's tape. Both sets will have extra states for each of A's state transitions to make the extra movements over T's tape and swap directions at the end of T's tape.

To cover the second point we note that for every A with an alphabet size n there exists a T which is equivalent except that it has an alphabet size 2. This can be achieved by using several symbols on T's tape to code one symbol on A's tape. Each of A's state transitions would be replaced by a small number of transitions in T for it to recognise the symbol and write the correct symbol in its place and move the read/write head the correct amount in the correct direction.

Some very small universal Turing machines have be designed; the smallest rely on mapping the functionality of the machine T first into a tag machine as described by Minsky [6] and then mapping that onto the tape of machine U. Minsky [6] described a 4 symbol 7 state universal Turing machine in this way. One machine designed like this, due to Rogozhin [9], is described below in Sect. 26.3.

26.3 Four State Six Symbol Universal Turing Machine

This is a description of one of the smallest known universal Turing machines. It has 4 states 6 symbols, and was designed by Rogozhin [9]. We will describe it by way of an example. The machine T must be coded into a 2-tag system. In Sect. 26.3.1 we look at a typical 2-tag system based on the work of Minsky [6]. The machine U will be described in Sect. 26.3.2.

26.3.1 Universal 2-Tag System

Tag systems were developed by Emil Post. In this subsection we describe the universal 2-tag system due to Minsky [6].

Tag systems manipulate strings of letters that make up a word by means of applying productions. Each production consists of two strings of letters. If the first part of the word to be transformed matches the first string of a production then the first n letters are removed from the word and the second string of the production is added to the end of remainder to make the transformed word. This process continues until none of the productions make any further change to the word. This is called a n-tag system.

We are only interested in 2-tag systems in which the first string of all the productions in just one letter and 2 letters are removed from the beginning of the word being transformed. This system has the property that only one production can apply to the word at any one time and therefore the result is deterministic. The application of productions stops when there is no production for the first letter of the word.

Let alphabet $A = \{a_1, a_2, \ldots, a_n, a_{n+1}\}$
Let the set of productions $P = \{p_1, p_2, \ldots, p_n\}$
Letter a_{n+1} is the letter that stops transformation when it appears at the front of the word
Let $\$$ be a string of letters such that the word is $a_i\$$
Then production P_i will apply and the transformation will be:

$$a_i \$ \rightarrow \$ p_i$$

In Minsky's scheme [6] the universal 2-tag system UP is made to perform the equivalent transformation as a Turing machine T. We will restrict T to 2 symbols without loss of generality, as shown in Sect. 26.2.3.

Let T's symbols be '0' and '1', with '0' being the symbol on blank tape. The contents of T's tape to the left of T's read/write head are a sequence of '0's and '1's which we will treat as a binary number value m. We will code this as part of UP's word using unary coding with pairs of letters. The part of T's tape to the right of its read/write head will be treated in the same way except that the bits are reversed, value n. That is to say in both cases the least significant bit is closest to the read/write head.

UP's word is of the form:

$$B_t a (b_t a)^m C_t a (c_t a)^n \quad (t \in \{1, \ldots, k\}),$$

k is the number of states transitions that T has and 'a' is any member of A. The transformation of UP will transform the above into:

$$B_{t'}a(b_{t'}a)^{m'}C_{t'}a(c_{t'}a)^{n'}$$

performing the equivalent of one step of T. The cycle of operation is:

- Writing a symbol and move the read/write head.
- Selecting the next state transition according to the value under the read/write head.

Skipping from one transition to another.

Moving Left

When moving the read/write head to the left we note that $m = 2*m' + v$ and $n' = 2*n + w$ where v is the value under the new position of the read/write head and where w is the value being written.

The first productions are:

$$B_t \rightarrow S_t$$
$$b_t \rightarrow s_t$$

Giving:

$$C_t a(c_t a)^n S_t(s_t)^m$$

The next productions will introduce the symbol to write w:

$$C_t \rightarrow D_{t1}D_{t0}(d_{t1}d_{t0})^w$$
$$c_t \rightarrow d_{t1}d_{t0}d_{t1}d_{t0}$$

Giving:

$$S_t(s_t)^m D_{t1}D_{t0}(d_{t1}d_{t0})^{2*n+w}$$

Substituting to replace m with m' and n with n' gives:

$$S_t(s_t s_t)^{m'}(s_t)^v D_{t1}D_{t0}(d_{t1}d_{t0})^{n'}$$

The effect of the next productions depends on v:

$$S_t \rightarrow B_{t1}B_{t0}$$
$$s_t \rightarrow b_{t1}b_{t0}$$

If $v = 0$ then the result is:

$$D_{t0}(d_{t0}d_{t1})^{n'-1}d_{t0}B_{t1}B_{t0}(b_{t1}b_{t0})^{m'}$$

We have an odd d_{t0} on the end to sort out with the next pair of productions:

$$D_{t0} \rightarrow a C_{t0} c_{t0}$$

$$d_{t0} \rightarrow c_{t0} c_{t0}$$

Giving:

$$B_{t0} b_{b1} (b_{t0} b_{t1})^{m'} C_{t0} c_{t0} (c_{t0} c_{t0})^{n'}$$

Mapping $B_{t0} \rightarrow B_{t'}$, $b_{t0} \rightarrow b_{t'}$, $C_{t0} \rightarrow C_{t'}$ and $c_{t0} \rightarrow c_{t'}$ gives us your result.

If $v = 1$ then the result is:

$$D_{t1} D_{t0} (d_{t1} d_{t0})^{n'} B_{t1} B_{t0} (b_{t1} b_{t0})^{m'}$$

Now just put in the final state:

$$D_{t1} \rightarrow C_{t'} c_{t'}$$

$$d_{t1} \rightarrow c_{t'} c_{t'}$$

to give:

$$B_{t1} B_{t0} (b_{t1} b_{t0})^{m'} C_{t'} c_{t'} (c_{t'} c_{t'})^{n'}$$

Mapping $B_{t1} \rightarrow B_{t'}$, $b_{t1} \rightarrow b_{t'}$, $B_{t0} \rightarrow a$ and $b_{t0} \rightarrow a$ gives us our result.

Moving Right

When moving the read/write head to the right we note that $n = 2 * n' + v$ and $m' = 2 * m + w$. The first productions putting in the value w are:

$$B_t \rightarrow D_t d_t (d_t d_t)^w$$

$$b_t \rightarrow d_t d_t d_t d_t$$

Giving:

$$C_t a (c_t a)^n D_t d_t (d_t d_t)^{2*m+w}$$

Substituting to replace m with m' and n with n' gives:

$$C_t a (c_t a)^{2*n'+v} D_t d_t (d_t d_t)^{m'}$$

The next pair of productions are:

$$C_t \rightarrow S_t$$

$$c_t \rightarrow s_t$$

Giving:

$$D_t d_t (d_t d_t)^{m'} S_t (s_t)^{2*n'+v}$$

Then:

$$D_t \rightarrow B_{t1} E_{t0}$$

$$d_t \rightarrow b_{t1} e_{t0}$$

Giving:

$$S_t (s_t s_t)^{n'} (s_t)^{v} B_{t1} E_{t0} (b_{t1} e_{t0})^{m'}$$

Now we pick up the value v with the next productions:

$$S_t \rightarrow C_{t1} F_{t0}$$

$$s_t \rightarrow c_{t1} f_{t0}$$

If $v = 0$:

$$E_{t0} (b_{t1} e_{t0})^{m'} C_{t1} F_{t0} (c_{t1} f_{t0})^{n'}$$

We have an odd f_{t0} on the end to sort out with the next pair of productions:

$$E_{t0} \rightarrow c_{t1} B_{t'} b_{t'}$$

$$e_{t0} \rightarrow b_{t'} b_{t'}$$

Giving:

$$F_{t0} (f_{t0} c_{t1})^{n'} B_{t'} B_{t'} (b_{t'} b_{t'})^{m'}$$

Now just put in the final state:

$$F_{t0} \rightarrow C_{t'} c_{t'}$$

$$f_{t0} \rightarrow c_{t'} c_{t'}$$

Giving our result:

$$B_{t'} b_{t'} (b_{t'} b_{t'})^{m'} C_{t'} c_{t'} (c_{t'} c_{t'})^{n'}$$

If $v = 1$ the outcome is:

$$B_{t1} E_{t0} (b_{t1} e_{t0})^{m'} C_{t1} F_{t0} (c_{t1} f_{t0})^{n'}$$

Mapping $B_{t1} \rightarrow B_{t'}$, $b_{t1} \rightarrow b_{t'}$, $C_{t1} \rightarrow C_{t'}$, $c_{t1} \rightarrow c_{t'}$, $E_{t0} \rightarrow a$, $F_{t0} \rightarrow a$ and $f_{t0} \rightarrow a$ gives us our result.

Tag Machine Example

The above scheme was tested using the Turing machine described in Fig. 26.8b as an example. The letters of the alphabet for the productions are made up of 4 characters

1 the production code letter B, b, C, c, D, d, $\ldots$
2 the state $1, \ldots, 6$ or stop state 7
3 the value under the read/write head
4 the value read where required and _ otherwise

It took 3,128 production cycles to convert:

$$B11\_,b11\_,C11\_,c11\_,c11\_,c11\_,c11\_,c11\_$$

into the final word:

B71_, E500, C71_, F500, c71_, f500, c71_, f500, c71_, f500, c71_, f500,
c71_, f500, c71_, f500, c71_, f500, c71_, f500, c71_, f500, c71_, f500,
c71_, f500, c71_, f500, c71_, f500, c71_, f500, c71_, f500, c71_, f500,
c71_, f500, c71_, f500, c71_, f500, c71_, f500, c71_, f500, c71_, f500,
c71_, f500, c71_, f500, c71_, f500, c71_, f500, c71_, f500, c71_, f500,
c71_, f500, c71_, f500, c71_, f500, c71_, f500, c71_, f500, c71_, f500,
c71_, f500, c71_, f500, c71_, f500, c71_, f500, c71_, f500, c71_, f500,
c71_, f500,

UP starts with *T* in state '1' with '1' under the read/write head, which is coded as '11' in the tag machine letters. There is no data to the left of the read/write head and the value to the right is 2, this is '01' in reverse order binary. Adding the '1' under the read/write head gives '101'. *T* interprets symbols in pairs so adding an extra '0' from blank tape gives '1010'. The machine *T* doubles the string of '10's and thus produces '10101010'. *UP* halts with *T* in the stop state '7' with the read/write head over the 1st '1' on the left, which is coded as '71' in the tag machine letters. Again there is nothing to the left of *T*'s read/write head, leaving '010101' to the right. This is of course 42 in reverse order binary. The productions for this are listed in Table 26.3.

26.3.2 Rogozhin's 2-Tag UTM

We will follow Minsky [6] and maintain that there is little point in trying to explain the machines structure as it is mixed up. We will go over the coding of the input and decoding of the output using a simple example.

Rogozhin's machine uses a 2-tag system similar to that described in Sect. 26.3.1 with one major difference. Rogozhin has added a constraint. This is based on the method of locating the productions.

Productions are located by indexing into a list structure using the code for the letter as the index. The machine starts writing the first letter of the production during the indexing procedure and has written the index value before looking at the first actual value. This early writing is corrected for in the coding by reducing the first value encoded for the first letter of the production by this amount.

There is unfortunately no algorithm which can code the letters and find an order for the productions, which satisfies this constraint, for all possible sets of productions. Rogozhin overcomes this by adding blank letters to the productions. These letters have no meaning and are discarded in the decoding of the result. The blank letter has the largest coding value of the alphabet and will have the identity production, i.e. for a 2-tag system with productions

Table 26.3 Productions for the tag machine version of Turing machine in Fig. 26.8b

B10_ → S10_	C10_ → D101D100	S10_ → B51_B50_
b10_ → s10_	c10_ → d101d100d101d100	s10_ → b51_b50_
B11_ → D11_d11_d11_d11_	C11_ → S11_	D11_ → B21_E110
S11_ → C21_F110	b11_ → d11_d11_d11_d11_	c11_ → s11_
d11_ → b21_e110	s11_ → c21_f110	B20_ → S20_
C20_ → D201D200d201d200	S20_ → B31_B30_	b20_ → s20_
c20_ → d201d200d201d200	s20_ → b31_b30_	B21_ → D21_d21_d21_d21_
C21_ → S21_	D21_ → B11_E210	S21_ → C11_F210
b21_ → d21_d21_d21_d21_	c21_ → s21_	d21_ → b11_e210
s21_ → c11_f210	B30_ → D30_d30_d30_d30_	C30_ → S30_
D30_ → B21_E300	S30_ → C21_F300	b30_ → d30_d30_d30_d30_
c30_ → s30_	d30_ → b21_e300	s30_ → c21_f300
B31_ → S31_	C31_ → D311D310d311d310	S31_ → B41_B40_
b31_ → s31_	c31_ → d311d310d311d310	s31_ → b41_b40_
B40_ → S40_	C40_ → D401D400d401d400	S40_ → B31_B30_
b40_ → s40_	c40_ → d401d400d401d400	s40_ → b31_b30_
B41_ → S41_	C41_ → D411D410d411d410	S41_ → B31_B30_
b41_ → s41_	c41_ → d411d410d411d410	s41_ → b31_b30_
B50_ → D50_d50_	C50_ → S50_	D50_ → B71_E500
S50_ → C71_F500	b50_ → d50_d50_d50_d50_	c50_ → s50_
d50_ → b71_e500	s50_ → c71_f500	B51_ → S51_
C51_ → D511D510	S51_ → B61_B60_	b51_ → s51_
c51_ → d511d510d511d510	s51_ → b61_b60_	B60_ → S60_
C60_ → D601D600	S60_ → B51_B50_	b60_ → s60_
c60_ → d601d600d601d600	s60_ → b51_b50_	B61_ → S61_
C61_ → D611D610d611d610	S61_ → B51_B50_	b61_ → s61_
c61_ → d611d610d611d610	s61_ → b51_b50_	D100 → c50_C50_c50_
d100 → c50_c50_	D101 → C51_c51_	d101 → c51_c51_
E110 → a20_B20_b20_	F110 → C20_c20_	e110 → b20_b20_
f110 → c20_c20_	D200 → c30_C30_c30_	d200 → c30_c30_
D201 → C31_c31_	d201 → c31_c31_	E210 → a10_B10_b10_
F210 → C10_c10_	e210 → b10_b10_	f210 → c10_c10_
E300 → a20_B20_b20_	F300 → C20_c20_	e300 → b20_b20_
f300 → c20_c20_	D310 → c40_C40_c40_	d310 → c40_c40_
D311 → C41_c41_	d311 → c41_c41_	D400 → c30_C30_c30_
d400 → c30_c30_	D401 → C31_c31_	d401 → c31_c31_
D410 → c30_C30_c30_	d410 → c30_c30_	D411 → C31_c31_
d411 → c31_c31_	E500 → a70_B70_b70_	F500 → C70_c70_
e500 → b70_b70_	f500 → c70_c70_	D510 → c60_C60_c60_
d510 → c60_c60_	D511 → C61_c61_	d511 → c61_c61_
D600 → c50_C50_c50_	d600 → c50_c50_	D601 → C51_c51_
d601 → c51_c51_	D610 → c50_C50_c50_	d610 → c50_c50_
D611 → C51_c51_	d611 → c51_c51_	

$$P = \{p_1, p_2, \ldots, p_n\}$$

letter a_{n+1} becomes the blank letter and letter a_{n+2} becomes the stop letter. In order to ensure that the blank letters do not interfere with the system they are added in pairs to each production, thus the productions of Sect. 26.3.1

$$a_i \rightarrow p_i$$

become

$$a_i \rightarrow a_{n+1} a_{n+1} p_i$$

The production for the blank letter is:

$$a_{n+1} \rightarrow a_{n+1} a_{n+1}$$

We will use a simpler example than Sect. 26.3.1 to explain the encoding. The example productions:

$$A \rightarrow AB$$
$$B \rightarrow BA$$

will transform the word $AABBH$, where 'H' is the stop letter, as follows:

$$A\ A\ B\ B\ H$$
$$B\ B\ H\ A\ B$$
$$H\ A\ B\ B\ A$$

Rogozhin's machine uses the alphabet '0' ,'1', 'b', '$\overleftarrow{b}$', '$\overrightarrow{b}$' and 'c'. The blank symbol on the tape is '0'. The tag machine letters are unary encoded using '1'. The end of the tag machine description is marked by '$\overleftarrow{b} b$'. Productions in the tag machine description are ended with '$b1b$'. Letters in the productions are separated by 'bb'. There is one 'b' between the tag machine description and the tag machine word. Letters in the tag machine word are separated by 'c'. The other letters in the alphabet are used for marking progress. U's tape is laid out with the tag word on the right and the productions on the left with the read/write head over the first symbol of the tag word. In normal operation there is used space between the productions and the tag word. As letters are deleted from the tag word the used space gets larger. When U finishes the start of the tag word is marked with 'c'. The productions are:

$$A \rightarrow b1b\ A\ bb\ B\ bb\ D\ bb\ D\text{-}A$$
$$B \rightarrow b1b\ B\ bb\ A\ bb\ D\ bb\ D\text{-}B$$
$$D \rightarrow b1b\ B\ bb\ D\text{-}D$$
$$H \rightarrow \overleftarrow{b} b$$

The coding for each letter is the number of 'b' symbols to the start of its production. The coding for the letters will be 'A' 1, 'B' 9, 'D' 17 and 'H' 21. With an initial tag word of $AABBH$ the initial coding of the tape will be:

$$\overleftarrow{b}\ bb1bbbb1b1bb1^9bb1^{17}bb1^8b1b1^9bb1bb1^{17}bb1^{16}b1c1c1^9c1^9c1^{21}$$

It takes U 43,971 transitions to complete the mapping resulting in the tape looking like:

$$\overleftarrow{b}\, bb\, \overleftarrow{b}\, bbbb\, \overleftarrow{b}\, b\, \overleftarrow{b}\, bb\, \overleftarrow{b}{}^9 bb\, \overleftarrow{b}{}^{17} bb\, \overleftarrow{b}{}^8 b\, \overleftarrow{b}\, b\, \overleftarrow{b}{}^9 bb\, \overleftarrow{b}\, bb\, \overleftarrow{b}{}^{17} bb\, \overleftarrow{b}{}^{16} b\, \overleftarrow{b}{}^{48}$$
$$c1^{17}c1^{17}c1c1^9 c1^{17}c1^{17}c1^9 c1c1^{21}$$

U stops when it reads the '$\overleftarrow{b}$' on the left of the tape. U has removed the 'c' separating 'H' from the used tape and adding the 'H' symbol on the end again leaving '$DDABDDBAH$' as the word. Removing the blank letter and the halt letter 'D' & 'H' leaves '$ABBA$'.

26.4 A Universal Turing Machine for the GoL Turing Machine

This machine does not have to be the smallest possible so the following criteria where chosen. It should be:

- Easily understandable.
- The simplest machine that would fit into the limitations of the author's Turing machine Built in Conway's Game of Life [7].
- Have the smallest possible description of machine T.
- Have the smallest possible description of machine T's tape.
- Have the shortest possible running time.

While these aims are almost totally mutually exclusive, some progress has been made towards each. The following subsections describe two machines which attempt to fit these criteria.

26.4.1 Eight Symbol Sixteen State Universal Turing Machine

This machine is derived from the simple machine described by Minsky [6] which has 8 symbols and 23 states. It was designed at the same time that the author designed the GoL Turing machine [7] with the objective of being more suitable for a demonstration model than the smallest machines and still fit the limitations of the GoL Turing machine.

UTM 8/16 Description

This universal Turing machine U directly simulates an arbitrary Turing machine T which has a single ended tape and just 2 states without loss of generality as shown in Sect. 26.2.3. There is a section of U's tape to represent T's tape. There is a section of U's tape to hold T's current state and the symbol which has been read. There is a section of U's tape which holds a list of T's state transition quintuples.

Table 26.4 Initial tape layout key

0^∞	is blank tape to the left and right
$a_1 a_2 \ldots a_{n-1}$	are T's tape contents to the left of T's read/write head using '0' and '1'
0	is the position of T's read/write head
$a_1' a_2' \ldots a_m'$	are T's tape contents to the right of T's read/write head using 'X' and 'M'
M	separates T's tape from T's working symbol and state
	and also marks the end of the quintuple list
k_v	is the symbol from under the read/write head using 'C' and 'D'
$k_{s1} \ldots k_{sj}$	is T's current state using 'C' and 'D'
Q_i	is $(k_{iv} k_{i1} \ldots k_{ij} k_{ij}^+ k_{ij-1}^+ \ldots k_{i1}^+ 0 v_i d_i)$, T's ith quintuple using '0' and '1'
X	marks the end of T's quintuples
k_{iv}	is the symbol read to match the ith quintuple
$k_{i1} \ldots k_{ij}$	is the state to match the ith quintuple
$k_{ij}^+ k_{ij-1}^+ \ldots k_{i1}^+$	is the next state for the ith quintuple
0	is the default value for the symbol read
v_i	is the value to write for the ith quintuple
d_i	is the direction to move the read/write head for the ith quintuple

U has alphabet $\{0, 1, A, B, C, D, X, M\}$.
U's tape is initially laid out as follows:

$$0^\infty a_1 a_2 \ldots a_{n-1} 0 a_{n+1}' \ldots a_m' M k_v k_{s1} \ldots k_{sj} Q_1 X Q_2 X \ldots X Q_t X M 0^\infty \quad (26.1)$$

These symbols are explained in Table 26.4.

U is described by 3 state transition diagrams (Fig. 26.7). It starts in state $F0$ in Fig. 26.7a with its read/write head over the k_v symbol in the initial layout (Eq. 26.1). k_v is replaced by the marked version using '0' and '1'. States $F1$ and $F2$ look for the matching value in the next quintuple, the first unmarked value. If this matches, it is marked using 'A' and 'B' and state $F5$ returns the read/write head to the left so that state $F0$ can continue to match the state value. If a match is not found, then state $F3$ is used to mark the rest of this quintuple and then state $F4$ moves the head back to the start position and unmarks the current state/symbol pair ready for state $F0$ to check the next quintuple. This will continue until either the 'M' symbol at the end of the quintuple list is encountered, causing U to halt, or a marked quintuple symbol is encountered indicating that the correct quintuple has been found. In the latter case state $C6$ initiates updating the current state.

State $C6$ in Fig. 26.7b moves the read/write head right and picks up data to copy in reverse order into the current state/symbol field. States $C7$ and $C8$ move the head back and write the data. This includes copying a '0' into the k_v position. This finishes when $C7$ or $C8$ encounter symbol 'M'. At this point the value they have read from the quintuple is the value to write. For state $C7$ this is '0' which is already

Fig. 26.7 UTM (**a**) Find, (**b**) Copy, (**c**) Move

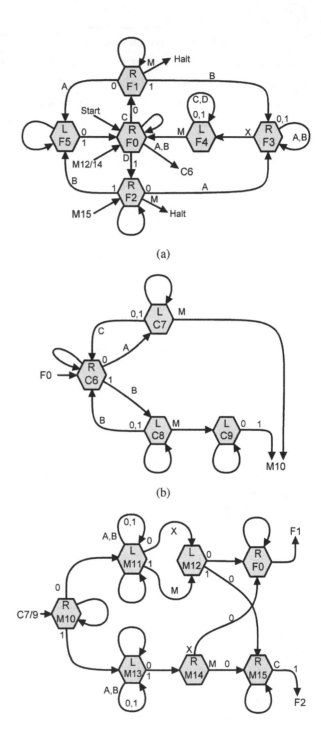

(a)

(b)

(c)

under the T's read/write head. For $C8$ state $C9$ is used to find the '0' marking the T's read/write head and write a '1'. State $M10$ in Fig. 26.7c then initiates moving the T's read/write head.

State $M10$ moves U's read/write head back to the next unread symbol in the current quintuple and $M11$ and $M12$ move the read/write head back to the '0' or '1' marking T's read/write head. Along the way these states reset the marked quintuples and current state ready for the next cycle.

State $M11$ moves T's read/write head left. It changes the '0' or '1' to 'X' or 'M' and $M12$ reads the value under T's read/write heads new position and writes a '0'. If the value is '0' this already matches the value copied from the quintuple and state $F0$ is selected to start the next cycle. If the value is '1' state $M15$ is used to move the read/write head right and correct it. It writes a '1' into the field as $F0$ would have done and then selects $F2$ to continue the next cycle.

State $M13$ moves T's read/write head right, it finds T's old read/write head position and state $M14$ reads the 'X' or 'M' under the new position. These become '0' as the new position and $F0$ or $M15$ are used as before to continue onto the next cycle.

UTM 8/16 Example T

We wish to demonstrate this universal Turing machine using the same example as before, the string doubler shown in Fig. 26.5 which we will call machine A. As this uses more than 2 symbols we must first convert it to a machine using just 2 which we will call machine T.

We will map A's alphabet into two symbols of T's alphabet as follows '0' $\rightarrow$ '00', '1' $\rightarrow$ '10' and '2' $\rightarrow$ '11'. A mechanical exercise results in the machine shown in Fig. 26.8a. A little more though produces a smaller machine as shown in Fig. 26.8b.

The Turing machine simulator [1] was modified to allow specification of the initial read/write head position to simulate this machine. The modified version is [8]. Figure 26.9 shows a screen shot of this simulator after completing the example program.

It took 78,512 transitions to transform the initial tape (Fig. 26.10a) to the final tape (Fig. 26.10).

The position of T's read/write head is the last '0' on the first part of tape. The value under this is coded by the 'C' (0) or 'D' (1) after the last 'M' in this section. The values on T's tape after its read/write head are coded using 'X' (0) and 'M' (1) The last 'M' is a separator.

The coding of the tape is quite straight forward. The order that the quintuples appear on the tape has a big impact on the time it takes. The most frequently used quintuples have been put on the left.

There is nothing in U from stopping T from moving its read/write head off the finite end of its tape and deleting the 'M' separator on the right. The consequence of this are left as an exercise for the reader.

Fig. 26.8 A two symbol version (**a**) and optimized version (**b**) of the program in Fig. 26.5

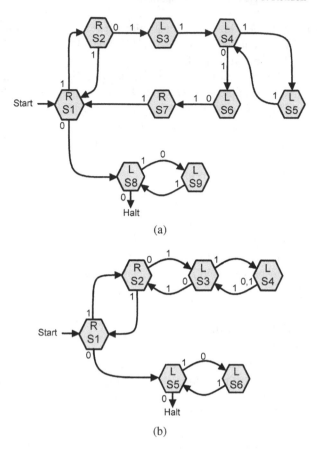

(a)

(b)

26.4.2 Eight Symbol Thirteen State Universal Turing Machine

This machine was designed meet the criteria in Sect. 26.4 a little better then the previous example.

UTM 8/13 Description

This universal Turing machine U directly simulates an arbitrary Turing machine T which has a single ended tape and just 2 states without loss of generality as shown in Sect. 26.2.3. There is a section of U's tape to represent T's tape and a section of U's tape to hold a description of T. This machine U uses a relative index system to locate T's transitions.

U has alphabet {'0', '1', 'A', 'B', 'C', 'D', 'X', 'M'}. U's tape is initially laid out as follows:

$$0^\infty a_1 a_2 \ldots a_{n-1} a_n a'_{n+1} \ldots a'_m X D'_1 X D'_2 X \ldots X D'_{i-1} X D'_i M D_{i+1} M \ldots M_t 0^\infty$$

$$(26.2)$$

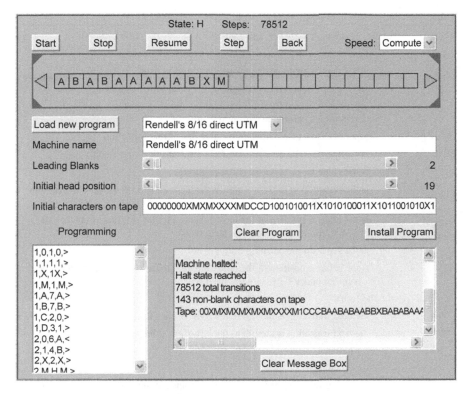

Fig. 26.9 Screen shot of the universal Turing machine running the Turing machine in Fig. 26.8b being simulated by [8]

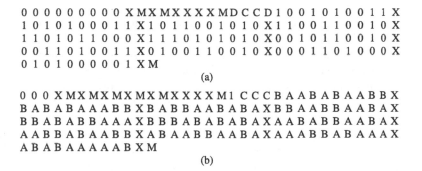

Fig. 26.10 UTM 8/16 (**a**) initial tape and (**b**) final tape for the example of Fig. 26.8b

These symbols are explained in Table 26.5.

U is described by 3 state transition diagrams (Figs. 26.11, 26.12 and 26.13). These diagrams are made slightly more general than those of Sects. 26.2.2 and 26.4.1 by putting the direction of movement with the symbol to write rather than in the state box. The states are shown as logical states. Where there is no con-

Table 26.5 Initial tape layout key

0^∞	is blank tape to the left and right
$a_1 a_2 \ldots a_{n-1}$	are T's tape contents to the left of T's read/write head using '0' and '1'
a_n	is T's tape contents under T's read/write head using '0' and '1'
$a'_{n+1} a'_{n+2} \ldots a'_m$	are T's tape contents to the right of T's read/write head using 'A' and 'B'
X	separates T's tape from T's description and also separates each of T's transitions before the current transition
'M'	separates each of T's transitions after the current transition
D_j	is $(v_j d_j t_{0j} C t_{1j})$ a description of T's jth transition (unmarked)
v_j	is the value to write for the jth transition
d_j	is the direction to move T's read/write head for the jth transition, 0 for left, 1 for right
t_{0j}	is the relative position of the next transition to the jth transition when the symbol under T's read/write head is '0'
C	is the separator between t_{0j} and t_{1j}
t_{1j}	is the relative position of the next transition to the jth transition when the symbol under T's read/write head is '1'
t_{0j} and t_{1j} take the form:	
0^n	the next transition is the nth to the right of the jth transition
1^n	the next transition is the nth to the left of the jth transition
10	for halt
nothing	for no change
D'_j	is the marked form of D_j using 'A', 'B' and 'D' instead of '0', '1' and 'C'. The marked form is used to the left of the current transition
D_i	is the current transition

flict the same real state number is used for 2 logical states. This occurs for *W2* and *N2*, *W3* and *N3* and *W4* and *N4*.

Initially *U*'s read/write head must be in the marked section of the tape between the '0' or '1' on the left which is T's read/write head position and the '0' or '1' on the right which is v_i the first part of *U*'s description for T's first transition. *U* starts in state *W1* (Fig. 26.11). States *W2* and *W3* are selected according the value to write on T's tape and they read the move direction. States *W4–W7* are selected accordingly and move *U*'s read/write head back to T's read/write head to perform these operations. In state *N8* *U*'s read/write head is over the new position of T's read/write head. If the value is '1' then state *N9* is used to skip passed t_{0i} by locating the 'C' symbol separating it from t_{1i}. The processing continues in common with *N8* '0' case with state *N10* handling both t_{0i} and t_{1i}.

Figure 26.12 shows the processing when the next transition is to the right of the current transition. Each '1' represents one transition to skip. This is changed to an 'M' symbol and state *N11* looks for a matching 'M' transition separator to the right. It marks this and all symbols up to it by substituting 'A' for '0', 'B' for '1', 'D' for 'C' and 'X' for 'M'. State *N4* moves *U*'s read/write head back to the 'M' of the last marked count and state *N2* checks to see if there is another count which *N11* will

Fig. 26.11 UTM part 1:
Write and Move T's
read/write head

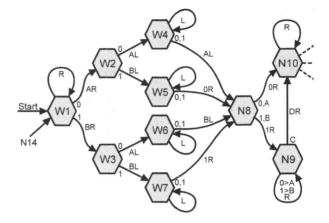

Fig. 26.12 UTM part 2: Next
transition to the right

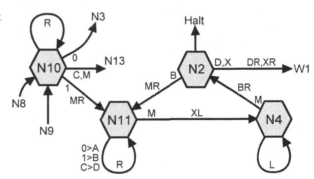

process. If state $N2$ finds either the 'D' or 'X' which mark the end of t_{0i} and t_{1i} respectively then the job is done and $W1$ will start the next cycle. State $N2$ may also find an 'A', this will be the marked '0' part of the halt transition '10'.

Figure 26.13 shows the processing when the next transition is to the left of the current transition or is the same transition. Each '0' represents one transition to skip. State $N3$ looks for an 'X' to the left and unmarks it by replacing it with an 'M'. State $N12$ looks for the next '0' count. The end of the list of '0's is either 'C' (t_{0i}) or 'M' (t_{1i}). In order prevent confusion between the different uses of 'M', states $N3$ and $N12$ convert 'M's for the previously the counted '0's to '1' and then back to 'M's again between them. The tidying up processing after counting is done by state $N4$. All marked symbols from U's read/write head up to the first 'X' on the right are unmarked by replacing 'A' with '0', 'B' with '1' and 'D' with 'C'. After that $W1$ is selected to start the next cycle. State $N4$ also handles resetting the current transition when state $N10$ detects either 'C' or 'M' implying that next transition is the current transition.

The Turing machine simulator [8] was used with this machine emulating the example in Sect. 26.2.2. Figure 26.9 shows a screen shot of this simulator after completing the example program.

Fig. 26.13 UTM part 3: Next transition to the left

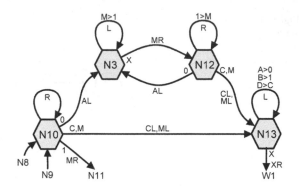

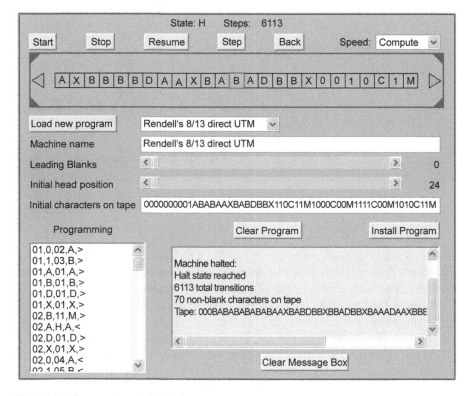

Fig. 26.14 Screen shot of the 8/13 universal Turing machine running the Turing machine in Fig. 26.8b being simulated by [8]

The modified Turing machine simulator [8] was used to transform the initial tape (Table 26.6) into the final tape (Table 26.7). It took 6,113 transitions as shown in the screen shot (Fig. 26.14).

The coding of the transitions is as shown in Table 26.8 in the order they appear on the tape. This order makes a big difference to the speed and was chosen based on the frequency of use of the transitions and the distance between them on the tape.

Table 26.6 UTM 8/13 initial tape for the example of Fig. 26.8b

```
0 0 0 0 0 0 0 0   0 1 A B A B A A   X B A B D B B X 1 1 0 C 1 1 M
1 0 0 0 C 0 0 M 1 1 1 1 C 0 0 M 1 0 1 0 C 1 1 M 0 0 1 0 C 1 M
0 0 1 0 C 0 0 M
```

Table 26.7 UTM 8/13 final tape for the example of Fig. 26.8b

```
0 0 0 B A B A B A B A B A B A A X B A B D B B X B B A D B B X
B A A A D A A X B B B B D A A X B A B A D B B X 0 0 1 0 C 1 M
0 0 1 0 C 0 0 M
```

Table 26.8 State transition for the example of Fig. 26.8b

Transition number	State	Value	Write	Move	Next for 0	Next for 1	Coding
T1	S2	0	1	L	T2	T3	101C11
T1	S4	0/1	1	L	T2	T3	
T2	S1	1	1	R	T1	T3	110C11
T2	S3	0	1	R	T1	T3	
T3	S3	1	1	L	T1	T1	1000C00
T4	S2	1	1	R	T6	T2	1111C00
T5	S6	1	1	L	Halt	T7	1010C11
T6	S1	0	0	L	Halt	T7	0010C1
T7	S5	1	0	L	–	T5	0010C00

26.5 Conclusion

The universal Turing machine presented in Sect. 26.4.2 is close to the optimum for a demonstration in the GoL Turing machine.

It required a run time of 6113 cycles and 69 tape cells to emulate the simple Turing machine in Sect. 26.4.1.

The GoL Turing machine with its small program performs one Turing machine cycle in 11,040 GoL generations. This will increase only linearly with the number of states and symbols. The open source Life program Golly [4] in Hashlife mode [5] can process this in seconds if not fractions of a second.

The GoL Turing machine uses two stacks for the tape; each stack cell is 87 Life cells wide. Thus having 69 stack cells rather than the current 5 will increase the pattern size much more than the increase in the size of the finite state machine. The current pattern width will increase from 1,714 life cells to 13,024, see Fig. 26.15 for a comparison of a mock-up of the Universal version at the same scale as the current version.

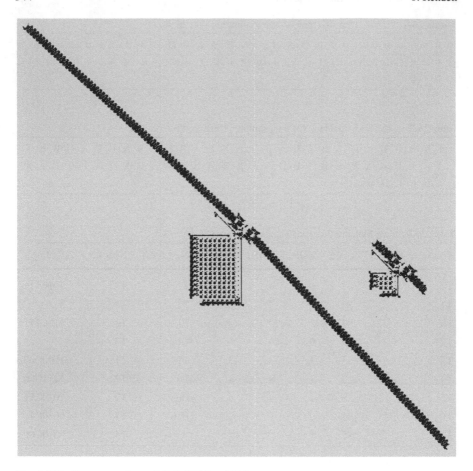

Fig. 26.15 Size comparison of GoL UTM and TM

We would therefore expect that Golly could run a universal version of the GoL Turing machine to complete the emulation of the current version in a small number or even fractions of hours with any modern personnel computer or laptop.

References

1. Britton, S.: Java applet Turing machine simulator. http://ironphoenix.org/tril/tm/
2. Gardner, M.: Mathematical Games articles in Scientific American: On cellular automata, self-reproduction, and the game "life" (February, 1971); The fantastic combinations of John Conway's new solitaire game "life" (October, 1970)
3. Gardner, M.: Wheels, Life and Other Mathematical Amusements. Freeman, New York (1983)
4. Golly, an open source, cross-platform application for exploring Conway's Game of Life and other cellular automata. http://golly.sourceforge.net/

5. Hashlife, an algorithm for computing the long-term fate of a given starting configuration in Conway's Game of Life and related cellular automata. http://en.wikipedia.org/wiki/Hashlife
6. Minsky, M.L.: Computation: Finite and Infinite Machines. Prentice Hall, New York (1967)
7. Rendell, P.: Conway's Game Life Turing machine. www.rendell-attic.org/gol
8. Rendell, P.: Java applet Turing machine simulator. http://www.rendell-attic.org/gol/TMapplet
9. Rogozhin, Y.: Small universal Turing machines. Theor. Comput. Sci. **168**, 215–240 (1996)
10. Turing, A.M.: On computable numbers, with an application to the Entscheidungsproblem. In: Proc. London Mathematical Society (1936)

Chapter 27
Computation with Competing Patterns in Life-Like Automaton

Genaro J. Martínez, Andrew Adamatzky,
Kenichi Morita, and Maurice Margenstern

We study a Life-like cellular automaton rule $B2/S2345$ where a cell in state '0' takes state '1' if it has exactly two neighbors in state '1' and the cell remains in the state '1' if it has between two and five neighbors in state '1.' This automaton is a discrete analog spatially extended chemical media, combining both properties of sub-excitable and precipitating chemical media. When started from random initial configuration $B2/S2345$ automaton exhibits chaotic behavior. Configurations with low density of state '1' show emergence of localized propagating patterns and stationary localizations. We construct basic logical gates and elementary arithmetical circuits by simulating logical signals with mobile localizations reaction propagating geometrically restricted by stationary non-destructive localizations. Values of Boolean variables are encoded into two types of patterns — symmetric (FALSE) and asymmetric (TRUE) patterns — which compete for the 'empty' space when propagate in the channels. Implementations of logical gates and binary adders are illustrated explicitly.

27.1 Introduction

Computational universality of Conway's Game of Life (GoL) cellular automaton [14] has been already demonstrated by various implementations. Most famous designs are realization of functionally complete set of logical functions [40], register machine [8], direct simulation of Turing machine [9, 39], and design of a universal constructor [17]. These implementations use principles of collision-based computing [2, 8] where information is transferred by localizations (gliders) propagating in an architecture-less, or 'free,' space. The theoretical results regarding GoL universality is only the first half-step in a long journey towards real-world implementation of the collision-based computers as unconventional computing [1, 42]. Controllability of signals is the first obstacle to overcome. Despite their mind-whirling elegance and complexity-wise efficiency of implementation the 'free-space' computing circuits are difficult to fabricate in physical or chemical materials [5] because propagating localizations (solitons, breathers,

A. Adamatzky (ed.), *Game of Life Cellular Automata*,
DOI 10.1007/978-1-84996-217-9_27, © Springer-Verlag London Limited 2010

kinks, wave-fragments) are notoriously difficult to manipulate, maintain and navigate.

The easiest way to control patterns propagating in a non-linear medium circuits is to constrain them geometrically. Constraining the media geometrically is a common technique used when designing computational schemes in spatially extended non-linear media. For example 'strips' or 'channels' are constructed within the medium (e.g. excitable medium) and connected together, typically using arrangements such as T-junctions. Fronts of propagating phase (excitation) or diffusive waves represent signals, or values of logical variables. When fronts interact at the junctions some fronts annihilate or new fronts emerge. The propagations in the output channels represent results of the computation.

The geometrical-constraining approach is far from being graceful (particularly, comparing to collision-based paradigm [2, 22, 35]) but simple and practical enough to be used by experimental scientists and engineers. The geometrical constraining is successfully applied in design of several laboratory prototypes of chemical (all based on Belousov–Zhabotinsky reaction) computing devices: logical gates [41, 43], diodes [13, 24, 36], counters [16], coincidence detectors [15], and memory [36].

What members of GoL family offer us most 'realistic' approximation of dynamical processes in spatially extended chemical computers? In its original form, GoL automaton is a discrete analog of sub-excitable chemical media [7, 23]. Localized wave-fragments (reaction–diffusion dissipative solitons) in sub-excitable Belousov–Zhabotinsky reaction [7, 23] are represented by gliders [6, 14, 19, 25, 37, 45] in GoL cellular automaton. See examples of direct comparisons between experimental chemical laboratory results and cellular automaton models in [12].

There is also a family of Life-life rules, where 'cells never die,' the state '1' is an absorbing state. This is the family of *Life without Death* (LwD), invented by Griffeath and Moore in [19]. In the LwD automaton we observe propagating patterns, formed due to rule-based restrictions on propagation similar to that in sub-excitable chemical media and plasmodium of *Physarum polycephalum* [3], but no complicated periodic structures or global chaotic behavior occurs. The LwD family of cell-state transition rules is an automaton equivalent of the precipitating chemical systems. This is demonstrated in our phenomenological studies of semi-totalistic and precipitating CA [6], where we selected a set of rules Life $2c22$, identified by periodic structures [26]. The clans closest to the family $2c22$ are *Diffusion Rule* (Life rule $B2/S7$) [28], all they also into of a big cluster named as Life $dc22$.[1]

In present chapter we study a Life-like cellular automaton, which somewhat imitates properties of both excitable and precipitating reaction–diffusion chemical systems, and who how to implement a sensible computation in such type of cellular automata. We develop a model where the channels are constructed using static patterns and computation is implemented by propagating patterns of precipitation. The re-

---

[1]http://uncomp.uwe.ac.uk/genaro/Life_dc22.html.

sults are based on our previous studies of reaction–diffusion analogs $B2/S2345678$, $B2/S23456$ [27, 30], and $B2/S2345$ [29].

27.2 Life Rule $B2/S2345$

Life rule $B2/S2345$ is described as follows. Each cell takes two states '0' ('dead') and '1' ('alive'), and updates its state depending on its eight closest neighbors:

1. Birth: a central cell in state 0 at time step t takes state 1 at time step $t + 1$ if it has exactly two neighbors in state 1.
2. Survival: a central cell in state 1 at time t remains in the state 1 at time $t + 1$ if it has two, three, four or five live neighbors.
3. Death: all other local situations.

Once a resting lattice is perturbed (few cells are assigned live states), patterns of states 1 emerge, grow and propagate on the lattice quickly.

A general behavior of rule $B2/S2345$ can be well described by a mean field polynomial and its fixed points. Mean field theory is a proved technique for discovering statistical properties of CA without analyzing evolution spaces of individual rules [10, 21, 32]. The method assumes that elements of the set of states are independent, uncorrelated between each other in the rule's evolution space. Therefore we can study probabilities of states in neighborhood in terms of probability of a single state (the state in which the neighborhood evolves), thus probability of a neighborhood is the product of the probabilities of each cell in the neighborhood.

The mean field polynomial for rule $B2/S2345$ is as follows:

$$p_{t+1} = 14p_t^2 q_t^3 \left(4p_t^4 + 2q_t^4 + 5p_t^3 q_t + 2p_t q_t^3 + 4p_t^2 q_t^2\right), \qquad (27.1)$$

where p_t is a probability of a cell being in state '1' at time step t, q_t is a probability of the cell to be in state '0' at time step t. Thus, knowing state of a cell at time step t and we can calculate probability of the cell's state at time step $t + 1$.

High densities of regions dominated by state 1 correspond to maximum point near $p = 0.35$. The average density is reached with one stable fixed point $p = 0.46$ when automaton find its stability around of 37% of cells in state '1.' Some interesting behavior can be found in extreme unstable fixed point when $p = 0.04$ (complex class, see mean field classification in [32]). Looking on configurations with less than 4% of cells in state '1' we can observer gliders, oscillators, and still life patterns. Thus unstable fixed points (Fig. 27.1) represent evidence of complex behavior as mobile localizations, gliders and small oscillators emerging in the automaton development. Figure 27.2 demonstrates a typical evolution of $B2/S2345$ starting with very small initial densities of state '1.' Mobile localizations emerge but do not survive for a long time.

A set of minimal particles, or basic periodic structures, in rule $B2/S2345$ include one glider, two oscillators (one blinker and one flip-flop configurations), and one still life configuration (see Fig. 27.3). The still life patterns [11, 31] represent precipitation of an abstract reaction–diffusion chemical system imitated by rule $B2/S2345$

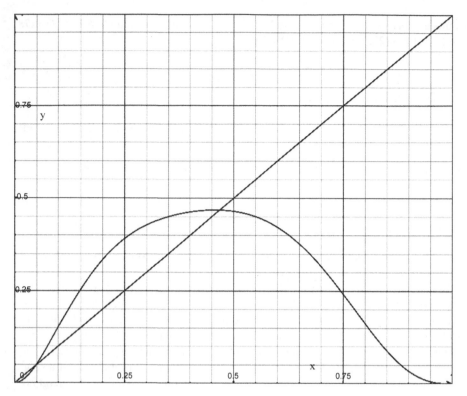

Fig. 27.1 Mean field curve for $B2/S2345$: p is a horizontal axis, and q is vertical axis

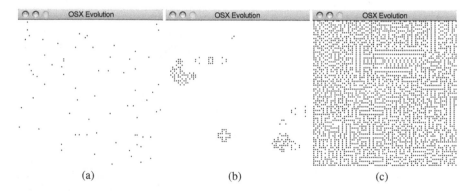

(a)   (b)   (c)

Fig. 27.2 Snapshots of automaton configurations of 300×300 cells showing activity of particles in $B2/S2345$. (**a**) Initial random configuration, density of state '1' is 1.2%. (**b**) Just after eight generations gliders and nucleation patterns emerge. (**c**) 540th generations

automaton. The still life blocks are not affected by their environment however they do affect their environment [27, 29, 30]. Therefore the still life patterns can be used to build channels, or wires, for signal propagation.

Fig. 27.3 Basic periodic
structures in $B2/S2345$:
(**a**) glider, (**b**) flip-flop
oscillator, (**c**) blinker
oscillator, and (**d**) still life
configuration

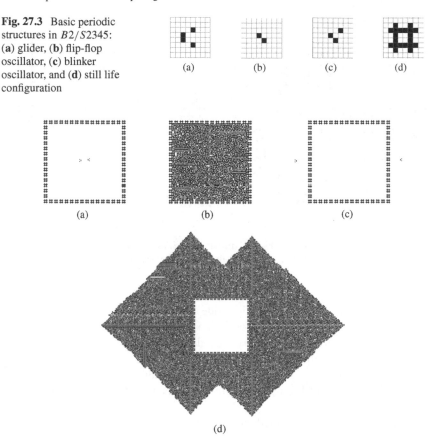

(a)   (b)   (c)   (d)

(a)   (b)   (c)

(d)

Fig. 27.4 Containment of growing pattern by indestructible localizations. (**a**) The explosion start
from a reaction between two gliders. (**b**) Final configuration of the container pattern. (**c**) Initial
positions of gliders outside the box walled by indestructible patterns. (**d**) Interior of the box is
protected from the growing pattern

27.2.1 Indestructible Pattern in B2/S2345

Some patterns amongst still life patterns in the rule $B2/S2345$ belong to a class
of *indestructible patterns* (sometimes referred to as 'glider-proof' patterns) which
cannot be destroyed by any perturbation, including collisions with gliders. A minimal indestructible pattern, still life occupying a square of 6×6 cells, is shown in
Fig. 27.3d.

The indestructible patterns discovered can be used to stop a 'supernova' explosions in Life-like rules. Usually a Life-like automaton development started at an
arbitrary configuration exhibits unlimited growth (generally related to nucleation
phenomenon [18]).

In rule $B2/S2345$ such an 'uncontrollable' growth can be prevented by a regular
arrangement of indestructible patterns. Examples are shown in Fig. 27.4. In the first

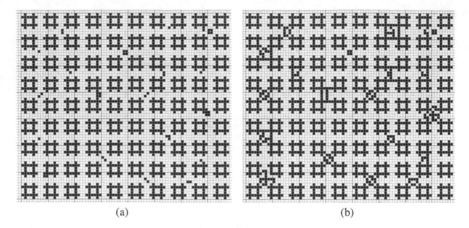

(a)   (b)

Fig. 27.5 Still life colony immune to local perturbations: (**a**) initial configuration, where 'viruses' are shown are irregular patterns of '1' states, (**b**) final state demonstrates that initial local perturbations remain contained

example (Fig. 27.4a) two gliders collide inside a 'box' made of still life patterns. The collision between the gliders lead to formation of growing pattern of '1' states. The propagation of the pattern is stopped by the indestructible localizations (Fig. 27.4b). In the second example, gliders collide outside the box (Fig. 27.4c) however interior of the box remains resting (Fig. 27.4c) due to impenetrable walls. Similarly, one can construct a colony of still life patterns immune to local perturbations. An example is shown in Fig. 27.5.

The indestructibility exemplified above allows us to use still life patterns to channel information in logical circuits.

27.3 Computing with Propagating Patterns

We built a computing scheme from channels — long areas of '0'-state cells walled by still life blocks, and T-junctions[2] — sites where two or more channels join together.

Each T-junction consists of two horizontal channels A and B (shoulders), acting as inputs, and a vertical channel, C, assigned as an output (Fig. 27.6). Such type of circuitry has been already used to implement XOR gate in chemical laboratory precipitating reaction–diffusion systems [5], and precipitating logical gates imitated in CA [27, 29, 30]. A minimal width of each channel equals three widths of the still life block (Fig. 27.3d) and width of a glider (Fig. 27.3a).

Boolean values are represented by reaction of gliders, positioned initially in the middle of channel, value 0 (Fig. 27.7a), or slightly offset, value 1 (Fig. 27.7c). The

---

[2] T-junction based control signals were suggested also in von Neumann [44] works.

Fig. 27.6 T-shaped system
processing information

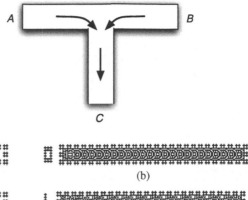

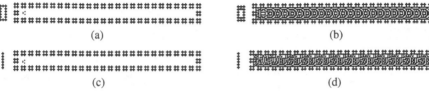

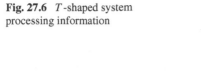

(a)

(b)

(c)

(d)

Fig. 27.7 Feedback channels constructed with still life patterns ((**a**) and (**c**)) show the initial state with the empty channel and one glider respectively. The symmetric pattern represent value 0 (**b**), and non-symmetric pattern represent value 1 (**d**) late of glider reaction

initial positions of the gliders determine outcomes of their reaction. Glider, corresponding to the value 0 is transformed to a regular symmetric pattern, similar to frozen waves of excitation activity (Fig. 27.7b). Glider, representing signal value 1, is transformed to transversally asymmetric patterns (Fig. 27.7d). Both patterns propagate inside the channel with constant, advancing unit of channel length per step of discrete time.

27.3.1 Implementation of Logic Gates and Beyond

When patterns, representing values 0 and 1, meet at T-junctions they compete for the output channel. Depending on initial distance between gliders, one of the patterns wins and propagates along the output channel. Figures 27.8, 27.9 and 27.11 show final configurations of basic logical gates.

Figure 27.8 shows two implementations of OR gate. Due to different locations of gliders in initial configurations of gates, patterns in both implementations of gates are different however, results of computation are the same. Configurations of AND gate are shown in Fig. 27.9.

Also we can implement a DELAY element as shown in Fig. 27.10.

The NOT gate is implemented using additional channel, where control pattern is generated, propagate and interfere with data-signal pattern. Initial and final configurations of NOT gate are shown in Fig. 27.11. Using delay elements we can construct serial channels with any number of NOT gates (Fig. 27.12). Number of control channels growth proportionally to number of gates in the circuit. Of course, it could not be the most elegant and efficient way of constructing NOT gate, but useful for our purposes in $B2/S2345$.

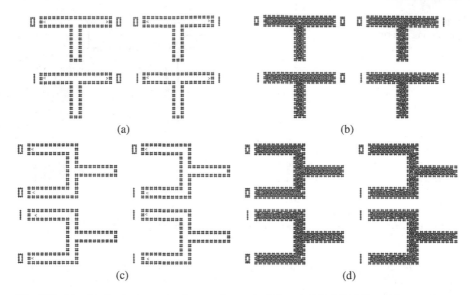

Fig. 27.8 Two kinds of OR gate implementation at the Life rule $B2/S2345$. Input binary values a and b they are represented as 'In/0' or 'In/1,' output result c is represented by 'Out/0' or 'Out/1.' (**a**) and (**c**) initial configurations of the gates, (**d**) and (**e**) final configurations of the gates

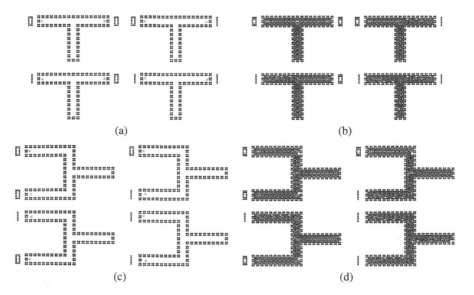

Fig. 27.9 Two kinds of AND gate implementation at the Life rule $B2/S2345$: (**a**) and (**c**) initial configurations of the gates, (**d**) and (**e**) final configurations of the gates

Fig. 27.10 Configurations of delay element for signal '0' (**ab**) and signal '1' (**cd**): (**a**) and (**c**) initial configurations, (**b**) and (**d**) final states

Fig. 27.11 NOT gate implementation for input '1' (**ab**) and input '0' (**cd**): (**a**) and (**c**) are initial configurations, (**b**) and (**d**) are final configurations

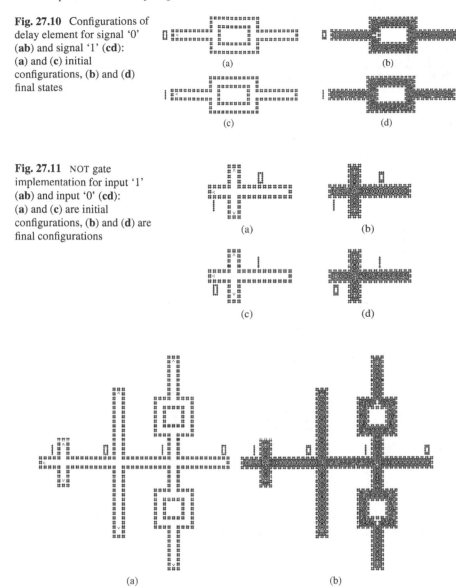

(a)

(b)

Fig. 27.12 Serial NOT gate with delays

27.3.2 Majority Gate

Implementation of MAJORITY gate is shown in Fig. 27.13. The gate has three inputs: North, West and South channels, and one output: East channel. Three propagating pattern, which represent inputs, collide at the cross-junction of the gate. The resul-

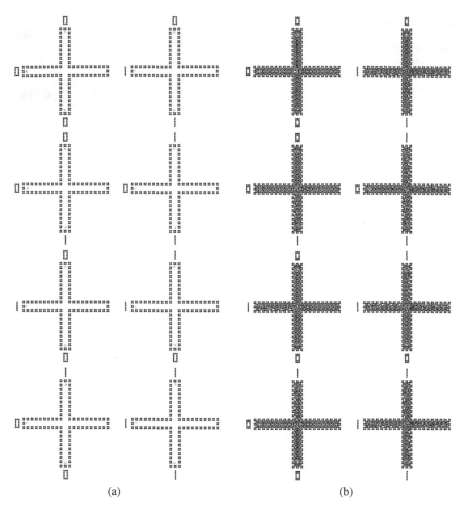

Fig. 27.13 MAJORITY gate implementation $(a \wedge b) \vee (a \wedge c) \vee (b \wedge c)$: (**a**) initial configuration: majority input values In/0 (*first column*), and majority input values In/1 (*second column*), and (**b**) final configurations of the majority gates

tant pattern is recorded at the output channel. Similarly gates in quantum-dot cellular automata are designed [38].

27.3.3 Implementation of Binary Adders

We represent two ways of implementing partial and full binary adders. First we can consider design based on cascading of logic gates, then second design employing only NOT-MAJORITY gates.

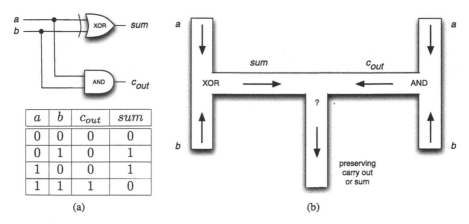

Fig. 27.14 (a) Half-adder circuit and its respective true table, and (b) scheme of half-adder implementation in geometrically constrained medium

Adder via Cascading of Logical Gates

Conventional logic circuit and truth table of a binary half-adder are shown in Fig. 27.14a (true tables were derived from [30]).[3] Schematic representation of the half-adder via T-junctions between channels is shown in Fig. 27.14b.

Final configurations of patterns in a one-bit half-adder are calculated for all inputs values. Figure 27.15a shows the initial configuration of inputs and outputs. Resultant patterns are shown in Fig. 27.15b where the last patterns is a carry-out operation for the next half-adder.

The circuit can be extended to a full binary adder via cascading of logical gates [20]. Configuration of the adder, built of still life blocks, and its description are shown in Fig. 27.16. The full adder consists of 16 T-junctions, linked together by channels and involve synchronization signals.

Working prototype of full adder is constructed in $1,118 \times 1,326$ cells lattice. In total, initial configuration has a population of 28,172 alive cells. The prototypes working cycle is 1,033 time steps with a final population of 63,662 alive cells. A data-area of the full adder is shown in Fig. 27.17.

Construction of Adder via NOT-MAJORITY Gates

The second model represents a binary adder constructed of three NOT-MAJORITY gates and two inverters working in $B2/S2345$ [29]. Such type of adder appears in several publications, particularly in construction of the arithmetical circuits in quantum-dot cellular automata [38]. Original version of the adder using NOT-MAJORITY gates was suggested by Minsky in his designs of artificial neural networks [33].

---

[3]http://uncomp.uwe.ac.uk/genaro/Diffusion_Rule/B2-S2345678.html.

Fig. 27.15 Half adder implemented in rule $B2/S2345$. Configurations represent sums $0+0$, $0+1$, $1+0$, and $1+1$ respectively, carry-out is preserved

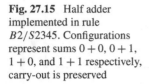
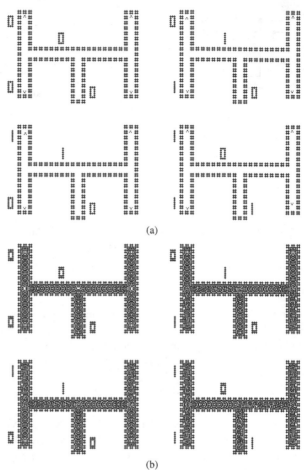

(a)

(b)

Figure 27.18 shows the classic circuit and the true table illustrating the dynamics of this adder. This way, Fig. 27.19 represents a scheme of the adder to implement in $B2/S2345$ (as was done in [29]). The scheme highlights critical points where some extra gates are necessary to adjust inputs and synchronize times of collisions.

Figure 27.20 presents most important stages of the full adder on $B2/S2345$ evolution space standing out DELAYS stages and NOT gates. The adder is implemented in $1,402 \times 662$ lattice that relates an square of 928,124 cells lattice with an initial population of 56,759 cells in state '1.' Final configurations of the adder for every initial configuration of inputs are shown in Figs. 27.21–27.28 with a final population of 1,439 alive cells on an average of 129,923 generations.

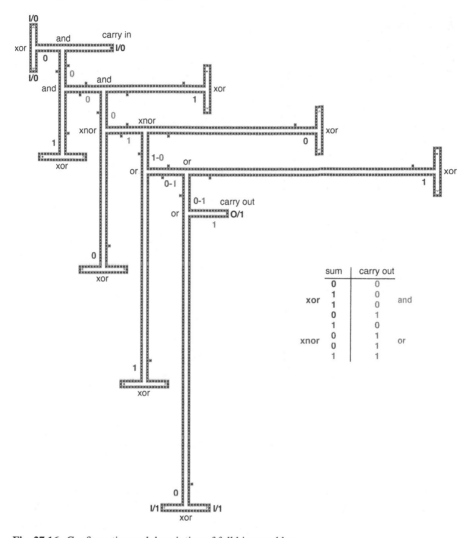

Fig. 27.16 Configuration and description of full binary adder

27.4 Conclusions

We studied a cellular automaton model — Life-like rule $B2/S2345$ — of a precipitating chemical system. We demonstrated that chaotic rule $B2/S2345$ supports stationary (still lifes) and mobile (gliders and propagating patterns) localizations. That relates another case where a chaotic rule contains non evident complex behavior and how such systems could have some computing information on its evolution [28, 34].

We have shown how construct basic logical gates and arithmetical circuits by restricting propagation of patters in a channels made of indestructible stationary localizations. Disadvantage of the approach presented is that computing space is

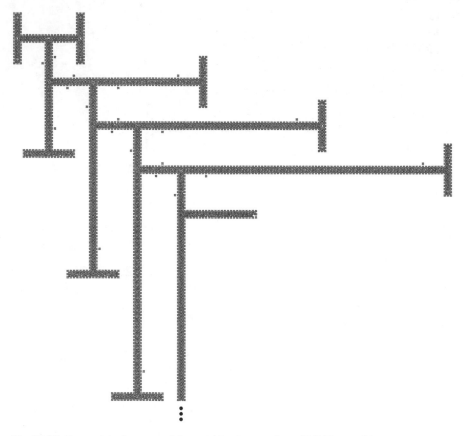

Fig. 27.17 Zoomed-in data-area of the serial implementation of full binary adder

geometrically constrained and not all cells of automaton lattices are used in computation. However, the geometrical constraining brings some benefits as well. Most computing circuits in Life-like automata are built using very complex dynamics of collisions between gliders and still lifes [9, 17, 39], in our case gliders are used

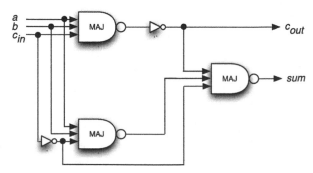

Fig. 27.18 Circuit and truth table of a full binary adder comprised of NOT-MAJORITY gates and inverters

Fig. 27.18 (continued)

a	b	c_{in}	$\sim maj_1$	$\sim maj_2$	$\sim maj_3$	c_{out}	sum
0	0	0	1	1	0	0	0
0	1	0	1	0	1	0	1
1	0	0	1	0	1	0	1
1	1	0	0	0	0	1	0
0	0	1	1	1	1	0	1
0	1	1	0	1	0	1	0
1	0	1	0	1	0	1	0
1	1	1	0	0	1	1	1

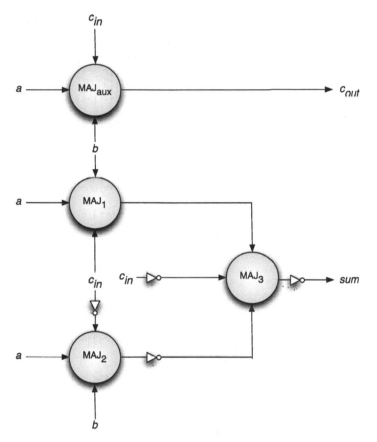

Fig. 27.19 Schematic diagram of a full binary adder comprised of NOT-MAJORITY gates. Delay elements are not shown

only to 'ignite' propagating patterns in the channels [5, 45]. In future studies we are planning to implement the computing architecture designed in the chapter to manufacture experimental prototypes of precipitating chemical computers; they will be based on crystallization of 'hot ice' [4].

Fig. 27.20 Full binary adder
designed with
NOT-MAJORITY gates, stages
and main circuit
implementation on
$B2/S2345$ evolution space

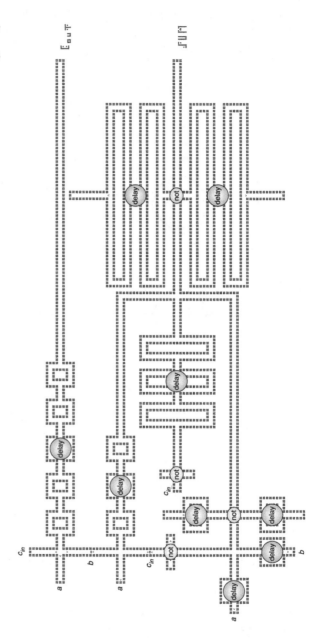

Fig. 27.21 Final configuration of the adder for inputs $a = 0$, $b = 0$ and $c_{in} = 0$, and outputs $c_{out} = 0$ and $sum = 0$

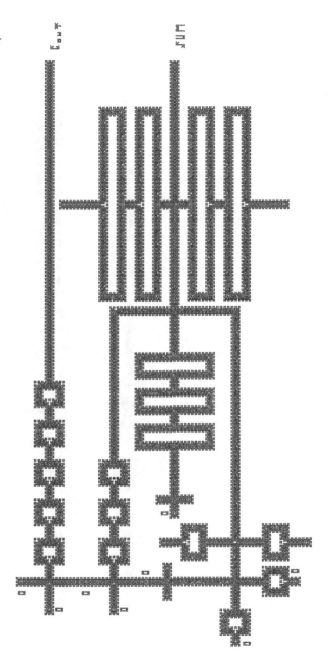

Fig. 27.22 Configuration of
the adder for inputs $a = 0$,
$b = 1$ and $c_{in} = 0$, and
outputs $c_{out} = 0$ and $sum = 1$

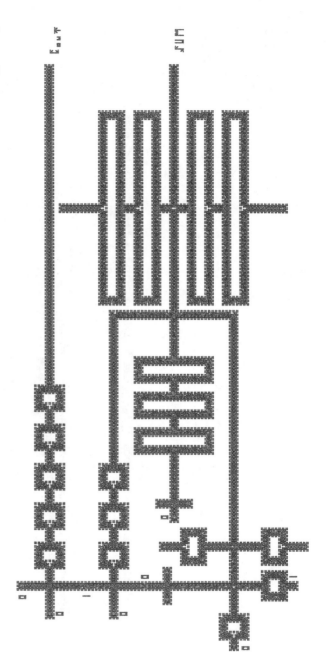

Fig. 27.23 Configuration of
the adder for inputs $a = 1$,
$b = 0$ and $c_{in} = 0$, and
outputs $c_{out} = 0$ and $sum = 1$

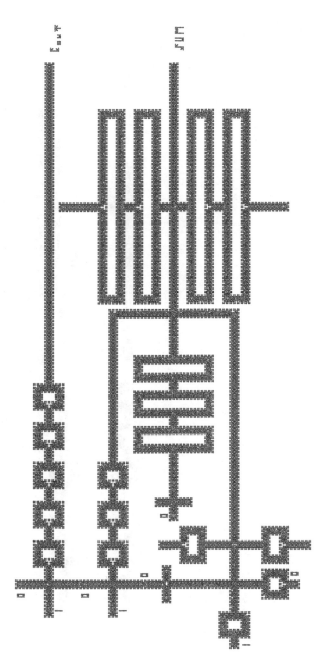

Fig. 27.24 Configuration of
the adder for inputs $a = 1$,
$b = 1$ and $c_{in} = 0$, and
outputs $c_{out} = 1$ and $sum = 0$

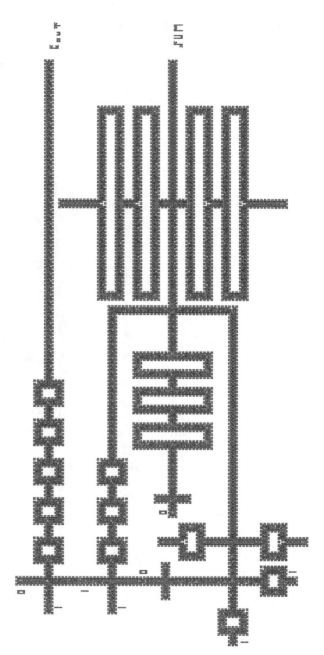

Fig. 27.25 Configuration of the adder for inputs $a = 0$, $b = 0$ and $c_{in} = 1$, and outputs $c_{out} = 0$ and $sum = 1$

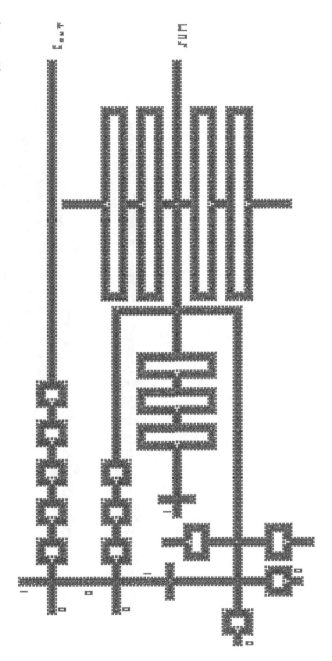

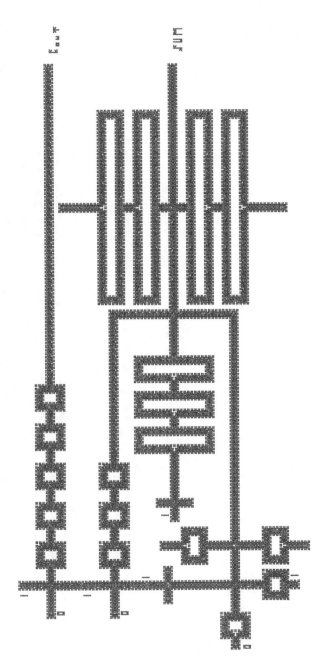

Fig. 27.26 Configuration of the adder for inputs $a = 0$, $b = 1$ and $c_{in} = 1$, and outputs $c_{out} = 1$ and $sum = 0$

Fig. 27.27 Configuration of the adder for inputs $a = 1$, $b = 0$ and $c_{in} = 1$, and outputs $c_{out} = 1$ and $sum = 0$

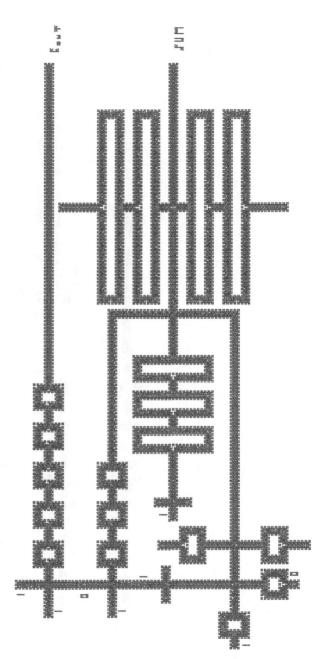

Fig. 27.28 Configuration of
the adder for inputs $a = 1$,
$b = 1$ and $c_{in} = 1$, and
outputs $c_{out} = 1$ and $sum = 1$

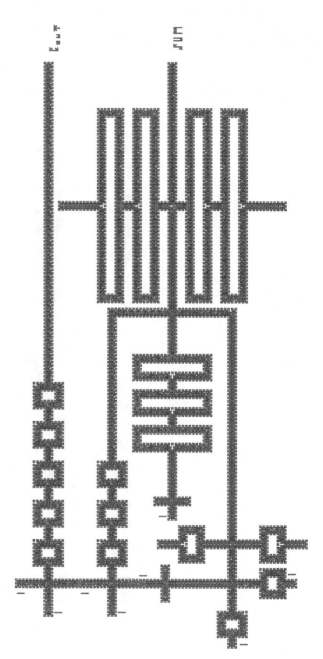

Implementations and constructions are done in *Golly system* (http://golly. sourceforge.net/). Source configurations and specific initial condition (RLE files) to reproduce the results are available in http://uncomp.uwe.ac.uk/genaro/Life_dc22. html.

Acknowledgements Genaro J. Martínez was partially funded by Engineering and Physical Sciences Research Council (EPSRC), United Kingdom, grant EP/F054343, and postdoctoral funding at the ICN and C3 of UNAM. Kenichi Morita was partially funded by Grant-in-Aid for Scientific Research (C) No. 21500015 from JSPS.

References

1. Adamatzky, A.: Computing in Nonlinear Media and Automata Collectives. Institute of Physics Publishing, Bristol and Philadelphia (2001)
2. Adamatzky, A. (ed.): Collision-Based Computing. Springer, Berlin (2002)
3. Adamatzky, A.: Physarum machines: encapsulating reaction–diffusion to compute spanning tree. Naturwissenschaften **94**, 975–980 (2007)
4. Adamatzky, A.: Hot ice computer. Phys. Lett. A **374**(2), 264–271 (2009)
5. Adamatzky, A., Costello, B.L., Asai, T.: Reaction–Diffusion Computers. Elsevier, Amsterdam (2005)
6. Adamatzky, A., Martínez, G.J., Seck-Tuoh-Mora, J.C.: Phenomenology of reaction–diffusion binary-state cellular automata. Int. J. Bifurc. Chaos **16** (10), 1–21 (2006)
7. Beato, V., Engel, H.: Pulse propagation in a model for the photosensitive Belousov–Zhabotinsky reaction with external noise. In: Schimansky-Geier, L., Abbott, D., Neiman, A., Van den Broeck, C. (eds.) Noise in Complex Systems and Stochastic Dynamics. Proc. SPIE, vol. 5114, pp. 353–362 (2003)
8. Berlekamp, E.R., Conway, J.H., Guy, R.K.: Winning Ways for Your Mathematical Plays, vol. 2, Chap. 25. Academic Press, San Diego (1982)
9. Chapman, P.: Life universal computer. http://www.igblan.free-online.co.uk/igblan/ca/ (2002)
10. Chaté, H., Manneville, P.: Evidence of collective behavior in cellular automata. Europhys. Lett. **14**, 409–413 (1991)
11. Cook, M.: Still Life theory. In: Griffeath, D., Moore, C. (eds.) New Constructions in Cellular Automata, pp. 93–118. Oxford University Press, London (2003)
12. Costello, B.L., Toth, R., Stone, C., Adamatzky, A., Bull, L.: Implementation of glider guns in the light sensitive Belousov–Zhabotinsky medium. Phys. Rev. E **79**, 026114 (2009)
13. Dupont, C., Agladze, K., Krinsky, V.: Excitable medium with left–right symmetry breaking. Physica A **249**, 47–52 (1998)
14. Gardner, M.: Mathematical Games — The fantastic combinations of John H. Conway's new solitaire game Life. Sci. Am. **223**, 120–123 (1970)
15. Gorecka, J., Gorecki, J.: T-shaped coincidence detector as a band filter of chemical signal frequency. Phys. Rev. E **67**, 067203 (2003)
16. Gorecki, J., Yoshikawa, K., Igarashi, Y.: On chemical reactors which can count. J. Phys. Chem. A **107**, 1664–1669 (2003)
17. Goucher, A.: Completed universal computer/constructor. http://pentadecathlon.com/lifeNews/2009/08/post.html (2009)
18. Gravner, J.: Growth phenomena in cellular automata. In: Griffeath, D., Moore, C. (eds.) New Constructions in Cellular Automata, pp. 161–181. Oxford University Press, London (2003)
19. Griffeath, D., Moore, C.: Life Without Death is P-complete. Complex Syst. **10**, 437–447 (1996)
20. Guan, Z., Qin, X., Zhang, Y., Shi, Q.: Network structure cascade for reversible logic. In: Proceedings of the Third International Conference on Natural Computation, vol. 3, pp. 306–310 (2007)

21. Gutowitz, H.A., Victor, J.D.: Local structure theory in more that one dimension. Complex Syst. **1**, 57–68 (1987)
22. Imai, K., Morita, K.: A computation-universal two-dimensional 8-state triangular reversible cellular automaton. Theor. Comput. Sci. **231**, 181–191 (2000)
23. Krug, H.J., Pohlmann, L., Kuhnert, L.: Analysis of the modified complete Oregonator (MCO) accounting for oxygen- and photosensitivity of Belousov–Zhabotinsky systems. J. Phys. Chem. **94**, 4862–4866 (1990)
24. Kusumi, T., Yamaguchi, T., Aliev, R., Amemiya, T., Ohmori, T., Hashimoto, H., Yoshikawa, K.: Numerical study on time delay for chemical wave transmission via an inactive gap. Chem. Phys. Lett. **271**, 355–60 (1997)
25. Magnier, M., Lattaud, C., Heudin, J.-K.: Complexity classes in the two-dimensional life cellular automata subspace. Complex Syst. **11**(6), 419–436 (1997)
26. Martínez, G.J., Méndez, A.M., Zambrano, M.M.: Un subconjunto de autómata celular con comportamiento complejo en dos dimensiones. http://uncomp.uwe.ac.uk/genaro/papers.html (2005)
27. Martínez, G.J., Adamatzky, A., Costello, B.L.: On logical gates in precipitating medium: cellular automaton model. Phys. Lett. A **1**(48), 1–5 (2008)
28. Martínez, G.J., Adamatzky, A., McIntosh, H.V.: Localization dynamic in a binary two-dimensional cellular automaton: the Diffusion Rule. J. Cell. Autom. **5**, 284–313 (2010)
29. Martínez, G.J., Adamatzky, A., Morita, K., Margenstern, M.: Majority adder implementation by competing patterns in Life-like rule $B2/S2345$. In: Calude, C.S., et al. (eds.) UC 2010. Lecture Notes in Computer Science, vol. 6079, pp. 93–104. Springer, Berlin (2010)
30. Martínez, G.J., Adamatzky, A., McIntosh, H.V., Costello, B.L.: Computation by competing patterns: Life rule $B2/S2345678$. In: Adamatzky, A., et al. (eds.) Automata 2008: Theory and Applications of Cellular Automata. Luniver Press, Beckington (2008)
31. McIntosh, H.V.: Life's Still Lifes. http://delta.cs.cinvestav.mx/~mcintosh (1988)
32. McIntosh, H.V.: Wolfram's Class IV and a Good Life. Physica D **45**, 105–121 (1990)
33. Minsky, M.: Computation: Finite and Infinite Machines. Prentice Hall, New York (1967)
34. Mitchell, M.: Life and evolution in computers. Hist. Philos. Life Sci. **23**, 361–383 (2001)
35. Morita, K., Margenstern, M., Imai, K.: Universality of reversible hexagonal cellular automata. Theor. Inform. Appl. **33**, 535–550 (1999)
36. Motoike, I.N., Yoshikawa, K., Iguchi, Y., Nakata, S.: Real-time memory on an excitable field. Phys. Rev. E **63**, 036220 (2001)
37. Packard, N., Wolfram, S.: Two-dimensional cellular automata. J. Stat. Phys. **38**, 901–946 (1985)
38. Porod, W., Lent, C.S., Bernstein, G.H., Orlov, A.O., Amlani, I., Snider, G.L., Merz, J.L.: Quantum-dot cellular automata: computing with coupled quantum dots. Int. J. Electron. **86**(5), 549–590 (1999)
39. Rendell, P.: Turing universality of the game of life. In: Adamatzky, A. (ed.) Collision-Based Computing, pp. 513–540. Springer, Berlin (2002)
40. Rennard, J.P.: Implementation of logical functions in the Game of Life. In: Adamatzky, A. (ed.) Collision-Based Computing, pp. 491–512. Springer, Berlin (2002)
41. Sielewiesiuk, J., Gorecki, J.: Logical functions of a cross junction of excitable chemical media. J. Phys. Chem. A **105**, 8189–8195 (2001)
42. Toffoli, T.: Non-conventional computers. In: Webster, J. (ed.) Encyclopedia of Electrical and Electronics Engineering, vol. 14, pp. 455–471. Wiley, New York (1998)
43. Tóth, A., Showalter, K.: Logic gates in excitable media. J. Chem. Phys. **103**, 2058–2066 (1995)
44. von Neumann, J.: Theory of Self-reproducing Automata. Edited and completed by Burks, A.W. University of Illinois, Urbana and London (1966)
45. Wainwright, R. (ed.): Lifeline – A Quarterly Newsletter for Enthusiasts of John Conway's Game of Life, Issues 1 to 11, March 1971 to September 1973

Index

A. Adamatzky (ed.), *Game of Life Cellular Automata*,
DOI 10.1007/978-1-84996-217-9, © Springer-Verlag London Limited 2010